GROUNDWATER ENGINEERING

GROUNDWATER ENGINEERING

ABDEL-AZIZ ISMAIL KASHEF

Geotechnical and Groundwater Engineering Consultant

McGRAW-HILL BOOK COMPANY

New York   St. Louis   San Francisco   Auckland
Bogotá   Hamburg   Johannesburg   London
Madrid   Mexico   Montreal   New Delhi
Panama   Paris   São Paulo   Singapore
Sydney   Tokyo   Toronto

Library of Congress Cataloging in Publication Data

Kashef, Abdel-Aziz I.
  Groundwater engineering.

  Bibliography: p.
  Includes index.
  1. Water, Underground. 2. Engineering geology.
I. Title.
TC176.K37 1986   627'.17   85-208
ISBN 0-07-033489-7

1234567890 DOC / DOC 898765

ISBN 0-07-033489-7

*The editors for this book were Joan Zseleczky and Susan
B. West, the designer was Naomi Auerbach, and the production
supervisor was Sally Fliess. It was set in Caledonia by Progressive
Typographers.*

Printed and bound by R. R. Donnelley & Sons Company.

Dedicated to the late
AWATEF SIDKY KASHEF
Sept. 28, 1924 to Jan. 30, 1979

About the Author

Dr. Abdel-Aziz Ismail Kashef is a geotechnical and groundwater engineering consultant. He was professor of civil engineering and head of the Earth Science Division at the American University of Beirut from 1956 through 1960. From 1962 to 1980, he was professor of civil engineering at North Carolina State University. He practiced engineering full time for 10 years and has worked as a consultant for 27 years in the United States and five other countries.

Dr. Kashef's degrees include a B.S. in irrigation engineering, an M.S. in structures, and a Ph.D. in soils from Purdue University. He was associated with the United Nations and also served as consultant for the High Aswan Dam of Egypt. From 1970 to 1972 he was editor of the *Water Resources Bulletin* of the American Water Resources Association and is a fellow both of that organization and of the American Society of Civil Engineers.

CONTENTS

PREFACE

Over the years of my professional life as an engineer, I have had the privilege of changing my specialization from that of a practicing professional engineer (irrigation, structures, and foundation engineering) to that of a teacher at various universities in the United States and the Middle East. From 1962 to 1980, I taught two graduate courses on groundwater at North Carolina State University. Perhaps because of this diversified career, I have observed two distinct groups of people whose approach to groundwater is quite different and between whom a gap exists: those who confine themselves to theory, and those who believe only in practice. In addition, the two main groups of people who deal with groundwater, engineers and geologists, have in the past been interested in different aspects of groundwater. The engineer's interest has been in seepage through and around dams, in dewatering of foundation sites, and in municipal groundwater supplies. Engineering geologists and geologists have been more concerned with the overall management of groundwater, particularly water-well production. However, this latter group has recently become aware of the importance of some aspects of geotechnical engineering, such as leakage, land subsidence, and seepage through dams, which is analogous to some conditions of saltwater intrusion. The intent of this book is to present the basic premises of groundwater flow and related subjects, with the hope of providing engineers and geologists with a text that combines both theory and practical solutions to groundwater problems in a single source that can be used by all.

Groundwater sciences have grown rapidly in the last 30 to 40 years. Formal courses in groundwater are presently taught in many universities. During the last decade, much attention has been focused on the use of numerical approaches, such as the finite-difference method, the finite-element method, the method of characteristics, and boundary integrals. By using these methods, one is able to solve complex equations on groundwater flow and the movement of contaminants if the natural and boundary conditions are known. These methods, however, have not proved to be totally effective for dealing with, for example, water-pollution problems. There are several specialized books and publications on numerical analysis, so that this and other topics, such as well drilling, water pollution, unsaturated flow, modeling techniques, and geophysical methods, have been excluded from this book. However, the correlations between these fields are explained and supplemented with selected references.

The main purpose of the book is to present the field of groundwater with a minimal amount of mathematics in a simple and clear manner, emphasizing the techniques of quantitative evaluations of groundwater flow, seepage through and around dams, water wells, saltwater intrusion, and groundwater management. The book can be used by students and professionals with little or no previous experience in this field and should be useful to those working in the fields of numerical analysis and modeling as a refresher course in basic fundamentals. The main features of the book are (1) simple presentation of complex topics, (2) avoidance of burdensome mathematical details that may disrupt the sequence of presentation, (3) emphasis on basic principles and the limitations of both theoretical and practical aspects, (4) introduction of reliable simple formulas that have been checked against more rigorous solutions, and (5) solved examples wherever it is felt that the material in the text needs further explanation. It is my hope that this book can be used by graduate and undergraduate students in geology, agriculture, and engineering, as well as by planners, environmentalists, water-resources managers, geologists, and professional engineers, such as geotechnical, sanitary, hydraulic, irrigation, agricultural, and construction engineers.

The book is divided into nine chapters. The first two chapters are devoted to a review of the basic fundamentals of groundwater occurrence and the properties of flow media. A discussion of groundwater quality is also included in these chapters. Chapter 3 explains the fundamentals of groundwater flow. The history of groundwater engineering is discussed in Chapter 4. Recent advances in groundwater engineering and a listing of sources of pertinent data and literature references are also included in Chapter 4. Two-dimensional flow systems are explained in Chapter 5, with emphasis on flow nets, seepage through earth dams (including new, simple approaches), and hydraulic design of solid dams or weirs. Chapter 6 discusses the various means of determining the hydrologic parameters of aquifers (other than by well pumping, which is presented in detail in Chapter 8). The environmental effects on these parameters as well as their inclusion in the hydrodynamic equations are also explained in Chapter 6. The main elements of groundwater management are briefly discussed in Chapter 7, supplemented by a relatively long list of references. Water-well hydraulics is discussed in detail in Chapter 8. The available techniques used in water wells are explained, and new techniques are introduced for overpumped artesian wells and gravity wells. Chapter 8 also includes a special section on land subsidence due to well pumping. Chapter 9 is devoted to an analysis of saltwater intrusion. Special attention is given to some recently introduced approaches: a modified version of the Ghyben-Herzberg curve, disturbance of the natural interface due to discharge wells, and a design method of controlling saltwater intrusion using recharge wells. All chapters (except Chapter 4) are followed by problems and discussion questions.

I am deeply grateful to the reviewers of the manuscript and to Mrs. Candace C. Morse for typing the manuscript and for her patience.

Finally, I would like to dedicate this book to the late Mrs. Awatef Sidky Kashef, who despite serious illness helped and encouraged me and to whom I am most grateful.

ABDEL-AZIZ I. KASHEF
Raleigh, North Carolina

1

PROPERTIES OF
ROCKS AND SOILS

The physical and mineralogic properties of the media through which groundwater percolates affect the quality of groundwater and its circulation. These properties can be better understood by studying the geologic origin of these media and how they were formed. The main types and properties of the media through which groundwater flows are briefly summarized in this chapter. These media include rocks and soils that originate from rocks known sometimes as regolith or unconsolidated rocks.

1.1 Rock Formations

Rocks were and are still being formed continuously (Legget, 1962) as a result of various natural processes, such as the cooling of molten rocks in the form of a hot liquid (magma) that percolates from considerable depths below the earth's surface, the precipitation of inorganic materials in water, the deposition of shells of various organisms, the condensation of gas containing mineral particles, the disintegration of other rocks due to various causes and the subsequent accumulation of the resulting minerals to form new types of rocks, and the action of intense heat and/or pressure on previously existing rocks. Rocks may be classified on the basis of their origin as *igneous, sedimentary,* or *metamorphic.* Each of these types has a characteristic general form, texture, structure, and mineral composition (U.S. Bureau of Public Roads, 1960; Legget, 1962; U.S. Bureau of Reclamation, 1977). Some of the main types of rocks are described in the following list:

Granite (igneous).   The color of granite varies from pale gray to deep red. It is primarily composed of feldspar (about 60 percent), quartz (about 30 percent), mica, and possibly hornblend. The granitic structure is usually massive. Joints in a massive granitic formation divide it into large blocks. However, closely spaced joints give a sheet appearance.

Volcanic rocks (igneous).   These may be hard or interbedded with loose permeable volcanic materials such as tuff, ashes, or sand. Basalt (traprock) is an example of a volcanic rock. Its color varies from dark gray to black, and it is characterized by the absence of quartz, the predominance of plagioclase, and the presence of considerable amounts of pyroxene and some olivine. In general, basalts are composed of fine-grained minerals. Columnar jointing (almost hexagonal) is one of the well-recognized features of basaltic structures. Weathered basalts become rusty in color.

Sandstone (sedimentary).   Sandstones have fine- to coarse-grained texture and are usually massive, although they may also be cross-bedded. Quartzite resembles limestone yet is harder and contains almost equal amounts of feldspar and quartz. Practically, sandstones may be looked on as cemented sand. Graywacke is a dark-gray to black sandstone cemented by silica or clay. Graywacke also may contain flakes of slate and shale.

Limestone (sedimentary).   This has a fine to coarse texture, and its color may be white, yellow, brown, gray, or a combination of these. Limestone is primarily calcium carbonate, and in the absence of fissures and solution channels, it is impervious; otherwise it is porous and has a loose texture. Marls and chalks are other forms of limestone.

Shales (sedimentary).   These are laminated rocks, mostly dark in color. Shales consist of clay-size particles and sometimes small percentages of sand or silt-size particles. The structural strength of shale varies from extremely soft to very hard, depending on the particle characteristics and the degree of their compaction and cementation.

Conglomorate (sedimentary).   The texture of conglomorates varies from very coarse to very fine, depending on the size of the cemented loose material. Usually 10 percent or more of the grains are coarse (larger than the size of sand).

Schist (metamorphic). This is a foliated rock, yet the foliation is not usually visible to the naked eye. Some schists are composed entirely of silica and form massive structures. Generally, the dip of the planes of schistosity is different from the dip of the whole formation.

Slate (metamorphic). This is a platy rock with an extremely fine texture. It is dark in color and susceptible to easy cleavage.

A simplified rock classification is given in Tables 1.1 to 1.3.

TABLE 1.1 Common Igneous Rocks

Color	Light		Intermediate	Dark	
Principal mineral	Quartz Feldspar Other minerals, minor	Feldspar	Feldspar Hornblende	Augite Feldspar	Augite Hornblende Olivine
Texture					
Coarse, irregular, crystalline	Pegmatite	Syenite pegmatite	Diorite pegmatite	Gabbro pegmatite	
Coarse and medium crystalline	Granite	Syenite	Diorite	Gabbro	Periodotite
			Dolerite		
Fine crystalline	Aplite			Diabase	
Aphanitic	Felsite			Basalt	
Glassy	Volcanic glass			Obsidian	
Porous (gas openings)	Pumice		Scoria or vesicular basalt		
Fragmental	Tuff (fine), breccia (coarse), cinders (variable)				

SOURCE: Naval Facilities Engineering Command, *Soil Mechanics*, Design Manual 7.1, Department of the Navy, NAVFAC, Alexandria, Va., 1982.

TABLE 1.2 Common Sedimentary Rocks

Group	Grain size	Composition			Name
Clastic	Mostly coarse grains	Rounded pebbles in medium-grained matrix			Conglomerate
		Angular coarse rock fragments, often quite variable			Breccia
	More than 50% of medium grains	Medium quartz grains	Less than 10% of other minerals		Siliceous sand-stone
			Appreciable quantity of clay minerals		Argillaceous sandstone
			Appreciable quantity of calcite		Calcareous sandstone
			Over 25% feldspar		Arkose
			25–50% feldspar and darker minerals		Graywacke
	More than 50% fine grain size	Fine to very fine quartz grains with clay minerals			Siltstone (if laminated, shale)
		Microscopic clay minerals	<10% other minerals		Shale
			Appreciable calcite		Calcareous shale
			Appreciable carbonaceous material		Carbonaceous shale
			Appreciable iron oxide cement		Ferruginous shale

Group	Grain size	Composition	Name
Organic	Variable	Calcite and fossils	Fossiliferous limestone
	Medium to microscopic	Calcite and appreciable dolomite	Dolomite limestone or dolomite
	Variable	Carbonaceous material	Bituminous coal
Chemical	Microscopic	Calcite	Limestone
		Dolomite	Dolomite
		Quartz	Chert, flint, etc.
		Iron compounds with quartz	Iron formation
		Halite	Rock salt
		Gypsum	Rock gypsum

SOURCE: Naval Facilities Engineering Command, *Soil Mechanics,* Design Manual 7.1, Department of the Navy, NAVFAC, Alexandria, Va., 1982.

TABLE 1.3 Common Metamorphic Rocks

Texture	Structure	
	Foliated	Massive
Coarse crystalline	Gneiss	Metaquartzite
Medium crystalline	Schist (sericite, mica, talc, chlorite, etc.)	Marble Quartzite Serpentine Soapstone
Fine to microscopic	Phyllite Slate	Hornfels Anthracite coal

SOURCE: Naval Facilities Engineering Command, *Soil Mechanics,* Design Manual 7.1, Department of the Navy, NAVFAC, Alexandria, Va., 1982.

Rocks undergo a geologic cycle that consists of three major actions: *denundation, deposition,* and *earth movement.* The structural form of the present earth crust results from these continuous actions, forming mountains, gullies, rivers, valleys, plains, lakes, sand dunes, and other physiographic features. Denundation is caused by severe temperature changes and the effects of wind and water. Destruction of rocks by these actions includes the solution of their minerals, decomposition, erosions, and disintegration of their grains. The products of this destruction are transported by water and/or wind and subsequently deposited in other locations. Geologically, some of these products were formed and moved by glaciers. Earth movement takes place because of uneven pressures produced by the gradation procedure caused by both denundation and deposition. The movement continues until a state of equilibrium is reached. Volcanism may be considered an action of earth movement.

1.2   Soil Formations

The products of rock destruction are known as unconsolidated rock or soil (known also as regolith). *Soil* may be defined as the material which disintegrates into individual grains by such gentle mechanical means as agitation in water or the application of low pressure (Peck, et al., 1974). A weathered rock may be structurally weaker in strength than soil, yet it is classified as a rock because it maintains all rock features except cementation. The exposed rock surface is known as an *outcrop,* even if it is later covered by a soil deposit.

Soils may be classified as either residual or transported, depending on the manner of their formation.

Residual soils Residual soils are encountered directly over the parent rock material from which they originated. Owing to weathering effects, the thicknesses of residual soils originating from igneous and metamorphic rocks are expected to be small because of the relatively high resistance of such rocks to weathering. However, the weak resistance of limestone to solution results in the formation of sinkholes, caverns, solution channels, and subterranean streams, as well as residual soils of deep thicknesses. The thickness of residual soils may be as little as 1.5 m or as large as 15 m depending on local climatic and physiographic conditions (Peck et al., 1974). Table 1.4 gives diagnostic features of weathering degrees (Naval Facilities Engineering Command, 1982).

Transported soils Transported soils are formed by the transportation of soils and disintegrated rocks and their redeposition in other locations. The transporting agents may be ice and glaciers (glacial soils), water

TABLE 1.4   Weathering Classification

Grade	Symbol	Diagnostic features
Fresh	F	No visible sign of decomposition or discoloration; rings under hammer impact.
Slightly weathered	WS	Slight discoloration inward from open fractures; otherwise similar to F.
Moderately weathered	WM	Discoloration throughout; weaker minerals such as feldspar decomposed; strength somewhat less than fresh rock but cores cannot be broken by hand or scraped by knife; texture preserved.
Highly weathered	WH	Most minerals somewhat decomposed; specimens can be broken by hand with effort or shaved with knife; core stones present in rock mass; texture becoming indistinct but fabric preserved.
Completely weathered	WC	Minerals decomposed to soil but fabric and structure preserved (saprolite); specimens easily crumbled or penetrated.
Residual soil	RS	Advanced state of decomposition resulting in plastic soils; rock fabric and structure completely destroyed; large volume change.

SOURCE: Naval Facilities Engineering Command, *Soil Mechanics,* Design Manual 7.1, Department of the Navy, Alexandria, Va., 1982.

(alluvial or fluvial soils), wind (aeolian soils), or gravitational forces that produce landslides and talus (colluvial soils). There are numerous types of each of these major groups, such as glacial drifts, a glacial outwash, glacial till, basal drift, ground moraines, lateral and terminal moraines, drumlines, eskers, kames, glacial lacustrines, glacial terraces, loess (primary and secondary), sand dunes of various shapes and sizes, river terraces, flood-plain sediments, natural levees, alluvial fans, delta formations, organic silts, openwork gravel, marshes, muskeg, peat, and shore deposits. These types are classified according to the nature of their deposition, their final geometric form, and their geographic location. The soil landform may also assume a rough topography, such as canyons, scablands, and badlands (Legget, 1962).

1.3   Soil Properties

The physical and chemical properties of soils are an area studied in several different fields, such as geology, engineering, and agronomy.

Therefore, it is not surprising that there exist various classification systems depending on the orientation and interest of the investigator. In an engineering study, for example, the major properties sought are in situ strength, deformation, and permeability. However, engineering classification systems presently in use are not based on these properties; if they were, these systems would be complex and impractical. At the present time, classification systems are used for preliminary identification of soils. They serve as a common language between field and laboratory technicians and geotechnical engineers. The available classification methods are based on textural and consistency properties. They may be considered as the first clues upon which further testing and investigations are planned.

Soil Identification and Basic Tests

A soil may be classified according to its texture as gravel, sand, silt, or clay. A natural soil consists of one or more of these general types and may contain a varying amount of organic material. Accordingly, if a silt is identified as "sandy silt," "gravelly silt," or "organic silt," it means that the soil is predominantly silt having somewhat similar engineering characteristics as pure silts. The adjectives are used for the soil ingredients of less predominance. The major types are differentiated according to their grain sizes. The size of gravel particles, for example, varies from about 3 to 75 mm, depending on the standard used (Table 1.5). Large-size grains that are coarser than gravel are called *cobbles, pebbles*, and *granules*. Very large sizes, over about 30 cm, are called *boulders*, such as those found in areas of glacial soils.

The individual soil particles of sands and gravels can be detected by the naked eye or a hand lens, whereas particles of silt and clay can be identified only by a microscope or more often by an electron microscope. Sands and gravels belong to the coarse-grained group (macroscopic), whereas silts and clays belong to the fine-grained group (microscopic). Coarse-grained soils are usually cohesionless, whereas fine-grained soils exhibit plasticity because of their cohesive properties (except for cohesionless silts such as the powdery "rock flour").

It was customary in the past for groundwater workers to overlook the examination of fine-grained soils. However, it has been found that these studies are needed in such practical problems as leakage and land subsidence due to water pumping (Chap. 8) as well as groundwater pollution problems (Sec. 4.3).

The major types of fine-grained soils are inorganic silts (plastic and nonplastic), inorganic clays, organic silts, and organic clays. Plastic silts contain minute flake-shaped (also platelike and sometimes needlelike)

TABLE 1.5   Comparison of Several Common Textural Classification Systems

Classification system	Grain size, mm						
	100	10	1	0.1	0.01	0.001	0.0001
Bureau of Soils, 1890–1895	Gravel		Sand	Silt		Clay	
			1	0.05	0.005		
Atterberg, 1905	Gravel		Coarse sand / Fine sand	Silt		Clay	
		2	0.2	0.02	0.002		
MIT, 1931	Gravel		Sand	Silt		Clay	
		2	0.06		0.002		
USDA, 1938	Gravel		Sand	Silt		Clay	
		2	0.05		0.002		
AASHO, 1970	Gravel		Sand	Silt		Clay / Colloids	
	75	2	0.075		0.002	0.001	
Unified, 1953 ASTM, 1967	Gravel	Sand		Fines (silt and clay)			
	75	4.75	0.075				

SOURCE: R. B. Peck, W. E. Hanson, and T. H. Thornburn, *Foundation Engineering*, 2d ed., Wiley, New York, 1974.

particles. Clays and plastic silts consist of microscopic or submicroscopic flake-shaped crystalline minerals that are characterized by their colloidal properties (plasticity, cohesion, and ability to adsorb ions). Some clays are noncrystalline. Organic materials contain finely divided particles of organic matters (shells, decayed plants, and animal organisms). The color of the soil varies from dark gray to black depending on the amount of organic matter and its stage of decay. Organic materials are usually detected by their odor. However, if only a small amount of organic material is present in the soil sample, the odor is detected by heating the sample slightly. Peat and muck are examples of organic clays. Usually it is not necessary to differentiate between organic silts and organic clays. Both types are usually referred to as "organic soils." Such soils have a high degree of compressibility.

Since inorganic silts and clays cannot be easily differentiated visually, simple manual field tests are usually performed (Table 1.6). These tests should be supplemented by standard physical tests in order to properly identify and classify these soils. The results of the standard physical soil tests supplement and/or correct the driller's field log. Highlights of such physical tests are briefly explained in the following section.

Laboratory determination of the average *specific gravity* of soil parti-

TABLE 1.6   Identification of Fine-Grained Soil Fractions from Manual Tests

Typical name	Dry strength	Dilatancy reaction	Toughness of plastic thread	Time to settle in dispersion test
Sandy silt	None to very low	Rapid	Weak to friable	30 s to 60 min
Silt	Very low to low	Rapid	Weak to friable	15 to 60 min
Clayey silt	Low to medium	Rapid to slow	Medium	15 min to several hours
Sandy clay	Low to high	Slow to none	Medium	30 s to several hours
Silty clay	Medium to high	Slow to none	Medium	15 min to several hours
Clay	High to very high	None	Tough	Several hours to days
Organic silt	Low to medium	Slow	Weak to friable	15 min to several hours
Organic clay	Medium to very high	None	Tough	Several hours to days

SOURCE: R. B. Peck, W. E. Hanson, and T. H. Thornburn, *Foundation Engineering*, 2d ed., Wiley, New York, 1974.

cles is done only for unusual soil types containing organic or heavy minerals or those known as extremely sensitive clays (Peck et al., 1974). Organic soils have variable specific gravities, but they are usually less than 2.0. Pure quartz sand has a specific gravity of 2.66, while the normal types of inorganic clays have a specific gravity ranging between about 2.3 and 2.9 (average 2.7). Table 1.7 includes the specific gravities of the most important soil constituents.

The distribution of the *grain sizes* is determined by means of gradation tests. The results of these tests do not indicate the physical properties of the soil, yet they are useful for (1) identifying and classifying the soils, (2) studying the proper means of stabilizing road subgrades, (3) designing filters and graded filters (Sec. 3.10), (4) identifying soil deposits of the same geologic origin (Terzaghi and Peck, 1967), and (5) determining the susceptibility of soils to frost action. These tests are either sieve mechanical analysis suitable for coarse-grained soils, wet

TABLE 1.7   Specific Gravity of Most Important Soil Constituents

Gypsum	2.32	Aragonite	2.94
Orthoclase	2.56	Biotite	3.0–3.1
Kaolinite	2.6	Augite	3.2–3.4
Chlorite	2.6–3.0	Hornblende	3.2–3.5
Quartz	2.66	Limonite	3.8
Talc	2.7	Hematite, hydrous	4.3±
Calcite	2.72	Magnetite	5.17
Muscovite	2.8–2.9	Hematite	5.2
Dolomite	2.87		

SOURCE: E. S. Larsen and H. Berman, *The Microscopic Determination of the Non-Opaque Minerals,* 2d ed., U.S. Department of the Interior Bulletin 848, Washington, 1934.

TABLE 1.8   U.S. Standard Sieves

Sieve number	Opening size		Sieve number	Opening size	
	Inches	Millimeters		Inches	Millimeters
3	0.2500	6.350	30	0.0232	0.590
4	0.1874	4.760	40	0.0165	0.420
6	0.1323	3.360	50	0.0117	0.297
8	0.0937	2.380	60	0.0098	0.250
10	0.0787	2.000	70	0.0083	0.210
16	0.0469	1.190	100	0.0059	0.149
20	0.0331	0.840	140	0.0041	0.105
			200	0.0029	0.074

analysis tests conducted on fine-grained soils, or a combination of both to analyze a soil mixture of coarse to fine texture. Practically, mesh sizes less than those of the no. 200 sieve (Table 1.8) are not used. The portion of the soil passing through this sieve (<0.074 mm) should be analyzed by a wet test, usually a hydrometer test (American Society for Testing and Materials, 1982; Bowles, 1978). No. 200 sieves and sieves of lesser mesh sizes are expensive and difficult to maintain.

The grain-size distribution is represented in graphic form on a semi-log plot in which the logarithm of the diameter D is plotted against the percent finer than D (Fig. 1.1). The size D corresponding to 10 percent finer is given the symbol D_{10}, which is known as the *effective size* (or *Allen-Hazen's effective size*) of the soil. Sizes larger than D_{10} constitute 90 percent of the aggregate, and those smaller than D_{10} constitute the balance. The effective size is used to compute the hydraulic conductivity of sands (Sec. 3.3). It is also used to find a qualitative value for the soil uniformity [Eq. (1.1)]. The effective size of clean sands (with about less than 3 percent fine material) may be regarded as that size of a uniform idealized material consisting of equal spherical grains that would produce the same hydraulic properties as that of the natural soil.

The effective size D_{10} is used to find the degree of uniformity of the soil. This is numerically determined by means of the uniformity coefficient C_u, defined as:

$$C_u = \frac{D_{60}}{D_{10}} \tag{1.1}$$

where D_{60} is the size of the particles at which 60 percent of the aggregate is finer. When C_u is less than about 4 or 5, the gradation exhibits a steep slope and the soil is then called *poorly graded* (practically uniform or

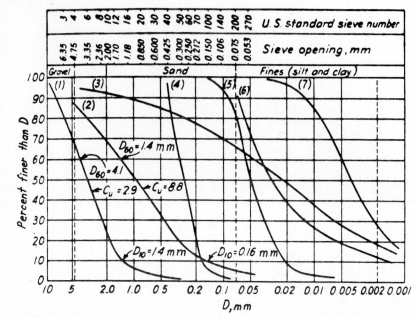

Figure 1.1 Typical particle-size distribution curves of natural soils. (1) Pea gravel, Castle Rock, Colorado; (2) river gravel, Denver, Colorado; (3) glacial till, Peoria, Illinois; (4) sand, Grenada, Mississippi; (5) glacial rock flour, Winchester, Massachusetts; (6) clayey silt, Smead, Montana; (7) silty clay, Marathon, Ontario, Canada. *(R. B. Peck et al., Foundation Engineering, 2d ed., Wiley, New York, 1974.)*

homogeneous). When the value of C_u is higher than about 4 or 5, the soil is considered *well graded* (heterogeneous), covering several sizes and/or different soil types. However, if the gradation curve is not continuous but rather is interrupted by an almost horizontal portion, the soil is identified as *gap-graded,* lacking intermediate soil sizes (e.g., a mixture of coarse sand and clay with no silt or a mixture of gravel and silt with no sand).

The hydrometer test is the most common laboratory method among the wet mechanical tests used to find the gradation of fine-grained soils. The method has several built-in uncertainties both theoretical and experimental (Bowles, 1978). It is based on Stoke's law, which determines the terminal velocity of a suspended sphere in a certain fluid as a function of the specific gravity of the solid grain and its diameter. Fine-grained soils with particle sizes smaller than about 0.005 mm, however, consist generally of flake-shaped (and sometimes needle-shaped) particles rather than of spheres. Moreover, a deflocculating agent should be added to the suspension before the test is made in order to separate the particles. However, the degree of effectiveness of such agents has not been satisfactorily established. The values of D_{10}, D_{15}, D_{30}, D_{50}, D_{60}, and C_u are usually reported in the results of a gradation test. These values are particularly useful in soil classification as well as in the design of filters (Sec. 3.10).

The natural moisture content w_n of a soil is the ratio of the weight of water W_w of a sample specimen of the natural soil to the dry weight W_s of the same specimen:

$$w_n(\%) = \frac{100 W_w}{W_s} \tag{1.2}$$

The total weight W of the soil specimen is determined as well as its weight after complete drying W_s. Then w_n is calculated as follows:

$$w_n(\%) = \frac{100(W - W_s)}{W_s} \tag{1.3}$$

where W is the sum of weight of the solid grains W_s and the weight of water W_w. Usually, the sample is dried in an oven for 24 h at a constant temperature of not more than 110°C; otherwise the mineralogic characteristics of the soil grains would change. Laboratory and field determinations of the water content may be rapidly obtained by using chemical and nutrons methods (American Society for Testing and Materials, 1982; Gardner, 1965).

The consistency of fine-grained soils is usually described in a driller's log by such terms as very soft, soft, firm (or medium), stiff, very stiff, and extremely stiff (or hard) (Table 1.9). Such descriptions are quantitatively verified by (Peck et al., 1974) the unconfined compressive strength (Table 1.9) and/or the Atterberg limits laboratory tests (Fig. 1.2).

TABLE 1.9   Qualitative and Quantitative Expressions for Consistency of Clays

Consistency	Field identification	Unconfined compressive strength, tons/ft²
Very soft	Easily penetrated several inches by fist	Less than 0.25
Soft	Easily penetrated several inches by thumb	0.25–0.5
Medium	Can be penetrated several inches by thumb with moderate effort	0.5–1.0
Stiff	Readily indented by thumb but penetrated only with great effort	1.0–2.0
Very stiff	Readily indented by thumbnail	2.0–4.0
Hard	Indented with difficulty by thumbnail	Over 4.0

SOURCE: R. B. Peck, W. E. Hanson, and T. H. Thornburn, *Foundation Engineering*, 2d ed., Wiley, New York, 1974.

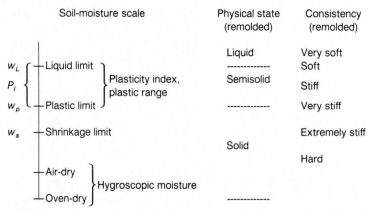

Figure 1.2 Soil-moisture scale showing Atterberg limits, corresponding physical state, and approximate consistency of remolded soil. *(R. B. Peck et al., Foundation Engineering, 2d ed., Wiley, New York, 1974.)*

The unconfined compressive strength laboratory tests are similar to the strength tests made on concrete cylinders. The soil sample should have the proper consistency so that it can stand vertically without lateral support. The sample is then loaded gradually until failure. These tests are also used to determine the degree of sensitivity of the soil material (explained later). After the test is made on an undisturbed sample, a similar test is made on a sample of the same specimen after it has been kneaded at an unaltered water content and formed into the same cylindrical size as the undisturbed sample. The ratio of the unconfined compressive strength of the undisturbed specimen to that of the disturbed (or re-molded) specimen is known as the *sensitivity* S_t. In normal clays, S_t varies from 2 to 4, whereas in the sensitive types, S_t varies from 4 to 8. If S_t is greater than 8, the clay is considered extrasensitive, such as clay with a high percent of montmorillonite (smectite). When S_t is less than unity, natural cracks or other weak features (joints and slickensides) are expected to exist in the natural soil structure; kneading the specimen in this case would improve the natural structure (Peck et al., 1974).

By means of the Atterberg limits test (Peck et al., 1974), the consistency of the soil in a saturated state at various moisture contents is determined and some limits are defined (Fig. 1.2). The highest limit is known as the *liquid limit* w_L, and it is determined by using the liquid-limit device. The lowest limit is the shrinkage limit w_s, which is determined by testing and/or calculation. Another intermediate limit is known as the *plastic limit* w_p, and it is determined by testing a small soil thread about 3.5 mm in diameter. These limits are nothing more than water contents at *complete saturation* that describe the soil consistency. If the natural water content w_n is higher than w_L, then the natural soil should be very soft and almost in a liquid state. However, if $w_s < w_n < w_L$, the natural soil should be within the plastic range. If the soil is in a semisolid state, then $w_n \leqslant w_s$. The solid state of the natural soil takes place when w_n decreases considerably below w_s. Within the plastic range, the consistency varies from stiff to medium (firm). When $w_n = w_L$, the soil in nature is soft, and when $w_n = w_p$, it is very stiff. At $w_n = w_s$, the soil is extremely stiff. When $w_n < w_s$, the soil is classified as hard. The range of water content within which the soil remains plastic is expressed by the plasticity index $P_I = w_L - w_p$.

The consolidation test (Terzaghi, 1943; Taylor, 1948; Terzaghi and Peck, 1967; Peck et al., 1974; Bowles, 1978) is used to determine the compressibility characteristics and the permeability properties of fine-grained material (Sec. 6.2). The results are extremely useful not only in the major engineering aspect of settlement analyses of structures and land subsidence due to well pumping, but also in the study of leakage problems (Secs. 6.4 and 8.3) in groundwater hydrology.

1.4 Volume-Weight Relationships

If the specific gravity of the soil grains of a saturated soil specimen is assumed or determined and if the water content is known, some valuable information such as the porosity and saturated unit weight of the soil can be calculated. This and other relationships can also be established by drawing what is called a *block* or *phase diagram* (Fig. 1.3) and indicating

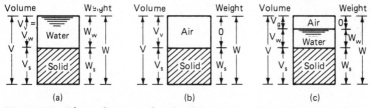

Figure 1.3 Phase diagrams for the three cases of saturation. *(a)* Completely saturated soil (two-phase diagram), $V_g = 0$, $V_v = V_w$, $V = V_v + V_s$, $W = W_w + W_s$; *(b)* completely dry soil (two-phase diagram), $V_v = V_v$, $V = V_v + V_s$, $W = W_s$, $W_w = 0$, weight of air $\cong 0$; *(c)* partially saturated soil (three-phase diagram), $V_v = V_w + V_g$, $V = V_v + V_s = V_g + V_w + V_s$, $W = W_w + W_s$, weight of air $\cong 0$.

the known parameters and calculating the unknown ones. In a block diagram, each material (air, water, and solids) is completely segregated from the others (in practice this is impossible, yet it simplifies the analysis without affecting the final results). The weights are indicated on one side and the volumes on the other. The general block diagram indicating the three material phases represents a partially saturated soil (moist). A two-phase diagram represents either a completely saturated soil (100 percent saturation) or a completely dry specimen (0 percent saturation). The weight of air is considered zero as compared with the weights of other materials.

In order to determine the weight-volume relationships, the following ratios are defined:

Percent porosity: $n(\%) = \dfrac{V_v}{V} 100$

where V_v and V are, respectively, the volume of voids and the total volume in a certain soil specimen.

Void ratio: $e = \dfrac{V_v}{V_s}$

where V_s is the volume of solids in the specimen.

also $n = \dfrac{e}{1+e}$ and $e = \dfrac{n}{1-n}$

It should be noted that the n value is always less than 1.0 [$n(\%) < 100$ percent], whereas the e value may be less, equal to, or greater than 1.0.

Percent water content: $w(\%) = \dfrac{100 W_w}{W_s}$

Percent degree of saturation: $S_r(\%) = \dfrac{100 V_w}{V_v}$

In sands, the S_r value describes their state (Table 1.10).

Bulk (mass) unit weight: $\gamma_m = \dfrac{W}{V}$

where γ_m is the ratio of the total weight to the total volume. When $S_r = 100$ percent, the bulk unit weight is called the *saturated unit weight* γ_{sat}, and when $S_r = 0$ percent, $\gamma_m =$ dry weight $\gamma_d = W_s/V$.

Unit weight of the soil particles: $\gamma_s = \dfrac{W_s}{V_s}$

Specific gravity of solid matters: $G = \dfrac{\gamma_s}{\gamma_w}$

where γ_w is the unit weight of water at $4°C$ ($\gamma_w = 1$ g/cm$^3 \simeq 62.4$ lb/ft^3),

TABLE 1.10 Degree of Saturation of Sand in Various States

Condition of sand	Degree of saturation, %
Dry	0
Humid	1–25
Damp	26–50
Moist	51–75
Wet	76–99
Saturated	100

SOURCE: K. Terzaghi and R. B. Peck, *Soil Mechanics in Engineering Practice*, 2d ed., Wiley, New York, 1967.

and $G = W_s/(V_s\gamma_w)$. If centimeter and gram units are used, γ_s is numerically equal to G.

Submerged unit weight: $\gamma' = (\gamma_{\text{sat}} - \gamma_w)$ or $\gamma' = \dfrac{W - V\gamma_w}{V}$

γ' is also known as the *buoyant unit weight*.

The specific gravity of water at a certain temperature is equal to the ratio of the unit weight of water at that temperature to γ_w at $4\,^\circ C$; practically, the variation of γ_w with temperature is negligible and the specific gravity of water may be considered unity at any temperature (Table 1.11).

Values of n, e, w, γ_d, and γ_{sat} for some natural deposits are given in Table 1.12 (Terzaghi and Peck, 1967).

Usually w is determined by testing, while γ_w and G may be assumed (Table 1.7). The values of γ_m, γ_{sat}, γ_d, γ', n, e, w, and S_r can be determined using block diagrams. The proper phase diagram is sketched and all weights and volumes of the materials are written in the appropriate spaces. If this cannot be completed, then by assuming any *one* unknown by a symbol (V_s, V_v, V_w, V, W_s, W_w, or W) or unity, other weights and volumes can be indicated on the block diagram in terms of that unknown. The assumed unknown value should be related to one or more of the known parameters, and all volumes and weights should be indicated in terms of that *one* unknown. Because the weight of air is negligible, $W = W_w + W_s$ whether the soil is partially or completely saturated.

EXAMPLE 1.1 If the water content of a certain saturated soil sample is found to be 23 percent and the specific gravity is assumed to be 2.7, determine γ_{sat}, γ_d, n, and e.

TABLE 1.11   Properties of Distilled Water

Temp., °C	Unit weight of water, g/cm^3	Viscosity of water, poises
4	1.0000	0.01567
16	0.9990	0.01111
17	0.9988	0.01083
18	0.9986	0.01056
19	0.9984	0.01030
20	0.9982	0.01005
21	0.9980	0.00981
22	0.9978	0.00958
23	0.9976	0.00936
24	0.9973	0.00914
25	0.9971	0.00894
26	0.9968	0.00874
27	0.9965	0.00855
28	0.9963	0.00836
29	0.9960	0.00818
30	0.9957	0.00801
40	0.9922	0.00654
50	0.9881	0.00529
60	0.9832	0.00470
70	0.9778	0.00407
80	0.9718	0.00357
90	0.9653	0.00317
100	0.9584	0.00284

Solution   Since the known values are $w = W_w/W_s$ and $G = W_s/(V_s\gamma_w)$, then the assumed unknown should be either W_s, W_w, or V_s, which are implied in w and G. Assuming the unknown value as V_s, which is shown circled in the phase diagram (Fig. 1.4), then

$$W_s = GV_s\gamma_w = 2.7 \times 1 \times V_s$$

and $\qquad w W_s = 0.23 \times 2.7 \times 1 \times V_s = 0.621 V_s$

Therefore,

$$V_w = \frac{W_w}{w} = 0.621 V_s$$

In addition, the total is

$$W = V_s \times 1 \times (2.7 + 0.621) = 3.321 V_s$$

and the volume is

$$V = V_w + V_s = V_s(1 + 0.621) = 1.621 V_s$$

Note that the components of weights and volumes in Fig. 1.4 are all written in

TABLE 1.12 Porosity, Void Ratio, and Unit Weight of Typical Soils in Natural State

Description	Porosity n, %	Void ratio e	Water content w, %°	Unit weight, g/cm³ $\gamma_d^\dagger$	Unit weight, g/cm³ $\gamma_{sat}^\ddagger$
1. Uniform sand, loose	46	0.85	32	1.43	1.89
2. Uniform sand, dense	34	0.51	19	1.75	2.09
3. Mixed-grained sand, loose	40	0.67	25	1.59	1.99
4. Mixed-grained sand, dense	30	0.43	16	1.86	2.16
5. Glacial till, very mixed grained	20	0.25	9	2.12	2.32
6. Soft glacial clay	55	1.2	45	—	1.77
7. Stiff glacial clay	37	0.6	22	—	2.07
8. Soft, slightly organic clay	66	1.9	70	—	1.58
9. Soft, very organic clay	75	3.0	110	—	1.43
10. Soft bentonite	84	5.2	194	—	1.27

° Water content when saturated, in percent of dry weight.
† Unit weight in dry state.
‡ Unit weight in saturated state.
SOURCE: K. Terzaghi and R. B. Peck, *Soil Mechanics in Engineering Practice*, 2d ed., Wiley, New York, 1967.

terms of just one unknown, V_s. Then

$$\gamma_{sat} = \frac{W}{V} = \frac{3.321}{1.621} = 2.049 \text{ g/cm}^3 \ (127.84 \text{ lb/ft}^3)$$

$$\gamma_d = \frac{W_s}{V} = \frac{2.7}{1.621} = 1.67 \text{ g/cm}^3 \ (103.94 \text{ lb/ft}^3)$$

$$n(\%) = \frac{100V_v}{V} = \frac{100 \times 0.621}{1.621} = 38.3\%$$

and
$$e = \frac{V_v}{V_s} = 0.621$$

or
$$e = \frac{n}{1-n} = \frac{0.383}{1-0.383} = 0.621$$

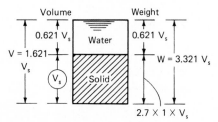

Figure 1.4 Phase diagram for Example 1.1.

The porosity of a soil is considered large if $n > 20$ percent ($e > 0.25$), as in Example 1.1. If n lies between 5 and 20 percent ($e = 0.05$ to 0.25), the porosity is said to be medium. When $n < 5$ percent ($e < 0.05$), the porosity would be very small. Exceptionally high values of porosities, i.e., between 80 and 95 percent, have been recorded in freshly deposited alluviums.

It should be noted that the assumed unknown value V_s in Example 1.1 canceled out in all cases. Also, the problem could be solved by assuming either W_w or W_s as the unknown. *Obviously, by assuming the unknown as unity, the problem also can be solved.*

If γ_d and γ_{sat} are known, G, w, n, and e can be calculated by following the same general procedure. Numerous relationships can be established (Jumikis, 1962), yet it is simpler to study each case by means of the block diagram rather than to use a formula.

EXAMPLE 1.2 A moist clay specimen weighs 54 g. After drying in the oven for 24 h, it weighs 50 g. The dry sample is immersed in a container full of mercury to determine its volume by weighing the spilled mercury. The volume is found to be 25 cm³. If the specific gravity of the clay is assumed to be 2.7, calculate the porosity, degree of saturation, original moist unit weight, and dry unit weight.

Solution   (See Fig. 1.5.)

$$W_s = 50 \text{ g}$$
$$W = 54 \text{ g}$$
$$W_w = (54 - 50) = 4 \text{ g}$$
$$V_s = \frac{50}{2.7} = 18.5 \text{ cm}^3$$
$$V = 38 \text{ cm}^3$$
$$V_w = 4 \text{ cm}^3$$
$$V_v = V - V_s = 38 - 18.5 = 20.5 \text{ cm}^3$$

Porosity:

$$n(\%) = \frac{20.5}{38} \times 100 = 53.9\%$$

Degree of saturation:

$$S_r(\%) = \frac{4}{20.5} \times 100 = 19.5\%$$

Moist unit weight:

$$\gamma = \frac{54}{38} = 1.42 \text{ g/cm}^3 \ (88.67 \text{ lb/ft}^3)$$

Dry unit weight:

$$\gamma_d = \frac{50}{38} = 1.32 \text{ g/cm}^3 \ (82.1 \text{ lb/ft}^3)$$

Note that in this example there was no need to assume an unknown; the given data were sufficient to solve the problem.

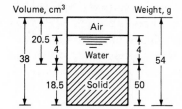

Figure 1.5 Phase diagram for Example 1.2.

1.5 Clay Minerals

Clay minerals have a limited number of crystalline components (Grim, 1953; Terzaghi and Peck, 1967). Clays are mainly hydrous aluminum silicates with magnesium or iron substituting wholly or partially for the aluminum in some minerals and with sometimes varying amounts of alkalies or alkaline earth. Clays may be composed of either a single clay mineral or, most often, a mixture of these minerals. Clays may also contain organic materials and water-soluble salts. The degree of perfection of the crystallinity varies in different types of clays, and in some rare cases (allophanes), the clays are noncrystalline. The study of clay mineralogy gives a better understanding of the characteristics of clays. However, clay mineral analyses cannot be a substitute for soil laboratory tests in engineering.

Clay particles are minute flakes consisting of atoms arranged in repeated units that form either silica or alumina sheets (Fig. 1.6). Clay 1 cm thick consists of approximately 200,000 of these sheets. The shape of a clay particle is determined using an electron microscope, and the arrangement of atoms is determined by x-ray diffraction techniques. Differential thermal analysis also may be used to identify clay minerals.

The three major groups of clay minerals are smectites (montmorillonites), kaolinites, and illites. *Kaolinites* consist of alternating sheets of silica and alumina. *Smectite* (Fig. 1.6) is composed of an alumina sheet between two silica sheets repeated indefinitely; the sheets are bound together somewhat loosely. The structure of *illite* is similar to that of smectite, with some changes in the chemical composition. Among the minor clay mineral groups are allophane, chlorite, vermiculite, attapulgite, palygorkite, and sepiolite (Grim, 1953).

Clay minerals have the property of adsorbing certain anions (negatively charged particles) and cations (positively charged particles). These adsorbed ions are retained in an exchangeable state, the process of which is known as *base exchange*. The softening of water by the use of zeolites, permutites, or carbon exchangers is a good example of an ion-exchange reaction. The exchange of ions takes place in a water solution or in a nonaqueous environment. The exchangeable ions are held around the outside of the silica-alumina sheets without affecting the structure of the sheets. For example, the sodium chloride (NaCl) in water dissociates

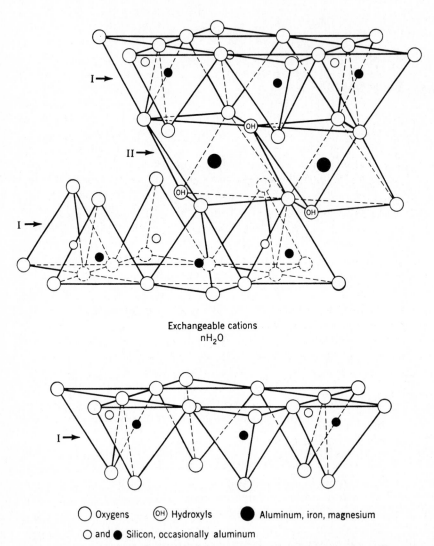

Exchangeable cations
nH_2O

○ Oxygens (OH) Hydroxyls ● Aluminum, iron, magnesium

○ and ● Silicon, occasionally aluminum

Figure 1.6 Diagrammatic sketch of the structure of smectite (montmorillonite). I = silica sheet; II = alumina sheet. *(R. E. Grim, Applied Clay Mineralogy, McGraw-Hill, New York, 1962.)*

into cations of sodium (Na^+) and anions of chloride (Cl^-). The negative, unbalanced charges at the surface of clay minerals attract the sodium cations. Adsorption is not permanent, and if the environment changes, for example, if NaCl is changed to potassium chloride, the sodium cations would be released and replaced by those of potassium. The exchange capacity varies in different minerals, and the rate of exchange depends not only on the mineral type but also on the nature and concentration of the cations and anions. Smectites are more active than illites. Kaolinites are the least active among the three major groups. Organic materials with high exchange capacities are restricted to recent soil sediments.

The water held directly on the surface of clay minerals is nonordinary water that has a physical state different from that of liquid water. Molecular layers of water can cover a distance as small as 8 Å (about 3 molecular layers) to one as high as 28 Å (about 10 molecular layers), where an angstrom (Å) equals one ten-millionth part of a millimeter. The transition from nonordinary water to ordinary water (free water) can be abrupt or gradual depending on the clay mineral type and the character of the adsorbed ions on the clay mineral surfaces. The thickness of the nonordinary water is smaller on the edges of rough clay particles than on smooth, flat particles (Grim, 1953). The density of this water is almost 1.0, and its quality varies from a solid to a plastic state (developing clay plasticity).

The structure of kaolinite resists the introduction of water molecules into its lattice; thus kaolinites are very stable clays and are not subject to excessive volume changes (heaving or swelling and shrinkage) when saturated or dried. Smectites, however, are unstable minerals whose sheets are loosely bound, which invites the insertion of adsorbed water molecules between the sheets. This causes swelling of smectite clays in water and excessive shrinkage after drying. The high plasticity of smectite clays is a result of this phenomenon. The attraction of water molecules between the sheets of smectite decreases their permeability. Illites do not expand or shrink as excessively as smectites, yet they suffer more volume changes than kaolinites.

The presence of fine-silt grains in all clays decreases the surface activity, but increases the clay sensitivity. Also, the accumulation of soluble salts as a result of groundwater movement or weathering alters the physical properties of the clay (including permeability). Sodium tends to disperse the clay mineral particles and thereby decreases the clay permeability, and in the case of highly sensitive clays, swelling increases. Bentonite, which is a well-known material in the engineering field, is essentially a smectite clay that is used as a drilling mud and in soil grouting to seal voids.

Although clays have porosities of 50 percent or more, they are impermeable as compared with sands of lesser porosity. The leakage properties of clay layers also depend on the properties of the clay minerals. Under pressure, clay layers lose only part of their free water. In practice, clay compressibility and leakage are usually determined through physical laboratory and field tests rather than mineralogically. However, an understanding of the mineralogic characteristics of clays can help explain some of their main features, such as their plastic characteristics and their low permeabilities and their high porosities.

1.6 Soil Classification Systems

There are numerous systems of soil classification, with each field developing its own system depending on its particular needs. Each of these classification systems is organized in such a way that identical or almost identical soils are grouped under specific categories. However, it immediately becomes obvious that no universal system can be found. The geologic classification systems do not satisfy the needs of agronomists or engineers. Instead, agronomists and road engineers use the pedologic classification system. Engineers devised the engineering classification system; these systems include few of the physical properties of the soil. Each system has numerous shortcomings. For example, in the engineering systems, soils are classified in their disturbed states, neglecting the more important physical properties of soils in situ. It is therefore recommended that one use more than one classification system to obtain sufficient preliminary information. Some of the main classification systems are summarized in the following section.

Textural Systems

The upper and lower limits of the sizes of grain particles in gravels, sands, silts, and clays are given by this system. The size groups are expressed as a percentage by weight of the total weight of the sample. As shown in the triangular chart in Fig. 1.7, for example, a soil aggregate is named on the basis of the percentage of sands, silts, and clay (excluding gravel). The soil is named as sand, silt, clay, loam, clay loam, sand clay loam, silty clay loam, sandy loam, or silty loam. In order to classify the soil according to this system, a gradation curve should also be used. The triangular chart is used by locating two coordinates in the directions shown. Other textural standards are given in Table 1.5.

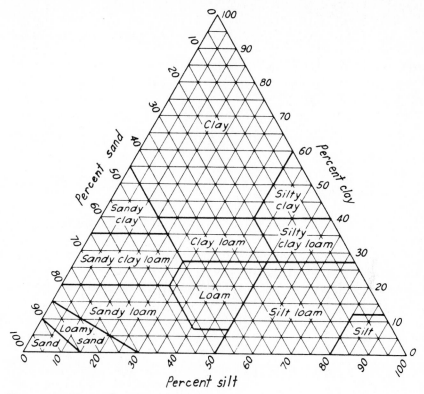

Figure 1.7 Triangular textural classification chart. *(After U.S. Department of Agriculture, Soil Classification: A Comprehensive System, USDA, Washington, 1960.)*

Unified Classification System

The system used mostly by engineers is known as the *Unified Classification System* (Table 1.13 and Fig. 1.8). The soil sample is screened through a no. 200 sieve. If more than 50 percent is retained on this sieve, the soil is considered coarse-grained and may be one of the following types: GW, SW, GC, SC, GP, SP, GM, or SM. The first letters G and S mean gravel and sand; letters W and P stand, respectively, for well graded and poorly graded; and letters C and M refer, respectively, to clayey and silty sands or gravels. However, if more than 50 percent passes the no. 200 sieve, the soil is considered fine-grained and may be one of the following types: ML, MH, CL, CH, OL, OH, or Pt. The letters M, C, and O refer to silts, clays, and organic soil, respectively, and Pt stands for peat.

TABLE 1.13 Unified Classification System

Major divisions	Group symbol	Typical names	Classification criteria[1]	
			Coarse-grained soils[2]	
Gravels[3,4] Clean gravels (little or no fines)	GW	Well-graded gravels, gravel-stone mixtures, little or no fines	$C_u = D_{60}/D_{10} > 4$ $C_c = 1 < D_{30}^2/D_{10} \times D_{60} < 3$	
	GP	Poorly graded gravels, gravel-sand mixtures, little or no fines	Not meeting all gradation requirements for GW	
Gravels with fines (appreciable amount of fines)	GM	Silty gravels, gravel-sand-silt mixtures	Atterberg limits below A line or $P_I < 4$	Above A line with $4 < P_I < 7$ are borderline cases requiring use of dual symbols
	GC	Clayey gravels, gravel-sand-clay mixtures	Atterberg limits above A line with $P_I > 7$	
Sands[4,5] Clean sands (little or no fines)	SW	Well-graded sands, gravelly sands, little or no fines	$C_u = D_{60}/D_{10} > 6$ $C_c = 1 < D_{30}^2/D_{10} \times D_{60} < 3$	
	SP	Poorly graded sands, gravelly sands, little or no fines	Not meeting all gradation requirements for SW	
Sands with fines (appreciable amount of fines)	SM	Silty sands, sand-silt mixtures	Atterberg limits below A line or $P_I < 4$	Limits plotting in hatched zone with $4 \le P_I \le 7$ are borderline cases requiring use of dual symbols
	SC	Clayey sands, sand-clay mixtures	Atterberg limits above A line with $P_I > 7$	

Fine-grained soils[6]

Silts and clays Liquid limit < 50	ML	Inorganic silts and very fine sands, rock flour, silty or clayey fine sands, clayey silts with slight plasticity
	CL	Inorganic clays of low to medium plasticity, gravelly clays, sandy clays, silty clays, lean clays
	OL	Organic silts, organic silty clays of low plasticity
Liquid limit > 50	MH	Inorganic silts, micaceous or diatomaceous fine sandy or silty soils, elastic silts
	CH	Inorganic clays of high plasticity, fat clays
	OH	Organic clays of medium to high plasticity, organic silts
Highly organic soils	Pt	Peat and other highly organic soils

[1] Coarse-grained soils can be classified on the basis of the percentage of fines (the fraction smaller than no. 200 sieve size) they contain. Soils with < 5 percent fines are classified as GW, GP, SW, or SP; those with > 12 percent, as GM, GC, SM, or SC; soils with 5 to 12 percent are borderline cases requiring dual symbols.
[2] More than half of material is larger than no. 200 sieve size.
[3] More than half of coarse fraction is larger than no. 4 sieve size.
[4] The percentage of sand and gravel can be determined from a grain-size curve.
[5] More than half of coarse fraction is smaller than no. 4 sieve size.
[6] More than half of material is smaller than no. 200 sieve size.

SOURCE: American Society for Testing and Materials *Annual Book of ASTM Standards: Soil and Rock, Building Stones*, D-2487, pt. 19, ASTM, Philadelphia, 1982.

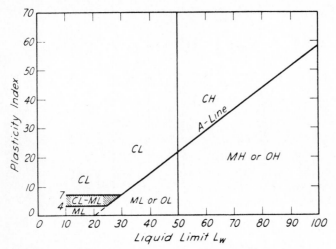

Figure 1.8 Plasticity chart for use together with Table 1.13. *(American Society for Testing and Materials, Annual Book of ASTM Standards: Soil and Rock; Building Stones, D-2487, pt. 19, ASTM, Philadelphia, 1982.)*

The fine group is classified by means of a plasticity chart (Fig. 1.8) on the basis of the results of the Atterberg limits test. The letter L stands for low to medium plasticity (or low to medium compressibility in nonplastic silts) when the liquid limit w_L is less than 50 percent. The letter H stands for high plasticity or high compressibility when w_L is greater than 50 percent.

It should be noted that the A line drawn on the plasticity chart (Fig. 1.8) and the vertical line passing through $w_L = 50$ percent divide the plasticity chart into the four zones for the indicated types. A dual classification is given for soils whose coordinates are close to the boundaries of the major zones (for example, CL-CH) or whose coordinates lie between the horizontal lines on the left-hand side of the chart (for example, CL-ML).

If the gravel and sand contain fine material and exhibit some plasticity, they should be classified as GM and GC or SM and SC depending on the classification criteria given in Table 1.13. If the material is cohesionless, with its C_u value (D_{60}/D_{10}) greater than 4 and its C_c value between 1 and 3 $[C_c = D_{30}^2/(D_{10} \times D_{60})]$, the soil may be classified as GW or SW. The letters G and S are selected on the basis of the percentage passing through a no. 4 sieve.

Many experienced engineers and technicians are able to classify soils visually (American Society for Testing and Materials, 1982) and manually by rubbing the specimens between their fingers to check their dry strength, the reaction of the specimens to shaking in the palm of the hand, and by other means (Tables 1.6 and 1.14). Tests would be required only when this determination does not provide clear results (Peck et al., 1974).

Pedologic Classification System

The parent-material groups in this system are subdivided on the basis of composition, drainability, and alkalinity or acidity. The pedologic concept is based on the premise that similar soil-forming materials (parent materials), if subjected to identical environmental conditions of climate, topography, and time, will develop identical soil profiles (Woods, 1960). In areas where drainage is possible, the soil profile consists of three distinct layers or horizons A, B, and C (and sometimes D) down to the parent rock:

A horizon is the zone of leaching.

B horizon is the zone of accumulation, and both A and B horizons are together called the *soil solum*.

C horizon is the unweathered parent material from which the upper layers were formed. It is described in the pedologic system by the same geologic nomenclature (Tables 1.1, 1.2, and 1.3).

D horizon is any stratum underlying the C horizon whose characteristics differ from C. It may or may not have an influence on the formation of the weathered portion of the profile. It may consist of hard rock or soil layers that are not parent material. If it is deep and has no influence, it is not usually recorded.

Further, pedologists use several features of the soil profile in identifying a soil unit: color, texture, structure, relative arrangement, chemical composition, thickness of the soil horizons, and geology of the soil mate-

TABLE 1.14   Classification of Fine-Grained Soils within the Unified Classification System

Group	Dry strength	Reaction to shaking test	Toughness at plastic limit
ML	None to very low	Rapid to slow	None
CL	Medium to high	None to very slow	Medium
OL	Very low to medium	Slow	Slight
MH	Very low to medium	Slow to none	Slight to medium
CH	High to very high	None	High
OH	Medium to high	None to very slow	Slight to medium

SOURCE: K. Terzaghi and R. B. Peck, *Soil Mechanics in Engineering Practice,* 2d ed., Wiley, New York, 1967.

rial. There are 36 major soil groups as well as soil series, soil types, and soil phases that have been identified. A reference key has been developed for soil series (Woods, 1960).

PROBLEMS AND DISCUSSION QUESTIONS

1.1 Define or explain the following:
   (a) Rock formation
   (b) Soil formation
   (c) Geologic cycle
   (d) Solum

1.2 What are the main differences between residual and transported soils? List four types of transported soils and indicate in each case the transporting agent.

1.3 How does one differentiate (visually and manually) between coarse-grained and fine-grained soils?

1.4 What is the predominant soil type in each of the following soils?
   (a) Silty clay   (c) Silty sand
   (b) Sandy silt   (d) Organic silt

1.5 Many engineers assume a value for the specific gravity of soil particles rather than determining it by testing.
   (a) Give the range of these assumed values for silts and clays.
   (b) When do you think it would be necessary to perform a specific gravity test?

1.6 (a) Does the soil-gradation test have any practical values other than classification purposes?
   (b) What are the main parameters that may be obtained from a soil-gradation curve?

1.7 Define or explain the following:
   (a) Effective size   (e) Dry unit weight
   (b) Uniformity coefficient   (f) Submerged unit weight
   (c) Hydrometer test   (g) Soil sensitivity
   (d) Poorly graded soil

1.8 Explain two major tests that may be used to determine the soil consistency of fine-grained soils. What is the practical significance of each determined parameter?

1.9 If the water content of a completely saturated soil sample is 28.7 percent and its specific gravity is assumed to be 2.65, determine the following by the aid of a phase diagram:
   (a) Saturated unit weight of the soil in grams per cubic centimeter
   (b) Dry unit weight in grams per cubic centimeter
   (c) Void ratio
   (d) Porosity in percent

1.10   A partially saturated soil has a dry density of 1.52 g/cm³ and a specific gravity of 2.67. If its moisture content is 10.5 percent, compute its degree of saturation.

1.11   The saturated unit weight of a clay sample is found to be 1.76 g/cm³. If its moisture content is 45.8 percent, find the specific gravity of the clay particles.

1.12   The weight of a completely saturated soil specimen is 155 g. The sample is dried in an oven for 24 h at 105°C and is weighed after drying. If its specific gravity is 2.67 and its dry weight is 110.7 g, find the following at the saturated state:
   (a) Water content
   (b) Total volume of the specimen before drying
   (c) Saturated unit weight
   (d) Void ratio
   (e) Porosity in percent

1.13   A sample of dry soil weighs 500 g. A sieve analysis test is performed on this sample and the results are as follows:

Sieve size, mm	Percent passing by weight
4.76	100.0
2.00	90.2
0.84	74.6
0.42	58.1
0.149	25.2
0.074	3.6
0.053	1.2

Plot the graduation curve on semilog paper and find the following:
   (a) Effective size and uniformity coefficient
   (b) Percentages of sand, silt, and clay, according to the U.S. Bureau of Standards, Atterberg, MIT, the American Geophysical Union, and the U.S. Department of Agriculture
   (c) The various different possibilities for classification of the sample type according to the Unified Classification System
   (d) The additional parameters required to exactly classify the soil (refer to part c)

1.14   What are the major three groups of clay minerals? Explain the differences in their structure and stability.

1.15   Explain briefly the ion-exchange reaction between soil particles and their environment. Also give an example for the process of base exchange.

1.16   What are the effects of water on each of the three main clay groups? What are the expected changes in the physical properties of these groups as a result of the intrusion of water molecules?

1.17   What is the importance of engineering classification systems in practice? Can you consider them satisfactory? Why?

1.18   What kind of soil parameters do you need to classify a soil by means of the following?
   (a) Textural systems
   (b) Unified Classification System

1.19   What are the main differences between the following classification systems?
   (a) Geologic system
   (b) Pedologic system
   (c) Engineering systems

REFERENCES

American Society for Testing and Materials: *Annual Book of ASTM Standards: Soil and Rock, Building Stones,* D-2487, pt. 19, ASTM, Philadelphia, 1982.

Bowles, J. E.: *Engineering Properties of Soils and Their Measurement,* 2d ed., McGraw-Hill, New York, 1978.

Gardner, W. R.: "Water Contents," in C. A. Black (ed.), *Methods of Soil Analysis,* pt. 1, Agronomy Monograph no. 9, American Society of Agronomy, Madison, Wisc., 1965, pp. 82–127.

Grim, R. E.: *Clay Mineralogy,* McGraw-Hill, New York, 1953.

Grim, R. E.: *Applied Clay Mineralogy,* McGraw-Hill, New York, 1962.

Jumikis, A. R.: *Soil Mechanics,* Van Nostrand, Princeton, N.J., 1962.

Larsen, E. S., and H. Berman: *The Microscopic Determination of the Non-Opaque Minerals,* 2d ed., U.S. Dept. of Interior Bulletin, 848, Washington, 1934.

Legget, R. F.: *Geology and Engineering,* 2d ed., McGraw-Hill, New York, 1962.

Naval Facilities Engineering Command: *Soil Mechanics,* Design Manual 7.1, Department of the Navy, NAVFAC, Alexandria, Va., 1982.

Peck, R. B., W. E. Hanson, and T. H. Thornburn: *Foundation Engineering,* 2d ed., Wiley, New York, 1974.

Taylor, D. W.: *Fundamentals of Soil Mechanics,* Wiley, New York, 1948.

Terzaghi, K.: *Theoretical Soil Mechanics,* Wiley, New York, 1943.

Terzaghi, K., and R. B. Peck: *Soil Mechanics in Engineering Practice,* 2d ed., Wiley, New York, 1967.

U.S. Bureau of Public Roads: *The Identification of Rock Types,* U.S. Department of Commerce, Washington, 1960.

U.S. Bureau of Reclamation: *Ground Water Manual,* U.S. Department of the Interior, Washington, 1977.

U.S. Department of Agriculture: *Soil Classification: A Comprehensive System,* U.S. Department of Agriculture, Washington, 1960.

Woods, K. B. (ed.): *Highway Engineering Handbook,* McGraw-Hill, New York, 1960, pp. 10-16 to 10-30.

2

WATER-BEARING FORMATIONS AND GROUNDWATER OCCURRENCE

The modes of groundwater occurrence are affected by the geologic development and properties, delineation, and boundary conditions of the soil and rock formations through which the water percolates. They depend also on ongoing activities and climatic and environmental conditions. These formations may be extensive or limited between natural and/or artificial boundaries. Groundwater management (Chaps. 4 and 7) requires careful investigations of the characteristics of the regional soil or rock formations, each of which represents a unit formed under similar geologic conditions. If water can be developed economically and in sufficient quantities for human and animal use, a formation is known as a *water-bearing formation* or an *aquifer* whether it is formed from rocks or soils or both. Thus a saturated clay layer cannot be called an aquifer, although it may contain more water than an equally thick sand layer.

The quality of water is an important aspect in groundwater utilization (Sec. 2.3). Water in nature is not pure; it contains dissolved and suspended matter that may affect the quality of the water to the extent that it is unsuitable for use as a water supply, in industry, and/or in agriculture. The quality of water in nature is affected by (1) the natural environment, such as the amount of rock and soil minerals dissolved in the groundwater and (2) the effects of human intervention, such as the injection of wastewater from an industrial complex into an aquifer. As a result of the past and present activities of human beings, the quality of water in nature cannot be easily attributed to only one of these factors.

2.1   Water-Bearing Formations

In groundwater studies, the rock structure of a formation is more important than its texture. The presence of small holes or vesicles throughout a formation (vesicular structure and vugs — Legget, 1962), such as those in pumice and some basalts, increases the permeability of these rocks. Open or closed rock fissures, in the form of joints and fractures, are important features in groundwater hydrology. Joints are fissures whose patterns tend to break the rock into cubes or regular blocks, whereas fractures may run in any direction. The rock may be practically impervious or impermeable° (that is, water cannot percolate through its microscopic voids). The presence of these deficient structures (vesicles, vugs, joints, and fractures) gives a rock formation a higher water-yielding capacity. These deficient features, which are desirable in groundwater hydrology, may be the source of failures in engineering structures. For example, noncoherent volcanic rocks may produce cavings in tunnels, slope failures, or piping and high seepage losses beneath dams.

The weathering of sedimentary rocks is discussed as an example to emphasize the fact that natural conditions are continuously changing. Weathering is affected by the resistance of cementing materials. The most durable cementing material is siliceous cement, and the least resisting cementing material is argillaceous cement (clay). Groundwater movement had been, and still is, slowly changing the characteristics of these formations and their cementing materials. Karstification, for example, is the result of natural processes in and on the earth's crust caused by the solution and leaching of limestone, dolomites, and other soluble rocks. There are numerous morphologic and hydrologic features in karsts (Colorado State University, 1976). Usually in groundwater hydrology, limestone (calcium carbonate) and dolomite (magnesium carbonate) are considered similar. However, dolomite has slightly higher microscopic (intrinsic) permeability than limestone. When limestone is partially converted to dolomite through the process of groundwater percolation carrying magnesium, its microscopic permeability increases. Legget (1962) believes that percolating water through limestone tends

---

° Usually the terms *impervious* and *impermeable* are used interchangeably to indicate that water (or any fluid) cannot percolate through. However, all nonmetallic structures are permeable. Thus the use of either term means *practically* impervious or impermeable.

to increase its yield over time. He estimated that for every 3784 m³ (1 million gallons) of water pumped out of the chalk of south England, about 1500 kg (3300 lb) of chalk would be removed, thus increasing pore space. The degree of permeability changes over time in this chalk. The process of weathering of sedimentary and other types of rocks illustrates that natural conditions are not static and that the water-bearing formations and the quality of groundwater are continuously changing.

Soil (unconsolidated rock) formations are usually bounded by rock formations. In many investigations, these rock layers were assumed impervious relative to the soil formations. However, the structural features of these rocks should not be ignored, especially in cases of leakage and saltwater intrusion (Chaps. 8 and 9). The main types of formations associated with groundwater occurrence are discussed in the following paragraphs.

Aquifers

Figure 2.1 shows a cross section of a pervious soil formation AB confined between relatively thin layers of compacted impervious soil (or massive rock) known as *confining beds.* These beds form more or less sealing boundaries that do not allow the easy penetration of water across them. The upper confining bed in this figure is overlain by a more or less uniform soil deposit that extends upward to the ground surface. All layers are exposed at their free end to an ocean or sea front. Figure 2.1 includes two aquifers AB and CD. The degree of water productivity of an

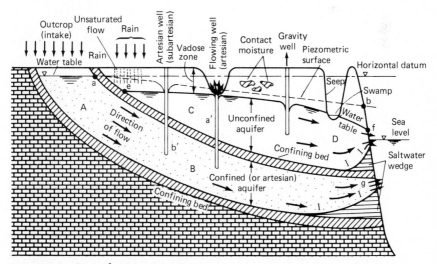

Figure 2.1 Groundwater occurrence.

aquifer depends on the degree of natural water recharge, its physical and chemical properties, climatic conditions, the characteristics of its geometric and hydrologic boundaries, vegetation, and human activities. The ease with which water penetrates through a soil formation defines the soil permeability. Therefore, aquifers should all have a relatively high degree of permeability.

Gravel aquifers are the most productive. Uniform sand aquifers (the uniformity coefficient $C_u < 6$) are less productive. Aquifers composed of well-graded dense soils, such as mixtures of sand and gravel, fine sand, loess, or mixed alluvial deposits containing small fractions of silt and products of rock disintegration or weathering that have never been transported, are among the least productive aquifers. The depth of an aquifer is also an important factor in its degree of productivity. It is obvious that a deep, saturated deposit of fine sand, for example, will yield more water to a well than a thin layer of clean coarse gravel. Aquifers of sand or gravel with little or no fine material are very productive. These generally occur as alluvium adjacent to and beneath streams and rivers. They also occur in buried valleys, abandoned river courses, and plains and valley fills between mountains.

Abundant water may be withdrawn from limestone as a result of the existence of solution channels, caverns, and fissures. A well drilled in a limestone aquifer therefore should intersect one or more of these openings. Similarly, in lava rocks, water is derived from the stored water in the interstitial spaces, open joints, and shrinkage cracks. The permeability and yield of rock aquifers depend on the degree of cementation of the rock minerals; the pattern and size of the cracks, fissures, solution channels, joints, and other openings; and the volume of shattered rock. The least productive rock aquifers are those of granite, quartzite, shale, slate, and schist. In a fault zone, crystalline rocks may yield water around 4 L/min in deep wells.

Examples of productive rocky aquifers are the Biscayne limestone, which provides the water supply for the Miami area; the Dakota and Nubian sandstone aquifers in Sudan, Egypt, and Saudi Arabia; and the Snake Plain aquifer, which is composed of a series of cracked lava flows and which supplies southern and eastern Idaho. In the latter formation, the water is suprisingly abundant, creating waterfalls from the cliffs of an area known as the Thousand Springs (Glover, 1974).

The natural velocities of groundwater are generally very small. They may be as low as 1.5 m per year and as high as 2.0 m per day. (Higher values up to 30 m per day have also been recorded.) Higher groundwater velocities are encountered as a result of artificial disturbances such as within zones in proximity to open channels or pumped wells.

Aquifers may be looked on as slow sand filters or underground reservoirs that perform better than surface reservoirs because of lesser degrees of pollution and evaporation and a lack of sedimentation. They also serve as conduits that transmit water from the intake zone to the withdrawal zone. Some aquifers are presently susceptible to pollution and contamination as a result of the disposal of wastes from industry and septic tanks, seepage from polluted streams, and leakage from landfills (Chap. 4). However, aquifers are recognized as water-quality control tools, especially for artificial recharge and wastewater.

In order to establish some comparative productivity indexes, aquifers have been classified according to various standards. One method of classification is based on an aquifer's ability to yield to a single well. For example, a certain aquifer is capable of yielding 23 m³/h (100 gal/min), and the pumped water contains no more than 1500 parts per million (ppm). Although this classification is extensively used by some geologists, it is definitely misleading. It is known that the yield of a well depends on several factors such as the effective radius of the well, its method of construction, its depth, and the present and future effects of water use such as pumping from neighboring wells. The quality of water may also change with time and/or as a result of the effects of environmental activities (Sec. 2.3). Another method of classifying aquifers is by reference to their source of replenishment. Some of the U.S. Geological Survey maps are based on this approach, and they should be helpful in water-resources planning. A third method of classification (which may be the best) is based on the geologic and engineering properties of the aquifer as well as the delineation of its boundaries. This method requires determination of the geometry of the aquifer, the properties of the rocks and/or soils and their state of densification, and their structural pattern. This last method of classification requires extensive and costly field and experimental investigations to determine hydrologic and other pertinent parameters of the aquifers.

Aquicludes and Aquitards

Aquicludes are practically impervious geologic formations that completely confine other strata and permit little water percolation across their boundaries. Aquicludes may themselves be completely saturated with water in a state of almost complete stagnation.

Aquitards are more pervious than aquicludes. They may be considered semiconfining beds through which water percolates at a very slow rate. The total amount of infiltrated water (known as *leakage* — see Secs. 6.4 and 8.3) may be very large if such aquitards occupy extensive areas. Aquitards also have different modes of occurrence. They may consist, for example, of thin clay layers interbedded in a sandy formation. In some cases, saucer-shaped aquicludes are encountered (Fig. 2.2). When water infiltrates downward from rainfall or other surface water sources, it is intercepted by these saucer-shaped aquicludes. The geometry of these aquicludes allows the formation of natural water reservoirs; this type of water is known as *perched water*. Sometimes these aquicludes flood (underground), and the water spills downward over the edges until it reaches the main groundwater table. In desert areas, perched water is used for limited water supplies. For example, perched water was the main water supply for the Allied troops during World War II along the coastal area of the Mediterranean Sea in the Western Desert of Egypt. The perched water table may be permanent or variable depending on the amount of water recharge.

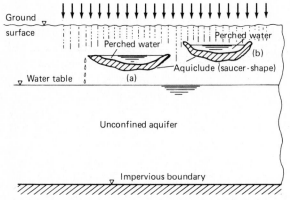

Figure 2.2 Perched water in aquicludes (saucer shaped) above the water table. *(a)* Water volume exceeds aquiclude capacity; *(b)* water volume is less than aquiclude capacity.

Confined and Unconfined Aquifers

There are two main idealized types of aquifers based on the nature of groundwater flow: *confined* (also known as *artesian* or *potentiometric* aquifers) and *unconfined* (also known as *water-table, ordinary,* or *gravitational* aquifers). In Fig. 2.1, aquifer *AB* is confined between two almost impervious beds and slopes gradually downward from its intake at the ground surface to its terminal boundary along the seashore. Confining beds allow practically no water to penetrate through. This is an idealization because water does flow across such beds in various amounts depending on the degree of permeability and the differences in water potentials above and below their boundaries. The groundwater level at the intake is exposed to the atmosphere and thus is known as a *ground-*

water table. However, as a result of continuous rainfall infiltration and other natural or artificial recharges and discharges, water moves within this aquifer, losing energy by friction through the pores as it travels. The line of pressure *ab*, which represents the distribution of water pressures along this aquifer, is the vertical projection of a surface known as the *piezometric (potentiometric) surface.* The head loss by friction while water is migrating from the intake point *a* to point *g* near the end is represented by the difference in elevations of points *a* and *b*. If an observation well (solid pipe with an open bottom) is drilled down to this confined aquifer at any location, the water will rise to the piezometric surface at that location. It should be emphasized that surface *ab* is not a groundwater table, yet it is a pressure diagram. The upper surface of water saturation within the aquifer *AB* coincides with the bottom surface of the upper confining bed.

The flow of groundwater through aquifer *AB* resembles to some extent the flow of water through a pipe conduit. If an observation well (or a productive well) is drilled at *a′* (Fig. 2.1), water will flow out of the well naturally without pumping because the original piezometric surface level at that particular location is above ground level. Such a well is known as a *flowing artesian well.* However, if another well is drilled at point *b′*, the piezometric surface level at this location is lower than the ground surface level so the water will spring up in the well to a level higher than the highest saturation level in the confined aquifer yet lower than the ground level. If one wishes to withdraw water from such a well, pumps must be used. Such a well is known as an *artesian well.* Actually both wells at *a′* and *b′* are of the artesian type; sometimes, however, the condition at *a′* is described as "artesian" and that at *b′* as "subartesian." Although any one of these two conditions depends on the level of the ground surface, *the flow pattern is totally independent of the surface topography.* Under artesian conditions, water pressures within the pores of the aquifer are always greater than atmospheric pressure unless water is depleted to levels lower than the upper boundary of the aquifer.

Since the upper confining bed of aquifer *AB* is overlain by aquifer *CD* (Fig. 2.1), water recharge from rainfall and other sources creates a saturated medium within the aquifer that produces a groundwater-table curve *ef* (known also as a *free surface* or *phreatic surface*). The water-table surface is, in fact, the locus of all points within the aquifer at which the pressure is atmospheric. The water table, however, is not the upper boundary of the saturated zone because it is well established that above the water table there should exist a vadose zone, which includes at its bottom a vertical range known as the *capillary fringe* (immediately above the water table) and at its top a soil-water zone (Fig. 2.3). The degree of saturation varies from 0 percent at the top of the fringe to almost 100

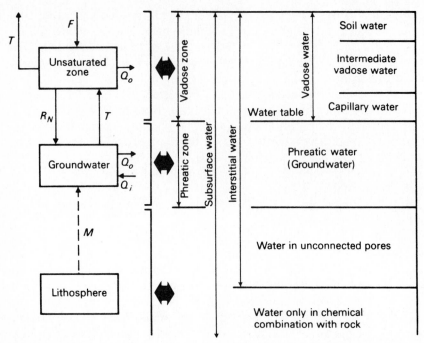

Figure 2.3 The groundwater subsystem and the water profile. *(P. A. Domenico, Concepts and Models in Groundwater Hydrology, McGraw-Hill, New York, 1972.)*

percent at the bottom close to the water table. The water pressure above the water table is always less than atmospheric pressure (the capillary fringe is not shown in Fig. 2.1). If an observation well, perforated throughout its entire depth, is drilled through the saturated zone of aquifer *CD* to a level slightly above the upper confining bed of aquifer *AB*, the water will rise to a level close to that of the water table. Such an aquifer is known as an *unconfined aquifer* (or ordinary, gravitational, or water-table aquifer). When the groundwater table intersects a depression, the water will seep to the side, forming a pond, swamp, lagoon, or lake, and the seeping water above the pond water level is known as a *seep* or *spring*. A seep is technically a spring with a very low rate of flow.

In Fig. 2.1, the right ends of all the aquifers and formations are in contact with a saltwater body (a sea or an ocean). In such a situation, the seawater invades both aquifers *AB* and *CD*, forming in each case a *saltwater wedge*. The boundary between the saltwater wedge and the fresh water is known as the *saltwater-freshwater interface*. The extent and shape of such an interface depend on many factors (Chap. 9). In an unconfined aquifer, fresh groundwater emerges below and above sea level. The exit surface is known as the *discharge* or *outflow face,* and usually the part of it that is above sea level is much smaller than that below. In an artesian aquifer, an outflow face through which fresh water flows to the ocean exists entirely below sea level. The zones of aquifers affected by saltwater intrusion from seas or oceans are identified as *coastal aquifers*. When dealing with saltwater intrusion problems, aquifers unaffected by saltwater are more precisely identified as *fresh-water aquifers*.

It should be noted that in the section shown in Fig. 2.1 and in similar geologic sections, the vertical scale is several times the horizontal scale. The actual slopes of confining beds are generally not as excessive as they appear in such drawings. The depth may be a few hundred meters as compared with a horizontal extent of, say, several kilometers. In most quantitative analyses of groundwater movements, the boundaries of the confining beds are assumed horizontal within the relatively small zones affected by local imbalances (such as pumped wells).

2.2 Groundwater Occurrence

Groundwater Origin and the Hydrologic Cycle

The *hydrologic cycle* (Fig. 2.4) represents the sequence of events when water drops from the atmosphere to the earth and hydrosphere (water

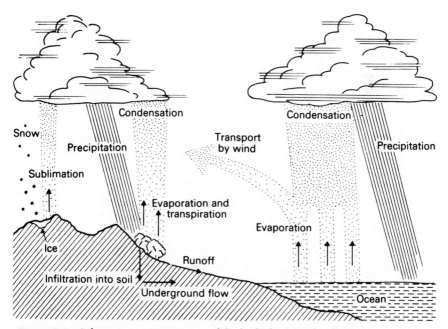

Figure 2.4 Schematic representation of the hydrologic cycle. *(P. A. Domenico, Concepts and Models in Groundwater Hydrology, McGraw-Hill, New York, 1972.)*

bodies such as rivers, oceans, and lakes covering the earth's surface) and then goes back to the atmosphere. Direct evaporation of water from oceans, seas, lakes, rivers (and other surface-water streams), wetlands, snow, and ice and the evapotranspiration of vegetation change water to particles of water vapor that are carried in the air. At high altitudes these particles form clouds which, by condensation, precipitate water (or snow) back to earth. As a result of wind action, the precipitation does not necessarily fall on the locations from which evaporation took place. Part of the precipitated water runs on the ground surface in the form of *surface runoff*, and this feeds streams, lakes, ponds, etc., or it drains to the seas and oceans before it evaporates and joins the hydrologic cycle again. Another portion of the precipitation is partially consumed by vegetation, and the rest infiltrates into the ground. This infiltrated portion may or may not reach the groundwater table. In addition, some of the precipitation evaporates before it reaches the ground surface.

The source of groundwater is precipitation and the infiltration of surface water from runoffs and streams. Although surface runoff represents water that was initially rejected by infiltration, after some time and under different physiographic and environmental conditions a portion of the runoff infiltrates to join the groundwater reservoir. Groundwater is obviously one of the phases of the hydrologic cycle.

The interaction between infiltrated water, precipitation, surface runoff, and groundwater is rather complex. It is affected by physiographic, geologic, meteorologic, and hydrologic factors as well as by human activities such as water use and pollution. Water-resource systems are interconnected by different water-resource subsystems (Fig. 2.5).

Surface runoff and evapotranspiration are considered water losses in terms of groundwater reservoirs. These losses may be permanent or temporary. Surface runoff, for example, may eventually discharge into a lake that recharges the groundwater reservoir.

Once rain droplets reach the ground, molecular and capillary forces begin to act, and this leads to a slow downward percolation of the water through zones of aeration known as *vadose zones* (Fig. 2.3), forming partially saturated flow systems before ultimately reaching the zone of complete saturation. The magnitude of infiltrated water depends primarily on rain intensity and duration, the type of soil and its presaturation conditions, vegetation, and climatic conditions. The continuous process of infiltration leads to an accretion to the zone of saturation that produces a rise in the water-table level. Water recharge through the outcrops of artesian aquifers and through direct and indirect leakage increases the amount of water in storage in these aquifers as well as the water pressures within the pores.

The relationship between monthly rainfall in centimeters and

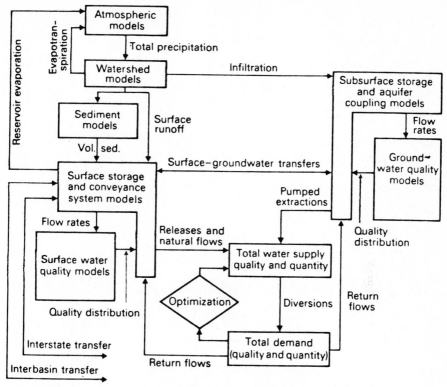

Figure 2.5 Water-resource systems. *(W. C. Ackermann, "Scientific Hydrology in the Unites States," The Progress of Hydrology, Internal Seminar for Hydrology Professors, 1st., University of Illinois, Champaign, 1969, pp. 563–571.)*

groundwater elevations and/or pressures is not easily established, although it does exist (Jacob, 1945). It depends on several factors, the most important of which are environmental, climatic, and physiographic conditions, the physical properties of the soil formations, and the nature of water usage. Such relationships vary with time in the same location (DeWiest, 1965).

A supply of water that raises the water-table level in an unconfined aquifer or increases the storage (and pressure) in an artesian aquifer is considered a *recharge* or *replenishment.* However, a *discharge* is the withdrawal of water out of an aquifer, depleting unconfined aquifers and decreasing storage (and pressure) in artesian aquifers. Both recharges and discharges may be either natural or artificial; their rates vary with time and depend on the geometric boundaries of flow media, the physical properties of their materials, climatic conditions, and water usage. A continuous circulation of groundwater within natural aquifers should therefore exist at all times.

Accretion in a groundwater reservoir may take place by *bank storage*, which implies seepage from streams, lakes, and flooded lands of higher water levels than that of the adjacent water table. The inflow components also may originate from the return flow of irrigation and sewage waters, lateral flow from adjacent groundwater basins, or any artificial recharge. The circulation of groundwater in certain zones changes its pattern as a result of changes in the natural or artificial conditions. Streams fed by an aquifer are known as *influent streams*. They may change to *effluent streams* and replenish the same aquifer. Usually, streams in arid regions are influent, whereas streams in humid areas are effluent. Bank storage and other water recharges are, in reality, indirect supplies from precipitation.

The depletion of groundwater levels in unconfined aquifers and the decrease in water pressures in confined aquifers are produced by the outflow of water due to several causes, such as artificial discharge (e.g., pumping out of wells); rejected inflow; natural outflow from flowing wells, springs, or seeps; evaporation; evapotranspiration of vegetation; and natural discharge into adjacent groundwater basins. During the process of evaporation, water vapor may completely fill the voids and migrate from regions of higher potential to regions of lower potential. In arid zones, aquifers do not usually reject any recharge, yet evaporation creates problems of water shortage. Arid zones that are not yet developed are invaded by natural plants known as *phreatophytes* (salt grass, greasewood, mesquite, and salt cedar) that consume enormous amounts of water. If such regions are developed, an abundant amount of water can be salvaged. It was estimated in 1958 that 64,750 km² (16 million acres) of the western lands in the United States were covered by these phreatophytes (Kashef, 1965) and that they consumed an annual amount of 30.8 km³ (25 million acre-feet) of water.[*] This is equivalent to about 8 percent of the entire surface storage in the United States (including the Great Lakes) or about 20 percent of the capacity of the largest artificial reservoir in the world[†] (Kashef, 1981*b*).

The evaluation of rates of water inflow and outflow in a certain aquifer or basin of known boundaries is essential to establish a water budget or groundwater inventory. The optimum condition exists when the maximum amount of water is withdrawn without producing any undesirable effect, such as excessive groundwater lowering (necessitating changes in pump capacities and depths or deepening of wells), land

---

[*] 1.0 km³ = 1 billion cubic meters.
[†] Water reservoir of the High Aswan Dam (in Egypt and Sudan).

subsidence (Chap. 8) due to excessive pumping, and saltwater intrusion (Chap. 9). If the annual rate of outflow does not exceed the annual rate of inflow, a safe but not an optimum (safe and economical) condition is reached. Aquifers are not only transmitting media, but are also reservoirs that yield water from storage. The problem of the "safe yield" of aquifers (Chap. 4) is one of the most complex phases of groundwater-resource management (Kazman, 1956, 1965).

Modes of Groundwater Occurrence

There are different modes of occurrence of water in soils (Fig. 2.4). *Hygroscopic moisture* is the moisture absorbed by dry soil from the atmosphere near the ground surface. When water infiltrates through the soil, it forms isolated moist patches as a result of the molecular attraction of soil grains for water. Such water cannot be separated by gravitational movement. It occupies the upper part of the vadose zone (Fig. 2.3). Immediately above the water table, a capillary fringe develops as a result of the upward movement of water under the action of capillarity. Generally, such water movement in the vadose zone is due to both capillary action and evaporation. Downward movement is due to infiltration of rain and surface water. *Vadose water* (Fig. 2.3) is mainly the unsaturated zone above the water table. It may be completely absent in humid areas when the water table is close to the ground surface, and it may be very deep in arid zones.

Below the water table, completely saturated soil is encountered. It should be realized that immediately above the water table, within the lower part of the capillary zone, the soil is also completely saturated even though the pore-water pressure is negative. The water table should therefore be thought of as the surface along which the pore-water pressures are atmospheric rather than as the upper level of saturation. In practical engineering problems, major attention is focused on completely saturated zones, whereas partially saturated soils are of prime concern to agronomists and highway and agricultural engineers. Only saturated flow systems are treated in this text, except in the very few instances where vadose zones and capillary action are discussed.

Genetically, water may be classified as follows (White, 1957*a*, *b*; Domenico, 1972):

> *Meteoric water* is water that has been recently involved in the hydrologic cycle. Its isotopic composition is similar to that of surface water. Meteoric water exists in the hydrosphere and occurs as relatively shallow groundwater in the aquifers used for water withdrawal and water recharge.

Marine water is water that has recently intruded (in a geologic sense) into coastal aquifers. It has an isotopic composition similar to that of seawater. This water is of interest to groundwater planners and engineers and should be recognized in water-resource management techniques (Chap. 9).

Connate water consists of water that migrated from where it was first buried with sedimentation. It is highly mineralized (Feth, 1965) and may have been derived from either oceanic or fresh water. Connate water is also known as *fossil water.* Sometimes it is not completely isolated from the hydrologic cycle, especially if it has been recently developed for use. Samples of fossil water taken from deep wells in the Egyptian Western Desert (Nubian sandstone aquifer) appear to be 20,000 to 30,000 years old (Kashef, 1981*a*). The tests were made using carbon 14 dating, which is capable of detecting ages of 20,000 to 30,000 years and more. The recent development (about 1960) of fossil water in Egypt encountered several problems, mainly due to the lack of water replenishment in the desert arid regions.

Magmatic water, such as the mineralized water of thermal springs, is derived from magmas. If these magmas are relatively shallow (3 to 5 km), the water is known as *volcanic water.* Deep thermal springs that produce water at very high temperatures are known as *volcanic springs* (Bryan, 1919). If the magmas occur at considerable depth, the water is known as *plutonic water.*

Metamorphic water is associated with rocks during their metamorphism.

Juvenile water is a recent water of magnetic or cosmic origin that has never been part of the hydrosphere.

Metamorphic and juvenile waters are not of interest to engineers. Magmatic water is of casual interest to engineers in its relation to springs. Interest in connate water is limited to its possible development in areas with no other water resources. Meteoric and marine waters are obviously of direct interest to engineers.

Metamorphic and juvenile waters are not part of the hydrologic cycle, while marine and meteoric waters join directly the hydrologic cycle. Connate (fossil) and magmatic waters may find their way to the ground surface and join the hydrologic cycle.

There are various types of water springs depending on their modes of occurrence and the media through which they flow (Bryan, 1919). They may be shallow springs or seeps, such as those developed as a result of gravitational movement of groundwater near the ground surface, or they may be very deep, such as volcanic springs associated with magmatic water (White, 1957b). If a spring discharges water through deep rock fractures, it is known as a *fissure spring*. In rocky coastal areas, submarine springs may develop and discharge through subsurface channels (Legget, 1962). Earthquakes produce changes in the discharges of springs and may develop new springs.

2.3 Groundwater Quality

Pure water (H_2O) does not generally occur in nature. Even rainwater is not pure, as was thought in the past. Natural water, from surface or ground sources, contains dissolved solids and gases as well as suspended matter. The quality and quantity of these constituents depend on geologic and environmental factors, and they are continuously changing as a result of the reaction of water with contact media and human activities. What is known as *natural water* may have already been polluted, and the term may be misleading. Natural water corresponds to the state of water at the time it was used or sampled for investigations and analyses.

In order to determine the acceptable water quality for recharge purposes or for use in agriculture and industry or for human use, the water has to undergo certain tests. Generally, these are chemical, physical, biologic, and radiologic tests. The results of these tests are then compared with the acceptable standard for any particular use. These standards vary from one discipline to another. For example, acceptable water quality for agriculture may be unacceptable for drinking. Even within a certain discipline such as industry, an acceptable water-quality standard for a certain operation may not be acceptable for another operation.

Water temperature is an important aspect of groundwater quality, especially for water used in industry, as a water supply, and as an environment for aquatic life. The temperature of groundwater in a certain location is usually uniform throughout the year. Groundwater is preferred to surface water as a water supply and in industry for this reason.

Study of the factors affecting groundwater quality may enable the prediction of its future quality as compared with its present quality. Groundwater-quality determinations may be important for reasons other than its suitability for certain uses. For example, Kunkle (1965) computed the groundwater runoff in a stream from measurements of the total dissolved solids (TDS) or ionic concentration of the groundwater and the total surface runoff of the stream. Study of the chemical constituents of water and their changes may also be useful in locating the source or sources of recharge, the direction of groundwater flow, the presence of aquifer boundaries (Hem, 1970; Walton, 1970; Todd, 1980), and the configuration of flow systems in these aquifers (Henningson, 1962; Toth, 1966; Maxey and Mifflin, 1966; Brown, 1967; Mifflin, 1968).

Changes in groundwater quality are due to the varying quality of the infiltrated precipitation, the reaction of groundwater to its environment, the length of the flow path, the residence time of water in a certain location, vegetative types, and human-determined features. In magmatic water, the quality also changes as a result of the absorption of gases.

Precipitation Effects

Groundwater quality is affected by the various phases of the hydrologic cycle. Atmospheric precipitation (rain and snow) consists of water (H_2O) that loses its purity while traveling through the atmosphere before reaching the earth's surface. The impurities and chemicals in the atmosphere originate from various natural and artificial sources, such as volcanic ashes and gases, dust and other windborne solids, airborne salts originating above the surfaces of salt water in coastal areas, reactions caused by lightning and cosmic rays, gas emissions from industrial plants, and radionuclides produced by nuclear explosions. Before nuclear weapons testing began in 1952, the tritium content (see the section "Radionuclides in Groundwater") of most precipitation ranged from about 1 to 10 tritium units (TU). After 1954, the tritium content increased from 10 TU to more than 20 TU in many areas (Walton, 1970) and then started to decay.

When rain reaches the ground, the impurities and chemicals in it have already been partially dissolved and they proceed to mix with surface water. The rainwater then reacts chemically with the minerals of soils and rocks. The amount and type of minerals that dissolve in the rainwater depend on the chemical composition and physical structure of these soils and rocks as well as the hydrogen-ion concentration (pH) and the redox potential (Eh) of water (Back and Hanshaw, 1965). The *hydrogen-ion concentration* (pH) refers to the effective concentration (activity) of hydrogen ions in the water. It is expressed as the negative logarithm to the base 10 of the H^+ activity in moles per liter (Bouwer, 1978). Distilled water at 25°C has an H^+ activity of 10^{-7} mol/L; therefore, its pH $= -\log 10^{-7} = 7$. The solution is considered neutral (equal number of H^+ and OH^- ions) when the pH $= 7$. When the pH < 7, the solution is considered acidic. Such water affects the life of pumps, well screens, and casings. Alkaline water has a pH > 7.0. Most natural waters have a range of pH between 6 and 8.5 (Hem, 1970). When pH values are greater than 8.5, water has usually come in contact with sodium carbonate; when pH values are below 4, water contains free acids. Moderately low values of pH indicate water that contains small amounts of mineral acids from sulfide sources or contains organic acids.

The *redox potential* (Eh) is a measure indicating the ability of a natural environment to promote or reduce a positive valence or loss in electrons (Domenico, 1972). Simply stated, it is a measure of the energy required to remove electrons from ions in a given chemical environment (Domenico, 1972).

Concentrations of chemicals in precipitation water vary seasonally as well as locally. Most of the ammonium and nitrate in water is soil-derived (Junge, 1958). The higher values are in the range of 2 mg/L in the United States.° The chloride (Cl) and potassium (K) concentrations reach, respectively, about 8 and 0.4 mg/L in coastal areas; the sodium concentrations vary from about 0.1 to 0.3 mg/L inland to about 4 mg/L along the coast; and calcium concentrations increase as a result of frequent dust storms and the presence of alkali soils (Junge and Werby, 1958).

Dissolved carbon dioxide (CO_2) from the atmosphere produces carbonic acid, which leads to rainwater with a pH that is less than 7. In industrial areas with severe air pollution, the pH of precipitation may be as low as 4.5 (Cogbill and Likens, 1974). Matcha (1983) believes that any precipitation with a pH below 5.6 should be viewed as acid rain and by implication caused by humans. It was surprising to find precipitation pH values near 4.3 in Hawaii at elevations of 3300 m in remote areas far from atmospheric pollution caused by industry. In addition, the precipitation had anions of sulfate rather than of nitrate. It is believed that the sulfate may have come from long-range transport of distant human-produced pollutants or from natural ocean biota (Matcha, 1983).

Effects of Soils and Rocks

When rainwater reaches the ground, it leaches the soil mantle and other materials decomposed from plants and animals. Through the root zones of vegetation, the chemical composition of water changes as a result of ion exchange between the constituents in the water and the soil, as well as because of the uptake of nutrients by plants. The infiltrating water is usually enriched with nitrate, phosphate, and potassium, which originate in chemical fertilizers. As plants and other living organisms produce

° The concentration of chemical elements or ions is expressed in milligrams per liter (ml/L) or parts per million (ppm) by weight. For example, if 1 L of a salt solution weighs 1.05 kg and the weight of the solid salt content (after drying) is 0.15 kg, then the concentration of salts or total dissolved solids (TDS) is 0.15×10^6 mg/L. If the TDS is expressed in parts per million, it is $0.15/1.05 \times 10^6 = 0.142 \times 10^6$ ppm. Generally, for concentrations less than about 7000 mg/L, the two expressions (milligrams per liter and parts per million) are numerically equal because the dissolved solids are generally a small fraction of 1 percent of the total weight of the groundwater sample. The concentration of different ions also may be expressed as a unit of chemical equivalence as milliequivalents per liter (mEq/L). The conversion between milligrams per liter and milliequivalents per liter is given in Table 2.1 (Hem, 1970).

TABLE 2.1 Conversion Factors for Chemical Equivalence*

Chemical constituent	Conversion factor
Aluminum (Al^{3+})	0.11119
Ammonium (NH_4^+)	0.05544
Barium (Ba^{2+})	0.01456
Beryllium (Be^{3+})	0.33288
Bicarbonate (HCO_3^-)	0.01639
Bromide (Br^-)	0.01251
Cadmium (Cd^{2+})	0.01779
Calcium (Ca^{2+})	0.04990
Carbonate (CO_3^{2-})	0.03333
Chloride (Cl^-)	0.02821
Cobalt (Co^{2+})	0.03394
Copper (Cu^{2+})	0.03148
Fluoride (F^-)	0.05264
Hydrogen (H^+)	0.99209
Hydroxide (OH^-)	0.05880
Iodide (I^-)	0.00788
Iron (Fe^{2+})	0.03581
Iron (Fe^{3+})	0.05372
Lithium (Li^+)	0.14411
Magnesium (Mg^{2+})	0.08226
Manganese (Mn^{2+})	0.03640
Nitrate (NO_3^-)	0.01613
Nitrite (NO_2^-)	0.02174
Phosphate (PO_4^{3-})	0.03159
Phosphate (HPO_4^{2-})	0.02084
Phosphate ($H_2PO_4^-$)	0.01031
Potassium (K^+)	0.02557
Rubidium (Rb^+)	0.01170
Sodium (Na^+)	0.04350
Strontium (Sr^{2+})	0.02283
Sulfate (SO_4^{2-})	0.02082
Sulfide (S^{2-})	0.06238
Zinc (Zn^{2+})	0.03060

* Concentration in milligrams per liter times the conversion factor yields concentration in milliequivalents per liter.

SOURCE: J. D. Hem, "Study and Interpretation of the Chemical Characteristics of Natural Water," 2d ed., U.S. Geological Survey Water Supply Paper 1473, 1970.

carbon dioxide (CO_2) and organic acids, the pH value of the infiltrating water decreases. Leachate from irrigated water therefore has salt concentrations much higher than those of surface water.

In arid and poorly drained zones, the rate of evaporation is usually higher than the rate of infiltration. This leads to the accumulation of salts near the ground surface, increasing the salt concentration above that produced by plants and fertilizers. An example of extreme salt accumulation is that of the palayas in the arid state of Nevada. Palayas are essentially dry lakes or depressions with relatively impermeable soil beds.

The acidity of rainwater and surface water (such as irrigation water) thus increases before and shortly after infiltrating into the ground. Accordingly, the rate of soil and rock weathering accelerates. Depending on the soluability and mineralogic characteristics of soils and rocks, the TDS of the infiltrating water vary. The degree of solubility is also affected by temperature and pressure. Seasonal variations in water recharge and discharge produce seasonal fluctuations in the chemical quality of water. The geochemistry of natural waters is therefore cyclic and influenced by the hydrologic cycle. The geochemical cycle is represented diagrammatically in Fig. 2.6.

The main soluble salts produced by the geochemical cycle have been divided into two main groups by Chebotarev (1955). The predominant ions that are present in relatively large proportions are Na^+, K^+, Ca^{2+}, Mg^{2+}, H^+, HCO_3^-, CO_3^{2-}, Cl^-, and SO_4^{2-}, and those infrequently encountered or of inconsistent occurrence are NH_4^+, Al^{3+}, Fe^{2+}, Fe^{3+}, NO_2^-, NO_3^-, SO_3^{2-}, OH^-, and SiO_3^{2-} (cations are positively charged ions and anions are negatively charged ions). Less commonly encountered constituents that may be important because of their beneficial or deleterious effects are given by the U.S. Bureau of Reclamation (1977) as boron (B), lead (Pb), arsenic (As), selenium (Se), barium (Ba), copper (Cu), zinc (Zn), hydrogen sulfide (H_2S), methane (CH_4), oxygen (O_2), and carbon dioxide (CO_2). Table 2.2, as presented by Durfer and Becker (1962) and modified by Todd (1980), gives more details. Table 2.3 (Davis and DeWiest, 1966) indicates the relative abundance of dissolved solids in potable water.

The salt concentration of meteoric groundwater increases with decreases in its velocity or when it has been isolated from the hydrologic cycle in locations with high mineral soils, such as connate and magmatic waters (Davis and DeWiest, 1966). Water velocities in aquifers commonly decrease with depth; the deep water layers in deep aquifers are almost stagnant. Thus water salinity increases with depth. In addition, bicarbonate waters occur near the top and chloride waters near the bottom of water-bearing formations.

Groundwaters (and surface waters) are classified according to their concentration of salinity as *fresh*, *brackish* (or moderately saline), *saline* (or very saline), and *briny waters*. The TDS of these various types of water are as follows: less than 1000 mg/L for fresh water, a range between 1000 and 10,000 mg/L for brackish or moderately saline water, a range between 10,000 and 100,000 mg/L for saline or very saline water,

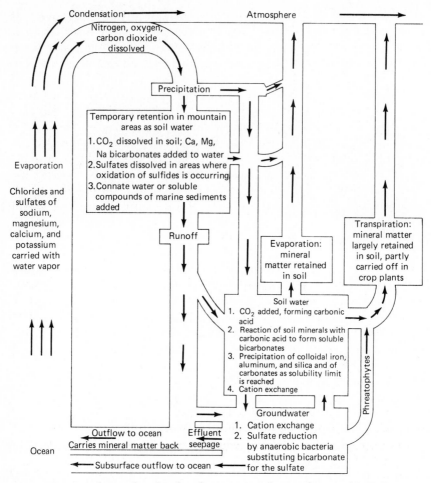

Figure 2.6 Geochemical cycle of surface water and groundwater. *(G. H. Davis et al., "Groundwater Conditions and Storage Capacity in the San Joaquin Valley, California," U.S. Geological Survey Water Supply Paper 1469, 1959.)*

and more than 1000,000 mg/L for brine (Carroll, 1962). These limits may change as a result of the arbitrary nature of classifications. For example, Hem (1970) gave the lower limit of moderately saline water as 3000 mg/L and the higher limit of very saline water (i.e., the lower limit of brines) as 35,000 mg/L.

Approximately two-thirds of the United States has saline waters with TDS > 1000 mg/L. Saline aquifers are undesirable as freshwater sources, yet they can be used as reservoirs for temporary storge of fresh water or for wastewater disposal. In industry, brackish water may be used for cooling purposes.

TABLE 2.2 Principal Chemical Constituents in Groundwater: Their Sources, Concentrations, and Effect on Usability

Constituent	Major natural sources	Concentration in natural water	Effect on usability of water
Silica (SiO_2)	Feldspars, ferromagnesium and clay minerals, amorphous silica, chert, opal	Ranges generally from 1.0 to 30 mg/L, although as much as 100 mg/L is fairly common; as much as 4000 mg/L is found in brines.	In the presence of calcium and magnesium, silica forms a scale in boilers and on steam turbines that retards heat; the scale is difficult to remove. Silica may be added to soft water to inhibit corrosion of iron pipes.
Iron (Fe)	Igneous rocks: amphiboles, ferromagnesian micas, ferrous sulfide (FeS), ferric sulfide or iron pyrite (FeS_2), magnetite (Fe_3O_4) Sandstone rocks: oxides, carbonates, and sulfides or iron clay minerals	Generally less than 0.50 mg/L in fully aerated water. Groundwater having a pH less than 8.0 may contain 10 mg/L; rarely as much as 50 mg/L may occur. Acid water from thermal springs, mine wastes, and industrial wastes may contain more than 6000 mg/L.	More than 0.1 mg/L precipitates after exposure to air; causes turbidity, stains plumbing fixtures, laundry, and cooking utensils, and imparts objectionable tastes and colors to foods and drinks. More than 0.2 mg/L is objectionable for most industrial uses.
Manganese (Mn)	Manganese in natural water probably comes most often from soils and sediments. Metamorphic and sedimentary rocks and mica biotite and amphibole hornblende minerals contain large amounts of manganese.	Generally 0.20 mg/L or less. Groundwater and acid mine water may contain more than 10 mg/L.	More than 0.2 mg/L precipitates upon oxidation; causes undesirable tastes, deposits on foods during cooking, stains plumbing fixtures and laundry, and fosters growths in reservoirs, filters, and distribution systems. Most industrial users object to water containing more than 0.2 mg/L.

TABLE 2.2 Principal Chemical Constituents in Groundwater: Their Sources, Concentrations, and Effect on Usability *(Continued)*

Constituent	Major natural sources	Concentration in natural water	Effect on usability of water
Calcium (Ca)	Amphiboles, feldspars, gypsum, pyroxenes, aragonite, calcite, dolomite, clay minerals	Generally less than 100 mg/L; brines may contain as much as 75,000 mg/L.	Calcium and magnesium combine with bicarbonate, carbonate, sulfate, and silica to form heat-retarding, pipe-clogging scale in boilers and in other heat-exchange equipment. Calcium and magnesium combine with ions of fatty acid in soaps to form soapsuds; the more calcium and magnesium, the more soap required to form suds. A high concentration of magnesium has a laxative effect, especially on new users of the supply.
Magnesium (Mg)	Amphiboles, olivine, pyroxenes, dolomite, magnesite, clay minerals	Generally less than 50 mg/L; ocean water contains more than 1000 mg/L, and brines may contain as much as 57,000 mg/L.	
Sodium (Na)	Feldspars (albite); clay minerals; evaporites, such as halite (NaCl) and mirabilite ($Na_2SO_4 \cdot 10H_2O$); industrial wastes	Generally less than 200 mg/L; about 10,000 mg/L in seawater; about 25,000 mg/L in brines.	More than 50 mg/L sodium and potassium in the presence of suspended matter causes foaming, which accelerates scale formation and corrosion in boilers.
Potassium (K)	Feldspars (orthoclase and microcline), feldspathoids, some micas, clay minerals	Generally less than about 10 mg/L; as much as 100 mg/L in hot springs; as much as 25,000 mg/L in brines.	Sodium and potassium carbonate in recirculating cooling water can cause deterioration of wood in cooling towers. More than 65 mg/L of sodium can cause problems in ice manufacture.
Carbonate (CO_3)	Limestone, dolomite	Commonly less than 10 mg/L in groundwater. Water high in sodium may contain as much as 50 mg/L of carbonate.	Upon heating, bicarbonate is changed into steam, carbon dioxide, and carbonate. The carbonate combines with alkaline earths — principally calcium and magnesium — to form a crustlike scale of calcium carbonate that retards flow of heat through pipe walls and restricts
Bicarbonate (HCO_3)	Limestone, dolomite	Commonly less than 500 mg/L; may exceed 100 mg/L	

Constituent	Source	Concentration	Significance
	in water highly charged with carbon dioxide.		flow of fluids in pipes. Water containing large amounts of bicarbonate and alkalinity is undesirable in many industries.
Sulfate (SO_4)	Oxidation of sulfide ores; gypsum; anhydrite	Commonly less than 300 mg/L except in wells influenced by acid mine drainage. As much as 200,000 mg/L in some brines.	Sulfate combines with calcium to form an adherent, heat-retarding scale. More than 250 mg/L is objectionable in water in some industries. Water containing about 500 mg/L of sulfate tastes bitter; water containing about 1000 mg/L may be cathartic.
Chloride (Cl)	Chief source is sedimentary rock (evaporites); minor sources are igneous rocks.	Commonly less than 10 mg/L in humid regions but up to 1000 mg/L in more arid regions. About 19,300 mg/L in seawater; and as much as 200,000 mg/L in brines.	Chloride in excess of 100 mg/L imparts a salty taste. Concentrations greatly in excess of 100 mg/L may cause physiological damage. Food-processing industries usually require less than 250 mg/L. Some industries — textile processing, paper manufacturing, and synthetic rubber manufacturing — desire less than 100 mg/L.
Fluoride (F)	Amphiboles (hornblende), apatite, fluorite, mica	Concentrations generally do not exceed 10 mg/L. Concentrations may be as much as 1600 mg/L in brines.	Fluoride concentration between 0.6 and 1.7 mg/L in drinking water has a beneficial effect on the structure and resistance to decay of children's teeth. Fluoride in excess of 1.5 mg/L in some areas causes "mottled enamel" in children's teeth. Fluoride in excess of 6.0 mg/L causes pronounced mottling and disfiguration of teeth.

TABLE 2.2 Principal Chemical Constituents in Groundwater: Their Sources, Concentrations, and Effect on Usability *(Continued)*

Constituent	Major natural sources	Concentration in natural water	Effect on usability of water
Nitrate (NO_3)	Atmosphere; legumes, plant debris, animal excrement	Commonly less than 10 mg/L.	Water containing large amounts of nitrate (more than 100 mg/L) is bitter tasting and may cause physiological distress. Water from shallow wells containing more than 45 mg/L has been reported to cause methemoglobinemia in infants. Small amounts of nitrate help reduce cracking of high-pressure boiler steel.
Dissolved solids	The mineral constituents dissolved in water constitute the dissolved solids.	Commonly contains less than 5000 mg/L; some brines contain as much as 300,000 mg/L.	More than 500 mg/L is undesirable for drinking and many industrial uses. Less than 300 mg/L is desirable for dyeing of textiles and the manufacture of plastics, pulp paper, rayon. Dissolved solids cause foaming in steam boilers; the maximum permissible content decreases with increases in operating pressure.

SOURCE: C. N. Durfer and E. Becker, "Public Water Supplies of the 100 Largest Cities in the United States," U.S Geologic Survey Water Supply Paper 1812, 1962; modified by D. K. Todd, *Ground Water Hydrology*, 2d ed., Wiley, New York, 1980.

TABLE 2.3 Dissolved Solids in Potable Water: Tentative Classification of Abundance

Major constituents, 1.0 to 1000 mg/L	Secondary constituents, 0.01 to 10.0 mg/L	Minor constituents, 0.0001 to 0.1 mg/L	Trace constituents, generally less than 0.001 mg/L
Sodium	Iron	Antimony°	Beryllium
Calcium	Strontium	Aluminum	Bismuth
Magnesium	Potassium	Arsenic	Cerium°
Bicarbonate	Carbonate	Barium	Cesium
Sulfate	Nitrate	Bromide	Gallium
Chloride	Fluoride	Cadmium°	Gold
Silica	Boron	Chromium°	Indium
		Cobalt	Lanthanum
		Copper	Niobium°
		Germanium°	Platinum
		Iodide	Radium
		Lead	Ruthenium°
		Lithium	Scandium°
		Manganese	Silver
		Molybdenum	Thallium°
		Nickel	Thorium°
		Phosphate	Tin
		Rubidium°	Tungsten°
		Selenium	Ytterbium
		Titanium°	Yttrium°
		Uranium	Zirconium°
		Vanadium	
		Zinc	

° These elements occupy an uncertain position in the list.
SOURCE: S. N. Davis and R. J. M. DeWiest, *Hydrogeology*, Wiley, New York, 1966.

Igneous and crystalline aquifers usually yield excellent-quality groundwater with salt concentrations lower than 100 mg/L and not exceeding about 500 mg/L. Good-quality water is encountered in the upper coastal aquifers and sand dunes. At lower depths in coastal aquifers, brackish to very saline waters are encountered as a result of saline water intrustion (Chap. 9). Groundwater of good quality is also encountered in volcanic and sedimentary aquifers with certain limitations. Water in sandstone aquifers may be high in sodium (Na^+) and bicarbonates (HCO_3^-); that in shale aquifers may be slightly acidic and high in iron, sulfate (SO_4^{2-}), and fluoride (F^-); and that in limestone aquifers may be slightly alkaline and contain calcium (Ca^{2+}) and magnesium (Mg^{2+}). Chemical analyses of groundwater for various geologic formations are given by White et al. (1963).

Hardness of Groundwater

When acidic rainwater infiltrates down to limestone or dolomite formations, it dissolves the calcium and magnesium carbonates, producing groundwater that is "hard." Hard water does not react with soap and thus is undesirable for household cleaning purposes. It produces streaking on chinaware, glassware, silverware, and car bodies. When hard water is transported and/or heated it produces an accumulation of scale and incrustations° in reservoirs and conduits. Hard groundwater is usually withdrawn from aquifers with dolomite, limestone, and gypsum overlain by thick topsoil layers. The continuous dissolving action of acidic water in such aquifers increases the hardness until the carbon dioxide content of the groundwater is exhausted, at which time carbonate hardness becomes stable. However, the sulfate content in groundwater depends on the dissolving action of water rather than on its carbon dioxide content. The dissolving action increases with the residence time of water; thus the sulfate content increases continuously as long as water is flowing. Sulfates constitute the largest dissolved solid, although carbonates exceed sulfates in these aquifers.

Softer water is thus closer to recharge zones than to discharge areas. When a river is hydraulically connected with an aquifer, the degree of hardness follows a seasonal pattern; the hardness and alkalinity of

---

° Incrustation may be considered the opposite of corrosion; it is characterized by an accumulation of minerals. In a water well, minerals are deposited primarily in and around the openings (circular holes or slots) of screens and in the voids of the formation surrounding the well. This accumulation of minerals blocks the passage of water into the well, eventually reducing its efficiency, i.e., decreasing the discharge and increasing the drawdown (Chap. 8).

groundwater increase gradually during long periods when the river is low (Jeffords, 1945) and decrease abruptly during periods when flooding occurs or when the river is high (Fig. 2.7). However, groundwater withdrawn as a result of induced infiltration from a stream exhibits a more or less uniform chemical quality in comparison with surface water from the same stream (Walton, 1970). Although most of the groundwater in this case is derived from the stream water, its quality is not affected by sudden changes in the quality of the surface water. Generally, ground-

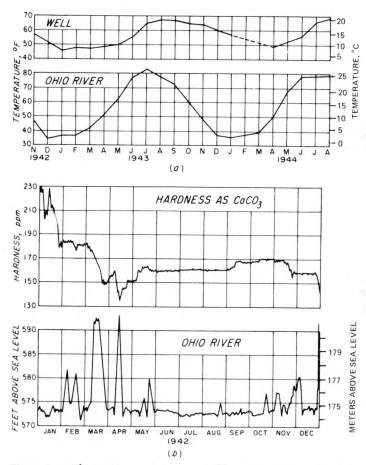

Figure 2.7 Fluctuations in average monthly temperature and in daily hardness. *(a)* Water pumped from a well at Point Pleasant, West Virginia; *(b)* water pumped from wells at Parkersburg, West Virginia. (*R. M. Jeffords, "Recharge to Water-Bearing Formations along the Ohio Valley," J. Am. Water Works Assoc., vol. 37, no. 2, 1945.*)

waters produced by induced infiltration from streams are free from taste, odor, turbidity, pathogenic bacteria, and organic matter unless the streams carry exceptionally heavy organic loads (Klaer, 1963).

Hardness of water H_r in general is numerically evaluated in terms of the concentrations of the calcium (Ca^{2+}) and magnesium (Mg^{2+}) ions. H_r is expressed in milligrams per liter as $CaCO_3$ (calcium carbonate) and is calculated from the equation $H_r = 2.5Ca + 4.1Mg$ (Todd, 1980). Water is classified as *soft* when H_r is less than 75 mg/L as $CaCO_3$, *moderately hard* when H_r ranges from 75 to 100 mg/L as $CaCO_3$, *hard* when H_r ranges from 150 to 300 mg/L as $CaCO_3$, and *very hard* when H_r exceeds 300 mg/L as $CaCO_3$ (Sawyer and McCarty, 1967).

Only soft water can be used domestically; otherwise water-softening processes are required. The effects of soft and hard water on health have been rather controversial (Crawford, 1972; Neri et al., 1975). Acceptable degrees of hardness in industry vary considerably depending on the type of industry. The proper locations of wells yielding water of appropriate degrees of hardness suitable for a certain industry should be carefully investigated. Such studies usually include drawing contour maps for groundwater hardness within the site. Potential changes in the magnitude of existing hardness due to environmental factors should also be evaluated.

Dissolved Gases

Most groundwaters contain dissolved gases that are included in the geochemical cycle (Fig. 2.6). These gases are derived either from the atmosphere (nitrogen, oxygen, and carbon dioxide) or from decaying organic material (which produces hydrogen sulfide and the flammable gas methane). Gases that occur less frequently are oxides of nitrogen, sulfur dioxide, and ammonia. Hydrogen sulfide is distinguished by its "rotten egg" odor at concentrations of less than 1.0 mg/L. Methane gas is produced in large amounts by the decomposition of buried organic materials, such as in solid-waste dumps (recent fills) and cemeteries. Methane gas is colorless and tasteless. Water containing as little as 1 to 2 mg/L of methane can produce an explosion in a poorly ventilated air space ("Gas in Ground Water," 1969). Methane is the cause of many accidents, including suffocations in dug wells and pump pits and explosions and fires in well pits. Methane gas may explode beneath such structures as basement and floor slabs in buildings and ground tanks because of its pressure unless these structures are properly ventilated.

Dissolved gases in groundwater are released by increasing temperature and/or reduced pressure. In most groundwaters, the concentration of dissolved gases varies from 1.0 to 100 ppm (Davis and DeWiest,

1966). Gases are damaging to pumps and well casings because of both corrosion and cavitation (U.S. Bureau of Reclamation, 1977).

Causes and Effects of Changes in Groundwater Quality

The interaction between groundwater and its natural environment tends to create a chemical equilibrium that leads to a stationary groundwater quality. However, such factors as chemical reactions, circulation of water of different quantities from various sources, and withdrawal of water and recharge of aquifers by polluted and clean water lead to changes in the chemical and other qualities of water.

The base exchange of ions (Chap. 1) produces chemical and physical changes in groundwater quality. Base exchange involves mainly certain cations, i.e., sodium, calcium, and magnesium. When high-sodium water infiltrates through soil, the sodium displaces calcium and other ions, leading to an accumulation of materials within the soil voids and thus reducing soil permeability. However, if gypsum ($CaSO_4$) is added to the soil, calcium will be the predominant cation, and by base exchange, the soil will become more permeable. These changes in permeability produce changes in the groundwater velocity that affect the water quality.

Base exchange is known to change water from hard to soft. The high amounts of calcium and magnesium in coastal aquifers are attributed to base exchange and cannot be the result of simple mixing of seawater and fresh water. The base-exchange phenomenon may help to explain certain properties that cannot be explained by simple-solution chemistry or the pattern of groundwater flow. For example, a salt content higher than that of seawater may be found in groundwaters that have not been in contact with evaporites (Hanshaw, 1972). Fine-grained materials such as clay or shale hold back certain ions through the process of base exchange and allow the passage of neutral substances. Aquicludes and aquitards, which are semipermeable soil layers, can thus produce differences in osmotic pressure, salt sieving (or ultrafiltration), and differences in electric potential (Hanshaw, 1972). A sandstone aquifer overlying shale indicated exceptionally low head pressures; this has been attributed to osmotic cross-formational withdrawal of water through shale (Hanshaw, 1972).

Changes in chemical quality are more intense in shallow aquifers than in deeper ones because shallow aquifers are more easily affected by seasonal variations and human activities. Chemical precipitation removes ions in solution by forming insoluble compounds, thus changing the chemical quality of water. The precipitation of calcium carbonate and the release of carbon dioxide result from pressure decreases and/or temperature increases.

Physical Properties of Groundwater

The physical analysis of groundwater includes the determination of its color, taste, odor, turbidity, and temperature. Water color is produced by minerals and organic matter in solution and is usually reported in milligrams per liter compared to a standard solution (Hem, 1970). Tastes and odors are usually defined by experience, although they can be qualitatively determined on the basis of the maximum degree of dilution in comparison with tasteless and odorless water (American Public Health Association, American Water Works Association, and Water Pollution Control Federation, 1975).

Turbidity is a measure of suspended and colloidal matter in water, such as organic matter, microscopic organisms, and particles of clay and silt. Its measurement may be determined on the basis of the length of the light path through the water that just causes the image of a standard candle to disappear (Todd, 1980). Groundwater withdrawn from sand and gravel aquifers is subject to natural filtration and is usually free of turbidity. Examples of water with high turbidity are water withdrawn from wells in fractured or cavernous rock, water from muddy springs that discharge shallow groundwater after rainy periods (Bouwer, 1978), and water with small particles of ferric oxide that result from rusty pump casings and columns. The concentration of all suspended matter should not exceed 5 ppm in production wells (Walton, 1970).

The temperature of shallow groundwater is affected by diurnal and seasonal fluctuations of surface temperature and to a lesser degree by the geothermal gradient from the interior of the earth. Owing to the geothermal gradient, water increases in temperature between 1.0 and 5°C (average about 2.5°C) for each 100 m of depth (White, 1973). The temperature of deep groundwater remains more or less constant. It has been found that the temperature of groundwater at depths between 10 and 20 m is almost constant and normally exceeds the mean annual air temperature by 1.0 to 1.5°C (Collins, 1925). The more or less stationary temperature of groundwater is an advantage in the use of groundwater in industry, as a water supply, and in such other applications as heat-pump temperature control in buildings for cooling in summer and heating in winter (Gass and Lehr, 1977).

The temperature of groundwater withdrawn from a production well by induced infiltration from a river should be expected to follow the same trend as that of the river water, yet it should be lesser in value and lagging in time (Walton, 1970) (Fig. 2.7). Deep-seated geothermal water can have a temperature above the boiling point of water (100°C); a range of temperatures between 200 and 300°C is not uncommon (Bouwer, 1978). Practical uses of geothermal waters from springs or wells include heating homes and greenhouses, industrial applications, fish culture, therapeutic purposes, and, if hot enough, steam generation for electric power (Koenig, 1973; Bouwer, 1978).

The temperature of water should be measured in the field immediately after water sampling. Immediate field analyses should also be conducted for turbidity, odor, and color.

Chemical and Biological Testing

It is usual practice to sample water from wells after allowing some pumping time in order to avoid nonrepresentative samples of stagnant or polluted water. Such a practice in fact results in sampling of water from various sources and layers. Consequently, the results of such tests should be given as a weighted average of the final product. In more refined analyses, multilevel water sampling using special sampling techniques is required (Chap. 4). In this way, a more precise distribution of water quality by depth is obtained. Special water sampling techniques are also required for the analysis of organic constituents.

Water samples should be of sufficient volume to conduct the required tests. These samples should be securely sealed in decontaminated bottles (sometimes amber bottles are required to protect the sample from light) and transported promptly to the laboratory for immediate testing; otherwise they should be stored in a cool place. However, storage may affect quality; for example, cations of Fe, Cu, Al, Mn, Cr, and Zn are subject to loss due to the ion exchange between the cations and the walls of the glass containers (Rainwater and Thatcher, 1960).

Before leaving the site, field analyses should be done to determine the pH, hardness, dissolved gases, specific electric conductance (explained later), and approximate chemical characteristics of the water. Although field tests are usually less accurate than laboratory tests, they give useful data for planning a reliable testing program. Portable laboratory kits for simple field chemical tests are commercially available.

Standard laboratory methods of water analysis are specified by the American Public Health Association, American Water Works Association, and Water Pollution Control Federation (1975) and other organizations (American Society for Testing and Materials, 1969). Chemical quality tests are usually presented either in tabular form or by means of graphic representation (Hem, 1970). The latter is convenient for comparing the results of various samples, identifying water of different constituents, and detecting chemical reactions that resulted from groundwater circulation. There are different methods of graphic representation (Schoeller, 1962; Todd, 1980). Schoeller (1962) has also shown that graphic methods can be used to determine the degree of saturation of $CaCO_3$ and $CaSO_4$ in groundwater.

The TDS can be approximately, yet rapidly, determined by measuring the electric conductance of groundwater. (The TDS includes all solid material in solution whether ionized or not; it does not include suspended materials or dissolved gases.) *Electric conductance* is a measure of the ability of water to conduct an electric current; it is a function of water temperature and TDS. The reciprocal of electric conductance is known as *resistance.* The electric conductance of 1 cm^3 of a solution at 25°C signifies its *specific electric conductance.* It is measured in microsiemens per centimeter (μS/cm), which is equivalent to micromhos per centimeter (mho/cm). In practice, the unit length (centimeters) is omitted. The value of the specific electric conductance increases with increases in salt content. It also increases by about 2 percent as a result of an increase in temperature of 1°C. The conductance of distilled water varies from 0.5 to 5.0 mho, that of normal groundwater varies from 30 to 2000 mho, and that of ocean water varies from 45,000 to 55,000 mho (Davis and De-Wiest, 1966). The relationship between milliequivalents per liter and milligrams per liter for most natural waters (range 100 to 5000 μS/cm) was given by Logan (1961) and Richards (1954) as 1 Meq/L of cations = 100 μS/cm and 1 mg/L = 1.56 μS/cm.

The interpretation of chemical test results of groundwater requires statistical analysis and the preparation of maps, diagrams, and graphs. Hydrographs of chemical quality are useful for studying the changes in water quality over time. Maps showing the areal distribution of TDS or individual constituents may be used to study the hydrologic and mineralogic characteristics of an aquifer system.

Microbial activity in aquifers is difficult to investigate because it is impossible to drill a sterile well (Willis et al., 1975). The origin of bacteria in aquifers is not known. It has been verified that varied and active microfloras exist in deep formations (McNabb and Dunlap, 1957; Smith et al., 1976; McCabe et al., 1970). Generally, bacteria do not migrate as far in fine-grained soils as they do in coarse-grained soils or fractured rocks (Romero, 1970). Disease-causing bacteria (pathogenic bacteria), viruses, and other microorganisms that are not native to the subsurface environment do not multiply and eventually die; these are, fortunately, rarely found in groundwater (Walton, 1970). Poor well construction and other sources of contamination necessitate biological analyses to isolate and identify bacteria of the coliform group. This test is required to verify the safety of drinking water (Walton, 1970; Todd, 1980). Coliform test results are reported as the most probable number (MPN) of coliform group organisms in a given volume of water using probability tables (Todd, 1980).

Organic materials in groundwater are usually tested by the carbon-chloroform extraction (CCE) method and the carbon-alcohol extraction (CAE) method (Bouwer, 1978). Biochemical oxygen demand (BOD) is a measure of the biodegradable material in water. Generally, BOD values in groundwater are essentially zero (Bouwer, 1978). Organic matter in peat and muck deposits are relatively stable, resulting in low BOD levels in groundwater.

Radionuclides in Groundwater

A *nuclide* is a particular constituent of an atomic nucleus; for example, the chlorine 37 nuclide has a nucleus of 17 protons and 20 neutrons. Unstable nuclides that tend to change spontaneously to other species of nuclides through various decay reactions (disintegration) are called *radioactive nuclides* or *radionuclides* (Walton, 1970). Tritium, for example, is an isotope product of cosmic-ray bombardment of ^{14}N; tritium is hydrogen 3 and has the symbol T or ^3H. Isotopes have the same chemical properties as their chemical elements but different atomic weights; thus tritium (^3H) is three times as heavy as hydrogen (H) (Walton, 1970). Radioactive substances give off alpha (α), beta (β), and gamma (γ) radiation as a result of nuclear decay reactions. The quantity of radionuclides in groundwater (or any other fluid) is measured indirectly by measuring the number of decay reactions during a given length of time. The basic unit used is the curie, which is defined as the number of decay reactions per second of 1 g of radium or 3.7×10^{10} disintegrations per second. Microcuries (μCi) and picocuries (pCi) are commonly used; 1 pCi/L = $10^{-9} \mu Ci/mL$. One tritium unit (TU) produces roughly 3.2×10^{-3} pCi/mL (picocuries per milliliter) of activity.

Radionuclides are found in natural groundwaters in several areas, especially where there is geothermal activity (Scott and Barker, 1962). Uranium concentrations in groundwater range from 0.05 to 10 parts per billion (ppb) (Davis and DeWiest, 1966). Most groundwater in sedimentary aquifers has a ^{226}Ra (radium) concentration of less than 1×10^{-9} $\mu Ci/mL$ (probable median of about $3 \times 10^{-6} \mu Ci/mL$); normal groundwater contains ^{222}Rn (radon) in a range of concentration between less than $1 \times 10^{-7} \mu Ci/mL$ and $3 \times 10^{-5} \mu Ci/mL$ and ^{40}K (potassium) in concentrations of about $2 \times 10^{-9} \mu Ci/mL$ (Walton, 1970). ^{226}Ra is a daughter product originating from the decay of ^{238}U (uranium), which in turn disintegrates to ^{222}Rn (a gaseous element). Radioactivity in groundwater is caused generally by the radionuclides ^{238}U (uranium), ^{232}Th (thorium), and ^{235}U (uranium) and to a lesser extent by ^{40}K and ^{87}Rb (robidium). Quantity standards of radionuclides in groundwater are expressed as the maximum permissible concentration (MPC) in microcuries per milliliter (Table 2.4).

Radionuclides in groundwater originate from three sources (Walton, 1970).

TABLE 2.4   Maximum Permissible Concentrations (MPC) of Radionuclides in Groundwater

Element	Mass no. of radioisotope	Half-life°	Radiation	MPC above natural background, μci/mL†
Barium	131	13 d	γ	2×10^{-4}
	140	12.8 d	β^-, γ	3×10^{-5}
Bromine	82	36 h	β^-, γ	3×10^{-4}
Calcium	45	153 d	β^-	9×10^{-6}
Carbon	14	5,600 y	β^-	8×10^{-4}
Cerium	144	290 d	β^-, γ	1×10^{-5}
Cesium	135	2.9×10^6 y	β^-	1×10^{-4}
	137	33 y	β^-, γ	2×10^{-5}
Chlorine	36	4×10^5 y	β^-	8×10^{-5}
Chromium	51	27.8 d	γ	2×10^{-3}
Cobalt	57	270 d	β^+, γ	5×10^{-4}
	60	5.3 y	β^-, γ	5×10^{-5}
Hydrogen	3	12.4 y	β^-	3×10^{-3}
Iodine	129	1.72×10^7 y	β^-, γ	4×10^{-7}
	131	8.04 d	β^-, γ	2×10^{-6}
Phosphorus	32	14.3 d	β^-	2×10^{-5}
Plutonium	238	92 y	α, γ	5×10^{-6}
	239	2.4×10^4 y	α, γ	5×10^{-6}
	240	6,580 y	α	5×10^{-6}
	242	5×10^5 y	α	5×10^{-6}
Radium	226[a]	1,620 y	α, γ	1×10^{-8}
	228	6.7 y	β^-	3×10^{-8}
Radon	222	3.83 d	α	A gas
Rubidium	86	18.7 d	β^-, γ	7×10^{-5}
	87	6×10^{10} y	β^-	1×10^{-4}
Ruthenium	103	40 d	β^-, γ	8×10^{-8}
	106	1 y	β^-	1×10^{-5}
Sodium	22	2.6 y	β^+, γ	4×10^{-5}
Strontium	89	51 d	β^-	1×10^{-5}
	90[a]	29 y	β^-	1×10^{-7}
Sulfur	35	88 d	β^-	6×10^{-5}
Uranium	235	7.1×10^8 y	α	3×10^{-5}
	238	4.5×10^9 y	α, γ	4×10^{-5}
Zinc	65	245 d	β^+, γ	1×10^{-4}

°Years, y; day, d; hours, h.
† In solution in water
SOURCE: Atomic Energy Commission, *Standards for Protection against Radiation*, pp. 10920–10924 [compiled by Walton (1970)], *Federal Register*, Nov. 17, 1960.

1.   Some radionuclides could have been part of the original material that formed the earth; in this case, they are known as *primordial nuclides* (Davis and DeWiest, 1966). This hypothesis is based on the long half-lives (Table 2.4) of such radionuclides as uranium and thorium 322 (which is used as a fuel in nuclear reactors). Table 2.4 compares the

half-lives of uranium and thorium with those of the daughter nuclides radium, radon, and tritium.

2. Some radionuclides were originally formed in the atmosphere by cosmic-ray bombardment, such as ^{14}N, ^{16}O, and ^{40}Ar (half-lives less than 1 million years). The best known products of these are tritium (from ^{14}N), which has half-life of 12.3 years, and radiocarbon or carbon 14 (^{14}C), which has half-life of 5730 years.

3. Some radionuclides originate from human-created nuclear explosions; the most hazardous of these in radioactive fallout are strontium (^{90}Sr) and cesium (^{37}Cs). These hazardous radionuclides can be adsorbed by clayey soils and organic material, yet they contaminate groundwater through coarse-grained aquifer outcrops or fractured and cavernous rocks exposed directly to the atmosphere.

Water-Quality Criteria

The quality of groundwater is continuously changing as a result of natural and human activities. The quality at a certain time may therefore be better or worse than at a future time. For this reason, frequent testing should be done to examine the suitability of groundwater for such applications as drinking, industrial, and/or agricultural use. The test results should meet the criteria for each type of water usage; otherwise the source location should be changed, the facility (such as an industrial complex) should be relocated, or the water should be treated to change its quality to the desired standards. It should be noted that it is rather complex to differentiate between the natural and human factors that cause the degradation of water quality. For example, rainfall in a certain area may be polluted as a result of industry or nuclear explosions in areas other than those where it precipitated, or it may be polluted after it reaches the ground from such sources as agricultural fertilizers.

The three main water-quality standards are water supply for humans, water for industry, and water for irrigation. If treatment of existing groundwater is planned in order to achieve better-quality water, the water at the source does not have to meet the required standard unless the cost of treatment is prohibitive.

The quality of a water supply that is to be used for drinking purposes is standardized in the United States (Table 2.5) by the U.S. Environmental Protection Agency (EPA). Gerba et al. (1975) suggested that drinking water should contain less than one virus per 400 to 4000 L. It is known that disinfection by chlorination or ozonation removes all viruses unless the water contains suspended solids that harbor and protect viruses against disinfection (White, 1975). As stated earlier, groundwaters are generally of low turbidity as compared with surface waters. Bacterial

TABLE 2.5   **Drinking Water Standards in the United States**

Physical characteristics		
Criterion	Recommended limit[a]	Tolerance limit[b]
Color, units	15	
Odor, threshold number	3, inoffensive	
Residue:		
Filtrable, mg/L	500	
Taste	Inoffensive	
Turbitity, units	5	

Inorganic chemicals		
Substance	Recommended limit,[a] mg/L	Tolerance limit,[b] mg/L
Alkyl benzene sulfonate (ABS)	0.5	
Arsenic (As)	0.01	0.05
Barium (Ba)	—	1.0
Cadmium (Cd)	—	0.01
Carbon chloroform extract (CCE)	0.2	
Chloride (Cl)	250	
Chromium, hexavalent (Cr^{6+})	—	0.05
Copper (Cu)	1.0	
Cyanide (CN)	0.01	0.2
Fluoride (F)	$0.8-1.7^{c,d}$	$1.4-2.4^{d}$
Iron (Fe)	0.3	
Lead (Pb)	—	0.05
Manganese (Mn)	0.05	
Mercury (Hg)	—	0.002
Nitrate (as N)	10	
Phenolic compounds (as phenol)	0.001	
Selenium (Se)	—	0.01
Silver (Ag)	—	0.05
Sulfate (SO_4)	250	
Zinc (Zn)	5	

Organic chemicals	
Substance	Tolerance limit, mg/L
Chlorinated hydrocarbons:	
Endrin	0.0002
Lindane	0.004
Methoxychlor	0.1
Toxaphene	0.005
Chlorophenoxys:	
2, 4-D	0.1
2, 4, 5-TP Silvex	0.01

Biological standards	
Substance examined	Maximum permissible limit
Standard 10-mL portions	Not more than 10% in 1 month shall show coliforms[e]
Standard 100-mL portions	Not more than 60% in 1 month shall show coliforms[e]

Radioactivity	
Substance	Recommended limit, pCi/L
Radium 226 (^{226}Ra)	3
Strontium 90 (^{90}Sr)	10
Gross beta activity	1000[f]

[a] Concentrations that should not be exceeded where more suitable water supplies are available.
[b] Concentations above this constitute grounds for rejection of the supply.
[c] Dependent on annual maximum daily air temperature.
[d] Where fluoridation is practiced, minimum recommended limits are also specified.
[e] Subject to further specified restrictions.
[f] In absence of strontium 90 and alpha emitters.
SOURCE: U.S. Environmental Protection Agency, *National Interim Primary Water Regulations*, vol. 40, no. 248, pp. 59566–59588 [compiled by Todd (1980)], *Federal Register*, Dec. 24, 1975.

contamination of groundwater does not generally occur within its source. Polluted surface water percolating into a water supply well or contamination of a distribution system due to structural failures in pipelines constitute other external polluting factors.

Water-quality standards for industrial use depend on the function of the industry. Recommended water criteria for selected industries are given in Table 2.6 (American Water Works Association, 1971). Water-quality requirements at point of use for fruit and vegetable processing and paper and textile plants are given in Table 2.7 (National Academy of Sciences and National Academy of Engineers, 1972). Standards for water quality for steam generation and cooling in heat exchanges are given in Table 2.8 (National Academy of Sciences and National Academy of Engineers, 1972). Some industries require water-quality standards similar to those of drinking water, such as the soft-drink industry. Additional information for the use of water in industry can be found in McKee and Wolf (1963), American Society for Testing and Materials (1966, 1967, and 1969), and the National Technical Advisory Committee (1968).

Groundwater quality for irrigation purposes is determined on the basis of the effects of the water on the quality and yield of the crops, as

TABLE 2.6 Ranges in Recommended Limiting Concentrations for Industrial Process Waters

Use	Turbidity, units	Color, units	Taste and odor threshold	Dissolved solids	Hardness, as CaCO3	Alkalinity, as CaCO3	pH, units	Chlorides, as Cl	Sulfates, as SO4	Iron, as Fe	Manganese, as Mn	Iron plus manganese	Hydrogen sulfide	Fluorides, as FL	Other requirements
Air conditioning			Low							0.5	0.5	0.5			Not corrosive or slime-promoting
Baking	10	10	None-low		[a]					0.2	0.2	0.2	0.2		Potable
Boiler feed															Potable if steam is used for food preparation
Brewing	0-10	0-10	None-low	500-1,500[b]	[c]	75-80[d]	6.5-7.0[e]	60-100		0.1	0.1	0.1	0.2	1.0	Potable; numerous other requirements
Carbonated beverages	1-2	5-10	None-low	850	200-250	50-130		250	250	0.1-0.2	0.2	0.1-0.4	0-0.2	0.2-1.0	Potable; COD, 1.5; organic matter, infinitesimal; algae and protozoa, none
Confectionery			Low	50-100	Soft					0.2	0.2				Potable
Dairy		None	None	500[f]	180		>7.0	30	60	0.1-0.3	0.03-0.1	0.2	0.2		Potable; NO3 as N, 5.5; NO2 as N, 0; NH3 as N, trace only; COD as KMnO4, 12
Drinking	5	15	3, inoffensive	500		30-250	>7.5	250	250	0.3	0.05			1.4-2.4[g]	Potable
Food canning and freezing	1-10		None-low	850	[h]					0.2	0.2	0.2-0.3	1.0	1.0	Potable; free from saprophytic organisms; NaCl, 1000-1500; NO3 as N, 2.8; NH3 as N, 0.4
Food equipment, washing	1	5-20	None	850	10	30-250		250	250			0.1	1.0	1.0	Potable; organic matter, infinitesimal
Food processing, general															
Ice manufacture	1-10	5-10	Low	850	10-250					0.2	0.2	0.2-0.3		1.0	Potable
Laundering	5	5	Low	170-1300	0-50	60	6.0-6.8[e]			0.2-1.0	0.2	0.2		[i]	Potable; SiO2, 10
Paper and pulp, fine	10	5		200	100[j]	75				0.1	0.05	0.2-1.0			Soluble SiO2, 20; free CO2, 10; residual Cl2, 2

Industry													Other
Paper, groundwood	50[k]	30		500	200	150			75	0.3	0.1			Soluble SiO_2, 50; free CO_2, 10
Paper, kraft, bleached	40	25		300	100	75		200	200	0.2	0.1			Soluble SiO_2, 50; free CO_2, 10
Paper, kraft, unbleached	100	100		500	200	150		200	200	1.0	0.5			Soluble SiO_2, 100; free CO_2, 10
Paper, soda and sulfate pulps	25[k]	5		250	100[l]	75		75	75	0.1	0.05			Soluble SiO_2, 20; free CO_2, 10
Rayon and acetate fiber pulp production	5	5	100[f]	8	50–75					0.05	0.03	0.05		Al_2O_3, 8; Si, 25, Cu, 5
Rayon manufacture	0.3		Low	55				20	20	0.0	0.0	0.0		
Sugar			Low	Low			7.8–8.3	20	20	0.1	0.0			Ca, 20; Mg, 10; bicarbonate, as $CaCO_3$, 100; sterile, no saprophytic organisms
Tanning	20	10–100		50–500	130		6.0–8.0		100	0.1–0.2	0.1–0.2	0.2		Bicarbonate hardness, low
Textile	0.3–25	0–70		0–50	75			100	100	0.1–1.0	0.05–1.0	0.2–1.0		COD, 8; heavy metals, none; Ca, 10; Mg, 5; bicarbonate, as $CaCO_3$, 200

Note: Units are milligrams per liter (mg/L) except as otherwise noted.

a Some calcium is necessary for yeast action. Too much hardness retards fermentation, but too little softens the gluten to produce soggy bread. Water of zero hardness is required for some cakes and crackers.

b Not more than 300 mg/L of any one substance.

c $CaSO_4$ less than 100 to 500 mg/L; $MgSO_4$ less than 50 to 200 mg/L.

d For dark beer, alkalinity as $CaCO_3$ may be 80 to 150 mg/L.

e Range, lower to upper limits.

f Total solids.

g Tolerance limit depends on annual average of maximum daily air temperatures for a minimum of 5 years.

h For legumes, 25 to 75; for fruits and vegetables, 100 to 200; for peas, 200 to 400.

i 1.5 mg/L of fluoride has been reported to cause embrittlement and cracking of ice.

j Calcium hardness, 50.

k No gritty material.

l Calcium hardness, 50; magnesium hardness, 50.

SOURCE: American Water Works Association, *Water Quality and Treatment*, McGraw-Hill, New York, 1971.

TABLE 2.7 Quality Requirements of Water at Point of Use for Fruit and Vegetable Processing, Paper Manufacturing, and Textile Plants*

	Fruits and vegetables	Paper	Textile
Acidity (H₂SO₄)	0		
Acidity (H_2SO_4)	0		
Alkalinity ($CaCO_3$)	250	75–150	50–200
Aluminum oxide (Al_2O_3)			8
Calcium (Ca)	100		
Calcium hardness ($CaCO_3$)		0–50	
Carbon dioxide (CO_2)		10	
Chloride (Cl)	250	0–200	100
Chlorine (Cl)	†	0–2	
Color, units	5	5–100	0–5
Copper			0.01–5
Fluoride (F)	1‡		
Hardness ($CaCO_3$)	250	100–200	0–50
Iron (Fe)	0.2	0.1–1	0–0.3
Magnesium hardness ($MgCO_3$)		0–50	
Manganese (Mn)	0.2	0.03–0.5	0.01–0.05
Nitrate (NO_3)	10‡		
Nitrite (NO_2)	0		
Organics (carbon tetrachloride extractables)	0.2		
Organic growths		0	
pH, units	6.5–8.5		
Silica (SiO_2)	50	20–100	25
Sulfate (SO_4)	250		100
Suspended solids	10	10–100	0–5
Total dissolved solids	500	200–500	100–200
Turbidity, units			0.3–5

° Numbers indicate milligrams per liter (mg/L) that normally should not be exceeded.
† Process water is chlorinated to prescribed levels. Unchlorinated water is used for canning syrups.
‡ Low values should be used for baby food.
SOURCE: Derived from National Academy of Sciences and National Academy of Engineering, *Water Quality Criteria 1972*, Washington, 1972.

well as the effects on drainage efficiency and characteristic changes in the soil (Richards, 1954; Wilcox, 1955). Extensive research has been done in irrigation and agriculture. Other detailed information regarding water quality for irrigation can be found in the following references: National Academy of Sciences and National Academy of Engineers (1972), van Schilfgaarde et al. (1973), Ayres (1975, 1977), and Bouwer

TABLE 2.8 Quality Requirements of Water at Point of Use for Steam Generation and Cooling in Heat Exchangers

| Characteristic | Boiler feedwater, quality of water prior to the addition of chemicals used for internal conditioning | | | | Cooling water | | | |
| | Industrial | | | Electric utilities 102 to 340 atm | Once through | | Makeup for recirculation | |
	Low pressure, 0 to 10 atm	Intermediate pressure, 10 to 48 atm	High pressure, 48 to 102 atm		Fresh	Brackish°	Fresh	Brackish°
Silica (SiO$_2$)	30	10	0.7	0.01	50	25	50	25
Aluminum (Al)	5	0.1	0.01	0.01	†	†	0.1	0.1
Iron (Fe)	1	0.3	0.05	0.01	†	†	0.5	0.5
Manganese (Mn)	0.3	0.1	0.01	0.01	†	†	0.5	0.02
Calcium (Ca)	†	0.4	0.01	0.01	200	420	50	420
Magnesium (Mg)	†	0.25	0.01	0.01	†	†	†	†
Ammonium (NH$_4$)	0.1	0.1	0.1	0.07	†	†	†	†
Bicarbonate (HCO$_3$)	170	120	48	0.5	600	140	24	140
Sulfate (SO$_4$)	†	†	†	†,‡	680	2 700	200	2 700
Chloride (Cl)	†	†	†	0.5	600	19 000	500	19 000
Dissolved solids	700	500	200	0.5	1 000	35 000	500	35 000
Copper (Cu)	0.5	0.05	0.05	0.01	†	†	†	†
Zinc (Zn)	†	0.01	0.01	0.01	†	†	†	†
Hardness (CaCO$_3$)	350	1.0	0.07	0.07	850	6 250	650	6 250
Alkalinity (CaCO$_3$)	350	100	40	1	500	115	350	115
pH (units)	7.0–10.0	8.2–10.0	8.2–9.0	8.8–9.4	5.0–8.3	6.0–8.3	†	†

TABLE 2.8 Quality Requirements of Water at Point of Use for Steam Generation and Cooling in Heat Exchangers (Continued)

Characteristic	Boiler feedwater, quality of water prior to the addition of chemicals used for internal conditioning — Industrial: Low pressure, 0 to 10 atm	Intermediate pressure, 10 to 48 atm	High pressure, 48 to 102 atm	Electric utilities 102 to 340 atm	Cooling water — Once through: Fresh	Once through: Brackish°	Makeup for recirculation: Fresh	Makeup for recirculation: Brackish°
Organics:								
Methylene blue active substances	1	1	0.5	0.1	†	†	1	1
Carbon tetrachloride	1	1	0.5	†, §	¶	¶	1	2
Chemical oxygen demand (COD)	5	5	1.0	1.0	75	75	75	75
Hydrogen sulfide (H_2S)	†	†	†	†	—	75	†	†
Dissolved oxygen (O_2)	2.5	0.007	0.007	0.007	Present	Present	†	†
Temperature					5 000	2 500	100	100
Suspended solids	10	5	0.5	0.05				

Note: Unless otherwise indicated, units are milligrams per liter (mg/L) and values that normally should not be exceeded. No one water will have all the maximum values shown.

° Brackish water—dissolved solids more than 1000 mg/L by definition 1963 Census of Manufacturers.

† Accepted as received (if meeting other limiting values); has never been a problem at concentrations encountered.

§ Controlled by treatment for other constituents.

‡ Zero, not detectable by test.

¶ No floating oil.

SOURCE: Derived from Academy of Sciences and National Academy of Engineering, *Water Quality Criteria 1972*, Washington, 1972.

TABLE 2.9 **Quality Criteria for Drinking Water for Farm Animals**

	Maximum concentration, mg/L
Total dissolved solids	3000
Aluminum	5
Arsenic	0.2
Boron	5
Cadmium	0.05
Chromium	1
Cobalt	1
Copper	0.5
Fluorine	2
Lead	0.1
Mercury	0.01
NO_3 (as N) plus NO_2 (as N)	100
NO_2 (as N)	10
Selenium	0.05
Vanadium	0.1
Zinc	25

SOURCE: Derived from National Academy of Sciences and National Academy of Engineering, *Water Quality Criteria 1972*, Washington, 1972.

(1978). Water-quality criteria for farm animals have also been established (Table 2.9) (National Academy of Sciences and National Academy of Engineers, 1972).

PROBLEMS AND DISCUSSION QUESTIONS

2.1   Define or explain the following:
   (a) Water-bearing formations
   (b) Aquifers (confined and unconfined)
   (c) Aquicludes
   (d) Aquitards
   (e) Flowing artesian well
   (f) Spring
   (g) Seep
   (h) Saltwater-freshwater interface
   (i) Coastal aquifer
   (j) Subartesian condition

2.2   What is the common range of natural water velocities?

2.3   Explain the various approaches to classifying aquifers.

2.4   By means of net sketches, explain the difference between a water table and a piezometric surface (potentiometric surface).

2.5   Construct your own flowchart for the hydrologic cycle on the basis of Fig. 2.3 and the explanations given in Sec. 2.2.

2.6   State the factors affecting the relationship between monthly rainfall and groundwater elevations and/or pressure. Explain the means of establishing such relationships by examining pertinent literature in your library. Refer to at least two studies in two different states.

2.7   By drawing appropriate freehand sketches, explain the terms *influent* and *effluent streams* and *bank storage.*

2.8   Study the appropriate sections in the works by Kazmann (1956 and 1965) and write three to four pages on the "safe yield of aquifers."

2.9   Indicate by a sketch the main three water systems within the vadose zone and explain differences in pore-water pressure in each system and in the saturated zone below the water table.

2.10   What would be the effect of the following factors on the vadose water?
(a) Excessive rainfall
(b) Excessive evaporation

2.11   Explain the relationship between the hydrologic cycle and the various genetic types of water.

2.12   List the main types of springs and explain the basis of their classification.

2.13   Why is rainwater not considered pure water?

2.14   List the main types of tests used to determine groundwater quality.

2.15   What was the effect of nuclear weapons testing in 1952 on the quality of precipitation?

2.16   Explain pH (hydrogen-ion concentration) and Eh (redox potential) of water.

2.17   If the chloride concentration of rainfall is 8 mg/L in a coastal area, find the same concentration in parts per million and milliequivalents per liter.

2.18   What is the effect of plants and living organisms on the quality of precipitation after it reaches the ground?

2.19   Explain the geochemical cycle and list the main soluble salts produced in water on the basis of this cycle.

2.20   Explain why groundwater salinity increases with depth.

2.21   What are the effects of carbon dioxide on groundwater hardness?

2.22   Why do sulfates form the largest TDS in an aquifer despite the fact that carbonate content exceeds that of sulfate?

2.23   From the water-quality point of view, what are the benefits of withdrawing groundwater induced by stream infiltration rather than withdrawing water directly from a stream?

2.24   What are the sources of dissolved gases in groundwater? What would be the effect of the changes in pressure and temperature on such gases?

2.25   What are the main factors that produce changes in groundwater quality? What are the effects of aquicludes and aquitards in this respect?

2.26 List the methods for measuring taste, odor, color, and turbidity of ground-water.

2.27 The concentration of all suspended material should not be more than 5 ppm in production wells. Explain the main reasons for this limitation.

2.28 What causes the high temperature of groundwater at great depth? Explain also the main reasons why the temperature of groundwater at depths between 10 and 20 m is more or less uniform throughout the year.

2.29 Define or explain the following:
(a) TOD
(b) BOD
(c) Specific electric conductance
(d) Pathogenic bacteria
(e) CCE
(f) CAE

2.30 Explain briefly the following:
(a) Radionuclides (primordial and daughter products)
(b) Curie unit
(c) MPC
(d) TU
(e) Half-life of a radionuclide

2.31 Explain the three probable sources of radionuclides in groundwater.

2.32 Prepare a list of references in which you can find the water-quality standards for
(a) Drinking water
(b) Industrial water
(c) Irrigation water

REFERENCES

Ackermann, W. C.: "Scientific Hydrology in the United States," The Progress of Hydrology, Internal Seminar for Hydrology Professors, 1st., University of Illinois, Champaign, 1969, pp. 563–571.

American Public Health Association, American Water Works Association, and Water Pollution Control Federation: *Standard Methods for the Examination of Water and Wastewater*, 14th ed., American Public Health Association, Washington, 1975.

American Society for Testing and Materials: *Manual on Industrial Water and Industrial Waste Water*, 2d ed., Philadelphia, 1966.

American Society for Testing and Materials: *Water Quality Criteria*. ASTM Special Technical Publication No. 1416, Philadelphia, 1967.

American Society for Testing and Materials: *Manual on Water*, ASTM Special Technical Publication No. 442, Philadelphia, 1969.

American Water Works Association: *Water Quality and Treatment*, McGraw-Hill, New York, 1971.

Ayres, R. S.: "Quality of Water for Irrigation," *Proceedings of the Irrigation and Drainage Division, Specialty Conference, American Society of Civil Engineers, August 13–15, Logan, Utah*, 1975, pp. 24–56.

Ayres, R. S.: "Quality of Water for Irrigation," *J. Irrig. Drain. Div., ASCE*, vol. 103, no. IR2, 1977, pp. 135–154.

Back, W. R., and B. Handshaw: "Chemical Geohydrology," in V. T. Chow (ed.), *Advances in Hydroscience*, vol. 2, Academic Press, New York, 1965, pp. 49–109.
Bouwer, H.: *Groundwater Hydrology*, McGraw-Hill, New York, 1978.
Brown, I. C.: Introduction and Chapter 1, in *Groundwater in Canada*, Canadian Department Mines Technical Survey, Geological Survey, Canadian Economic Geological Service 24, 1967, pp. 1–30.
Bryan, K.: "Classification of Springs", *J. Geol.* vol. 27, 1919, pp. 522–561.
Carroll, D.: "Rainwater as a Chemical Agent of Geologic Processes: A Review," U.S. Geological Survey Water Supply Paper 1535-G, 1962.
Chebotarev, I. I.: "Metamorphism of Natural Waters in the Crust of Weathering," *Geochim. Cosmochim. Acta*, vol. 8, 1955, pp. 22–48, 137–170, and 198–212.
Cogbill, C. V., and G. E. Likens: "Acid Precipitation in Northeastern United States," *Water Resources Res.*, vol. 10, 1974, pp. 1133–1137.
Collins, W. D.: "Temperature of Water Available for Industrial Use in the United States," U.S. Geological Survey Water Supply Paper 520-F, 1925, pp. 97–104
Colorado State University: "Karst Hydrology and Water Resources," *Proceedings of the U.S.-Yugoslavian Symposium, Dubrovnik, June 2–7, 1975*, Water Resources Publication, Fort Collins, Colorado, 1976.
Crawford, M. D.: "Hardness of Drinking Water and Cardiovascular Disease," *Proc. Nutr. Soc.*, vol. 31, 1972, pp. 347–353.
Davis, G. H., et al.: "Groundwater Conditions and Storage Capacity in the San Joaquin Valley, California," U.S. Geological Survey Water Supply Paper 1496, 1959.
Davis, S. N., and R. J. M. DeWiest: *Hydrogeology*, Wiley, New York, 1966.
DeWiest, R. J. M.: *Geohydrology*, Wiley, New York, 1965.
Domenico, P. A.: *Concepts and Models in Groundwater Hydrology*, McGraw-Hill, New York, 1972.
Durfer, C. N., and E. Becker: "Public Water Supplies of the 100 Largest Cities in the United States," U.S. Geological Survey Water Supply Paper 1812, 1962.
Feth, J. H.: "Selected References on Saline Ground Water Resources of the United States," U.S. Geological Survey Circular 499, 1965.
"Gas in Ground Water," *J. Am. Water Works Assoc.*, vol. 61, 1969, pp. 413–414.
Gass, T. E., and J. H. Lehr: "Ground Water Energy and the Ground Water Heat Pump," *Water Well J.*, vol. 31, no. 4, 1977, pp. 42–47.
Gerba, C. P., C. Wallis, and J. L. Melnick: *Viruses in Water: The Problem, Some Solutions*, vol. 9, Environmental Science & Technology, 1975, pp. 1122–1126. Published by the American Chemical Society, Easton, Pa.
Glover, R. E.: *Transient Ground Water Hydraulics*, Colorado State University, Fort Collins, Colorado, 1974.
Hanshaw, B. B.: "Natural-Membrane Phenomena and Subsurface Waste Emplacement," in T. D. Cook (ed.), *Underground Waste Management and Environmental Implications*, vol. 18, American Association of Petroleum Geologists Memoires, Tulsa, Oklahoma, 1972, pp. 308–317.
Hem, J. D.: "Study and Interpretation of the Chemical Characteristics of Natural Water," 2d ed., U.S. Geological Survey Water Supply Paper 1473, 1970.
Henningsen, E. R.: "Water Diagenesis in Lower Cretaceous Trinity Aquifers of Central Texas," Baylor Geological Studies Bulletin 3, 1962.
Jacob, C. E.: "Correlation of Groundwater Levels and Precipitation on Long Island, N.Y.," New York Department Conservation Bulletin GW-14, Water Power and Control Commission, 1945.
Jeffords, R. M.: "Recharge to Water-Bearing Formations along the Ohio Valley," *J. Am. Water Works Assoc.*, vol. 37, 1945, pp. 144–154.
Junge, C. E.: "The Distribution of Ammonia and Nitrate in Rain Water Over the United States," *Trans. Am. Geophys. Union*, vol. 39, 1958, pp. 241–248.
Junge, C. E., and R. T. Werby: "The Concentration of Chloride, Sodium, Potassium, Calcium, and Sulfate in Rainwater Over the United States," *J. Meteorol.*, vol. 15, 1958, pp. 417–425.
Kashef, A. I.: "Evaluation of Drainage Problems in Waterlogged Areas and Bolsons in

Nevada," Center of Water Resources Research, Desert Research Institute, University of Nevada, Reno, Report No. 1, 1965.

Kashef, A. I.: "The Nile: One River and Nine Countries," *J. Hydrol.*, vol. 53, 1981*a*, pp. 53–71.

Kashef, A. I.: "Technical and Ecological Impacts of the High Aswan Dam," *J. Hydrol.*, vol. 53, 1981*b*, pp. 73–84.

Kazmann, R. G.: " 'Safe Yield' in Ground-Water Development, Reality or Illusion?" *J. Irrig. Drain. Div., Am. Soc. Civil Engineers*, vol. 82, no. IR3, 1956, pp. 1103–1 to 1103–12.

Kazmann, R. G.: *Modern Hydrology*, Harper & Row, New York, 1965.

Klaer, F. H., Jr.: "Bacteriological and Chemical Factors in Induced Infiltration," *Ground Water*, vol. 1, no. 1, 1963, pp. 38–43.

Koenig, J. B.: "Worldwide Status of Geothermal Resources Development," in P. Kruger and C. Otte (eds.), *Geothermal Energy*, Stanford University Press, Stanford, California, 1973, pp. 15–58.

Kunkle, G. R.: "Computation of Groundwater Discharge to Streams During Floods, or to Individual Reaches during Base Flow, by Use of Specific Conductance," U.S. Geological Survey Professional Paper 525-D, 1965.

Legget, R. F.: *Geology and Engineering*, 2d ed., McGraw-Hill, New York, 1962.

Logan, J.: "Estimation of Electrical Conductivity from Chemical Analysis of Natural Waters," *J. Geophys. Res.*, vol. 66, 1961, pp. 2479–2483.

McCabe, L. J., J. M. Symons, R. D. Lee, and G. G. Robeck: "Survey of Community Water Supply Systems," *J. Am. Water Works Assoc.*, vol. 62, 1970, pp. 670–687.

McKee, J. E., and H. W. Wolf (eds.): "Water Quality Criteria," Publication No. 3-A, California State Water Resources Control Board, Sacramento, 1963.

McNabb, J. F., and W. J. Dunlap: "Subsurface Biological Activity in Relation to Ground-Water Pollution," *Ground Water*, vol. 13, no. 1, 1975, pp. 33–44.

Matcha, L.: "Acid Rain: Controllable?" *Trans. Am. Geophys. Union*, vol. 64, 1983, p. 953.

Maxey, G. B., and M. D. Mifflin: "Occurrence and Movement of Groundwater in Carbonate Rocks of Nevada," *Nat. Speleol. Soc. Bull.*, vol. 28, no. 3, 1966, pp. 141–157.

Mifflin, M. D.: "Delineation of Groundwater Flow Systems in Nevada," Desert Research Institute, Technical Report Service H-W, no. 4, Reno, Nevada, 1968.

National Academy of Sciences and National Academy of Engineering: *Water Quality Criteria 1972*, Report prepared by Committee of Water Quality Criteria at request of U.S. Environmental Protection Agency, Washington, 1972.

National Technical Advisory Committee: *Water Quality Criteria*, Federal Water Pollution Control Administration, Washington, 1968.

Neri, L., C. D. Hewitt, G. B. Schreiber, T. W. Anderson, J. S. Mandel, and A. Zdrojewski: "Health Aspects of Hard and Soft Waters," *J. Am. Water Works Assoc.*, vol. 67, 1975, pp. 403–409.

Rainwater, F. H., and L. L. Thatcher: "Methods for Collection and Analysis of Water Samples," U.S. Geological Survey Water Supply Paper 1454, 1960.

Richards, L. A. (ed.): *Diagnosis and Improvement of Saline and Alkali Soils*, Agricultural Handbook 60, U.S. Department of Agriculture, Washington, 1954.

Romero, J. C.: "The Movement of Bacteria and Viruses Through Porous Media," *Ground Water*, vol. 8, no. 2, 1970, pp. 37–49.

Sawyer, C. N., and P. L. McCarty: *Chemistry for Sanitary Engineers*, 2d ed., McGraw-Hill, New York, 1967.

Schoeller, H.: *Les Eaux Souterraines*, Masson & Cie, Paris, 1962.

Scott, R. C., and F. B. Barker: "Data on Uranium and Radium in Ground Water in the United States, 1954–1957," U.S Geological Survey Professional Paper 426, 1962.

Smith, D. B., R. A. Downing, R. A. Monkhouse, R. L. Otlet, and F. J. Pearson: "The Age of Groundwater in the Chalk of London Basin," *Water Resources Res.*, vol. 12, 1976, pp. 392–404.

Todd, D. K.: *Ground Water Hydrology*, 2d ed., Wiley, New York, 1980.

Toth, J.: "Groundwater Geology, Movement Chemistry, and Resources Near Olds, Alberta," Research Council of Alberta (Canada), Geologic Division, Bulletin 17, 1966.

U.S. Bureau of Reclamation: *Ground Water Manual*, U.S. Department of Interior, Washington, 1977.

van Schilfgaarde, J., L. Bernstein, J. D. Rhoades, and S. L. Rawlins: "Irrigation Management for Salt Control," *Proceedings of the Irrigation and Drainage Division, Specialty Conference, American Society of Civil Engineers, April 22–24*, Fort Collins, Colorado, 1973, pp. 647–672.

Walton, W. C.: *Groundwater Resource Evaluation*, McGraw-Hill, New York, 1970.

White, D. E.: "Magmatic, Connate, and Metamorphic Waters," *Bull. Geolog. Soc. Am.*, vol. 68, 1957a, pp. 1659–1682.

White, D. E.: "Thermal Waters of Volcanic Origin," *Bull. Geolog. Soc. Am.*, vol. 68, 1957b, pp. 1637–1658.

White, D. E.: "Characteristics of Geothermal Resources," in P. Kruger and C. Otte (eds.), *Geothermal Energy*, Stanford University Press, Stanford, California, 1973, pp. 69–94.

White, D. E., et al.: "Data on Geochemistry — Chemical Composition of Subsurface Waters," 6th ed., U.S. Geological Survey Professional Paper 440-F, 1963.

White, G. C.: "Disinfection: The Last Line of Defense for Potable Water," *J. Am. Water Works Assoc.*, vol. 67, 1975, pp. 410–413.

Wilcox, L. V.: "Classification and Use of Irrigation Waters," U.S. Department Agricultural Circular 969, Washington, 1955.

Willis, C. J., G. H. Elkan, E. Horvath, and K. R. Dail: "Bacterial Flora of Saline Aquifers," *Ground Water*, vol. 13, no. 5, 1975, pp. 406–409.

3

FUNDAMENTALS OF GROUNDWATER FLOW

Soil layers consist of deposits of soil that vary in their physical properties (Chap. 1). Soil units may vary from boulders to colloids. Some deposits are a more or less homogeneous formation (or practically so), such as sands, clays, silts, silty sands, sandy silts, etc. The homogeneity of a formation may be interrupted by various soil pockets or lenses such as open-work gravel and clay pockets. The soil may be completely saturated, completely dry, or partially saturated. Thus there is a difference between soils and other engineering materials such as steel. Voids exist in soils, and these voids are continuous and connected together, forming irregular capillary tubes in random patterns. The two major engineering properties of metallic materials are their deformation (deflection of the members, for example) and strength. In soil structures, the major properties are also deformation (compressibility of soils and related settlement or land subsidence) and strength (related to the bearing power of the soil), as well as a third major property, permeability, which results from the existence of voids. Insofar as the voids in soils are continuous, the soil is said to be pervious or permeable. Any other material that contains continuous voids should also be permeable; this includes concrete, bricks, neat cement after setting, sound granite, and very stiff clays. In practice, these materials are called impermeable or impervious, but in fact they are permeable, although yet to an insignificant degree from a practical point of view.

3.1 Permeability

Qualitatively, *permeability* is defined as the ease with which water or any other fluid is able to seep through the interconnected pores that form an intricate net of irregular capillary tubes. Accordingly, permeability depends on the properties of the soil medium as well as those of the fluid. Quantitatively, permeability is expressed in the following ways:

1. As K, the coefficient of permeability, transmission constant, or hydraulic conductivity. K depends on properties of both the medium and the fluid. Most groundwater hydrologists prefer the term *hydraulic conductivity*, whereas geotechnical engineers prefer the term *coefficient of permeability*.

2. As K_p, the specific permeability, intrinsic permeability, physical permeability, or simply permeability (Taylor, 1948). K_p depends on the soil properties only and is related to K as explained later. The most common term used is the *physical permeability*.

In practice, groundwater problems that include the study of permeability are numerous; examples are found in irrigation practice, drainage of agricultural land, drainage of highways and runways, temporary or permanent drainage of sites for the construction of foundations, water supply and sanitary engineering, compressibility of soil layers underneath engineering structures, land subsidence due to well pumping, and seepage through and underneath dams, weirs, cutoffs, and several other hydraulic structures. A sound study of the hydraulic properties of soil and rock formations (Chaps. 6 and 8) can minimize construction costs and is necessary for efficient management, planning, design, and performance.

In most groundwater problems, the fluid is generally water (fresh, brackish, or salty) and the flow medium may be soil or rock or both. The laws that govern flow are the same whether the medium is solid or unconsolidated rock (soil). However, the rate of water seepage through the flow medium depends on the physical properties of the medium. The downstream face of many solid dams may appear dry because the rate of evaporation from the face is higher than the rate of water percolation through the body of the dam. The same is seen in damp cuts in apparently dry quarries. The steady flow pattern through two different media of the same geometric and hydraulic boundaries is theoretically the same (see discussion of flow nets, Sec. 5.1).

3.2   General States of Flow

Laminar and Turbulent Flow

The flow of fluid may be laminar or turbulent. In *laminar* flow, each fluid particle travels along a well-defined path that never intersects another flow path (Taylor, 1948). However, *turbulent* flow paths are irregular; they vary with time and have no definite pattern. Reynolds observed (Taylor, 1948) that at low velocities of flow in a glass pipe, flow is laminar (Figs. 3.1 and 3.2). If the velocity of the fluid in a glass pipe is increased

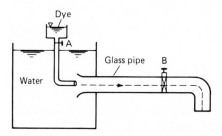

Figure 3.1 Reynold's experiment to visually find the critical velocity stage. *A*, valve to regulate dye; *B*, valve to regulate the velocity of water flow.

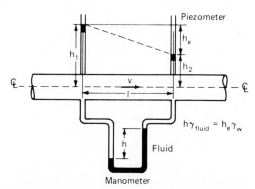

Figure 3.2 Piezometer or manometer to measure the head loss in Reynold's experiment.

gradually, the relationship between the hydraulic head h and the velocity v remains linear, but when h is increased to a point beyond a certain critical velocity v_c, the flow passes through turbulence states. If the test is repeated by decreasing h, a lower critical velocity is obtained (Fig. 3.3).

In the case of turbulence, the relationship between velocity v and the hydraulic gradient i_g (loss in head h per unit length) in a pipe is given by

$$v^n = C'i_g \tag{3.1}$$

where $2.0 > n \geqslant 1.79$ (Taylor, 1948); the value of n (dimensionless)

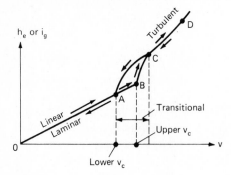

Figure 3.3 Velocity versus head in Reynold's experiment.

depends on the degree of turbulence, and C' is a constant. In the laminar flow state in a pipe, $n = 1.0$ and

$$v = Ci_g < v_c \tag{3.2}$$

where C is a constant that has a velocity dimension. If the flow is laminar, the average velocity is directly proportional to the hydraulic gradient. The same relationship holds true in soils, as indicated by Darcy's law [Eq. (3.5)].

Steady and Transient Flow

If at any specific point within the flow medium the velocity of flow is constant at all times (magnitude and direction), the flow is said to be *steady* or under *steady-state conditions*. When the groundwater level does not change, the natural flow is steady. However, such a case of equilibrium may suffer an imbalance due to natural and artificial factors such as recharge from rainfall, artificial discharge or recharge by wells, and seepage losses by evaporation. These nonequilibrium conditions produce what is called a *transient condition* (or *unsteady state of flow*), where the velocity of flow at a specific point changes in magnitude and/or direction with time. For example, when the pressure changes suddenly from a condition of steady-state flow, such as when aquifer water is pumped by a well, the steady natural flow will be disturbed and the flow toward the well will change over time and develop transient conditions of flow. If the well is pumped continuously, a state of equilibrium is reached after a long time and the steady-state flow will prevail. In such cases, the steady state represents a theoretically ultimate condition (time $t = \infty$) as a special case of the transient state. Theoretically speaking, this steady-state case does not develop unless there is aquifer recharge or leakage (Chap. 8). From a practical point of view, whenever insignificant changes take considerable time to develop, flow may be considered approximately steady. Transient flow systems also may be laminar or turbulent depending on the water velocities and the characteristics of the flow medium.

3.3 Darcy's Law

Darcy developed experimentally in 1856 the following empirical formula for the flow of water through soils (mainly sands) (Fig. 3.4):

$$Q = \frac{KA}{l}\left(\frac{p}{\gamma_w} + l \pm \frac{p_o}{\gamma_w}\right) \tag{3.3}$$

where Q = the rate of flow, cm³/s

A = the overall cross-sectional area of the specimen, normal to the flow direction, cm³

l = the height of the soil specimen, cm

$\dfrac{p}{\gamma_w}, \dfrac{p_o}{\gamma_w}$ = the water-pressure heads at the top and bottom sections of the sample, cm

The level of the water at the starting section of the flow is known as the *upgradient* or *upstream level*, and the tail water level (Fig. 3.4) is known as the *downgradient* or *downstream level*. Thus the superficial velocity $v = Q/A$ is given by

$$v = \frac{K}{l}\left(\frac{p}{\gamma_w} + l \pm \frac{p_o}{\gamma_w}\right) = K\frac{h_e}{l} \tag{3.4}$$

or

$$v = Ki_g = K\frac{h_e}{l} \tag{3.5}$$

The superficial velocity v is also known as the *darcian, average,* or *discharge velocity*. It should be noted that v does not give the average seepage velocity v_s, which is the actual velocity of the fluid within the

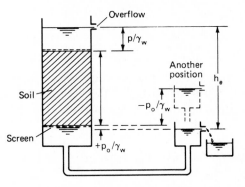

Figure 3.4 Darcy's experiment on soils.

pores of the sample. In a homogeneous soil, porosity n is given by $n = A_v/A$, where A_v is the area of the voids within the overall cross-sectional area A. Then the rate of flow Q is given by $Q = A_v v_s = Av$, and thus the relationship between the darcian velocity v and the seepage velocity v_s is given by

$$v_s = \frac{v}{n} \tag{3.6}$$

Factors Affecting Hydraulic Conductivity

Darcy's law expresses the loss in total head as a result of friction in a certain formation along a certain flow path. The constant K of proportionality between the average macroscopic velocity v and the hydraulic gradient i_g indicate the degree of permeability of the soil. Strictly speaking, *K is not a constant* because it depends on the properties of the flow medium as well as those of the fluid. The unit weight and absolute viscosity of any fluid vary with temperature and pressure (see Table 1.11). Thus changes in the unit weight and/or viscosity of the fluid are accompanied by changes in the hydraulic conductivity K.

Although Darcy's law was developed experimentally, several attempts have been made to derive it analytically (Taylor, 1948). Such derivations made it possible to fully understand the factors on which K depends. In one approach, Poiseuille's law[*] for the flow of fluids through capillary tubes, which was worked out between 1840 and 1841, was modified for use in soils by introducing certain equivalent soil parameters (Taylor, 1948). The derived equation is as follows:

$$K \text{ (cm/s)} = C_s \frac{\gamma_w}{\mu} \frac{e^3}{1+e} D_s^2 \tag{3.7a}$$

$$K \text{ (cm/s)} = C_s \frac{\gamma_w}{\mu} \frac{n^3}{1-n} D_s^2 \tag{3.7b}$$

where γ_w = unit weight of water, g/cm^3
  μ = absolute or dynamic viscosity, $g \cdot s/cm^2$
  e = void ratio of soil
  n = porosity of the soil = $e/(1+e)$
  C_s = shape factor that depends on the pattern of the capillary tube system in the soil and on the tortuosity of the flow path
  D_s = diameter of an *ideal* flow medium which consists of spherical grains of equal size and which behaves collectively (with

---

[*] This is also known in the literature as the Poiseuille-Hagen law because it was discovered that Hagen developed a similar law in 1839.

regard to such hydraulic characteristics as head loss) as the actual soil under consideration

Sometimes D_s is assumed as the average size of a real soil and is approximately equal to D_{10} of the natural soil. Equation (3.7a and b) cannot be used as such to calculate K because of the complexity involved in finding reasonable and acceptable values for C_s and D_s. However, these equations clarify the factors that affect the value of K. It is clear from these equations that K varies as the square of the average grain size. For filter sands, Hazen (1911) developed the following empirical formula for K (in centimeters per second) in terms of the effective size D_{10} (in centimeters):

$$K = C_h D_{10}^2 \qquad (3.8)$$

The factor C_h is approximately equal to 100, and because it has a dimension $(s^{-1}cm^{-1})$, the units of K and D_{10} are centimeters per second and centimeters, respectively. Moreover, Hazen's equation is restricted to values of D_{10} between 0.1 and 3 mm and should not be used whenever the uniformity coefficient $C_u = D_{60}/D_{10}$ exceeds 5. In Hazen's tests (Taylor, 1948), the constant C_h varied from 81 to 117 (average 100), and some individual values were as low as 41 and as high as 146. It should be noted that γ_w, μ, and the void ratio or porosity do not appear in Eq. (3.8). This leads to an unacceptable conclusion that loose sand would have the same K after densification and that K would not be affected by temperature changes, which do affect the values of γ_w and μ.

If Eqs. (3.7a) and (3.8) are compared, it follows that C_h depends on the properties of the fluid and the soil structure, or

$$C_h = C_s \frac{\gamma_w}{\mu} \frac{e^3}{1+e} \qquad (3.9)$$

Temperature changes affect the value of K in a specific soil medium. Thus if K_1 and $(\gamma_w/\mu)_1$ correspond to temperature t_1 and K_2 and $(\gamma_w/\mu)_2$ correspond to temperature t_2, then if the *soil structure of the soil medium remains unchanged,*

$$\frac{K_2}{K_1} = \frac{(\gamma_w/\mu)_2}{(\gamma_w/\mu)_1} \qquad (3.10)$$

It is known that the γ_w of water changes slightly with temperature, whereas μ changes considerably (see Table 1.11). For example, at $4°C$, μ is about 5 times its value at $100°C$, while γ_w changes only about 1.04 times. Accordingly, Eq. (3.10) may be written as

$$\frac{K_2}{K_1} \simeq \frac{\mu_1}{\mu_2} \qquad (3.11)$$

Equation (3.10) or (3.11) can be used to determine the effects of changes in the fluid properties due to changes in temperature, pressure, and/or water quality. For example, the K values in coastal aquifers in the fresh-water and saltwater zones would be different.

Equation (3.7a) may be written in the following form:

$$K = K_p \frac{\gamma_w}{\mu} \tag{3.12}$$

where $K_p = C_s D_s^2 e^3/(1 + e)$ and is known as the *intrinsic* or *physical permeability*. Its use is common in the petroleum industry, where the occurrence of various types of fluids within the same soil medium is common. K may be expressed in square centimeters or square feet, or by using a unit called *darcy*, which is defined as a rate of flow of 1.0 cm³/s percolating through a cross-sectional area of 1.0 cm² and a pressure gradient i_{gp} of 1 atm/cm when the viscosity of the flowing fluid is 1 cP, where

> 1.0 darcy = 0.987×10^{-8} cm² = 1.062×10^{-11} ft²
> 1.0 atm = 1.0132×10^6 dyn/cm2
> 1.0 cP = 0.01 P = 10 mP = 0.01 dyn $\cdot$ s/cm²
> 1.0 g $\cdot$ s/cm² = 980.7×1000 mP
> i_{gp} = pressure gradient = γ_w times the hydraulic gradient = $i_g \gamma_w$
> (μ for water $\simeq 10^{-5}$ g $\cdot$ s/cm² at room temperature)

Referring again to Eq. (3.7a), a change in the void ratio e (by compaction, compression, or vibration) while the fluid remains unchanged is governed by the relationship between the hydraulic conductivities K_1 and K_2 at two states of soil density 1 and 2 by

$$\frac{K_1}{K_2} = \frac{[e^3/(1 + e)]_1}{[e^3/(1 + e)]_2} \tag{3.13}$$

It is assumed that the shape factor C_s [Eq. (3.7a)] is independent of the density changes in a certain soil formation. From a practical point of view (Taylor, 1948), it has been found that Eq. (3.11) can be simplified as (Fig. 3.5)

$$\frac{K_1}{K_2} \simeq \frac{e_1^2}{e_2^2} \tag{3.14}$$

Equations (3.8), (3.13), and (3.14) are restricted to cohesionless soils and *should not* be used for cohesive soils such as plastic silts and/or clays. It has been proved experimentally (Taylor, 1948), however, that there is a linear relationship between e and log K for cohesive soils (Fig. 3.6).

Water in nature as well as distilled water includes air either in an entrapped form or in a dissolved state. The presence of entrapped air and

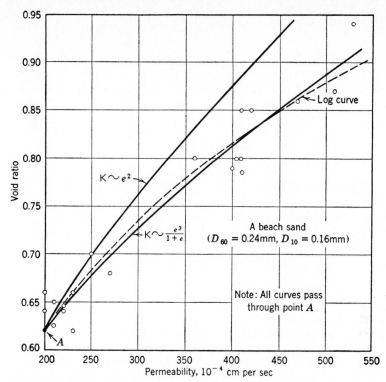

Figure 3.5 Experimental checks on theoretical proportionalities between K and e in a beach sand. *(D. W. Taylor, Fundamentals of Soil Mechanics, Wiley, New York, 1948.)*

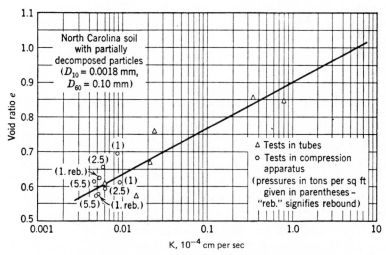

Figure 3.6 Relationship between K and e in a fine-grained soil. *(D. W. Taylor, Fundamentals of Soil Mechanics, Wiley, New York, 1948.)*

the release of dissolved air (by increases in temperature and/or decreases in water pressure) affect the value of K determined from laboratory or field tests. In the presence of organic materials, water can free some gases as a result of chemical reactions, and this also affects the magnitude of the average hydraulic conductivity.

Range of Validity of Darcy's Law

In the case of flow of a fluid through a circular pipe of diameter D, the critical velocity v_c beyond which the flow starts to change from laminar to turbulent is grouped with other parameters by Reynold's number Re.

$$\text{Re} = v_c D \frac{\gamma_w}{\mu g} = v_c D \frac{\rho}{\mu} = v_c \frac{D}{v'} \qquad (3.15)$$

where v_c = lower critical velocity (Fig. 3.3)
    μ = absolute viscosity of the fluid
    v' = kinematic viscosity of the fluid, cm^2/s
    γ_w = unit weight of the fluid

Re is a dimensionless number that defines the practical limit of laminar flow before it starts to change to turbulent. The ranges of Re for various sizes and types of conduits can be found in any textbook on hydraulics. Re is about 2000 in circular tubes, 1000 for the flow between parallel plates considering D as the spacing between the plates, 500 for flow through a trapezoidal conduit considering D as the depth of the fluid, and 1.0 for flow around a sphere of diameter D. These values are conservative; for example, in pipes with circular cross sections, turbulent flow may take place whenever Re exceeds 4000 rather than 2000 because of special testing techniques. Since the values of μ and v' vary with temperature, the product $v_c D = v'\text{Re}$ should be constant at a certain temperature. This indicates that at a certain temperature, the critical velocity of a fluid is higher in small-diameter tubes than in large-diameter tubes. The flow is turbulent in most pipelines.

Reynold's number may be approximated for any conduit of irregular cross section. Soil may be looked on as a medium consisting of an infinite number of irregular, tortuous, and continuous capillary tubes. Attempts made to determine the Reynold's number Re for flow through soils (Muskat, 1937; Polubarinova-Kochina, 1962) gave the following range:

$$\text{Re} \geqslant v_c D_s \frac{\gamma_w}{\mu g} = 1.0 \text{ to } 10 \qquad (3.16)$$

The symbols in the preceding equation are the same as those in Eq. (3.15) except for D_s, which represents the soil size as expressed by Eq. (3.7a).

Darcy's law is valid only when flow is laminar, and it cannot be applied to soils with large-size grains, even in some clays, when the gradient is very small (Polubarinova-Kochina, 1962). Muskat (1937) found that Re [Eq. (3.16)] varies between 1.0 and 5, while Polubarinova-Kochina (1962) found a range between 3.0 and 10. The discrepancies are obviously due to the fact that Eq. (3.16) does not include any parameter related to the structure of the soil medium.

The flow of groundwater is not laminar in soils with particle sizes greater than those of coarse sands. If the upper limits of K and i_g in coarse sands are assumed to be 1 cm/s and 1.0, respectively, the Reynold's number would be about 10 ($\gamma_w \simeq 1$ g/cm^3, $\mu \simeq 10^{-5}$ g · s/cm^2, $g = 1000$ cm/s^2, and $D_s = D_{10} = 0.1$ cm), which is the upper limit given by Eq. (3.16).

3.4 Laboratory Permeability Tests

Hydraulic conductivity is best determined by field tests (see Chaps. 6 and 8). There are several types of laboratory permeameters for measuring K, yet the operating principles are almost the same. The major laboratory tests are given briefly in the following paragraphs. Detailed experimental procedures and techniques can be found elsewhere (Taylor, 1948; Bowles, 1978; American Society for Testing and Materials, 1982).

Constant-Head Permeameters

A diagrammatic sketch of this permeameter is given in Fig. 3.7. A steady-state flow is established by maintaining the upgradient and downgradient water levels. The sample is a cylindrical specimen of over-

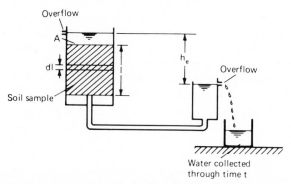

Figure 3.7 Diagrammatic sketch of contant-head permeameter.

all cross-sectional area A and length l. Usually the soil grains are maintained in place by confining the specimen between two thin layers of coarse, clean, and homogeneous sand (such as Ottawa sand), which in turn is held in place by a circular wire mesh at each end. The volume of the flowing water V_w is collected over a period of time t. Then by direct application of Darcy's law, K is calculated from

$$K = \frac{V_w l}{h_e A t} \tag{3.17}$$

At the beginning of the test, at least two measurements V_{w1} and V_{w2} should be made during time periods t_1 and t_2 in order to check that the flow reached the steady state when $V_{w1}/t_1 = V_{w2}/t_2 = $ constant.

Strictly speaking, the sample should be undisturbed in order to represent the actual field conditions.[*] Unless the permeameter is specifically designed to accommodate the same sampling spoon used in the field, extrusion of the sample from the spoon and insertion of the spoon into the permeameter probably would disturb the original structure of the soil. In practice, it may be adequate to prepare a sandy specimen (disturbed) in such a way that the dry weight of the specimen would be equal to that in the field in order to alleviate the effects of disturbance.

The test is usually conducted on completely saturated samples using distilled water. However, in the field, the soil may be partially saturated and the water may contain different types and amounts of minerals and other constituents (Chap. 2) depending on the location. The water may also contain certain amounts of entrapped as well as dissolved air. In order to be able to obtain a *unique* value of K for a certain soil, the test has to be standardized. The sample may first be subjected to a vacuum and then completely saturated with distilled water. Because entrapped and dissolved air exists in distilled water, the water flow could be obstructed by the entrapped air and/or by the release of air bubbles. In order to avoid obstruction from these bubbles of air in a relatively small sample, de-aired (or degassed) distilled water may be used. Air-free water is obtained by spraying the distilled water in a vacuum (Taylor, 1948). The sample should be protected during its percolation from contact with the atmosphere. Most laboratories are not equipped with devices for this purpose. It may be sufficient to warm the distilled water slightly above room temperature (Taylor, 1948); the water cools off during the test period, reducing the possibility of air release. In addition to these draw-

---

[*] In geotechnical engineering, undisturbed samples are obtained by using sampling tubes that preserve, as much as possible, the natural state of the sample structure and its field water content. Split-spoon samples are usually disturbed more than undisturbed samples, but less than a truly disturbed sample, in which the natural structure is destroyed.

backs, a "filter skin" may develop because of the presence of fine foreign material in the sample. This would also obstruct the flow.

The results of the tests should thus be considered very approximate. They may be used to find the *relative* permeabilities of several different samples, but *not* the absolute values of these permeabilities. In nature, water is not distilled and contains air and other foreign materials. However, any laboratory test should be standardized in order to get a unique result for a certain soil. This means that in situ permeability tests should give more realistic K values than laboratory tests (Chaps. 6 and 8).

Falling-Head Permeameters

The hydraulic conductivity of fine-grained soils such as silts may be determined by a falling-head permeameter (Fig. 3.8). This permeameter is designed to avoid collecting V_w, which would be very small for these soil types and would demand a very long period of time to collect a measurable amount. A small-diameter burette is connected to the upper section of the sample. The lower section is connected to a container in which the water level is maintained constant at all times. The flow of

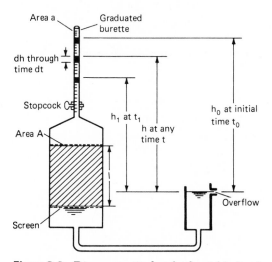

Figure 3.8 Diagrammatic sketch of variable-head permeameter.

water is of the transient type in this permeameter (Sec. 3.2). The records of the test should include measurement of any two water heads h_0 and h_1 at times t_0 and t_1, respectively (Fig. 3.8), where $t_1 - t_0$ is the test period. If a and A are the cross-sectional areas of the burette and the soil specimen, respectively, then K is determined from the following equation (Terzaghi and Peck, 1967):

$$K = \frac{al}{A(t_1 - t_0)} \ln \frac{h_0}{h_1} \tag{3.18}$$

The variable rate of flow $Q = AK\,(h/l)$ does not appear in this equation (h is also variable).

The drawbacks of this permeameter are the same as those for the constant-head permeameter. However, the release of air due to decreases in water pressure are more likely to be pronounced in this test because of the change in water pressures.

Consolidation Tests

Consolidation tests are designed primarily to determine the compressibility of clay layers in order to evaluate the anticipated settlements of structures built above clay layers from which the samples are obtained (Sec. 6.2). The determined value K is considered more reliable than that determined from the falling-head permeameter. Although consolidation

TABLE 3.1   Hydraulic Conductivity K of Common Natural Soil Formations

Formation	Value of K, cm/s
River deposits:	
Rhone at Genissiat	Up to 0.40
Small streams, eastern Alps	0.02–0.16
Missouri	0.02–0.20
Mississippi	0.02–0.12
Glacial deposits:	
Outwash plains	0.05–2.00
Esker, Westfield, Mass.	0.01–0.13
Delta, Chicopee, Mass.	0.0001–0.015
Till	Less than 0.0001
Wind deposits:	
Dune sand	0.1–0.3
Loess	0.001±
Loess loam	0.0001±
Lacustrine and marine offshore deposits:	
Very fine uniform sand, $C_u = 5$ to 2	0.0001–0.0064
Bull's liver, Sixth Ave., N.Y., $C_u = 5$ to 2	0.0001–0.0050
Bull's liver, Brooklyn, $C_u = 5$	0.00001–0.0001
Clay	Less than 0.0000001

SOURCE: K. Terzaghi and R. B. Peck, *Soil Mechanics in Engineering Practice*, 2d ed., Wiley, New York, 1967.

tests have not as yet been extensively used in the field of groundwater hydrology, their use should be very beneficial in leakage problems (Secs. 6.4 and 8.3) and in evaluating land subsidence due to water pumping (Sec. 8.11). According to the consolidation theory (Peck et al., 1974), K is given as

$$K = c_v m_v \gamma_w \tag{3.19}$$

where K = hydraulic conductivity, cm/s
  c_v = coefficient of consolidation cm²/s
  m_v = vertical compressibility of the soil, cm²/g

Once the coefficients c_v and m_v are determined from consolidation tests, K is easily determined from Eq. (3.19). The coefficient m_v is also known among geotechnical engineers as the *coefficient of volume compressibility*, and it is the same as the soil compressibility implied in the coefficient of storage S, which is explained later in Secs. 6.2 and 6.3.

The K values for common natural formations are given in Table 3.1 (Terzaghi and Peck, 1967). Ranges of K for soils and rocks are given in Fig. 3.9 (U.S. Bureau of Reclamation, 1977). Usually, m_v and K are found from testing. In either case, c_v can be determined from Eq. (3.19).

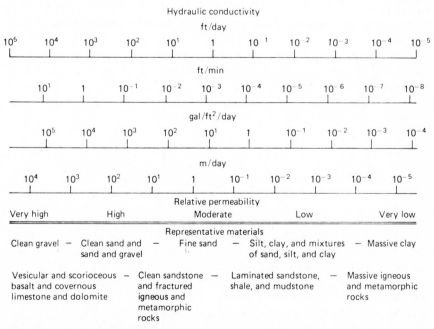

Figure 3.9 Range of hydraulic conductivities for representative aquifer materials. (See App. B for conversion factors.) *(U.S. Bureau of Reclamation, Ground Water Manual, U.S. Department of Interior, Washington, 1977.)*

3.5   Hydraulic Heads

In surface-water hydraulics, the total hydraulic head h at two different points 1 and 2 along a flow line is given by Bernouilli's equation.

$$h_1 = \frac{p_1}{\gamma_w} + y_{e1} + \frac{v_1^2}{2g} \tag{3.20}$$

$$h_2 = \frac{p_2}{\gamma_w} + y_{e2} + \frac{v_2^2}{2g}$$

where p/γ_w = pressure head, cm
$\qquad p$ = water pressure, g/cm^2
$\qquad v$ = water velocity, cm/s
$\qquad y_e$ = elevation head with respect to any selected horizontal datum, cm
$\qquad g$ = acceleration due to gravity, cm/s^2

In groundwater flow systems, Eq. (3.20) is also used, neglecting the velocity head $v^2/2g$ due to the small seepage velocities. It was pointed out in Chap. 2 that natural velocities vary from about 0.4 cm per day (about 5 ft per year) to about 2.0 m per day (about 6.6 ft per day) and may occasionally be as high as 30 m per day (about 100 ft per day). If exceptionally high values of K and i_g are assumed, for example, 5 cm/s and 1.0, respectively, then seepage velocity v_s corresponding to a porosity n of 30 percent would be 16.7 cm/s; velocity head = $v_s^2/2g = 0.14$ cm, which is a negligible fraction compared with other heads. Therefore, in groundwater hydrology, Bernouilli's equation is written as

$$h = \frac{p}{\gamma_w} + y_e \tag{3.21}$$

Then the energy equation between two points would be

$$h_e = h_2 - h_1 = \Delta h = \Delta\left(\frac{p}{\gamma_w} + y_e\right) \tag{3.22}$$

which is the loss in head between points 1 and 2.

The pressure head p/γ_w at some point in a flow region is the elevation above this point to which water rises in a piezometric tube installed at the point. The pressure head is positive in saturated zones below the groundwater table (or zero at the water table) and negative in capillary zones. Position head y_e is positive if the point under consideration is above the selected datum and negative if it is below. The horizontal datum is arbitrarily selected, and its location has no effect on p/γ_w or the value of $\Delta h = h_e$.

EXAMPLE 3.1   As an example, see Fig. 3.10, in which the flow starts from section A and ends at B. Figure 3.10 is similar to a constant-head permeameter. If it is imagined that a drop of dye is inserted at A, it would migrate along the flow line AB. It is also imagined that piezometric tubes are installed at A and B.

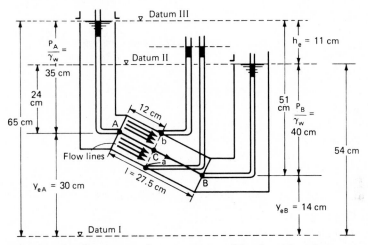

Figure 3.10   Illustration of various types of heads with reference to arbitrary selected datums in a unidirectional flow system (Example 3.1).

With reference to datum I:

At A: $\dfrac{p_A}{\gamma_w} = +35.0$ cm $y_{eA} = +30.0$ cm $h_A = \dfrac{p_A}{\gamma_w} + y_{eA}$

$h_A = +35 + 30 = +65.0$ cm

At B: $\dfrac{p_B}{\gamma_w} = +40$ cm $y_{eB} = +14$ cm $h_B = \dfrac{p_B}{\gamma_w} + y_{eB}$

$h_B = +40 + 14 = +54$ cm

As long as $h_A > h_B$, water should flow from A to B. It should be noted that in Fig. 3.10, $p_B/\gamma_w = 40$ cm $> p_A/\gamma_w = 35$ cm. This means that *water may flow from a point of low pressure to a point of high pressure;* and this indicates that the total energy includes the potential energy implied in the position head y_e.

Applying Eq. (3.22);

$$h_e = \Delta h = +65 - 54 = +11.0 \text{ cm}$$

Therefore, the hydraulic gradient

$$i_g = \frac{h_e}{l} = \frac{h_e}{AB}$$

$$= \frac{11}{27.5} = 0.4 \qquad \text{(dimensionless)}$$

The values of h_e and i_g are independent of the location of the datum, as indicated in the following solutions with respect to other datums.

With reference to datum II:

At A: $\dfrac{p_A}{\gamma_w} = +35.0$ cm $y_{eA} = -24.0$ cm $h_A = \dfrac{p_A}{\gamma_w} + y_{eA}$

$h_A = +35.0 - 24.0 = +11.0$ cm

At B: $\dfrac{p_B}{\gamma_w} = +40.0$ cm $y_{eB} = -40.0$ cm $h_B = \dfrac{p_B}{\gamma_w} + y_{eB}$

$h_B = +40 - 40 = 0$

$$h_e = \Delta h = +11.0 - 0 = +11.0 \text{ cm}$$

With reference to datum III:

At A: $\dfrac{p_A}{\gamma_w} = +35$ cm $y_{eA} = -35$ cm $h_A = \dfrac{p_A}{\gamma_w} + y_{eA}$

$h_A = +35 - 35 = 0$

At B: $\dfrac{p_B}{\gamma_w} = +40$ cm $y_{eB} = -51$ cm $h_B = \dfrac{p_B}{\gamma_w} + y_{eB}$

$h_B = +40 - 51 = -11$ cm

$$h_e = \Delta h = 0 - (-11) = +11 \text{ cm}$$

Discussion of Example 3.1   Although such analysis may be necessary in some cases, it is apparent that h_e is simply *the difference in water elevations in the piezometers installed at the two points.* Section aCb (Fig. 3.10) is normal to the flow line AB. The water levels in the piezometric tubes installed at both a and b should be the same because the traveled path of any water particle from the section at A to section aCb should have the same length and consequently the same dissipation in h (the gradient i_g is constant in this case.) Line ab is known as an *equipotential line* because all points on ab have the same h (but not the same p/γ_w). The loss in head h_e between A and $C = i_g \times 12.0$ cm $= 0.4 \times 12.0 = 4.8$ cm.   Therefore, $h_c = h_A - 4.8 = 11.0 - 4.8 = +6.2$ cm (with reference to datum I).

3.6   Generalization of Darcy's Law

Although Darcy's law was developed experimentally, it is useful to express it in a mathematical form, especially in two-dimensional flow systems (Fig. 3.11), where the hydraulic gradient changes along any specific flow line. The velocity $v_{s,t}$ at any point a in the flow medium has the same direction as the tangent to the flow line. Its vector may therefore be analyzed to $v_{x,a}$ and $v_{y,a}$ in the x and y directions, respectively. Therefore, at any point a,

$$v_{s,t} = -K\frac{\delta h}{\delta s_t} \qquad v_{x,a} = -K\frac{\delta h}{\delta x} \qquad v_{y,a} = -K\frac{\delta h}{\delta y} \qquad (3.23)$$

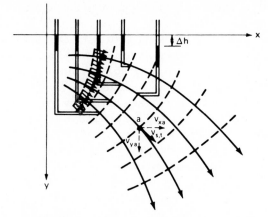

Figure 3.11 Families of stream lines (———) and equipotential lines (– – – –) in a two-dimensional flow system.

where s_t is the tangential direction of the flow lines, and x and y are, respectively, the horizontal and vertical axes. The negative signs in Eq. (3.23) mean that the total head h decreases whenever s_t, x, or y increases. Positive signs are possible depending on the directions of the axes with respect to the flow direction.

The generalization of Darcy's law also may be expanded to three-dimensional cases with cartesian or cylindrical coordinates. However, in common practical problems, most of the three-dimensional cases are approximated to two-dimensional ones. Solutions of three-dimensional problems have been attempted by means of numerical analyses.

3.7 Effects of Soil Stratification on Hydraulic Conductivity

Soil formations are generally more or less stratified. The hydraulic conductivity K_{max} in the direction of the stratification is the maximum possible as compared with any other K along any other direction at a specific point in the formation. The minimum possible value K_{min} takes place in the direction normal to the stratification. The ratio K_{max}/K_{min} ranges between 2 and 3 and almost infinity (Terzaghi and Peck, 1967). In the direction of angle α to the bedding, the value of K_α is intermediate between K_{max} and K_{min} and is given by (Casagrande, 1937)

$$K_\alpha = \frac{K_{max}K_{min}}{K_{max}\sin^2\alpha + K_{min}\cos^2\alpha} \qquad (3.24)$$

The values of K_{max} and K_{min} cannot be routinely determined by laboratory or field tests. The methods are generally too complex and expensive for everyday use.

When the soil structure has a certain pattern and texture that prevails throughout the formation, the soil is called *homogeneous*. When the hydraulic properties are the same in all directions at any point in the medium (that is, $K_{max} = K_{min} = K_\alpha = K$), the soil is called *isotropic*. If the soil is stratified but the values of K_{max} and K_{min} are the same at any point in the formation, the soil is known as *anisotropic* and *homogeneous*. If these conditions are not satisfied, the soil is *heterogeneous*.

In theory and practice, soil is commonly assumed homogeneous and isotropic. The solutions for groundwater patterns in these latter cases can be expanded for homogeneous and anisotropic soils. Theoretically speaking, solutions can be found for certain heterogeneous soils (using numerical analyses, for example). However, these solutions require such preliminary data as the determination of K in all directions at various points and the delineation of all layers or any existing soil pockets in the heterogeneous formations, and these are not practically feasible to determine.

Loess formations are the only exceptions (Terzaghi and Peck, 1967) because vertical permeability is the maximum possible in loessial soils owing to the occurrence of vertical vesicles in these deposits.

3.8 Flow Configuration

When all the flow lines are parallel and have the same direction (horizontal or vertical or inclined), the flow system is known as a *unidirectional flow system* (Fig. 3.10). Laboratory permeameters all belong to this system (Figs. 3.7 and 3.8).

The radial flow system shown in Fig. 3.12 represents the idealized case of a water well in which the initial water level or piezometric surface is horizontal and so are the aquifer boundaries. The flow in such a case is also two-dimensional. Other examples of two-dimensional flow systems are given in Fig. 3.13.

If drops of dye are inserted at selected entrance points *a* and *b* in a laboratory model to represent the case shown in Fig. 3.13*b*, the full line curves can be traced. These curves indicate the flow lines (or stream lines). When the flow is steady and the medium is homogeneous and isotropic, the curves orthogonal to these lines represent the equipotential lines. All points lying on a certain equipotential line should have the same total head. The two sets of these curves (Figs. 3.11 and 3.13) are known as the *flow net*. There are, of course, an infinite number of these curves within a flow medium. However, a certain number of these curves are selected to trace a flow net that indicates the flow configuration. Different methods are used to trace these flow nets, whether experimen-

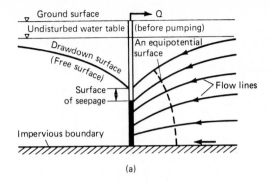

(a)

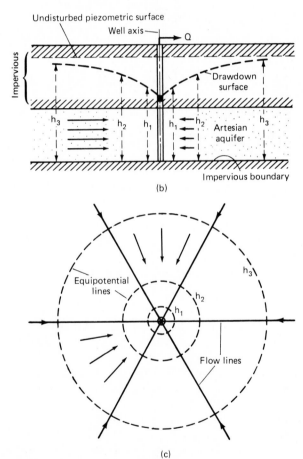

(b)

(c)

Figure 3.12 Radial flow systems. *(a)* Gravity well; *(b)* artesian well (vertical section); *(c)* artesian well (horizontal section).

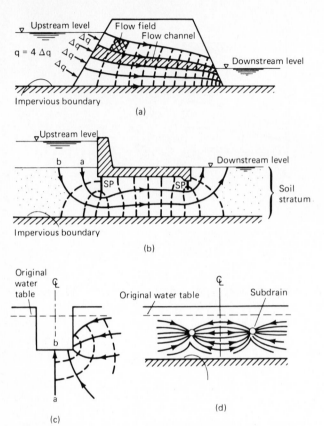

Figure 3.13 Examples of two-dimensional flow systems. *(a)* Seepage through an earth embankment or an earth dam; *(b)* seepage beneath a weir or solid dam (SP, sheet piles); *(c)* flow toward a trench; *(d)* flow toward subdrains.

tally or analytically (see Sec. 5.1). The region bounded between two consecutive flow lines is known as a *flow channel* (Fig. 3.13a), through which a portion Δq of the total rate of flow q passes (q is the rate of flow per unit length, normal to the cross section; it is in square meters per minute). The difference in intensity Δh between the total heads of two consecutive equipotential lines is known as the *equipotential drop* (Fig. 3.12). A zone bounded by two consecutive flow lines and two consecutive equipotential lines is known as a *field* (Fig. 3.13a). The geometric shape of a very small field is a square if the Δh's and Δq's are equal; otherwise it would be a rectangle. Because the sides of a field are curved, its shape is known as a curvilinear square or curvilinear rectangle (Casagrande, 1937; Taylor, 1948; Terzaghi and Peck, 1967). Similar large flow nets are obtained in the field for several human activities, such as in a well field (Fig. 3.14; see also Sec. 5.1).

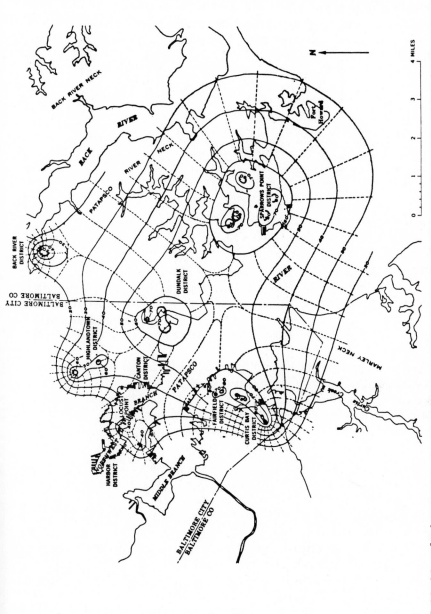

Figure 3.14 Map of Baltimore, Maryland, industrial area showing altitude of piezometric (potentiometric) surface (———) and generalized flow lines (– – –) in the Patuxent Formation in 1945. The contour interval is 10 ft, and the datum is mean sea level. The symbol ● indicates wells for which the altitude of the head was obtained. (R. R. Bennett and R. R. Meyer, "*Geology and Ground-Water Resources of the Baltimore Area (Maryland)*," *Maryland Dept. of Geology, Mines and Water Resources Bull. 4, 1952.*)

Under the steady-state conditions of uniform flow, the stream lines in a unidirectional flow system are parallel straight lines with a constant hydraulic gradient. The pattern of the system is affected by imposed geometric boundaries, as in the case of permeameters. In practice, such a flow configuration *rarely* takes place. However, a flow zone may include straight lines along which the gradient is variable, such as the vertical line of symmetry *ab* in Fig. 3.13*c*, and along the horizontal boundaries, such as the impervious boundaries in Fig. 3.13*a*, *b*, and *d*.

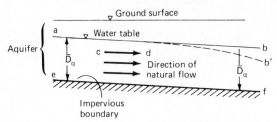

Figure 3.15 Uniform flow in unconfined aquifers.

Figure 3.15 indicates a unidirectional uniform flow in direction *cd*, assuming that water table *ab* has the same slope as the lower impervious boundary *ef*; that is, the depth $\overline{D}_q$ of water is constant. In small reaches, this assumption may be valid. In fact, the water table *ab* should be gently curved to *ab'* (Fig. 3.15), changing the flow system to two-dimensional (Chap. 5).

The piezometric surface *ab* in the idealized artesian aquifer in Fig. 3.16 is above the water table *cd* of the aquitard overlying the aquifer. The water should therefore percolate upward from the aquifer to the aquitard. This condition of water percolation is known as *leakage* (see Sec. 6.4 and Chap. 8), and the flow may be assumed uniform with a

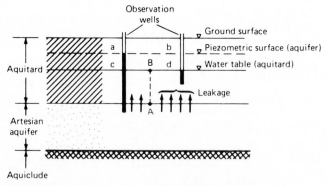

Figure 3.16 Water leakage from artesian aquifers.

constant gradient along any vertical section such as *AB* (Fig. 3.16). The rate of flow is so small that it has no appreciable effect on the levels of the water table or piezometric surface. These levels are subjected to periodic fluctuations due to climatic effects and human activities.

Despite the assumptions implied in a unidirectional flow system, its study is useful for understanding more complex actual cases. Analysis of these systems clarifies such important phenomena and parameters as effective stresses, seepage forces, and "quicksand conditions" which also occur in other general flow systems.

3.9 Effective Stresses

The soil profile shown in Fig. 3.17 consists of an aquitard of thickness D_a overlying a deep artesian aquifer. The top surface of the aquitard coincides with the bottom of a lake that has a more or less stationary water level of height h_{e1} above its bed. If a piezometer is installed at point B_1 along the upper boundary of the artesian aquifer, the piezometric surface can be recorded. In this analysis, three different piezometric levels I, II, and III are considered to occur at three different times (Fig. 3.17). When the water level is at I, a condition of equilibrium is established. At a different time when the water level reaches stage II, there would be an excess head h_{e2} (in excess of the static head at I) causing an upward flow (leakage) from the main artesian aquifer through the aquitard. At another time when the level reaches stage III, the excess head h_{e3} (negative) would produce a downward flow from the aquitard to the aquifer. It should be noted that the water level in the lake is, in fact, the water table in the aquitard and that a piezometer installed at A_1 would give the same

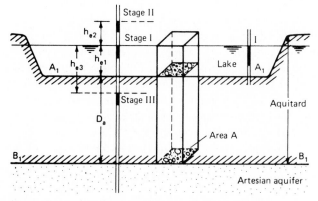

Figure 3.17 Unidirectional (vertical) flow system.

level at I. Whether the flow is at rest or moving upward with a uniform gradient $h_{e2}/D_a = i_{gII}$ or downward with a gradient $i_{gIII} = h_{e3}/D_a$, the total stress at plane $B_1 B_1$ of the prism shown in Fig. 3.17 does not change. The total stress σ_t is equal to the weight of the material above B_1 in that prism of a cross-sectional area A; that is,

$$\sigma_t = \frac{\gamma_{sat} A D_a + \gamma_w h_{e1} A}{A} = \gamma_{sat} D_a + \gamma_w h_{e1}$$

The force $\sigma_t A$ at B_1 is supported by both the water and the solids of the section. Then,

$$\sigma_t A = p A_v + \bar{\sigma}(A - A_v) \qquad (3.25)$$

where $\quad A_v$ = the average area occupied by water only within area A
$\quad\quad (A - A_v)$ = the area occupied by solids
$\quad\quad\quad \bar{\sigma}$ = the effective stress, which represents the intergranular pressure (the grain-to-grain pressure)

Once the water-pressure head at B_1 is measured by a piezometer, the pressure p is known. Therefore $\bar{\sigma}$ may be determined from Eq. (3.25), if A_v is known. However, refined tests (Bishop, 1960) indicate that Eq. (3.25) can be approximately written as

$$\bar{\sigma} \simeq \sigma_t - p \qquad (3.26)$$

eliminating A_v and A.[°]

The preceding equation is very important in engineering practice and was originally given by Terzaghi (1943). The effective stresses are responsible for volume changes in the soil, such as land subsidence due to water pumping and settlement of enginering structures. If σ_t is constant, as in the case shown in Fig. 3.17, then $\bar{\sigma}$ increases when p decreases, and vice versa. When comparing upward and downward flow effects with the hydrostatic condition, if the flow of water is upward, $\bar{\sigma}$ is less than that corresponding to the static condition. If the excess head h_e increases in such a way that $p \simeq \sigma_t$, then $\bar{\sigma}$ reaches zero. This condition is known as the *quicksand phenomenon*, in which the pressures between the soil grains disappear and the grains barely touch each other. If the flow is downward, $\bar{\sigma}$ increases over that corresponding to the static condition.

Water withdrawal from an artesian aquifer lowers the piezometric surface and increases $\bar{\sigma}$ within the aquifer. The effective stresses there-

---

[°] It will be obvious later in Section 3.11 that Eq. (3.26) is theoretically correct and is not an approximation.

fore compress the soil skeleton, reducing the volume of the pores. Consequently, water is released from storage. This phenomenon is expressed analytically by the coefficient of storage S (see Sec. 6.3), which is the main factor in withdrawing water from aquifers with no recharge.

Equation (3.26) is valid at any other section within a soil formation. Instead of computing $\bar{\sigma}$ at various points, it may be convenient to plot a diagram whose horizontal ordinates give $\bar{\sigma}$ at any point. In the case shown in Figure 3.17, the gradient is uniform, yielding linear pore-water pressure diagrams (p diagrams). The σ_t diagram is also linear, with increasing values of σ_t as depth increases. Therefore a $\bar{\sigma}$ diagram, which is determined by superimposing the positive σ_t diagram over a negative p diagram [Eq. (3.26)], should be also a linear diagram (Fig. 3.18).

The effective stress diagrams corresponding to the three water levels I, II, and III (Fig. 3.17) are shown in Fig. 3.18. The three cases I, II, and III are superimposed on each other to indicate the deviations from the static condition I (Fig. 3.18) in the other two moving conditions II and III.

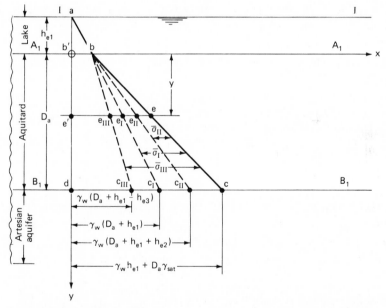

Figure 3.18 Diagrams for pore-water pressure, total stress, and effective stress corresponding to the three stages shown in Fig. 3.17. Effective-stress diagrams: case I = $bcc_{\mathrm{I}}b$, case II = $bcc_{\mathrm{II}}b$, case III = $bcc_{\mathrm{III}}b$. Water-pressure diagrams: case I = $abc_{\mathrm{I}}da$, case II = $abc_{\mathrm{II}}da$, case III = $abc_{\mathrm{III}}da$. Total stress diagram: all cases = $abcda$.

At level B_1 at the bottom:

$$\sigma_t = \overline{dc} = \gamma_w h_{e1} + D_a \gamma_{sat}$$
$$p_I = \overline{dc_I} = \gamma_w (D_a + h_{e1})$$
$$p_{II} = \overline{dc_{II}} = \gamma_w (D_a + h_{e1} + h_{e2}) = \gamma_w (D_a + h_{e1} + i_{gII} D_a)$$
$$p_{III} = \overline{dc_{III}} = \gamma_w (D_a + h_{e1} - h_{e3}) = \gamma_w (D_a + h_{e1} - i_{gIII} D_a)$$
$$\bar{\sigma}_I = \overline{cc_I} = \sigma_t - p_I = D_a (\gamma_{sat} - \gamma_w) = D_a \gamma'$$
$$\bar{\sigma}_{II} = \overline{cc_{II}} = \sigma_t - p_{II} = (\gamma_w h_{e1} + D_a \gamma_{sat}) - (\gamma_w D_a + \gamma_w h_{e1}$$
$$+ \, i_{gII} \gamma_w D_a) = D_a \gamma' - i_{gII} \gamma_w D_a$$

and $\quad \bar{\sigma}_{III} = \sigma_t - p_{III} = D_a \gamma' + i_{gIII} \gamma_w D_a$

At level y below level A_1:

$$\sigma_t = \overline{ee'} = \gamma_w h_{e1} + y \gamma_{sat}$$
$$p_I = \overline{e'e_I} = \gamma_w (y + h_{e1})$$
$$p_{II} = \overline{e'e_{II}} = \gamma_w (y + h_{e1} + i_{gII} y)$$
$$p_{III} = \overline{e'e_{III}} = \gamma_w (y + h_{e1} - i_{gIII} y)$$
$$\bar{\sigma}_I = \overline{ee_I} = y \gamma'$$
$$\bar{\sigma}_{II} = \overline{ee_{II}} = y \gamma' - i_{gII} \gamma_w y$$
$$\bar{\sigma}_{III} = \overline{ee_{III}} = y \gamma' + i_{gIII} \gamma_w y$$

The forces are obtained by multiplying the stresses by the area A of the cross section of the prism. The upward water movement reduces the effective force $yA\gamma'$ of the static condition by a force $i_{gII}\gamma_w yA$, whereas the downward movement increases the effective force by an additional force $i_{gIII}\gamma_w yA$. These additional forces are in fact seepage forces F_s produced by the flow. When the flow is downward, the effective stresses increase, and when it is upward, they decrease. Generally,

$$F_s = i_g \gamma_w V \tag{3.27}$$

where V is the volume of the soil prism (or any soil mass under consideration). The seepage force per unit volume is known as the pressure gradient i_{gp} (Terzaghi and Peck, 1967).

$$i_{gp} = i_g \gamma_w \tag{3.28}$$

3.10   Quicksand and the Design of Filters

In stage II (Fig. 3.17) if the upward flow has a high gradient in such a way that $\sigma_t = p_{II}$ and consequently $\bar{\sigma} = 0$ [Eq. (3.26)], then

$$\bar{\sigma} = \sigma_t - p_{II} = \gamma_w h_{e1} + y \gamma_{sat} - \gamma_w (y + h_{e1} - i_{gII} y) = 0$$

Thus $\qquad y\gamma' = y\gamma_w i_{gII} \qquad$ or $\qquad i_{gII} = \dfrac{\gamma'}{\gamma_w}$

In such a case, quicksand develops when the hydraulic gradient i_{gc} is critical; that is,

$$i_{gc} = \frac{\gamma'}{\gamma_w} \tag{3.29}$$

Writing γ' in terms of e and G gives

$$i_{gc} = \frac{\gamma'}{\gamma_w} = \gamma_w \frac{(G-1)/(1+e)}{\gamma_w} = \frac{G-1}{1+e} \tag{3.30}$$

[Equation (3.30) may be obtained by the use of a block diagram as explained in Chap. 2.] Under normal circumstances, γ_{sat} is roughly twice γ_w (see Table 1.12); therefore,

$$\gamma_{sat} - \gamma_w \approx \gamma' \approx \gamma_w$$

or

$$i_{gc} \approx 1.0 \tag{3.31}$$

The maximum limit of an upward gradient in laminar flow should then be about unity. If the upward exit gradient is higher than i_{gc}, erosion will start to develop at the exits. These exit surfaces have to be protected by means of a filter.

Theoretically, the quicksand phenomenon can occur in any type of soil once i_{gc} is reached. However, in cohesive soils, such as clays and plastic silts, failure by erosion when $i_g \geq i_{gc}$ may be prevented by the inherent cohesive forces (Sec. 1.3). In gravels, quicksand cannot be developed under normal conditions because of the large pores, which reduce excessively the hydraulic gradients and consequently decrease head losses. Quicksand is therefore a phenomenon rather than a sand type, the prevailing misconception notwithstanding. It produces "soil boiling," which leads to the washing of fine material as water reaches an exit.

Figure 3.19 presents various types of filters for various cases: Fig. 3.19a is a solid dam or weir with a horizontal filter known as a *filter blanket*, Fig. 3.19b is an earth dam with a rock toe acting as a filter, Fig. 3.19c is an earth dam with a horizontal filter blanket, Fig. 3.19d is a trench with vertical filters at the sides and graded or layered filters at the bottom, and Fig. 3.19e is a water well with a *gravel pack* that is essentially a filter. In each case, a specific stream line *ae* is selected to indicate the exit point *e*. If a filter were not constructed for the dam shown in Fig. 3.19a, the wash of fines at *e* would progress, forming cavities along *ae* in a backward direction, which is known as *piping*. Piping along the base of the dam is known as *roofing* (Taylor, 1948). Eventually, piping leads to failure of the structure. Thus such filters are provided to prevent failures by piping and are known as *protective filters*. All types of filters have the

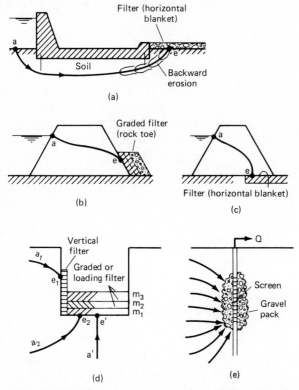

Figure 3.19 Filters in various structures. *(a)* Horizontal blanket in weirs and solid dams; *(b)* rock toe (graded filter) in earth dams; *(c)* horizontal blanket downstream from an earth dam; *(d)* graded or loading filters in an excavation and vertical filter along its side; *(e)* gravel pack around a well screen.

same design, but they differ in name and method of construction. For example, the filter shown in Fig. 3.19*d* is known as a *loading graded filter;* its downward vertical weight should be much greater than the upward seepage forces in order to counterbalance their effects with a good margin of safety. Well filters (Fig. 3.19*e*) are known as *gravel packs,* and so on.

A filter should be designed to satisfy the following two conditions (Taylor, 1948):

1. It must be fine enough to prevent passage of the fine material in the natural soil (to be protected) through its pores; otherwise it will become clogged.

2. At the same time, it should be coarse enough that the water flow is not obstructed. In other words, the loss in head through the filter should be very small (corresponding to small velocities, hydraulic gradients, and seepage forces).

The filter should be of an optimum size and gradation to satisfy these conditions. The significance of the first condition is theoretically verified by assuming that the soil and filter are each composed of equal spherical grains with radii R_s and R_f, respectively (Fig. 3.20a). In this idealized arrangement, R_f should not be larger than the limiting size that would barely allow a soil sphere with radius R_s to touch the spheres of the filter. The geometry of the densest possible arrangement of filter particles gives the size of the spherical soil particle that barely touches the walls of the filter particles (Taylor, 1948).

$$\cos 30° = \frac{\sqrt{3}}{2} = \frac{R_f}{R_f + R_s}$$

then
$$R_f + R_s = \frac{2R_f}{\sqrt{3}}$$

or
$$R_f \leqslant 6.5\, R_s \qquad (3.32)$$

The preceding equation satisfies the first condition of a successful filter design for idealized soil and filter, each made up of equally spherical soil particles. In a uniform homogeneous natural soil, the particles cannot have equal sizes and are not spherical in shape. It has been found by testing that the average grain size of a homogeneous filter may be 10 times as large as the average grain size $D_s \simeq D_{10}$ of a homogeneous soil before any appreciable wash-through of fine particles takes place. For any soil type, Terzaghi developed in 1922 an empirical formula (Terzaghi and Peck, 1948) for the design of a successful filter meeting the two previously cited conditions. His method has since been checked and reinvestigated by several agencies (e.g., U.S. Bureau of Reclamation, 1960, 1963; Hunter Blair, 1970), and a few refinements were added.

Terzaghi's criterion was written as

$$\frac{D^f_{15}}{\text{Finest } D_{85} \text{ of soil}} < 4 \text{ to } 5 \text{ times} < \frac{D^f_{15}}{\text{coarsest } D_{15} \text{ of soil}} \qquad (3.33)$$

where D^f_{15} is the D_{15} of the filter. The gradation curve from which D_{15} and D_{85} are obtained cannot be a single curve representing the results of one soil sample only. In practice, several samples are tested; the various curves produce a range (Terzaghi and Peck, 1948) such as the shaded area in Fig. 3.20b, and the values of the "finest D_{85}" and "coarsest D_{15}" included in Eq. (3.33) are obtained from the range. No restrictions were

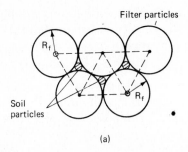

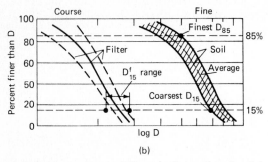

Figure 3.20   Design of filters and gravel packs. *(a)*
Dense state of spherical particles (soil particles have
a radius R_s, and filter particles have a radius R_f); *(b)*
relationship between the gradations of soil and
filter.

given in Eq. (3.33) with regard to the extent of the upper limb of the
filter's gradation curve above its D_{15}^f. It was recommended, however,
that the curve be as parallel as possible to the average gradation curve of
the soil to be protected. Equation (3.33) means that the gradation curve
of the filter material is exactly the same as that of the soil, yet the grain
sizes are enlarged four to five times.

It would be fortunate if the filter gradation curve determined by this
procedure were as close as possible to some natural deposit near the
construction site. However, if such a natural deposit is not available or
requires hauling from far locations, the filter must be constructed from
other materials available in the neighborhood of the site. Such a procedure is expensive, and therefore, the costs should be compared with
those of hauled natural deposits. The construction of gravel packs in
water wells is also complex and expensive.

Gravel pack is a prevailing conventional term that does not necessarily mean that it is composed of all gravel. A gravel pack is nothing more

than a filter that is designed on the basis of the gradation of the natural soil in the same manner explained above.

In cases where a filter is required to protect a relatively fine soil, it has to be graded into successive filters. Construction of horizontal filter layers (Fig. 3.19a, c, and d) is not as complex as construction of vertical or inclined layers, such as shown in Fig. 3.19b and e. The graded filter shown in Fig. 3.19a is designed in several layers m_1, m_2, and m_3 in such a way that filter layer m_1 has the same gradation as that for a filter protecting the natural soil. Filter layer m_2 is then designed to protect layer m_1, as if m_1 were the soil to be protected. Layer m_3 is then designed to protect layer m_2, and so on. The number and thicknesses of these layers are determined on the basis of both practical considerations and the required weights to counterbalance the upward seepage forces (Sec. 3.11).

The range of D_{15}^f varies from the maximum limit $[(4 \text{ to } 5)D_{85}]$ to the minimum limit $[(4 \text{ to } 5)D_{15}]$. If it is found that $(4 \text{ to } 5)D_{15} > (4 \text{ to } 5)D_{85}$, then it signifies a wide range of soil gradation; i.e., the shaded area in Fig. 3.20b will be relatively large. This means that the soil to be protected is almost heterogeneous and has to be categorized as belonging to two or more groups of somewhat similar types and two or more filters of different gradations have to be used.

Another method for designing filters has been recommended by the U.S. Bureau of Reclamation (1963). In this method, two filter ratios are identified.

$$R_{50} = \frac{D_{50}^f}{D_{50}} \quad \text{and} \quad R_{15} = \frac{D_{15}^f}{D_{15}} \tag{3.34}$$

When the soil to be protected is uniform ($C_u = 3$ to 4), the range of R_{50} lies between 5 and 10, and any R_{15} value is acceptable. When the soil is nonuniform ($C_u > 4$) and its gradation indicates well-graded to poorly graded soil, then the ranges of R_{50} and R_{15} are as follows:

Nonuniform soil	R_{50}	R_{15}
Subrounded grains	12–58	12–40
Angular particles	9–30	6–18

If the soil to be protected ranges from gravel (over 10 percent larger than a no. 4 sieve) to silt (over 10 percent passing a no. 200 sieve), the ratios R_{15} and R_{50} should be based on the fraction passing the no. 4 sieve (4.76 mm). The maximum size of the filter material should not exceed 3 in (7.62 cm). In this method, it is also recommended that the gradations of filter and soil to be protected should be approximately parallel, especially in the finer range of sizes.

Sherard et al. (1984a) conducted extensive experimental research on sand and gravel filters. They stated that the U.S Bureau of Reclamation method based on the parameters given by Eq. (3.34) is not supported by experiments or theory, and they recommended that the method be abandoned. However, they concluded that the filter criterion given by Eq. (3.33) is conservative, but not excessively, and that its use should be continued for sand and gravel filters with D_{15}^f larger than about 1.0 mm. For the same filters, Sherard et al. (1984a) found that it is not necessary for the gradation curves of the soil and filter to be similar and that angular particles of crushed rock are as satisfactory as filters as rounded alluvial particles.

In an accompanying paper, Sherard et al. (1984b) studied filters for silts and clays. They concluded that

1. For cores of dams of fine-grained clays, a sand filter with D_{15}^f of 0.5 mm is satisfactory.

2. The filter criterion given by Eq. (3.33) is satisfactory (conservative and reasonable) if it is used to design filters to protect cores of dams made up of sandy clays and silts.

3. Quantitative filter criteria are not necessary for filters placed upstream of cores of dams made up of clay and coarse gravel or small quarried rock can be used, even for a very fine clay core.

Sherard et al. (1984a) also found that the hydraulic conductivity of dense filters ranges between 0.2 and $0.6(D_{15}^f)^2$; the average is $K = 0.35(D_{15}^f)^2$, where K is in centimeters per second and D_{15}^f is in millimeters.

The design of filters has been developed experimentally by choosing materials and designs that do not lead to clogging of the filter. This satisfies the first requirement for a successful filter. In order to prove that the design criterion does not allow any obstruction to the flow (the second requirement), Terzaghi's method and Eqs. (3.33), (3.34), and (3.8) (Hazen's Equation) are used. If it is assumed that $D_{10} \simeq D_{15}$, then from Eq. (3.33), $D_{10}^f/D_{10} \simeq 4$ to 5, and from Eq. (3.34), and $D_{10}^f/D_{10} \simeq 6$ to 40. The smaller ratio of $D_{10}^f/D_{10} = 4$ to 5 [Eq. (3.33)] is conservative for our purpose. Considering the flow line $a'e'$ in Fig. 3.19d, the rate of flow at the boundary between the soil and the filter should be the same in both media. Therefore, $v = Ki_g$ (of filter) $= Ki_g$ (of soil). Applying Eq. (3.8),

$$\frac{i_g \text{ (of filter)}}{i_g \text{ (of soil)}} = \frac{(D_{10})^2}{(D_{10}^f)^2} = \frac{1}{16} \text{ to } \frac{1}{25}$$

The gradient through the filter would therefore be just a small fraction of that through the soil, indicating negligible head losses through the filter. In other words, filters designed in this manner would not obstruct the flow, satisfying the second requirement.

After completing the well construction, the well should be pumped several times at various rates and each time the pump should be shut down. This leads to surging action, which forces the fine particles to move out or in and settle down. Such a procedure is known as *well development.* The most common methods of well development are by pumping, surging, and the use of compresed air. Well development can also be accomplished by hydraulic jetting, by the addition of chemicals, by hydraulic fracturing, and even by the use of explosives (Walton, 1970; Johnson Division, 1972; U.S. Bureau of Reclamations, 1977; Todd, 1980). When the efficiency of a productive water well decreases over time, rehabilitation of the well is needed to increase the efficiency (Walton, 1970; Todd, 1980). Rehabilitation methods include development techniques as well as other procedures.

In water wells, gravel packs surround perforated screens whose perforations may be either circular or slotted. In small-capacity wells, screens may be used without gravel packs. The screen openings should be designed hydraulically (U.S. Bureau of Reclamation, 1977), and their sizes should satisfy certain requirements to prevent clogging by the fines of the gravel packs or surrounding natural soil. The gradation curves of the soil surrounding the screen are used as a basis for designing the perforation size. As explained earlier, these gradation curves usually cover a range (Fig. 3.20b) because they are based on various soil samples. The most common recommendations for finding slot size are as follows:

1. If the groundwater is noncorrosive, the slot size should be equal to or less than D_{60} of the soil surrounding the screen (Johnson, 1972); the C_u of the soil should not exceed 3.0.

2. In the case of corrosive groundwater, the slot size should be equal to or less than D_{50} of the soil surrounding the screen; the C_u of the soil should also not exceed 3.0 (Johnson, 1972).

3. When the uniformity coefficient C_u is greater than 6.0, the size of the slot should vary between D_{50} and D_{70} of the soil surrounding the screen (Ahrens, 1957).

4. The slot size should be smaller than approximately D_{60} of the surrounding material (Terzaghi and Peck, 1967).

The preceding recommendations give a slot size that should not exceed a narrow range between D_{50} and D_{70}. It is suggested, therefore, that the size of the screen slot should be limited between the finest values of D_{50} and D_{70} of the gradation of the material surrounding the screen.

3.11   Analysis of Hydraulic Forces

Assuming the soil prism shown in Fig. 3.17 with a height D_a, then from equilibrium the downward forces should be equal to the upward forces throughout D_a. The relationship between these forces is analyzed for three cases (Taylor, 1948; Kashef, 1965) as follows:

Case I: Hydrostatic Conditions

Downward forces within the prism $\downarrow W = \downarrow V \gamma_{sat}$, where $V = AD_a$. The resultant F_B of hydraulic forces acting on the top and bottom boundaries of the prism are

$$F_B = \uparrow (h_{e1} + D_a)A\gamma_w - \downarrow (h_{e1})A\gamma_w$$
$$= \uparrow D_a A\gamma_w = \uparrow V\gamma_w = \text{total buoyancy} \uparrow U$$

The effective force at the bottom section of the prism (resultant of all forces) therefore is

$$\bar{\sigma}A = V\gamma_{sat} - V\gamma_w = V\gamma' = \text{submerged weight} \downarrow$$

The total buoyancy $U = V\gamma_w$ is the weight of water occupying a volume equivalent to that of the total volume of the prism. Since $V\gamma_w = V_v\gamma_w + V_s\gamma_w$, the total buoyancy U is numerically equal to the sum of the weight of water W_w within the pores, and the buoyancy on the grains $U_g = V_s\gamma_w$.

Case II: Upward Flow

Gradient $i_{gII} = h_{e2}/D_a$. The downward forces $\downarrow W = \downarrow V\gamma_{sat}$. And the resultant F_B of hydraulic forces acting on the boundaries of the prism are

$$F_B = \uparrow (h_{e1} + D_a + h_{e2})A\gamma_w - \downarrow h_{e1}A\gamma_w$$
$$= \uparrow V\gamma_w + \uparrow (h_{e2})A\gamma_w \frac{D_a}{D_a}$$
$$= \uparrow V\gamma_w + \uparrow V\gamma_w i_{gII}$$
$$= \text{total buoyancy} + \text{seepage force}$$

or $$F_B = U + F_s$$

and $\bar{\sigma}A = \text{effective force} = \text{resultant of all forces} = W - F_B$
$$= V\gamma_{sat} - V\gamma_w - V\gamma_w i_{gII} = V\gamma' - F_s$$

Dividing by A, we get

$$\bar{\sigma} = D_a\gamma_{sat} - D_a\gamma_w - \gamma_w h_{e2}$$
$$= (D_a\gamma_{sat} + h_{e1}\gamma_w) - (D_a + h_{e1} + h_{e2})\gamma_w = \sigma_t - p$$

that is, $\qquad \sigma = \sigma_t - p \qquad$ [compare with Eq. (3.30)]

This proves that Eq. (3.30) is correct and not approximate.

Case III: Downward Flow

Gradient $i_{gIII} = h_{e3}/D_a$. Following the same analysis as in case II,

$$\bar{\sigma}A = \text{effective force} = V\gamma_{sat} - V\gamma_w + V\gamma_w i_{gIII}$$
$$= W - (U - F_s)$$

where $\qquad\qquad (U - F_s) = F_B$

Generally, therefore,

$$F_B = U \pm F_s = U_g + W_w \pm F_s \qquad (3.35)$$

and $\qquad$ Effective force $= W - F_B = V\gamma' \pm F_s$
$$= W - (U \pm F_s) \qquad (3.36)$$
$$= W - (U_g + W_w \pm F_s)$$

Equations (3.35) and (3.36) are valid in two-dimensional cases where F_B and F_s are generally inclined; however, F_B and F_s are added vectorially (see Sec. 5.5).

In steady-state cases, F_s should vanish in Eqs. (3.35) and (3.36). When the quicksand condition occurs, the effective force should be zero, and from Eq. (3.36), $V\gamma' = F_s$ numerically; that is, $V\gamma' = V\gamma_w i_{gc}$ or $i_{gc} = \gamma'/\gamma_w$, as previously determined in Eq. (3.29).

Also from Eq. (3.36), $V\gamma' = W - U = W - U_g - W_w = W_s - U_g =$ weight of the solid grains minus the buoyancy on the grains.

EXAMPLE 3.2 Referring to Fig. 3.17, if $h_{e1} = 1.0$ m, $D_a = 10$ m, $h_{e2} = 5.0$ m, and $\gamma_{sat} = 2.08$ g/cm³, determine $\bar{\sigma}$ and the various forces U, $V\gamma'$, U_g, F_s, and F_B. The specific gravity G of the solid material is 2.7, and the cross sectional area A is 1.0 m² ($\gamma_w = 1$ g/cm³ = 1000 kg/m³).

Solution

$$i_g = \tfrac{5}{10} = 0.5$$

Since $\gamma_{sat} = \gamma_w(G + e)/(1 + e)$, then $2.08 = 1.0(2.7 + e)/(1 + e)$, or $e = 0.574$. Therefore,

$$W = 10 \times 1 \times 2.08 \times 1000 = 20{,}800 \text{ kg}\downarrow$$

$$V_s = \frac{V}{1 + e} = \frac{10}{1.574} = 6.353 \text{ m}^3$$

$$W_s = W - W_w = W - (V - V_s)\gamma_w = 20{,}800 - (10 - 6.353)1000$$
$$= 17{,}153 \text{ kg}\downarrow$$

$$U = D_a\gamma_w A = 10 \times 1 \times 1000 = 10{,}000 \text{ kg}\uparrow$$

$$V\gamma' = W - U = 20{,}800 - 10{,}000 = 10{,}800 \text{ kg}\downarrow$$

$$U_g = \gamma_w V_w = 1000 \times 6.353 = 6353 \text{ kg}\uparrow$$
$$F_s = iD_a A\gamma_w = 0.5 \times 10 \times 1 \times 1000 = 5000 \text{ kg}\uparrow$$
$$F_B = A\gamma_w(D_a + h_{e1} + h_{e2}) - A\gamma_w h_{e1} = AD_a\gamma_w + A\gamma_w h_{e2} = 1000(10 + 5)$$
$$= 15,000 \text{ kg}\uparrow$$

Effective force $= \bar{\sigma}A = W - F_B = 20,800 - 15,000 = 5800$ kg

or $= V\gamma' - F_s = 10,800 - 5000 = 5800$ kg

or $= W - U - F_s = 20,800 - 10,000 - 5000 = 5800$ kg

or $= W_s - U_g - F_s = 17,153 - 6353 - 5000 = 5800$ kg

Therefore,

$$\bar{\sigma} = \frac{5800}{A} = 5800 \text{ kg/m}^2$$

PROBLEMS AND DISCUSSION QUESTIONS

3.1   What is the main property of soil that does not exist in steel materials? Does this property exist in dry cement mortar?

3.2   State the main engineering characteristics of soils and indicate the practical aspects related to these properties.

3.3   What are the other terms for the following?
(a) Hydraulic conductivity
(b) Physical permeability
Give the SI units for each.

3.4   What are the factors upon which hydraulic conductivity depends?

3.5   Give other terms used for darcian and seepage velocities and indicate how to compute the seepage velocity in terms of the darcian velocity.

3.6   A sample of sand was placed in a vertical cylindrical tube 15 cm long and 5 cm in diameter and then was tested in a constant-head permeameter. The excess head was maintained at 25 cm. The volume of water collected after 3, 6, 12, and 24 min was, respectively, 25, 70, 164, and 328 cm³. Determine the steady-state rate of flow and compute the hydraulic conductivity of the sand. (The flow is assumed downward.)

3.7   Taking a vertical section in the sample in Problem 3.6, plot a *square* flow net to a scale of 1:1 consisting of four flow lines and the appropriate number of equipotential lines. Indicate the intensities of the equipotential lines assuming that the datum coincides with the downgradient water level and that this level coincides with the bottom of the sample. Count the number of fields in your drawing.

3.8   A consolidation test was performed on a clay specimen, and the coefficient of consolidation was found to be 1.06×10^{-3} cm²/s and the coefficient of volume compressibility was found to be 4.06×10^{-5} cm²/s. Calculate the hydraulic conductivity of the clay.

3.9   A soil sample was tested in a variable-head permeameter. After 5 min, the excess head was 30 cm. After another 15 min, the excess head was 25 cm. If the length of the sample was 12 cm and the ratio of the cross-sectional area of the sample to that of the burette was 100, find the hydraulic conductivity of the sample. Also find the ratio of the rates of flow at the beginning and the end of the test.

3.10 By means of a field pumping test, the hydraulic conductivity of a sand below the water table was found to average 2×10^{-2} cm/s. The average porosity of the sand was 32 percent, and the water temperature was about $25°C$. Later on, the sand was densified by vibration and it is believed that its porosity reached an average of 25 percent. What was the average hydraulic conductivity of the densified sand at a temperature of $25°C$ and at a temperature of $36°C$? [Hint: Use Eq. (3.7b), assuming that C_s remains unchanged.]

3.11 From the answers to Problem 1.13, find the approximate value of the hydraulic conductivity of the soil using Hazen's equation. Are the limitations of using Hazen's equation satisfied in this case? Why?

3.12 (a) Determine the magnitude of the critical hydraulic gradient of a soil that has a natural void ratio of 0.60. The average specific gravity of the soil grain is 2.67.
      (b) Explain why the quicksand phenomenon does not occur in gravels or clay.

3.13 (a) What are the conditions validating the use of Darcy's law?
      (b) Define or explain laminar, turbulent, steady-state, and transient flow.

3.14 Explain why Reynold's number cannot be used satisfactorily in flow through soils.

3.15 The stratifications of a formation makes an angle of $10°$ with the horizontal. If $K_{max} = 2 \times 10^{-2}$ cm/s and $K_{min} = 8 \times 10^{-3}$ cm/s, determine the hydraulic conductivities along the directions at $8°$, $55°$, and $90°$ to horizontal. Make a sketch indicating these directions and the corresponding K values.

3.16 In Fig. 3.10, equal pressure heads at A and B were imposed at 55 cm. These were the only changes made in the setup shown in Fig. 3.10. Determine the total pressure and position heads at each of the points A, C, and B corresponding to each of the datums I, II, and III (at the same locations).

3.17 In Fig. 3.17, the values of h_{e1} and D_a are, respectively, 0.6 and 9.0 m. The pore-water pressure at B_1 is 0.95 kg/cm². Plot the following diagrams (use $B_1 B_1$ as datum):
      (a) Position head
      (b) Pressure head
      (c) Total head
      (d) Total stress
      (e) Effective stress

3.18 If the soil in Problem 3.17 has the properties $K = 9.8 \times 10^{-3}$ cm/s, $e = 0.6$, and $G = 2.65$, calculate the following for the prism shown:
      (a) Total buoyancy
      (b) Buoyancy on the grains
      (c) Rate of flow
      (d) Seepage force
      (e) Effective force
      (f) Pressure head at B_1 corresponding to the quicksand condition

3.19 Redo Problems 3.17 and 3.18 if the pressure head at B_1 is 9.6 m.

3.20 Redo Problems 3.17 and 3.18 if the pressure head at B_1 is 7.9 m.

3.21 Design a filter by means of Terzaghi's method if the soil to be protected is sand with the following ranges of gradation:

Size, mm	Percent passing	Size, mm	Percent passing
9.525	100	0.59	25–75
4.76	95–100	0.297	8–30
2.38	80–100	0.149	0.5–10
1.19	45–95	0.074	0.0–3.0

Plot the filter gradation on a semilog plot and check the design by the method used by the U.S. Bureau of Reclamation (1963).

REFERENCES

Ahrens, T. P.: "Well Design Criteria," *Water Well J.*, vol. 11, nos. 9 and 11, 1957.

American Society for Testing and Materials: *Annual Book of ASTM Standards, Soil and Rock; Building Stones*, pt. 19, 1982.

Bishop, A. W.: "The Principle of Effective Stress," Norweigian Geotechnical Institute Publication 32, Oslo, 1960, pp. 1–5.

Bowles, J. E.: *Engineering Properties of Soils and Their Measurement*, 2d ed., McGraw-Hill, New York, 1978.

Casagrande, A.: "Seepage through Dams," *J. N. Engl. Water Works Assoc.*, vol. 51, 1937, pp. 131–172.

Darcy, H.: *Les fontaines publiques de la ville de Dijon*, V. Dalmont, Paris, 1856.

Hazen, A.: Discussion of "Dams on Sand Foundations, by A. C. Koenig," *Trans. Am. Soc. Civ. Eng.*, vol. 73, 1911, p. 199.

Hunter Blair, A.: "Well Screens and Gravel Packs," *Ground Water*, vol. 8, no. 1, 1970, pp. 10–21.

Johnson Division, UOP Inc.: *Ground Water and Wells*, 2d ed., Edward E. Johnson, St. Paul, Minn., 1972.

Kashef, A. I.: "Exact Free Surface of Gravity Wells," *Proc. Am. Soc. Civ. Eng.*, vol. 91, no. HY4, 1965 pp. 167–184.

Muskat, M.: *The Flow of Homogeneous Fluids through Porous Media*, McGraw-Hill, New York, 1937 (2d printing by J. W. Edwards, Ann Arbor, Mich., 1946).

Peck, R. B., W. E. Hanson, and T. H. Thornburn: *Foundation Engineering*, 2d ed., Wiley, New York, 1974.

Poiseuille, J. L. M.: "Recherches expérimentales sur le mouvment des liquides dans les tubes de très petits diamètres," *Acad. Sci., Paris, Comptes Rendus*, 961–967, 1041–1048 (1840); 112–115 (1841).

Polubarinova-Kochina, P. Y.: *Theory of Groundwater Movement* (translated by R. J. M. DeWiest from the 1952 Russian edition), Princeton University Press, Princeton, N.J., 1962.

Sherard, J. L., L. P. Dunningan, and J. R. Talbot: "Basic Properties of Sand and Gravel Filters," *J. Geotech. Eng.* vol. 110, no. GT6, 1984*a*, pp. 684–700.

Sherard, J. L., L. P. Dunningan, and J. R. Talbot: "Filters for Silts and Clays," *J. Geotech. Eng.* vol. 110, no. GT6, 1984*b*, pp. 701–718.

Taylor, D. W.: *Fundamentals of Soil Mechanics*, Wiley, New York, 1948.

Terzaghi, K.: *Theoretical Soil Mechanics*, Wiley, New York, 1943.

Terzaghi, K., and R. B. Peck: *Soil Mechanics in Engineering Practice*, 1st ed., Wiley, New York, 1948.

Terzaghi, K., and R. B. Peck: *Soil Mechanics in Engineering Practice*, 2d ed., Wiley, New York, 1967.

Todd, D. K.: *Groundwater Hydrology*, 2d ed., Wiley, New York, 1980.

U.S. Bureau of Reclamation: *Design of Small Dams*, Washington, 1960.

U.S. Bureau of Reclamation: *Earth Manual*, U.S. Department of Interior, Washington, 1963.

U.S. Bureau of Reclamation: *Ground Water Manual*, U.S. Department of Interior, Washington, 1977.

Walton, W. C.: *Groundwater Resource Evaluation*, McGraw-Hill, New York, 1970.

4

EVOLUTION OF GROUNDWATER ENGINEERING

Groundwater engineering is defined as "the art and science of investigating, developing, and managing groundwater for the benefit of man" (Water and Power Resources Service, 1981). In general, water (surface water and groundwater) is used in industry, in agriculture, and for human consumption. Owing to the continuous increase in water demands and the near exhaustion of surface waters, groundwater has become a major water source. Investigation of the occurrence of groundwater is an important part of groundwater technology for planning adequate and economical means of groundwater development whether for direct use or for supplementing surface water. The efficient planning, developments, and management of groundwater resources (Chap. 7) would prevent or alleviate the detrimental side effects of groundwater misuse, such as groundwater pollution, land subsidence due to excessive water pumping (Chap. 8), and groundwater mining.

Up until about 50 years ago, groundwater development was based mainly on the local practices of well drillers, some geologic data, and the then very limited procedures of groundwater hydrology. The visibility of surface water was probably the main reason for neglect of the invisible and relatively unknown sources of groundwater. However, the continuously increasing demands for water resulting from advances in agriculture and industry and the accelerating growth of global population made it clear that surface-water resources were not adequate. This realization shifted attention to invisible groundwater resources and necessitated reevaluation of the old practices in this relatively new direction.

Groundwater engineering at the present time encompasses an intricate net of highly scientific and technologic disciplines. These include geology; geophysics; mathematics; hydrology; hydraulics; meteorology; agronomy; physics; agricultural engineering; soil sciences; biological sciences; chemistry and chemical engineering; modeling and laboratory techniques; computer sciences; electric, mechanical, civil, sanitary, nuclear, and geotechnical engineering; statistics; systems analysis; drilling and metallurgical technology; and several other disciplines. Sound managerial approaches include part of or all these fields in addition to certain social, health, economic, and legal considerations. Sampling and testing of soils and water necessary to evaluate present or potential pollution have accelerated development in field and laboratory techniques, including refinements in sophisticated models for regional studies. The interpretation of aerial photographs and the use of remote sensing techniques have also been successful in certain aspects of groundwater hydrology and pollution evaluation. It is obvious that it is impossible for anyone to master all these related sciences. However, a successful groundwater manager should be aware of the influence of these disciplines and should be capable of comprehending and synthesizing the results of the investigations in these related technical and scientific areas.

In this chapter the evolution in our understanding of the origins of groundwater and its utilization as a water source are traced. Practical problems of water misuse are also discussed. A list of publications containing information about groundwater and related fields appears at the end of the chapter.

4.1 Historical Notes

Origin of Groundwater

Although groundwater has been used since early times, an understanding of the origin of groundwater as related to the hydrologic cycle was established only in the latter part of the seventeenth century. Many incorrect explanations of the origin of surface water and groundwater were given by Greek and Roman philosophers. Sporadic correct explanations appeared from time to time long before the seventeenth century, but these explanations were either not accepted or were overlooked.

Several incorrect hypotheses explaining the occurrence of groundwater were given by such early Greek philosophers and historians as

Homer (about 1000 B.C.), Anaxagoras and Herodotus (fifth century B.C.), Plato (427–347 B.C.), and Aristotle (384–322 B.C.). For details of these hypotheses, the reader is referred to Meinzer (1934, 1942), Bury (1929), Rawlinson (1942), and Biswas (1970).

Plato thought that one huge underground cavern in the earth was the source of all rivers and that water flowed back from the ocean to this cavern. Surprisingly, however, Plato's "Critias" includes an accurate description of the hydrologic cycle (Krynine, 1960). Aristotle (a student of Plato) taught that groundwater occurred in an intricate spongelike system of underground openings that discharged their water into springs. However, he thought that the major part of spring water originated from water vapor in the interior of the earth. Despite these incorrect hypotheses, Aristotle recognized that some cavern water originated from direct rainfall infiltration (Baker and Horton, 1936; Adams, 1938).

The Roman philosophers followed the Greek teachings and contributed little to the subject. These hypotheses were unquestioned until the end of the seventeenth century. During this period, correct explanations for the hydrologic cycle appeared, but they were denied or unnoticed.

The Roman architect Marcus Vitrivius, who lived from about 15 B.C. to A.D. 58, was probably the first in recorded history to have a correct grasp of the hydrologic cycle (DeWiest, 1965). He realized that the mountains receive large amounts of water from melting snow that seeps through the rock strata and emerges as springs at lower elevations (Walton, 1970). During the same period, the famous philosopher Lucius Anneus Seneca (who lived from 4 B.C. to A.D. 65) denied the reality of infiltrating rainwater and maintained essentially the same hypothesis as Aristotle. Seneca's idea that rainfall was an insufficient source to develop springs remained acceptable for more than 1500 years (Baker and Horton, 1936; Adams, 1938).

Al-Biruni (A.D. 973–1048) accurately explained the mechanics of groundwater movement as well as the occurrence of natural springs and artesian wells "on the principle of water finding its own level in communicating channels" (Dampier, 1949). Al-Biruni was a philosopher, astronomer, and geographer who, among other achievements, measured the specific gravity of precious stones (Dampier, 1949). He was one of several Moslem scholars who participated in the advancement of science during the Arab civilization that reached its peak during the second half of the eighth century. These scholars were well versed in the contents of the Holy Quran, and most probably Al-Biruni's hypothesis was based on some verses of the Quran unknown to the Western world. It is of interest to note that verse 21 of the thirty-ninth chapter of the Quran (recorded between A.D. 611 and 622) is very explicit in stating that rainwater infiltrates into the ground and appears as springs (Kashef, 1955).

Bernard Palissy (1509–1589) is recognized as the first in modern history to explain the hydrologic cycle, the origin of springs, and the relationship between wells and rivers (Cap, 1961). The first field measurements were made by Pierre Perrault (1608–1680). He studied evaporation and capillary rise and measured the rainfall and runoff of the upper drainage basin of the Seine River in France. Perrault reported that the precipitation was about six times as much as the river discharge (DeWiest, 1965). This was a major change from the early Greek and Roman teachings, which claimed that rainfall was not adequate to feed springs and rivers. The findings of Perrault were verified several years later by Edmé Mariotte (1620–1684), whose report appeared in 1686 after his death (Baker and Horton, 1936; Adams, 1938; Meinzer, 1942). Mariotte estimated the flow of the Seine River at the Pont Royal in Paris to be 5600 m³/min, which was less than about one-sixth the estimated total precipitation on the water shed of the river upstream of Paris. His work also examined the properties of fluids, the origin of flowing wells, winds, storms, and hurricanes, and several other subjects, and his work was collected and published by Leiden (Mariotte, 1717). The English astronomer Edmund Halley (1656–1742) verified the work of Perrault and Mariotte. He measured evaporation in the Mediterranean Sea and reported in 1693 that the amount of evaporation was sufficient to supply rivers and springs.

Outstanding documents on the subject of artesian wells were written in 1715 by Antonio Vallisnieri, President of the University of Padua, Italy (DeWiest, 1965). His paper was illustrated with one of the earliest known geologic sections drawn by Johann Schenchzer (Hagen, 1853).

One of the astonishing methods of groundwater prospecting was the use of a forked twig or so-called divining rod. The forks of the rod are usually held in the hands in such a manner that the butt end points upward (Ellis, 1917). The supposition is that when the rod is carried to a place beneath which water or other minerals lie, the butt end will be attracted downward or, according to some diviners, will whirl round and round. The origin of the method is lost in antiquity. Publications on the subject date back to the sixteenth century. Unfortunately, this method is still used at the present time in the United States and other countries, although it is based only on superstition (Ellis, 1917).

In the nineteenth century, quantative measurements were initiated by Darcy (1856) and supplemented by the analytical work of Dupuit (1863), Thiem (1906), and Forchheimer (1886). This work stimulated groundwater research in the twentieth century and shifted groundwater hydrology from a descriptive subject to a more rigorous analytical science. It would be a rather difficult task to compile a complete list of researchers who made contributions to the field of groundwater from the latter part of the nineteenth century to the present time. The major achievements between 1856 and 1955 have been compiled by Ferris and Sayre (1955). In the author's opinion, the outstanding investigators in groundwater and/or seepage in this century who stimulated others are Oscar Meinzer, Charles V. Theis, M. King Hubbert, J. Kozeny, Philip Forchheimer, Karl Terzaghi, Arthur Casagrande, Morris Muskat, C. E. Jacob, and Mahdi Hantush.

Groundwater Utilization

Some of the old methods for developing groundwater are still practiced in parts of the world. The present techniques for groundwater with-drawal are rather limited; wells are the most common source (and to a lesser extent infiltration galleries), and subdrains (known also as horizontal wells) are sometimes used to skim thin aquifers or fresh water overlying saline water (Chap. 9). Subdrains are generally used for subsurface drainage; however, they are occasionally used for supplying water by pumping the water through vertical shafts.

The utilization of groundwater goes back to early times, but the first application has been lost in antiquity. Springs and wells are mentioned in several verses in the Old and New Testaments as well as in the Holy Quran. Withdrawal of groundwater was accomplished by water wells as well as by means of longitudinal channels known as *qanats;* in general, these qanats (explained later) are somewhat similar to subdrains. Old practices are therefore generally the same as the present, except for the method of construction and the type of energy used to lift the water. Many of these old wells and qanats still exist in several places, and some of them are still used.

The Roman wells along the coast of the Mediterranean Sea in the Western Desert (west of the Nile) in Egypt are circular wells dug to intercept the groundwater and to collect the little rain that drops in the wintertime (about 25 mm per year). The intensity of rainfall decreases southward until it becomes almost nil near Cairo, which is about 250 km south of the coast. The frequency of appearance and diameter of these Roman wells were planned in accord with rain intensity; large-diameter wells were dug along the coast and a lesser number of smaller-diameter wells were dug inland. It is believed also that Mary's Well in the Cairo suburb of Mataryiah was the same well that existed about 30 B.C. It is also believed that the dug wells that exist along a belt connecting the oases in the Western Desert in Egypt were used during pharanoic times. These wells were clogged by sandstorms because of a lack of maintenance until they were reopened sometime after 1882. It is known that this belt of lowland was very fertile during pharaonic times. More recently (after

1882), these old wells were repaired, but they deteriorated again at the turn of the twentieth century (Kashef, 1981).

Generally, in the Middle East, the depth of dug wells rarely exceeds about 50 m (DeWiest, 1965). Recent advances in well-drilling techniques were unknown in early times, although the Egyptians used perforated core drilling in stone quarry operations as early as 3000 B.C. (Brantly, 1961a and b). It is believed that the art of constructing wells and infiltration galleries started in the dry regions of Asia (Bowman, 1911). The ancient Chinese developed a churn drill for water wells that was almost identical in principle to modern machines (Brantley, 1961a and b), but used wood rather than steel and human power rather than mechanical energy. Through patience over several years and probably decades, some very deep wells (1200 to 1500 m) were drilled for brine and gas (Bowman, 1911; Tolman, 1937). The ancient Chinese methods of churn drilling are still used in rural areas in Laos, Cambodia, and Thailand, as well as China, with slight modification (DeWiest, 1965). The ancient methods of lifting water by human and animal power using hoists and primitive tools are still practiced in some countries in Asia and Africa. In order to lift deep groundwater from dug wells, a water wheel is used that consists of two toothed timber wheels geared together, one horizontal and the other vertical. The horizontal wheel is rotated by an animal, which causes the vertical wheel to move. If a chain of buckets is mounted on the vertical wheel, each bucket gets filled with water when it reaches the groundwater in the well and is then emptied into a small ditch when it reaches the higher ground level. The same method is used to lift water from surface streams. Surface water in shallow channels and ditches is also lifted manually by rotating an inclined wooden cylinder with an interior Archimedes' screw.

After the discovery of flowing wells (artesian) in Flanders about A.D. 1100 and in eastern England and northern Italy a few decades later (DeWiest, 1965), percussion methods of drilling were developed. The early artesian wells were tapped about A.D. 1126 from the fractured chalk formations outcropping in the higher plateau of the Province of Artois in France. The name of this province was the origin of the name *artesian* (Norton, 1897). The fast-growing search for artesian wells stimulated a rapid development of drilling techniques. Authors, inventors, well drillers, and those who drilled wells in unexplored zones were encouraged and rewarded with medals and prizes for a number of years by the Royal and Central Society of Agriculture in France (Norton, 1897). With the introduction of the drilling mud in about 1890, the method of rotary drilling has steadily gained popularity (Bowman, 1911; Brantly, 1961a and b). Wells exceeding 300 m in depth became feasible near the end of the nineteenth century. The efficiency of deep wells was increased after the perfection of turbine pumps between 1910 and 1930 (DeWiest, 1965).

Intricate systems of water channels known as *qanats*° (Beaumont, 1968) were used in several locations between Iran and Afghanistan in southwestern Asia and in Morocco in northwestern Africa. These qanats were dug below the ground surface to intercept groundwater. They resemble underdrains or culverts. Qanats are still used in certain areas of Iran and in the Hijaz Province in Saudi Arabia. They were usually dug in alluvial materials and sometimes in rocks. The bed levels of the initial segment had a slope and then dropped below the groundwater level. At the end of the qanat, water discharged by gravity into an irrigation canal. Within the intermediate zone, closely spaced vertical shafts were constructed to lift water for water supply purposes (Wulff, 1968). The first upstream vertical shaft was known as the "mother well" and commonly its depth was less than 50 m, but occasionally it was as deep as 250 m. Most qanats were less than 5 km long, but some reached a maximum of 30 km (Beaumont, 1971). The discharge from qanats fluctuated seasonally but seldom exceeded 100 m^3/h (Todd, 1980). Although qanats are not presently found in Egypt, an extensive one was reported to have been built about 500 B.C. in order to irrigate 3500 km^2 of the then fertile land in the desert west of the Nile (Tolman, 1937).

4.2 Seepage and Groundwater Hydrology

Until the early fifties, few references on groundwater were available; the most common were the books by Meinzer (1942), Tolman (1937), and Muskat (1937) and the excellent chapter on groundwater by Jacob (1950). A few outstanding papers published during the same time period are still considered masterpieces, such as those written by Theis (1935), Hubbert (1940), and Casagrande (1937). During the two decades from 1930 to 1950, groundwater was of interest to three major groups:

1. Geologists, who limited most of their work to water-bearing formations, springs, and wells.

2. Hydrologists, who were mainly engineers interested in the hydrologic cycle in relation to surface water. Their interest in groundwater was essentially due to its role in the hydrologic cycle and the estimates of base flow in rivers and streams.

3. Geotechnical and civil engineers, who were interested in the impact of groundwater on the characteristics of soils and their bearing capacity and settlement as well as the design of hydraulic structures. The

---

° Means "canal" in Arabic; *qanah* or *qanat* is singular and *qanawat* is plural.

term *seepage* has been used to characterize the water flow beneath structures (such as solid dams or weirs) or the water flow through and below the main body of an earth dam impounding surface water.

Groundwater sciences and technology were developed to a great extent by people of various backgrounds and goals, including engineers (civil, chemical, sanitary, irrigation, and geotechnical), geologists, meteorologists, geographers, chemists, environmentalists, economists, and legislators. Specialists in groundwater are sometimes called *hydrogeologists* or *geohydrologists;* both terms are obviously inaccurate because they indicate that groundwater is confined only to hydrology and geology.

In engineering practice, the analysis of groundwater and seepage is a very important phase of vital engineering projects, such as water supply schemes, dams and hydraulic structures, irrigation and drainage systems, and dewatering of construction sites. Engineering practices in groundwater and seepage management have preceded the theoretical development of groundwater movement in general and water-well design in particular, which was initiated by Muskat (1937) and Theis (1935). Field tests and empirical rules were developed for the design of the length of weir floors by Bligh (1910) and Lane (1935) (Chap. 5). The pattern of flow in soils was developed by graphic sketching using the method of flow nets (as proved by the Laplace equation; see Chap. 5). The method was developed by Richardson (1908), Lehman in 1909 (Kashef, 1951), and Forchheimer in 1917, whose work was introduced into this country by Terzaghi (1943), the founder of the soil mechanics field. Seepage through earth dams was also well explained on the basis of flow nets by Casagrande (1937). This approach was based on the work of Kozeny, who gave a rigorous mathematical analysis of earth dams with horizontal downstream blankets (Chap. 5). The use of seepage methods in irrigation engineering was extensively explained by Leliavsky (1955), who was probably the first author to introduce in English some Russian investigations in groundwater. This was followed by two publications that appeared in 1962. The first was a translation by Roger DeWiest of a book by Polubarinova-Kochina (1962) that was published by Princeton University. The original Russian version was published 10 years earlier in 1952. The second book, which included many Russian references, was authored by Harr (1962). In all these references, very little information was given about water wells.

In engineering, drainage by wells and subdrains has been practiced efficiently since the beginning of the twentieth century. Most of the analyses were based on steady-state conditions, whether for wells, subdrains, or seepage below and through dams. Advances in the analysis of transient flow systems initiated by Theis (1935) and Muskat (1937) went unnoticed by many engineers, except perhaps in the field of petroleum engineering. Only very recently have these advances started to find their way into engineering practice.

From the early fifties to around 1965, the theoretical development of groundwater movement and water-well hydraulics continued to expand through the efforts of Jacob (1950) and his student Mahdi Hantush. The new theories stimulated others working in three major places: universities, the U.S. Bureau of Reclamation, and the U.S. Geological Survey.

In the teaching of hydrology, little was said about groundwater. It was mentioned casually in discussions of the hydrologic cycle and infiltration and in estimates of base flow (Chap. 7). Special groundwater courses were taught in a few universities, mainly in departments of geology (or geosciences), civil engineering, and agricultural engineering. In some small universities and colleges, geology departments are still teaching groundwater on a descriptive basis. In agricultural engineering, emphasis is placed on unsaturated flow and drainage of agricultural land. In civil engineering, not only hydrology is taught but also groundwater courses, including almost all aspects of groundwater sciences. Geotechnical engineers should be credited with the introduction of special groundwater courses, such as the first course taught at Harvard by Casagrande, whose students realized its importance and introduced it at other universities after they left Harvard. As early as 1945, Purdue University was among the few universities to introduce a similar course. At the present time, groundwater sciences and technology are taught in almost all well-known universities, mostly at the graduate level as separate courses or as part of undergraduate courses in hydraulics, hydrology, environmental engineering, sanitary engineering, and geotechnical engineering.

Drainage of agricultural land is closely related to seepage and groundwater. In foundation engineering, site drainage of an excavation is extremely important. Drainage of highways and airports is another important aspect of transportation engineering. Groundwater problems are also associated with the construction of tunnels, caisson foundations, and the bearing capacity of cohesionless soils, coffer dams, and sheet piles. Special departments for water resources have been established in some western universities in the United States. These offer most of the courses related to the present broad fields of surface water and groundwater.

4.3 Recent Advances in Groundwater Engineering

Since the Water Resources Act of 1965, 50 water-resource institutions have been established in universities to stimulate research in water resources. The impetus of this act was so great that the public became aware of the urgent need for clean (unpolluted) water. The establishment of the Environmental Protection Agency added another dimension to the field of water resources and emphasized all kinds of pollution problems. Water-resource investigations increased tremendously. State governments, companies, legislators, universities, technical associations, the federal government, and others are presently associated in these activities. Since 1965, groundwater engineering has started to deviate from the traditional practice of solving one problem at a time. The management of water resources (Chap. 7) encompasses various technologic, sociologic, and economic fields related to both surface water and groundwater.

The recent advances in groundwater engineering have been the outcome of several factors, such as

1. The increase in water demand due to the continuously increasing population and the subsequent expansion of industry and agriculture to meet their needs.

2. The continuously rising standard of living; this created many water problems that did not exist in the past.

3. Public awareness of the deterioration of the environment.

4. Research contracts awarded to universities and other organizations.

5. Recent laws and regulations that control water pollution and water withdrawal.

6. The detrimental effects of the misuse of water, such as the mining of aquifers, land subsidence, and water pollution.

7. Engineering advances in field and laboratory testing equipment and techniques.

8. The inadequacy of available nonpolluted surface waters to meet the increasing demands for water.

9. The impact of computers on the enhancement of mathematical models used in water-resource planning, especially for future predictions of the impact of groundwater use.

10. The successful use of some of the recent advancements, such as remote sensing and improved methods of field monitoring.

A brief summary of the highlights of these factors is given below, and the technical details of some of these factors are given in Chap. 7.

Water Demand

The average water consumption per capita per day for domestic use varies considerably according to the standard of living. In primitive areas with no industry, this average does not exceed about 30 L per person per day. In industrial areas with high standards of living, the average consumption may exceed 600 L per person per day (DeWiest, 1965). This means that in these latter areas, water shortages are common. The additional water requirements of industry and agriculture add more problems in these advanced areas and necessitate intelligent methods of water-resource management. Surface water cannot satisfy these high demands for water, and therefore, the proper development of groundwater has become increasingly important.

Murray and Reeves (1977) indicated that groundwater withdrawal in the United States increased from about 0.12 km³ per day to over 0.3 km³ per day[*] during a period of 25 years (1950–1975). The largest demand on groundwater is irrigation; it amounts to 71 percent of the groundwater used for all other purposes. About 90 percent of all irrigation water is used in the western states, where the climate varies from arid to semiarid. The largest industrial users of groundwater are the following (in descending order of usage): oil refineries, paper manufacturers, metalworking plants, chemical manufacturers, air-conditioning and refrigerating units, and distilleries (MacKichan, 1957).

Murray and Reeves (1977) listed the percentages of groundwater use for various purposes as compared with the total amount of surface and groundwater combined as follows:

Use	Percent of total
Rural domestic supplies	96
Rural livestock	57
Irrigation	41
Public water supply	37
Self-supported industries	25
Thermoelectric power utilities	1

A breakdown of the amount of water used in agriculture and certain industries is given by DeWiest (1965) as follows:

> 10 L of water to produce 1 L of cow's milk
> 10 L of water to process 1 kg of meat
> 100 L of water to produce 1 kg of paper
> 200 L of water to produce 1 kg of steel
> 400 L of water to produce 1 kg of crop (average)
> 1000 L of water to obtain 1 kg of produced dry alfalfa hay

The water consumptions of a horse, sheep, and laying hen are, respectively, 14,600, 2200, and 180 to 360 L per year (De Wiest, 1965).

[*] $1 \text{ km}^3 = 10^9 \text{ m}^3 = 1$ billion m³; this unit is sometimes called a *milliard*.

Groundwater use is not restricted to arid zones. As a matter of fact, in some metropolitan areas located in humid regions, groundwater consumption is higher per capita than in arid zones. However, in some arid zones, surface water is the main water source for irrigation. For example, the water of the Nile River in Egypt and northern Sudan comes from humid areas in central Africa (Kashef, 1981), and as a result, the groundwater resources were left intact for a very long time. Hawaii, where the soil has a very high infiltration capacity, is an example of a very humid area that uses groundwater as its primary source of water. As a result of social, cultural, and economical problems, areas in central Africa that have huge amounts of groundwater rely only on rainfall (Kashef, 1981). These areas have been suffering from droughts in the recent years, although their water problems can be solved technically by developing the available groundwater.

Control of Groundwater Withdrawal

The individual discharge of wells varies from as little as 5.5 m³ per day (about 1.0 gal/min) in domestic wells to more than 27,000 m³ per day (about 5000 gal/min) in large-capacity wells used for irrigation, by industry, or as a municipal water supply. Small-capacity wells rarely cause problems if they are properly designed and constructed. However, heavy water withdrawal from large-capacity wells produces undesirable effects if the wells have not been well planned, managed, and operated. The most common problems are

1. *Aquifer mining.* This practice continuously depletes groundwater. Some of the wells may dry out, and others may require redrilling to reach the lowered water levels. The pumps may also have to be changed and/or deepened to accommodate the new situation.

2. *Encroachment on the water rights of others.* Encroachment is caused by the expansion of interfering wells and can lead to legal disputes.

3. *Land subsidence due to excessive pumping.* Land subsidence (see Chap. 8) can lead to failures or cracking of such structures as pipelines and buildings.

4. *Induced infiltration.* If wells are close to a stream, they may withdraw large amounts of the stream surface water (induced infiltration). In most arid zones, this cannot be permitted, expecially if the stream water is allocated to others downstream from the sites of these large-capacity wells.

5. *Saltwater intrusion.* In coastal plains, large amounts of water withdrawal can lead to saltwater intrusion (Chap. 9). Ultimately, the

pumped water may become brackish or salty. The intrusion may expand, affecting other neighboring wells.

6. *Crustal uplift*. The removal of large amounts of groundwater sometimes produces an elastic expansion of the lithosphere (the solid part of the earth excluding the hydrosphere), leading to a crustal uplift (Holzer, 1979), which is a tectonic upheaval of land. Crustal uplift is the reverse process of land subsidence. The pumping of 43.5 km³ of ground-water within an area of 8070 km³ over a period of 19 years (1948 – 1967) in the Santa Cruz River Basin in Arizona produced a crustal uplift of 6 cm (Todd, 1980).

Several investigations are required for planning large-capacity wells. These include evaluation of the aquifer characteristics, determination of appropriate well spacing, and design of the well pattern to avoid well interference. Wells must also be properly designed and their pumps properly selected to achieve the most economic and trouble-free opera-tion (Water and Power Resources Service, 1981). In the process of well operation, the drawdowns should be continuously monitored and the water should occasionally be tested for quality (Chap. 2).

Water Pollution

Groundwater pollution results from the degradation of water quality. Surface-water and groundwater pollution may be the result of recent and/or past activities. The source of pollution may be municipal (sewer leakage and liquid and solid wastes), industrial (tank and pipeline leak-age, disposal of liquid wastes, mining activities, and oil-field brines) or agricultural (fertilizers, pesticides, animal wastes, and irrigation return water). It can also come from such sources as spills and surface dis-charges, water infiltration through stockpiles, leakage of septic tanks and cesspoles, roadway deicing, saltwater intrusion, polluted surface water infiltrating into the soil or surface water being polluted during its infil-tration, and infiltration of water in the sites of cemeteries.

Investigations of water pollution are relatively recent, although water pollution has been recognized by engineers for a long time. Do-mestic wells were usually specified to be dug or drilled at safe distances from septic tanks and sewage plants. (Engineers in the past were even aware of air pollution. Industrial complexes were located on the outskirts of cities and towns in such a way that the prevailing winds would drive away the smoke from inhabited areas.) However, water pollution in industrial countries is becoming highly complex, and its control highly sophisticated. Much attention has been given to surface-water pollution. Groundwater pollution has received attention only within the last few years. In some areas it is still permissible to dispose of the water wastes into aquifers through injection wells. Few states regulate such means of disposal (Miller, 1980). Many people still believe that there is no danger of groundwater pollution because the wastewater is usually localized around the source. This belief still persists because of the following reasons:

1. Groundwater is commonly free from pathogenic organisms and needs little or no treatment.

2. Turbidity and color are generally absent in groundwater.

3. Although groundwater contains larger amounts of dissolved solids than surface water, its chemical composition is almost stationary.

Despite these facts, groundwater *is* susceptible to pollution. The elements of water pollution appear in the aquifers a very long time after they leave their source because of the very slow percolation of fluids into the ground.

The study of groundwater pollution is one of the major recent advances in groundwater engineering. Such a study requires expertise in the related subjects mentioned in the early part of this chapter. Investigations of groundwater pollution should include at least the following (Walton, 1970; Bouwer, 1978; Miller, 1980; Todd, 1980):

1. *Identification of the source.*   It should be determined whether the source is municipal, industrial, agricultural, or one or more of the miscellaneous sources cited earlier.

2. *Disposal of pollutants.*   The various techniques that have been suggested include spreading the wastes on the ground surface or placing them in seepage pits and trenches, in dry streambeds, or in disposal and injection wells. The migration of pollutants and their interaction with groundwater should be traced using modeling techniques together with a well-designed field-monitoring system.

3. *Attenuation of pollution.*   This requires evaluation of filtration, sorption, chemical processes, microbiological decomposition, and dilution. It is obvious that each of these methods requires a series of tests and adequate knowledge of chemistry, biology, microbiology, soil physics and mineralogy, and the theories of dispersion and diffusion.

4. *Monitoring of groundwater quality.*   Although modeling techniques have become highly sophisticated, they usually do not give definite answers. They are useful tools for determining the trends of the migration of certain pollutants and approximate predictions of their future trends. For this reason, field-monitoring systems are used irrespective of their exhorbitant costs. These systems should be well planned, and

the equipment should be well selected to delineate the pollution boundaries (known as *plumes*).

5. *Control of polluting.* Once the pollutants are defined as to their type, source, present delineation, and mode of propagation, some means to control them should be devised, such as prevention, reduction, elimination, or the imposition of legal sanctions. A good example of the control of saltwater intrusion by reduction is given in Chap. 9.

Additional references on water pollution listed in Walton (1970), Bouwer (1978), and Todd (1980) should be checked for greater detail. Up-to-date investigations are published by the Environmental Protection Agency (EPA).

Regional Investigations

Until very recently, selected instances of groundwater problems were solved without any consideration of their effects on the environment. Each groundwater problem was confined within the hydrologic aspects necessary to solve it. Examples include the determination of flow patterns around wells, aquifer parameters, seepage from surface reservoirs and streams, stability of slopes and bank storage, base flow in streams (groundwater runoff), drainage of agricultural land using wells and/or subdrains, temporary and permanent drainage of highways, airports, and sites of building foundations, and seepage below and through dams. Some side effects, such as land subsidence, slope failures, and saltwater intrusion, were recognized. However, an efficient and successful plan of water-resource management necessitates consideration of a *hydrologic unit* rather than analysis on a case-by-case basis. Usually, such a hydrologic unit is called a *groundwater basin* (Domenico, 1972), which consists of a large aquifer or several interrelated and connected aquifers capable of providing an adequate water supply of a specified water quality with little or no undesirable side effects at the minimum possible cost.

Regional investigations stimulated the use of models of various types. They are becoming more complex because of the introduction of such additional factors as water pollution and the interaction between surface water and groundwater (Chap. 7).

Legal Aspects

The rights of ownership and use of water are regulated by the courts and legislative bodies of several states. In the United States, there are no federal statutes under which a water right can be acquired (Thomas, 1953). There are two different doctrines for acquiring water rights:

1. *Doctrine of riparian rights.* This doctrine stems from the ownership of land contiguous with a natural water source, such as a stream or lake (Water and Power Resources Service, 1981). Under this doctrine, the ownership of land overlying an aquifer is sufficient to establish groundwater rights. This doctrine is recognized in 31 predominantly eastern states (Water and Power Resources Service, 1981) and is often referred to as the English rule of unlimited use.

2. *Doctrine of prior appropriation.* Under this doctrine, water is considered the common property of the people, and thus the ownership of water is vested in the state. The first water user, not the landowner, has a prior right to use the water. In some states, water rights can be acquired by using the water beneficially for some number of consecutive years, even if the landowners or prior appropriators have rights to the same water. Usually, water for domestic use is not subject to the need for appropriation. This doctrine is recognized in 17 western states.

There are certain statutes and regulations in some states, in addition to the preceding two doctrines, that control and administer the use of groundwater in order to protect the public interest. Examples of these include certain construction practices to prevent contamination, restrictions on the disposal of pollutants, securing permits for drilling wells or rehabilitating old wells, keeping geologic records of new wells, and licensing and bonding of well drillers.

The conjunctive use of surface water and groundwater has usually caused legal conflicts. Such use should be well planned and managed to avoid these conflicts.

Water pollution was first regulated by the federal government in 1948 (Water Pollution Control Act). The National Environmental Policy Act (NEPA) of 1969 (Public Law 91-190) requires that all federal agencies prepare environmental impact statements (EIS) on major federal or federally regulated actions significantly affecting the quality of the environment. The EPA put in force regulations for the implementation of NEPA that listed groundwater protection in determining the need for an EIS. The Federal Water Pollution Control Act Amendments of 1972 (Public Law 92-500) dealt with waste-treatment management (Sec. 208) and indicated in Sec. 304(e) that the role of the EPA is to provide guidance and information, but gave the EPA no implementation authority. The Safe Drinking Water Act of 1974 (Public Law 93-523) requires the regulation of underground injection that may endanger underground drinking water resources and includes "deep and shallow waste disposal wells, oil-field brine disposal wells and secondary recovery wells, and engineering wells." Some of the major polluting activities may not be subject to the requirements of Public Law 93-523 (Miller, 1980). In general, federal and state regulations give low priority to groundwater quality as compared with that of surface water. However, the present regulations should be looked on as moving in the right direction, especially since advances in the groundwater field as a whole can be considered relatively recent.

The preceding discussion indicates some of the more recent trends in groundwater investigation. It is also obvious that the problems involved are becoming more and more intricate over time and need continuous research and planning. The groundwater field has now moved from consideration of individual local problems to a more broad, regional analysis. In other words, the recent advances in the groundwater field are centered around the proper planning and management of water resources, including water pollution, among other factors. Advances in the technical aspects of management will also continue to be useful in enhancing the quality of the data input in modeling systems.

4.4   Groundwater Data and Literature

Publications on groundwater and related fields have increased tremendously in the last two decades. The sources of these data and literature are usually the following organizational publications.

U.S. Geological Survey Publications

The results of the U.S. Geological Survey are published as circulars, professional papers, and water-supply papers (Clarke et al., 1978). Most of the field work is done on a cooperative basis with the states. Groundwater levels in various parts of the country are published periodically. The publications may contain solutions to field problems in certain areas, the reporting of field data, or purely theoretical analyses. In 1971, the U.S. Geological Survey established The National Water Data Storage and Retrieval (WATSTORE) System to provide a large-scale computerized system for the storage and retrieval of water data, including water quality (analysis of biological, chemical, physical, and radiochemical characteristics) and groundwater site inventory (geohydrologic characteristics, field measurements, and history of groundwater sources). These data are all available to the public through any of the district offices of the U.S. Geological Survey (Clarke et al., 1978).

Government Publications

Other than the U.S. Geological Survey, various governmental agencies publish material related directly or indirectly to groundwater and seepage. The U.S. Corps of Engineers, the U.S. Navy, and the U.S. Bureau of Reclamation, Department of the Interior, publish bulletins and manuals from time to time related to water resources in general and groundwater in particular. Most of these manuals are designed for use by practicing engineers. However, many other publications are the result of extensive research. The recent publications of the EPA are very useful references in water pollution. These should be considered as invaluable sources of information on this subject.

University Publications

Most of the research conducted in universities is published in professional and scientific journals (national and/or international). These journals usually present summaries of the voluminous reports published by universities. Some of these reports may be found in the publications of local water-resource research institutes (or equivalent agencies) or as separate bulletins published by universities. The reports usually contain the details of investigations and other data that may be useful to research workers. However, some of the university bulletins and reports have never been published as papers in national or international scientific journals for various reasons.

Another great source of information is the master and doctorate theses completed every year. Although the results in these theses are supposed to be published in professional and scientific journals, many of the better theses have never been published. Usually students and their advisors are reluctant to publish the results of master's theses, although some of these have proved to be useful additions to the field. University libraries usually compile lists periodically of all the theses in the United States and Canada. Recently, scientific journals have started to summarize these theses (see, for example, *Ground Water,* the journal of the National Water Well Association).

Professional and Scientific Journals

There are various associations and societies in the world that publish journals and bulletins related directly or indirectly to the groundwater field. Most of the journals published in the United States are also distributed abroad. Many foreign scientists publish their papers in U.S. journals. Also, international conferences and congresses are held periodically in various locations throughout the world. Occasionally, national and local conferences and symposia with specific themes are also held, and the presented papers are compiled in their proceedings.

The American Society of Civil Engineers publishes about 18 journals in their various divisions. (The number changes over time. In the old system papers, called "proceeding separates," were compiled and published each year as proceedings and transactions.) Groundwater papers are found occasionally in journals of the divisions of hydraulics, irrigation and drainage, environmental engineering, and water resources planning and management and less frequently in journals of the divisions of construction, engineering mechanics, and geotechnical engineering.

The American Geophysical Union has published *Water Resources Research* since 1965. This journal contains contributions in groundwater (and other water-resource research papers). Before 1965, papers pertinent to groundwater were published in the *Journal of Geophysical Research* and the *Transactions of the American Geophysical Union*. Also in 1965, the American Water Resources Association published the *Water Resources Bulletin*, which deals with all phases of water-resource planning and management, including the social, economic, and environmental aspects, etc., with occasional papers on groundwater (especially during 1970 and thereafter).

Pertinent papers also appear in journals of physics, chemical engineering, and engineering mechanics, as well as foreign journals and those of other U.S. associations. The most important publications are those of the International Association of Hydrological Sciences (IAHS), the *Journal of Hydrology*, which is published in Amsterdam and distributed worldwide, and of the American Water Works Association and other geologic, geophysical, and agricultural associations.

The National Water Well Association publishes *Water Well Journal*, which is confined to water-well technology and practical aspects of well drilling and related equipment. Since 1963, the same association has published *Ground Water*, which is devoted exclusively to groundwater.

State Publications

Details of local groundwater issues are published in some states. The technical standard of these publications varies widely from state to state. Extensive groundwater investigations have been published by state agencies in California and Illinois, where the water-resource agencies are better equipped than in other states. Bibliographies of all water-resource publications of the state agencies can be found in Giefer and Todd (1972 and 1974) or by contacting these agencies.

Books on Groundwater

There are specialized books on the flow of fluids through porous media or such subjects as geochemistry, pollution, and water rights. General books on groundwater have appeared mainly in recent years, except Tolman's book, which was published in 1939, and the first edition of

Todd's book, published in 1959. The references in these books are usually very important sources for greater detail. Most publishing companies request the replacement of old references with more recent ones. The reader should be aware of this fact (the author of the reference may be the compiler rather than the originator of a certain theory). It is difficult to include *all* books published worldwide. The name of the author or authors and the year of publication of major books (or chapters) on groundwater and seepage are listed in the following table in chronologic order. With the exception of a few, these books are published in the United States. The complete titles of these publications are listed among the references at the end of this chapter.

Muskat, 1937	Heath and Trainer, 1968
Tolman, 1937	DeWiest (ed.), 1969
Meinzer (ed.), 1942	Verruijt, 1970
Jacob, 1950	Walton, 1970
Thomas, 1951	Bear, 1972
Todd, 1959 (2d ed., 1980)	Brown et al. (eds.), 1972
Scheidegger, 1960 (3d ed., 1974)	Cooley et al., 1972
Collins, 1961	Domenico, 1972
Harr, 1962	Huisman, 1972
Polubarinova-Kochina, 1962 (Russian version, 1952)	Randkivi and Callander, 1976
	Cedergren, 1977
Schoeller, 1962 (French)	McWhorter and Sunada, 1977
Castany, 1963 (French)	Bouwer, 1978
DeWiest, 1965	Bear, 1979
Davis and DeWiest, 1966	Freeze and Cherry, 1979
Bear et al., 1968	Marino and Luthin, 1982

The preceding list does not include the publications of the organizations mentioned previously. The list also does not include United Nations publications, industrial publications, and books on general hydrology and/or geotechnical engineering. There is good material in these publications related directly or indirectly to groundwater and seepage. Reference has already been made in this text to some of these publications.

REFERENCES

Adams, F. D.: *The Birth and Development of the Geological Sciences*, Dover, New York, 1938, Chap. 12, pp. 426–460

Baker, M. N., and R. E. Horton: "Historical Development of Ideas Regarding the Origin of Springs and Ground Water," *Trans. Am. Geophys. Union*, vol. 17, 1936, pp. 395–400.

Bear, J.: *Dynamics of Fluids in Porous Media*, Elsevier, New York, 1972.

Bear, J., et al.: *Physical Principles of Water Percolation and Seepage*, UNESCO, Paris, 1968.

Bear, J.: *Hydraulics of Groundwater*, McGraw-Hill, New York, 1979.

Beaumont, P.: "Qanats on the Varamin Plain, Iran," *Trans. Inst. Br. Geogr.*, vol. 45, 1968, pp. 169–179.

Beaumont, P.: "Qanat Systems in Iran," *Bull. Int. Assoc. Sci. Hydrol.*, vol. 16, 1971, pp. 39–50.

Biswas, A. K.: *History of Hydrology*, Elsevier, New York, 1970.

Bligh, W. G.: "Dams, Barrages and Weirs on Porous Foundations," *Engineering News*, vol. 64, 1910, pp. 708–710.

Bouwer, H.: *Groundwater Hydrology*, McGraw-Hill, New York, 1978.

Bowman, I.: "Well-Drilling Methods," U.S. Geological Survey Water-Supply Paper 257, 1911, pp. 23–30.

Brantly, J. E.: "Percussion Drilling System," in D. V. Carter (ed.), *History of Petroleum Engineering*, American Petroleum Institute, New York, 1961a, pp. 133–269.

Brantley, J. E.: "Hydraulic Rotary-Drilling System," in D. V. Carter (ed.), *History of Petroleum Engineering*, American Petroleum Institute, New York, 1961b, pp. 271–452.

Brown, R. H., et al. (eds.): *Ground-Water Studies*, UNESCO, Paris, 1972.

Bury, R. G.: Translation of Plato's "Critias" and "Timaeus," in *Plato's Works in the Loeb Classical Library*, vol. 7, Harvard University Press, Cambridge, Mass., 1929.

Cap, P. A.: *Les oeuvres complète de Bernard Palissy — Des eaux et fontaines*, Albert Blanchard, Paris, 1961, pp. 436–483.

Casagrande, A.: "Seepage through Dams," *J. N. Engl. Water Works Assoc.*, vol. 51, 1937, pp. 131–172.

Castany, G.: *Traité pratique des eaux souterraines*, Dunod, Paris, 1963.

Cedergren, H. R.: *Seepage, Drainage, and Flow Nets*, 2d ed., Wiley, New York, 1977.

Clarke, P. F., et al.: "A Guide to Obtaining Information from the USGS, 1978," U.S. Geological Survey Circular 777, Washington, 1978.

Collins, R. E.: *Flow of Fluids through Porous Materials*, Reinhold, New York, 1961.

Cooley, R. L., et al.: "Principles of Ground-Water Hydrology," in *Hydrologic Engineering Methods for Water Resources Development*, vol. 10, U.S. Army, Corps of Engineers, Davis, Calif., 1972.

Dampier, Sir William Cecil: *A History of Science and Its Relations with Philosophy and Religion*, 4th ed., Macmillan, New York, 1949.

Darcy, H.: *Les fontaines publiques de la ville de Dijon*, V. Dalmont, Paris, 1856.

Davis, S. N., and R. J. M. DeWiest: *Hydrogeology*, Wiley, New York, 1966.

DeWiest, R. J. M.: *Geohydrology*, Wiley, New York, 1965.

DeWiest, R. J. M. (ed.): *Flow through Porous Media*, Academic Press, New York, 1969.

Domenico, P. A.: *Concepts and Models in Groundwater Hydrology*, McGraw-Hill, New York, 1972.

Dupuit, J.: *Etudes théoriques et pratiques sur le mouvement des eaux dans les canaux découverts et à travers les terrains perméables*, Dunod, Paris, 1863.

Ellis, A. J.: "The Divining Rod: A History of Water Witching," U.S. Geological Survey Water-Supply Paper 416, Washington, 1917.

Ferris, J. G., and A. N. Sayre: "The Quantative Approach to Groundwater Investigations," *Economic Geology*, 50th anniversary volume, 1955.

Forchheimer, Ph.: "Uber die Ergebigkeit von Brunnen Anlagen und Sickerschlitzen," *Zeitschrift des Architekten-und Ingenieur Vereins zu Hannover*, vol. 32, 1886, pp. 539–564.

Freeze, R. A., and J. A. Cherry: *Groundwater*, Prentice-Hall, Englewood Cliffs, N.J., 1979.

Giefer, G. J., and D. K. Todd (eds.): *Water Publication of State Agencies*, Water Information Center, Port Washington, N.Y., 1972.

Giefer, G. J., and D. K. Todd (eds.): *Water Publications of State Agencies, First Supplement, 1971–1974*, Water Information Center, Huntington, N.Y., 1976.

Hagen, G.: *Handbuch der Wasserbaukunst*, vol. 1, Bornträger, Koenigsberg, 1853, p. 87.

Harr, M. E.: *Groundwater and Seepage*, McGraw-Hill, New York, 1962.

Heath, R. C., and F. W. Trainer: *Introduction to Ground-Water Hydrology*, Wiley, New York, 1968.

Holzer, T. J.: "Elastic Expansion of the Lithosphere Caused by Groundwater Depletion," *J. Geophys. Res.*, vol. 84, 1979, pp. 4689–4698.

Hubbert, M. K.: "Theory of Ground-Water Motion," *J. Geol.*, vol. 48, no. 8, 1940, p. 785.

Huisman, L.: *Groundwater Recovery*, Winchester Press, New York, 1972.

Jacob, C. E.: "Flow of Ground Water," in H. Rouse (ed.), *Engineering Hydraulics*, Wiley, New York, 1950, Chap. 5, pp. 321–386.

Kashef, A. I.: "Numerical Solutions of Steady State and Transient Flow Problems. Ph.D. thesis at Purdue University, West Lafayette, Ind., 1951.

Kashef, A. I.: "On the History of Ground Water Research," *Civil Engineering Magazine*, vol. 3, pp. 113–126, April 1955.

Kashef, A. I.: "The Nile: One River and Nine Countries," *J. Hydrol.* (in Arabic), vol. 53, 1981, pp. 53–71.

Krynine, P. D.: "On the Antiquity of Sedimentation and Hydrology," *Bull. Geol. Soc. Am.*, vol. 71, 1960, pp. 1721–1726.

Lane, E. W.: "Security from Under-Seepage Masonry Dams on Earth Foundations," *Trans. Am. Soc. Civ. Eng.*, vol. 100, 1935, pp. 1235–1351.

Leliavsky, S.: *Irrigation and Hydraulic Design*, vols. 1–3, Chapman & Hall, London, 1955.

MacKichan, K. A.: "Estimated Use of Water in the United States, 1955," *J. Am. Water Works Assoc.*, vol. 49, 1957, pp. 369–391.

McWhorter, D. B., and D. K. Sunada: *Ground-Water Hydrology and Hydraulics*, Water Resources Publishers, Fort Collins, Colo., 1977.

Marino, M. A., and J. N. Luthin: *Seepage and Groundwater*, Elsevier, Amsterdam, 1982.

Mariotte, E.: *Oeuvres de Mr. Mariotte*, vols. 1 and 2, P. Van der Aa, Leiden, 1917.

Meinzer, O. E.: "The History and Development of Ground-Water Hydrology." *J. Wash. Acad. Sci.*, vol. 24, 1934, pp. 6–32.

Meinzer, O. (ed.): *Physics of the Earth*, vol. 9, McGraw-Hill, New York, 1942.

Miller, D. W.: *Waste Disposal Effects on Ground Water*, Premier Press, Berkeley, Calif., 1980.

Murray, C. R., and E. B. Reeves: "Estimated Use of Water in the United States, 1975," U.S. Geological Survey Circular 765, Washington, 1977.

Muskat, M.: *The Flow of Homogeneous Fluids through Porous Media*, McGraw-Hill, New York, 1937.

Norton, W. H.: "Artesian Wells of Iowa," *Iowa Geol. Surv.*, vol. 6, 1897, pp. 122–134.

Polubarinova-Kochina, P. Y.: *Theory of Groundwater Movement*, Princeton University Press, Princeton, N.J., 1962.

Randkivi, A. J., and R. A. Callander: *Analysis of Groundwater Flow*, Wiley, New York, 1976.

Rawlinson, G.: "Persian Wars," in F. R. B. Godolphin (ed.), *The Greek Historians Complete and Unabridged Historical Works of Herodotus*, vol. 1, Random House, New York, 1942.

Richardson, L. F.: "A Freehand Graphical Way of Determining Streamlines and Equipotentials," *Philos. Mag. Ser. VI* (London), vol. 15, 1908, pp. 237–250.

Scheidegger, A. E.: *The Physics of Flow through Porous Media*, University of Toronto Press, Toronto, 1962; 3rd ed., 1974.

Schoeller, H.: *Les eaux souterraines*, Masson & Cie, Paris, 1962.

Terzaghi, K.: *Theoretical Soil Mechanics*, Wiley, New York, 1943.

Theis, C. V.: "The Relation between the Lowering of the Piezometric Surface and the Rate and Duration of Discharge of a Well Using Groundwater Storage," *Am. Geophys. Union Trans.*, vol. 16, 1935, pp. 519–524.

Thiem, A.: *Hydrologische Methoden*, Gebhardt, Leipzig, 1906.

Thomas, H. E.: *The Conservation of Ground Water*, McGraw-Hill, New York, 1951.

Thomas, H. E.: "Ground Water Law," transcript lecture presented at Ground Water Short Course, U.S. Geological Survey and Bureau of Reclamation, Fort Collins, Colo., April 6–17, 1953.

Todd, D. K.: *Ground Water Hydrology*, 2d ed., Wiley, New York, 1980.

Tolman, C. F.: *Ground Water*, McGraw-Hill, New York, 1937.

Verruijt, A.: *Theory of Groundwater Flow*, Gordon and Breach, New York, 1970.

Walton, W. C.: *Groundwater Resource Evaluation*, McGraw-Hill, New York, 1970.

Water and Power Resources Service, U.S. Department of Interior: *Ground Water Manual*, Washington, 1981; first edition was published in 1977 by the United States Bureau of Reclamation before its name change.

Wulff, H. E.: "The Qanats of Iran," *Sci. Am.*, vol. 218, 1968, pp. 94–100, 105.

5

TWO-DIMENSIONAL
STEADY-STATE FLOW

Two-dimensional flow systems are similar to those shown in Figs. 3.13 and 3.19. The solution of any of these problems should result in a flow net similar to those shown in Fig. 3.11. The flow toward a discharge well or away from a recharge well is radial but two-dimensional (Fig. 3.12). Well problems are usually idealized by assuming wells to be vertical, aquifer boundaries to be horizontal, water tables and piezometric surfaces to be initially horizontal, and wells to be completely penetrated and screened throughout the intercepting zone. The problem is then called *axisymmetric;* i.e., vertical planes passing through the well axis exhibit the same flow pattern. In idealized artesian wells (Fig. 3.12*b* and *c*), any horizontal plane within the aquifer should also have the same pattern as shown.

Generally, there are two main classes of flow: *confined flow* and *unconfined flow.* In confined flow, the uppermost flow line is under a pressure that is higher than atmospheric pressure. For example, Fig. 3.13*b* shows the flow beneath a solid dam or weir. Artesian wells also fall into this category (Fig. 3.12*b*). The uppermost flow line coincides with the upper impervious boundary of the artesian aquifer.

In unconfined flow (gravity flow), the uppermost flow line is in balance with atmospheric pressure. The pressure along this line should be atmospheric ($p = 0$) at any point. Examples of this class of flow are natural flow systems in unconfined aquifers (Fig. 3.15), seepage through earth dams (Figs. 3.13a and 3.19b and c), flow toward subdrains (Fig. 3.13d), flow toward trenches or cuts (Figs. 3.13c and 3.19d), and flow toward a gravity well (Fig. 3.12a). The uppermost flow line in an unconfined system is known also as a *phreatic line, free surface,* or *line of seepage.*

The general hydrodynamic equation that governs flow is the same for all steady-state flow systems (Sec. 5.1). However, solutions for the various systems differ as a result of the effects of geometric and hydraulic boundaries.

The governing equation for two-dimensional systems is known as the *Laplace equation* (Sec. 5.1), which may be transformed to cylindrical coordinate systems (Chap. 6) suitable for radial flow cases. The mathematical solutions are in most cases complex, except in highly idealized cases in which the boundaries are simplified. The solutions become even more complex in unconfined flow systems than in confined systems. The free surface is in fact an unknown boundary of the flow region until an analytical solution is found; the free surface has to be determined with this solution. However, the boundary conditions of the free surface can be expressed both physically and mathematically. Advanced mathematical techniques have been developed to solve special cases of gravity flow systems (Muskat, 1937; Polubarinova-Kolchina, 1962; DeWiest, 1969). Such methods are not attractive to practicing engineers and professionals who are accustomed to closed-form solutions. In many instances mathematical approximations have been introduced, and the results have been compared in some cases with the more rigorous analyses. In other instances, mathematical solutions are not feasible, and computerized numerical procedures have to be used (Zienkiewicz, 1971; Rushton and Redshaw, 1979).

In addition to mathematical approximations, various modeling techniques have been used (Todd, 1980). Sand models, in which dyes are introduced to trace the flow lines, are used for both confined and unconfined flow systems. In unconfined systems, these models have capillary flow within a region of the same size as that of the prototype and cannot be scaled down to the model size (Cahill, 1967). The diffusion of dyes is another drawback, among several others. These sand models are now generally used only for illustrative purposes.

The analogy between groundwater flow as expressed by the Laplace equation [Eq. (5.3)] and other fields of physics made it possible for models simulating groundwater flow systems to be used. These models include thermal, viscous flow (see Hele-Shaw models in Chap. 9), stretched membrane or soap film (Hansen, 1952), optical, electric conduction, electrolytic tank (Boothroyd et al., 1949; Debrine, 1970), and magnetic models. In each of these models, the flow net is traced using various techniques. For example, in magnetic models, the flow pattern is determined by using fibers of dielectric material, such as iron fillings, fragments of carbon, glass wool, gypsum, mica dust, tourmaline, mahogany dust, and even tea leaves (Hague, 1929). Most of these models have several drawbacks, except perhaps electrical models built to solve confined flow systems. Electric conduction models were developed to solve gravity flow problems by a trial-and-error procedure (Muskat, 1937). Electric resistance models have been used successfully to solve steady-state problems. The model is composed of an electronic analyzer coupled with an analog model that consists of an array of only resistors (Todd, 1980). An understanding of these models requires a knowledge of numerical analysis.

Numerical analysis was introduced about four decades ago (Southwell, 1940, 1946; Maasland and Bittinger, 1963; Yang, 1949; Kashef, 1951) to solve engineering and groundwater problems using various techniques (mainly relaxation and iteration techniques). Numerical analysis implies changing the governing mathematical equation to a set of algebraic equations. More recently, computerized numerical analysis (Prickett, 1979) has been performed using mainly the finite-difference and finite-element techniques (Kashef, 1951, 1963; Kashef and Safar, 1975). These are probably the best techniques available at the present time.

The basic assumptions of the major mathematical equations involved in flow are explained in this chapter and in Chap. 6 in order to emphasize the limitations of the solution methods, including numerical analysis. The solutions presented in this book can be used by practitioners or can be used to find general trends before seeking more rigorous solutions. The method of flow nets using freehand sketching is explained in detail. Although the method appears to be approximate, it is a useful and powerful approach that serves its purpose in practice (Cedergren, 1977). The principles of this method cannot be fully understood without a knowledge of mathematics. Its results may be sufficient for many practical purposes, and they also can be used by computer programmers to speed up their computerized solutions by setting these simple solutions as the initial conditions.

In many cases, steady-state flow is never reached; however, its investigation is useful in understanding the more realistic transient states. In water wells, for example, transient states of flow are presently treated by several techniques (Chap. 8), but the investigated cases have characteristics that are similar to those of the steady state. Several useful principles have been obtained from the study of steady-state solutions. Moreover, some transient flow solutions are somewhat similar to steady-state solutions (Chap. 8). Finally, in engineering, extreme conditions that are not expected to occur are imposed as a measure of safety, and these justify the use of steady-state solutions. In dams across rivers, for example, the upstream levels are considered at their highest possible level, while downstream levels are assumed to be dry, even though these rivers may never dry out. Steady-state flow represents, in these cases, the most critical conditions that might be expected to occur that would test the stability and safety of the structures.

5.1 The Laplace Equation and Flow Nets

In Fig. 5.1a, the region *abcd* lies within a flow medium. Under steady-state conditions, the rate of flow crossing the line *abc* is equal to the rate of flow leaving the region across line *adc*. While water is traveling from *abc* to *adc*, the losses in total head produce decreases in water pressures *p* at all points within *abcd*. Thus the effective stresses $\bar{\sigma}$ increase [Eq. (3.26)], and the soil skeleton suffers compressibility, releasing water from storage within the region *abcd*. The decreases in water pressures also lead to expansion of water in the pores within the same region,

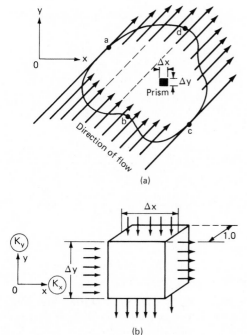

(a)

(b)

Figure 5.1 Water flow through soil media. *(a)* Region *abcd*; *(b)* infinitesimal prism $[(\Delta X)(\Delta Y)(1.0)]$ within region *abcd* (enlarged).

releasing an additional amount of water. The total heads vary over time, and the flow is thus transient.

When the water and the soil skeleton are incompressible, steady-state flow is reached. In a practical sense, steady-state flow occurs after the soil and water have reached their ultimate possible compressibility or when the discharge is equal to the recharge. Mathematically, the analysis is made on a prism $[(\Delta x)(\Delta y)(1.0)$ in Fig. 5.1$b]$, and the rate of flow entering the prism is equated to the rate of flow leaving it. Thus

$$v_y(\Delta x) + v_x(\Delta y) = v_y'(\Delta x) + v_x'(\Delta y) \tag{5.1}$$

If v is written in terms of the total head h and Darcy's law [Eq. (3.5)] for a homogeneous and anisotropic medium is used, then as the prism approaches a point, the following equation is obtained:

$$K_x \frac{\delta^2 h}{\delta x^2} + K_y \frac{\delta^2 h}{\delta y^2} = 0 \tag{5.2}$$

where K_x and K_y are, respectively, the hydraulic conductivities along the x axis (direction of stratification) and the y axis. In homogeneous and isotropic media, $K_x = K_y = K$; therefore,

$$\frac{\delta^2 h}{\delta x^2} + \frac{\delta^2 h}{\delta y^2} = 0 \tag{5.3}$$

Equation (5.3) is known as the *Laplace equation,* and it governs flow in homogeneous and isotropic media under steady states of flow. The solution to this equation for certain boundary conditions is unique (no other solution is possible). The solution gives all data for plotting a flow net consisting of a set of flow lines orthogonal to a set of equipotential lines, as shown in Figs. 5.2a and 5.3. Any specific flow line (or stream line) has a certain intensity ψ (in square centimeters per second) known as the *stream function.* If, for example, the uppermost flow line $BB'CDEF$ in Fig. 5.2a is given an intensity of zero ($\psi_1 = 0$), then the lowest flow line $A'G'$ has an intensity of $\psi_5 = q$. The difference in the ψ values of any two consecutive flow lines forming the boundaries of a flow channel is equal to $\Delta\psi = \Delta q$, which is the rate of flow through that channel. If the flow net is constructed in such a way that Δq has the same value for all flow channels, then

$$q = n_F(\Delta q) = n_F(\Delta\psi) \tag{5.4}$$

where n_F is the number of flow channels and q is the total rate of flow per unit length (in square centimeters per second). In Fig. 5.2a, there are five flow lines ($n_F = 4$). The intensities ψ_1, ψ_2, ψ_3, ψ_4, and ψ_5 can be assigned, respectively, as 0, Δq, $2\Delta q$, $3\Delta q$, and $4\Delta q$, or as q, $q - \Delta q$,

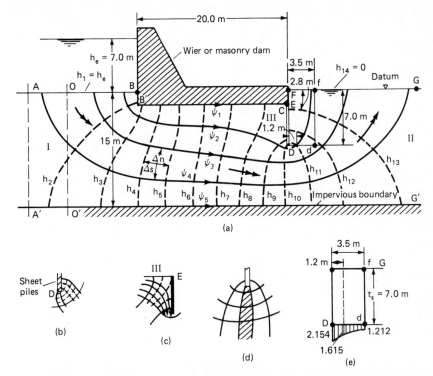

(a)

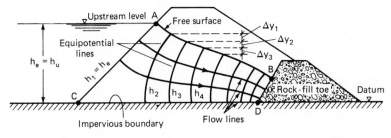

Figure 5.2 Seepage beneath a solid dam (or weir). CD, impervious sheet piles. *(a)* Flow net beneath the weir; *(b)* details at bottom of sheet piles; *(c)* details at corner C; *(d)* residual portion of a square flow net; *(e)* analysis of upheaval.

$q - 2\Delta q$, $q - 3\Delta q$, and $q - 4\Delta q$, or as any other set of values that maintains $\Delta\psi = \Delta q = $ constant ($\Delta q = \frac{1}{4}q$ in the case shown in Fig. 5.2a).

An equipotential line with intensity h is the locus of all points within the flow medium that have the same total head. The intensity of the equipotential lines can be given by h or by the velocity potential ϕ,

Figure 5.3 Flow net in an earth dam with a downstream rock-fill toe.

defined as

$$\phi = -Kh + \text{constant} = -K\left(\frac{p}{\gamma_w} + y\right) + \text{constant} \qquad (5.5)$$

The constant may be positive, negative, or zero depending on the location of the selected datum with reference to which the total heads are measured. (The location of the datum should of course remain fixed in a certain problem.) Other choices of datum location would affect the absolute value of h but would have no effect on the equipotential drop Δh or $\Delta \phi$, where

$$\Delta \phi = K(\Delta h) \qquad (5.6)$$

Use of the functions ϕ and ψ is helpful in solving Eq. (5.3) in other fields of physics as well as in groundwater. The relationships between ψ, ϕ, and h are expressed by

$$v_x = -\frac{\delta \phi}{\delta x} = -\frac{\delta \psi}{\delta y} = -K\frac{\delta h}{\delta x} \qquad (5.7a)$$

and

$$v_y = -\frac{\delta \phi}{\delta y} = +\frac{\delta \psi}{\delta x} = -K\frac{\delta h}{\delta y} \qquad (5.7b)$$

The ϕ and ψ curves should be orthogonal in isotropic and homogeneous media, as indicated by the preceding equations.

Solution of the Laplace equation [Eq. (5.3)] is possible for certain simple cases (Muskat, 1937). In homogeneous and anisotropic media, Eq. (5.2) may be transformed to a laplacian form [Eq. (5.3)] as follows:

$$\frac{\delta^2 h}{\delta(x\sqrt{K_y/K_x})^2} + \frac{\delta^2 h}{\delta y^2} = 0 \qquad (5.8)$$

or

$$\frac{\delta^2 h}{\delta X^2} + \frac{\delta^2 h}{\delta y^2} = 0 \qquad (5.9)$$

where

$$X = x\sqrt{K_y/K_x} \qquad (5.10)$$

If the original x axis is shortened to the new axis X given by Eq. (5.10), a transformed section is obtained in which the governing equation is laplacian [Eq. (5.9)]. The flow net for the transformed section is obtained as if it were homogeneous and isotropic, and the real flow net is projected back to the original section (Fig. 5.4) by multiplying each X coordinate by $\sqrt{K_x/K_y}$ while y remains the same. In evaluating the rate of flow q, the hydraulic conductivity K should be replaced by $\sqrt{K_xK_y}$ (Taylor, 1948).

Referring to Fig. 5.2a, where the medium is assumed homogeneous and isotropic, there are 14 equipotential lines and 13 equipotential drops. Considering the "field" bounded by ψ_3, ψ_4, h_4, and h_5 and apply-

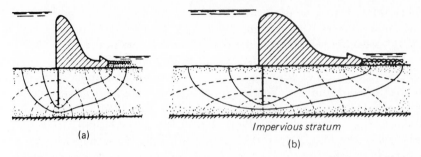

(a)

Impervious stratum

(b)

Figure 5.4 Construction of flow net in homogeneous and anisotropic medium. (a) Transformed section, $K_x = 4K_y$ and $\text{scale}_{\text{horiz}} = \text{scale}_{\text{vert}} \times \sqrt{K_y/K_x} = \text{scale}_{\text{vert}} \times \frac{1}{2}$; (b) true section, natural scale. (K. Terzaghi and R. B. Peck, *Soil Mechanics in Engineering Practice*, Wiley, New York, 1967.)

ing Darcy's law [Eq. (3.5)], the rate of flow Δq through the flow channel between ψ_3 and ψ_4 is given by

$$\Delta q = \Delta \psi = \psi_4 - \psi_3 = K \frac{h_4 - h_5}{\Delta s} \times \Delta n \times 1.0$$

per unit dimension normal to the cross section shown. Δn and Δs are, respectively, the average width and length of the shown field (Δs is measured along the flow path).

Setting Δh equal to $h_4 - h_5$, we get

$$\Delta q = K(\Delta h)\frac{\Delta n}{\Delta s} \qquad (5.11)$$

If Δh is maintained constant between any two consecutive equipotential lines (usually this can be achieved in theoretical and experimental solutions) and Δq is the same in any other flow channel, then the ratio $\Delta n/\Delta s$ for various sizes of fields should be the same. If Δs is twice as much as Δn, then this ratio should remain $\frac{1}{2}$ for *all* other fields (curvilinear rectangles). From Eqs. (5.4) and (5.11),

$$q = Kh_e\frac{n_F}{n_E} \times \frac{\Delta n}{\Delta s} \qquad (5.12)$$

where h_e is the excess head, that is, the difference between the intensities of the first and last equipotential lines or simply the difference between the upstream and downstream water levels, and n_E is the number of equipotential drops. If the ratio $\Delta n/\Delta s$ is selected as 1.0 ($\Delta n = \Delta s$), the flow net consists of curvilinear squares and

$$q = Kh_e\frac{n_F}{n_E} \qquad (5.13)$$

The flow net may be drawn by freehand sketching when it consists of curvilinear squares (Richardson, 1908; Casagrande, 1937; Cedergren, 1977), observing the mathematical conditions expressed by Eq. (5.7). Freehand sketching of flow nets is very popular among engineers, especially in complex cases where analytical solutions are complex, nonexistent, or time-consuming. The method of flow nets has been attributed to Lehman in 1909 (Kashef, 1951), Prášil in 1913 (Harr, 1962), and Forchheimer in 1917 (Terzaghi and Peck, 1967). However, according to Kashef (1951), the first known publication of freehand sketching for flow nets was that of Richardson (1908). However, its use was well explained by Casagrande (1937) and Cedergren (1977).

A flow net is drawn by trial and error until it satisfies all requirements. Once these requirements are met, the resulting net is the *unique* solution of the Laplace equation [Eq. (5.3)]. It is simpler to sketch a net composed of curvilinear squares rather than of rectangles. Referring to Fig. 5.2, the following conditions should be observed:

1. The flow lines should be orthogonal to the equipotential lines (the tangents at the points of intersection should be normal). This requirement should also be observed at the boundaries. In Fig. 5.2, *all* flow lines should be normal to the equipotential lines AB and FG, and *all* equipotential lines should be normal to the boundary flow lines $BB'CDEF$ and $A'G'$. In the same figure, CD is an impervious sheet pile (usually steel) of relatively small thickness; water percolates downward along the left-hand side of CD and then turns upward along the right-hand side of DEF.

2. The flow and equipotential lines should be drawn as curves that are as smooth as possible with no kinks. The resulting fields should be checked to be sure that they are curvilinear squares ($\Delta n \simeq \Delta s$ in all fields).

3. No flow lines should cross each other, and no equipotential lines should cross each other. The exceptions to this condition are as follows: In a location of high velocity, such as point D in Fig. 5.2a, the field becomes small and appears to be of a shape deviating from a curvilinear square. However, by enlargement (Fig. 5.2b), the field may be subdivided into any number of small squares, indicating that the equipotential lines are not intersecting. Mathematically, a point such as D is known as a singular point where the gradient $i_g = \infty$ (infinite velocity). Also in Fig. 5.2, fields I, II, and III do not appear as squares; however, field III can be subdivided into small squares, as shown in Fig. 5.2c.

4. The end fields I and II should extend to infinity (theoretically speaking), where the water velocities are zero. Graphically, equipotential and flow lines bordering such fields cannot be checked. Therefore, at least two of these boundaries (ψ_4 and h_2 in field I and ψ_4 and h_{13} in field II)

should be checked with the rest of the interior fields to form the required squares. Once all fields other than I and II are squares, the flow net should be correct. Therefore, it is imperative in sketching a flow net to observe that all interior flow lines are completed to the boundaries; for example, if the left end of the drawing paper were at OO' rather than at AA' (Fig. 5.2), then incomplete lines would result and it would be difficult to check the correctness of the flow net.

Guidelines for sketching flow nets are available (Taylor 1948; Casagrande, 1937). However, there is no doubt that drawing a good flow net demands a special artistic talent. From the author's experience in the classroom with people of varying artistic talent, it has been found that one should start by drawing a coarse flow net consisting only of *three* flow channels (two interior flow lines other than the two boundary flow lines). The equipotential lines should then be sketched observing the previous given conditions. Corrections should be made by adjusting both the flow and equipotential lines to satisfy the main conditions: orthogonality and square fields. Starting with four flow channels (three interior flow lines would make it difficult to visualize the large number of the resulting squares and would take more time to correct. However, starting with two flow channels would make it difficult to construct a reasonable flow net. Once the initial flow net is corrected, each flow channel should then be subdivided into two or more channels and new equipotential lines added accordingly. This subdivision would consist of smaller fields that would enable the detection of errors that were unnoticeable in the initial coarse net.

In the suggested method of sketching, n_F should be a whole number. The resulting n_E cannot be guaranteed to also be a whole number. In order to clarify this point, let us assume that the problem shown in Fig. 5.2 was solved mathematically and that q was found to be equal to $0.30 K h_e$. By freehand sketching, the flow net shown in Fig. 5.2 has $n_E = 13$ and $n_F = 4$. Then q [Eq. (5.13)] is

$$q = K h_e \frac{4}{13} = 0.3077 K h_e$$

Since the mathematical solution is considered exact, the ratio n_F/n_E is found to be 0.30 (rather than 0.3077 by sketching). Assigning the number of n_F as 4, the number of equipotential drops n_E is about $13\frac{1}{3}$ if the sketch of the flow net is very precise. If such differences are noticed by the naked eye, the fractional rectangular fields ($\Delta s = \frac{1}{3}\Delta n$) should be drawn in the interior of the flow net (Fig. 5.2d). These fractions are hardly noticeable in a freehand sketch. From an engineering point of view, these fractions do not have appreciable effect on the final result if the flow nets are drawn as fine nets.

In unconfined flow systems (Fig. 3.13a, c, and d and Fig. 5.3), free-hand sketching of flow nets becomes more difficult. In Fig. 5.3, the free surface shown by flow line AB is the upper boundary of the flow medium in the earth dam and its location is unknown before the solution is found. In this type of system, freehand sketching entails more work in adjusting the unknown upper boundary AB to the rest of the flow net. The boundary conditions of the free surface AB should also be observed (Sec. 5.2).

In these systems, the approximation theories to locate the free surface may be used to minimize the number of trials, thus saving time in sketching the flow nets (see Sec. 5.5).

In anisotropic cases, the flow nets are drawn in the transformed section [Eq. (5.9)] exactly as in homogeneous and isotropic sections, but q is calculated as follows (Taylor, 1948; see App. C for proof):

$$q = \sqrt{K_x K_y}\, h_e \frac{n_F}{n_E} \tag{5.14}$$

5.2 Boundary Conditions

Impervious Boundaries

The impervious surfaces in Fig. 5.2a are $BB'CDEF$ of the structure (including the sheet piles CD) and the natural impervious surface $A'G'$. These surfaces prevent the flow of water across them and should thus constitute flow lines of certain intensities ψ. The boundary conditions are thus expressed by

$$d\psi = 0 \qquad \text{(i.e., } \psi \text{ is constant)}$$

and
$$\frac{\delta\psi}{\delta s} = 0$$

where s is the direction along the boundary and

$$\frac{\delta\phi}{\delta n} = 0$$

where n is the normal direction at any point along the boundary (i.e., the velocity in a direction normal to the boundary is zero).

Free Surfaces

In unconfined flow systems such as that shown in Fig. 5.3, the free surface AB is a stream line and has the same boundary conditions as an impervious boundary. Another condition is that the pressure along AB is atmospheric:

$$h = \frac{p}{\gamma_w} + y + \text{constant} = 0 + y + \text{constant} \qquad (5.15a)$$

$$\phi + Ky = \text{constant (or zero)} \qquad (5.15b)$$

Accordingly, for a square flow net (Fig. 5.3),

$$\Delta y_1 = \Delta y_2 = \Delta y_3 = \cdots = \Delta h$$

A line along which the pressure is the same everywhere is known as an *isobar*. Thus a free surface is a flow line *and* an isobar.

Entrance Surfaces

The upstream boundary surfaces *AB* in Fig. 5.2a and *AC* in Figs. 5.3 and 5.5 are the entrance surfaces at which the percolation of water through the media starts. These surfaces are equipotential lines; that is, $\phi =$ constant $= -Kh$. Note that in Fig. 5.2a, y is constant along *AB* and thus p is constant, while in Figs. 5.3 and 5.5, y varies linearly along *AC* and therefore p also decreases linearly [Eq. (5.15)] whenever y increases.

Exit Surfaces

The water leaves the medium at some surface such as *FG* (Fig. 5.2a), *BD* (Fig. 5.3), or *BFD* (Fig. 5.5). These exit surfaces may be equipotential lines (*FG* in Fig. 5.2a) or isobars (*BD* in Fig. 5.3) or a combination of both (*BFD* in Fig. 5.5). In the latter case, the portion *FD* is an equipotential line and the portion *BF* is an isobar with atmospheric pressure intensity.

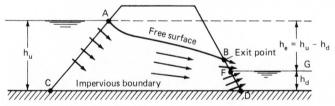

Figure 5.5   Flow through an earth dam (or embankment).

When the exit surface (or part of it) is an equipotential line, it has the condition that $\phi =$ constant $= -Kh_d$. If there is no tailwater, then $h_d = 0$ (e.g., along *FG* in Fig. 5.2a).

The exit surface along which pressure is atmospheric is known in the literature as the *surface of seepage*, the *outflow surface*, or the *discharge face*. In reality, these descriptions should apply to the entire exit surface, which may include a portion of an equipotential line. These terms would then conventionally be considered only to represent the portions of the exit surface *above* the downstream level (or tailwater). In gravity wells, it is the portion above the water level in the well. Along such surfaces,

$$\phi + Ky = 0 \qquad \text{or} \qquad h = y \tag{5.16}$$

The outflow surfaces are isobars, and they are neither flow lines nor equipotential lines.

The flow lines in a flow medium are normal to the equipotential line portion of the downstream surface. Flow lines crossing an outflow surface make angles α with that surface varying between zero at its upper limit (point B in Fig. 5.5) and 90° at its lower limit (point F in Fig. 5.5). The terminal point B (Figs. 5.3 and 5.5) along the free surface coincides with the upper point of the outflow surface. This point is known as the *exit point.*

5.3   Analysis of Forces within a Flow Field

Assuming a field in a flow net (Fig. 5.6a) between two flow lines MN and OP and two equipotential lines OM and PN, the pressure distribution along the sides is almost linear if the field is sufficiently small, represented by a perfect square of sides Δs. There is a single value of pressure head at any corner point of the prism (Fig. 5.6b). Within the prism, the total buoyancy $U = \gamma_w(\Delta s)^2$, which is represented by an upward force vector. The seepage force F_s is equal to $F_s = i_g \gamma_w(\Delta s)^2$. The resultant pressure head diagrams are indicated in Fig. 5.6c as two rectangular distribution diagrams. Similar to the analysis of forces in a unidirectional flow system (Sec. 3.11), there is a relationship between the boundary hydraulic forces, the buoyancy, and the seepage force. In two-dimensional systems, forces F_s, F_{B1}, and F_{B2} (Fig. 5.6c) are sloping rather than vertical. A force polygon diagram can thus be drawn (Fig. 5.6d) including all possible forces [see proofs in Taylor (1948) and Kashef (1965a)]; the vectors in this diagram are as follows:

$AB \downarrow$ = saturated weight of the prism = $\gamma_{sat}(\Delta s)^2$

$AD \downarrow$ = weight of the solid grains = $\gamma_{dry}(\Delta s)^2$

$DB \downarrow$ = weight of water within the pores = $(\gamma_{sat} - \gamma_{dry})(\Delta s)^2$

$CB \uparrow$ = total buoyancy $U = -\gamma_w(\Delta s)^2$

$AC \downarrow$ = submerged weight = $\gamma'(\Delta s)^2 = (\Delta s)^2(\gamma_{sat} - \gamma_w)$

$CD \uparrow$ = buoyancy on the grains $U_g = -(AD - AC) = -(\Delta s)^2(\gamma_w - \gamma_{sat} + \gamma_{dry})$

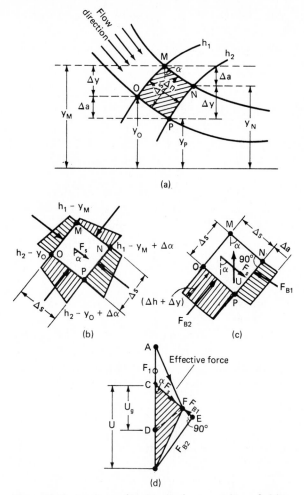

Figure 5.6 Analysis of hydraulic forces within a field in a flow net. (a) Field $OMNP$: $y_M - y_O = \Delta y$, $y_N - y_P = \Delta y$, $h_1 - h_2 = \Delta h$, $\Delta s \cong \Delta n$. (b) Pressure diagrams on each side of the field. (c) Resultant pressure diagrams. $F_{B1} = \Delta a \Delta s \gamma_w$, $F_{B2} = (\Delta h + \Delta y) \gamma_w$, $F_s = i \gamma_w (\Delta s)^2$, $U = \gamma_w (\Delta s)^2$. (d) Triangle of filtration: F_1 corresponds to vertical flow.

CF = seepage force $F_s = i_g \gamma_w (\Delta s)^2$ along the direction of flow

BE = boundary hydraulic force (resultant of water pressures) on sides OP and $MN = F_{B2} = \gamma_w (\Delta h + \Delta y)$

EF = resultant boundary hydraulic force on sides NP and $OM = F_{B1} = \gamma_w (\Delta s \Delta a)°$

° Note that force BE is normal to force EF and force F_s has the same direction as force EF.

BF = resultant of BE and EF, which constitutes the resultant boundary water pressures

DF = resultant of F_s and U_g (CB and CD)

AF = resultant body force, which is either $AC \leftrightarrow CF$, $AD \leftrightarrow DF$, $AB \leftrightarrow BF$, or $AB \leftrightarrow BE \leftrightarrow EF$

In the case of a unidirectional upward flow, point F (Fig. 5.6d) lies at some point F_1 vertically above point C on vector CA (force BE is equal to BF_1, EF equals zero, and F_s equals $CF_1 \uparrow$). When the values of $F_s = CA$, which is numerically equal to the submerged weight AC, the quicksand condition is reached.

The triangle BFC is known as the *triangle of filtration* (Polubarinova-Kochina, 1962). It indicates that the horizontal component of the seepage force is numerically equal to the horizontal component of force BF representing the resultant boundary water pressures. It shows also that the algebraic sum of the vertical components of BF and the seepage force are numerically equal to the total buoyancy U. This analysis was used to derive some gravity well formulas (Sec. 8.10).

In the case of a hydrostatic condition in which the total head is the same at all points M, N, O, and P (Fig. 5.6a), it can easily be shown that BE in the force polygon shown in Fig. 5.6d remains unchanged, while point F moves to point C ($F_s = 0$); the total buoyancy is then numerically equal to the resultant of the boundary water pressures. The resultant of all forces in the static case is thus equal to AC (the submerged weight).

5.4   Hydraulic Design of Weirs or Solid Dams

Weirs or dams (Figs. 3.13b, 3.19a, 5.2a, and 5.7) are constructed for many purposes, such as flood control, water storage, and recreation. Excessive seepage beneath these structures is considered a loss if they are used for water storage and recreation. However, for flood control purposes, excessive seepage should be avoided only when it is detrimental to the safety of the structure or its stability. In the design of such structures, the following should be evaluated:

1. Rate of seepage losses.

2. Exit gradients and heaving near the downstream to prevent scour action (piping and roofing, see Sec. 3.10), which may necessitate providing sheet piles, horizontal filter blankets, and/or elongating the floor.

3. Uplift pressures on the floor base to check the overall stability, which may necessitate increasing the dead loads of the structure.

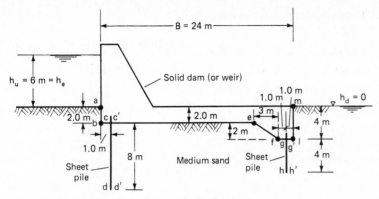

Figure 5.7   A weir with a sheet pile (Example 5.1).

4. Structural resistance and safety against soil bearing and overturning. This last requirement is outside the scope of this book because it
belongs to structural and geotechnical engineering.

Preliminary Dimensioning of the Floor

In order to investigate the hydraulic features of a weir, a tentative cross
section should be assumed, and hydraulic data, such as the maximum
upstream depth of the dammed-up water, should be available. If the
selected cross section proves to be unsafe or too conservative, it has to be
changed. The design, therefore, is based on trial-and-error procedures
similar to those in the design of any other engineering structure. If the
tentative dimensions are closer to those of the final design, the number of
trials is reduced. The selection of the initial trial should therefore be
based on experience, judgment, and/or available approximate guidelines, such as those explained in the following paragraph.

Extensive field studies were made by Bligh (1910) on weirs in Egypt
and India.[*] In order to prevent scour action at the downstream end of the
floor, a safe length L was determined by defining empirical weighted
creep ratios L/h_e that have a constant value for each type of soil. The
creep length L is the length of the uppermost flow line. Once the design
value of h_e is known (Fig. 5.7), the safe length $L = abcdd'c'efghh'g'lm$ is
determined by multiplying h_e by a proper weighted creep ratio, which is
the reverse of a safe *uniform* gradient. The length L is supposed to be the
minimum possible safe length that should be used. The thickness of the
floor (Fig. 5.7) has a minor effect on L and is tentatively assumed before it

[*] These studies were carried out after the catastrophic failure in 1898 of the Narora Dam
on the Ganges River in India.

TABLE 5.1   **Recommended Weighted Creep Ratios**

Material	Safe weighted creep-head ratios L/h_e
Very fine sand or silt	8.5
Fine sand	7.0
Medium sand	6.0
Coarse sand	5.0
Fine gravel	4.0
Medium gravel	3.5
Coarse gravel, including cobbles	3.0
Boulders with some cobbles and gravel	2.5
Soft clay	3.0
Medium clay	2.0
Hard clay	1.8
Very hard clay or hardpan	1.6

SOURCE: E. W. Lane, "Security from Under-Seepage Masonry Dams on Earth Foundations," *Trans. Am. Soc. Civil Eng.*, vol. 100, 1935, pp. 1235–1351.

is structurally checked later against uplift pressures. The design value of h_e is in fact the maximum excess head $h_e = h_u - h_d$ corresponding to $h_d = 0$ (no tail water). Bligh's method was overconservative and was later modified by Lane (1935) on the basis of a study of about 280 existing dams, 24 of which had failed. The safe weighted creep ratios L/h_e corresponding to Lane's method are given in Table 5.1. The L value is computed as follows, making the assumption that $K_{max}/K_{min} = 3$ (Sec. 3.7):

$$L = \text{sum of lengths of vertical flow paths} \qquad (5.17)$$
$$+ \tfrac{1}{3} \text{ (sum of lengths of horizontal paths)}$$

The preceding equation does not take into account the wide range of K_{max}/K_{min}. Any inclined surface is considered vertical if its slope is greater than 45° and horizontal if its slope is less than 45°.

EXAMPLE 5.1   In Fig. 5.7, a concrete weir is to be constructed in medium sand. A toe is to be constructed at the downstream end of the floor and two rows of steel sheet piles are to be provided at c ang g. Find the safe L.

Solution

$$L = (ab + cd + d'c + gh + h'g + lm) + \tfrac{1}{3}(bc + c'ef + fg + g'e)$$
$$= (2 + 8 + 8 + 4 + 4 + 4) + \tfrac{1}{3}(1 + 21 + 1 + 1) = 38 \text{ m}$$

$L/h_e = 38/6 = 6.3$, which is greater than 6.0 for medium sand (Table 5.1); therefore, it is safe. If the toe and sheet piles are eliminated, then $L = 4 + 4 + \tfrac{24}{3} = 12$ m and $L/h_e = 2$ (unsafe). In this case, the minimum safe L would be $L = 6h_e = 36 = 4 + \tfrac{1}{3}B$ or $B = 96$ m, which is a very excessive and uneconomical length if sheet piles are not used.

The hydraulic gradient h_e/L is considered uniform in Lane's method along the length L, which is the uppermost flow line. This is not the case if flow nets are drawn (Fig. 5.8). The spacings of the equipotential lines at the floor base are not uniform. Moreover, in Lane's method the safe gradients are assumed to depend on the soil type. In exact methods of steady-state flow based on the Laplace equation (Sec. 5.1), the hydraulic conductivity K does not appear in the equation. This indicates that the flow configuration and gradients are not affected by the soil type in homogeneous and isotropic media under steady-state flow. The gradients are affected by the magnitude of h_e, and so are the uplift pressures. Despite these drawbacks, Lane's method is still useful for finding a *tentative* value for the length of the floor that can be used for preliminary dimensioning.

Seepage Losses, Gradients, and Uplift Pressures

Once preliminary dimensioning is made on the basis of Lane's method, a flow net has to be drawn using mathematical or experimental data or freehand sketching. By counting the numbers n_F and n_E, the rate of flow q (seepage losses) can be determined for a unit length of the structure normal to the cross section by applying Eq. (5.13) or, in the case of anisotropy, Eq. (5.14). The hydraulic gradient within any field in a flow net is determined from $\Delta h/\Delta s$, where Δs is the average length of that field along the flow direction, scaled from the drawing.

In square flow nets, where Δh is constant, it is obvious that the gradients get higher as the fields get smaller. High gradients occur mostly at the bottom ends of the sheet piles and at the exits. In most cases, it is sufficient to find the upward gradients at the downstream exits. This is needed in order to design the extent and thickness of the downstream filter blankets and serves to check the safety of the weir or dam. A decision has to be made whether to elongate the floor or provide a toe and/or construct more than one row of sheet piles. The high gradients within the interior of the flow medium can assume any direction depending on the configuration of the stream lines. If they dip downward, they do not present a danger, whereas if they are directed upward, the potential of soil upheaval should be checked, as explained in the following section.

The water pressures acting on the floor base have an upward resultant force known as the *uplift force* that should be less than the total downward weights of the structure. The ratio of the downward forces to the uplift force should be greater than 1.0 by an appropriate margin of safety. This ratio varies from 4 to 5. The factor of safety in hydraulic

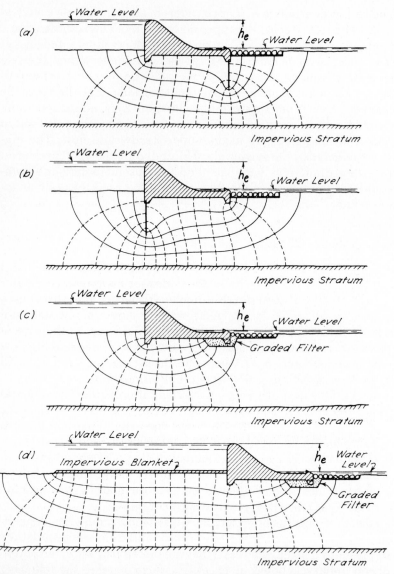

Figure 5.8 Flow nets beneath a concrete dam. *(a)* Sheet pile in downstream; *(b)* sheet pile in upstream; *(c)* downstream filter and no sheet piles; *(d)* downstream filter, upstream impervious blankets, and no sheet piles. *(A. Casagrande, Discussion of "Security from Under-Seepage Masonry Dams on Earth Foundations," Trans. Am. Soc. Civil Eng., vol. 100, 1935.)*

structures should not generally be less than 4. In Fig. 5.2a, for example, the total heads at the points of intersection of the equipotential lines with the floor base are already known from the intensities h of these lines. At each of these points, the pressure head can be obtained. At such points as B' and C, the heads are determined by subdividing the fields where the points lie. The uplift diagram is obtained by drawing the ordinates of the pressure head p/γ_w at the base and joining them by straight lines. More precision can be obtained by using finer flow nets. The area of this diagram (which is the volume per unit length of the weir) multiplied by γ_w gives the magnitude of the uplift force.

Elimination of the toe and sheet piles (Fig. 5.7) decreases the lengths of all flow lines, thus increasing the gradients. The downstream sheet pile gh helps in decreasing the exit upheaval of the downstream soil mass (see following section). It has the same purpose as the toe, but it may be more economical and practical to have both the downstream sheet piles and the toe. The row of sheet piles at the upstream (cd in Fig. 5.7) serves to elongate the flow paths, thus decreasing the gradients and reducing drastically the uplift pressures due to dissipation in head along the path $cdd'c'$. The increase in flow path also may be achieved by constructing an impervious blanket upstream (Fig. 5.8d). The decision to use one or more of these provisions is based on experience and/or carrying out several solutions for different layouts. Upstream impervious blankets, for example, have little effect in anisotropic soils (Taylor, 1948).

The effect of the location of sheet piles on the magnitude of uplift forces can be approximated by assuming a uniform gradient along the uppermost flow line beneath the floor of the dam (Fig. 5.9). Selecting the datum at surface abm, the total and pressure heads at the main points are given in Table 5.2 for each of the two positions for the sheet piles. The uniform gradient in each case is $i_g = 6/(2 + 19 + 8 + 8 + 1 + 2) = 6/40 = 0.15$. The uplift diagram corresponds to the water-pressure diagram along the base $b'cgm'$. The uplift diagram $bdefolgcb$ corresponds to location of the sheet piles at c, whereas $bde'f'folgcb$ corresponds to location of the sheet piles at g (Fig. 5.9b). The increase in the uplift force is thus $2.4 \times 18 \times \gamma_w = 2.4 \times 18 \times 1000 = 43,200$ kg ($= 43.2$ metric tons) if the row of sheet piles is shifted from c to g. This analysis gives approximate values of the uplift forces and can be used as a tentative guide to determine the depths and locations of the sheet piles. In stability analysis, the position of the resultant uplift force should be determined. The line of action of this force should coincide with the centroid of the uplift-pressure diagram.

A flow net beneath a dam in a stratified soil where $K_x \neq K_y$ is shown in Fig. 5.4 ($K_x = 4K_y$). The flow net is drawn in the transformed section as if the soil were homogeneous and isotropic. However, the rate of flow q is

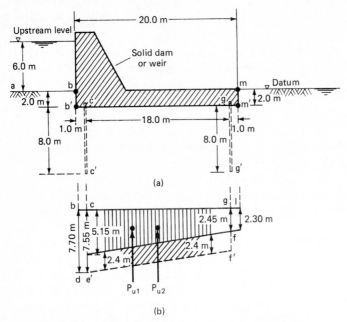

(b)

Figure 5.9 Effect of sheet-pile location on the magnitude of uplift forces.. *(a)* Two locations of sheet piles *(cc′* and *gg′)*; *(b)* uplift diagrams corresponding to the two locations of the sheet piles.

TABLE 5.2   Uplift Diagrams*

| Points | Total head h and pressure head p/γ_w at designated points, m | | | |
| | Sheet pile at c | | Sheet pile at g | |
	h	$\dfrac{p}{\gamma_w}$	h	$\dfrac{p}{\gamma_w}$
b	6.00	6.00	6.00	6.00
b'	5.70	7.70	5.70	7.70
Left of c	5.55	7.55	5.55	7.55
c'	4.35	—	—	—
Right of c	3.15	5.15	5.55	7.55
Left of g	0.45	2.45	2.85	4.85
g'	—	—	1.65	—
Right of g	0.45	2.45	0.45	2.45
m'	0.30	2.30	0.30	2.30
m	0.00	0.00	0.00	0.00

° See Fig. 5.9*b*.

determined from Eq. (5.14). The gradients and the uplift-pressure diagram are determined from the natural (true) section (Fig. 5.4b) after plotting the flow net back from the transformed section. If only q is required, Eq. (5.14) is used by counting n_F and n_E directly from the transformed section.

Safety against Heaving

With reference to Fig. 5.2a, there is a possibility that the upward seepage forces immediately to the right of the sheet piles CD will produce soil upheaval leading to a rise in the horizontal surface at F. Heaving was investigated by Terzaghi (Terzaghi and Peck, 1967) for a case of one row of sheet piles damming up water from one side. The mass of soil prism most likely to failure has a depth $t_s = \overline{FD}$ and width $\frac{1}{2}\overline{FD}$ (Fig. 5.2e). Assuming that the datum passes through line FG and that the average total head along the base of the prism Dd is h_{av}, then the factor of safety (FS) against heaving is given by

$$FS = \frac{t_s \gamma'}{h_{av}\gamma_w} = \frac{\gamma'/\gamma_w}{i_{g,av}} = \frac{i_{gc}}{i_{g,av}} \tag{5.18}$$

Since $i_{gc} \simeq 1$ [Eq. (3.31)], then

$$FS = \frac{1}{i_{g,av}} \tag{5.19}$$

The factor of safety against upheaval can be thought of as the ratio of downward to upward forces in the prism $FDdf$ (Fig. 5.2e). According to the analysis of forces given in Sec. 3.11, the downward force $= W = V\gamma_{sat} = \frac{1}{2}t_s^2\gamma_{sat}$. The upward force would be equal to the resultant of the pore-water pressures acting on the base of the prism $Dd = (h_{av} + t_s)$ $(t_s/2)\gamma_w$. Therefore,

$$FS = \frac{\frac{1}{2}t_s^2\gamma_{sat}}{(h_{av} + t_s)(t_s/2)\gamma_w}$$

or $\qquad$ $$FS = \frac{t_s\gamma_{sat}}{h_{av}\gamma_w + t_s\gamma_w} = \frac{t_s\gamma' + t_s\gamma_w}{h_{av}\gamma_w + t_s\gamma_w} = \frac{W}{F_B} \tag{5.20}$$

Equation (5.20) is more realistic than Eq. (5.18). If the factor of safety is less than 4 to 5, a filter should be constructed in such a way that

$$FS = \frac{W + W_{fi}}{F_B} \tag{5.21}$$

where W_{fi} is the dry weight of the filter. If a value for FS is assigned, the thickness of the filter t_{fi} can be determined from $W_{fi} = t_{fi}\gamma_{fi}$, where γ_{fi} is the unit weight of the filter material.

In Eqs. (5.18) through (5.21), the frictional resistance along the sides of the prism $FDdf$ (Fig. 5.2e) is neglected in order to be on the side of safety. The resistence consists of the adhesion along DF and the shearing strength along df (Fig. 5.2a and e).

EXAMPLE 5.2   Find the factor of safety against upheaval in the dam shown in Fig. 5.2a.

Solution   The values of h_1 through h_{15} and the values of ψ_1 through ψ_5 in terms of K are given in Table 5.3.

$$n_E = 13 \qquad n_F = 4 \qquad h_e = 7.0 \text{ m}$$

$$q = Kh_e \frac{n_F}{n_E} = 2.15K \text{ m}^2/\text{s} \qquad (K \text{ in m/s})$$

$$\Delta h = \frac{7}{13} = 0.538$$

$$\text{Exit gradient} \cong \frac{\Delta h}{2.8} = \frac{0.538}{2.6} = 0.19 < \text{critical}$$

TABLE 5.3   Values of h^* and ψ in Fig. 5.2a

Equipotential h_n	Intensity, m
h_1	7.000
h_2	6.462
h_3	5.923
h_4	5.385
h_5	4.846
h_6	4.308
h_7	3.769
h_8	3.231
h_9	2.692
h_{10}	2.154
h_{11}	1.615
h_{12}	1.077
h_{13}	0.538
h_{14}	0.000
ψ_n	m^2/s
ψ_1	0
ψ_2	0.54K
ψ_3	1.08K
ψ_4	1.62K
ψ_5	2.15K

° Datum at tailwater level.

The prism height $t_s = 7.0$ m and its base width $= 3.5$ m. This base intersects the two equipotential lines h_{10} and h_{11} of intensities 2.154 and 1.615 m, respectively; h at point $d \simeq 1.212$ m (see Fig. 5.2). Therefore,

$$h_{av} = \frac{(2.154 + 1.615)0.6 + (1.615 + 1.212)1.15}{3.5} = 1.57 \text{ m}$$

If $\gamma_{sat} = 2160$ kg/m³ and $\gamma_w = 1000$ kg/m³, then $\gamma' = 2160 - 1000 = 1160$ kg/m³, $W = 7 \times 3.5 \times 2160 = 52{,}920$ kg/m, and

$$F_B = \frac{t_s}{2}\gamma_w(t_s + h_{av}) = 1000(7 + 1.57)3.5 = 29{,}995 \text{ kg/m}$$

Applying Eq. (5.18),

$$FS = \frac{7.0 \times 1160}{1.57 \times 1000} = 5.17$$

which is greater than 5 and therefore safe. Applying Eq. (5.20),

$$FS = \frac{52{,}920}{29{,}995} = 1.76$$

which is less than 5 and therefore unsafe.

In this example, Eq. (5.20) gave a lesser FS value than Eq. (5.18), indicating that a filter is needed. Equation (5.20) is more logical than Eq. (5.18). It is also recommended that these equations be applied at other levels shallower than t_s because the gradients cannot be uniform within prism *FDdf* and the factor of safety may be less for shorter prisms.

Selected Mathematical Solutions

A few idealized cases that have mathematical solutions are given below. Only the results are given; the procedures are outside the scope of this book. The reader may check the results against the simpler methods previously cited. The mathematical solutions become more complex once these idealized cases change to *practical* cases in which slight deviations are imposed, such as having dam embedments, providing a toe, or having an impervious bed that is not horizontal.

Soil of infinite depth with no embedment and no sheet piles (Fig. 5.10a)
The flow pattern is symmetrical with respect to the y axis, which passes through the midpoint of the base of the dam. The total rate of flow q approaches infinity, and the ψ values of the flow lines are given by (Harr, 1962)

$$\frac{4x^2}{B^2 \cosh^2{(\pi\psi_n/Kh_e)}} + \frac{4y^2}{B^2 \sinh^2{(\pi\psi_n/Kh_e)}} = 1 \qquad (5.22)$$

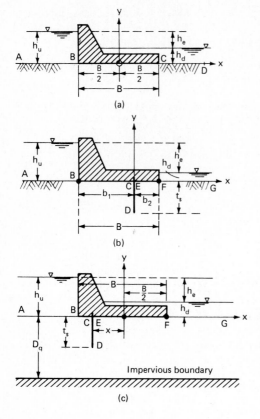

(a)

(b)

(c)

Figure 5.10 Analysis of seepage beneath weirs under various conditions. *(a)* No embedment, no sheet piles, and aquifer of infinite depth; *(b)* no embedment, one row of sheet piles, and aquifer of infinite depth; *(c)* no embedment, one row of sheet piles, and aquifer of finite depth.

where h_e = excess head = $h_u - h_d$, in which h_u is the upstream and h_d the downstream water depth

B = the base width of the dam

ψ_n = stream functions $(n = 0, 1, 2, \ldots, n)$

At the base of the dam, $n = 0$ and ψ is ψ_0.

The flow lines are represented by a family of ellipses with foci at $x = \pm B$ [Eq. (5.22)]. The equipotential lines consist of a family of confocal hyperbolas (with the same foci as the flow lines) given by

$$\frac{4x^2}{B^2 \cos^2 (\pi\phi_n/Kh_e)} - \frac{4y^2}{B^2 \sin^2 (\pi\phi_n/Kh_e)} = 1 \qquad (5.23)$$

where ϕ_n = velocity potentials $(n = 0, 1, 2, \ldots, n)$

h = total head at any point with respect to a *datum along the tailwater level*

Along AB, the velocity potential ϕ is at its maximum, ϕ_n; along CD, it is at its minimum, ϕ_0.

In order to apply Eqs. (5.22) and (5.23) to plot a flow net, a certain number of equipotential drops n_E between the maximum $\phi_n = -Kh_e$ along AB and the minimum $\phi_0 = 0$ along CD are assumed; then

$$\Delta h = \frac{Kh_e - 0}{n_E K} = \frac{h_e}{n_E}$$

In a flow net consisting of curvilinear squares,

$$\Delta q = K\Delta h$$

If ψ_0 corresponds to the stream line beneath the base of the dam, then $\psi_1 = \Delta q$, $\psi_2 = 2\Delta q$, etc. For a particular ψ, the stream line can be obtained from Eq. (5.22) by computing the y values corresponding to assigned x values.

Applying Eq. (5.23) to any specific value ϕ, that is, $\phi = \phi_0$ along CD, $\phi_1 = K\Delta h$, $\phi_2 = 2K\Delta h$, $\phi_3 = 3K\Delta h$, . . . , $\phi_n = Kh_e$, any equipotential line can be plotted by computing y corresponding to assigned x values.

The horizontal velocity v_x along base BC is given by

$$v_x = \frac{2Kh_e/\pi}{(B^2 - 4x^2)^{1/2}} \tag{5.24}$$

At points B and C, $v_x = \infty$.

Total uplift force P_u on base BC *in excess of the effect of the tailwater* is

$$P_u = \tfrac{1}{2}h_e\gamma_w B \tag{5.25}$$

Equation (5.25) was determined mathematically (Harr, 1962). Using the approximate method, where the gradient is assumed uniform, the value of the total uplift is $\gamma_w[(h_u + h_d)/2]B$. The effect of tailwater is $\gamma_w h_d B$, and thus

$$P_u = \tfrac{1}{2}\gamma_w(h_u + h_d)B - Bh_d\gamma_w = \tfrac{1}{2}h_e\gamma_w B$$

which is the same result given by Eq. (5.25). The distribution of the uplift pressures on the base is therefore uniform in this case.

Soil of infinite depth with no embedment and one row of sheet piles This case was solved by Pavlovsky (Harr, 1962) by means of the velocity hodograph. The x axis coincides with the base, and the sheet pile is assumed at various locations at point C (Fig. 5.10b). The y axis passes through C (or E), changing the x values for each new pile position.

The pressure p_b acting on the base is

$$p_b = \gamma_w \left(\frac{h}{\pi} \cos^{-1} \frac{B_1 t_s \pm \sqrt{t_s^2 + x^2}}{B_2 t_s} + h_d \right) \qquad (5.26)$$

and p_s along the sheet pile $(x = 0)$ is

$$p_s = \gamma_w \left(\frac{h}{\pi} \cos^{-1} \frac{B_1 t_s \pm \sqrt{t_s^2 - y^2}}{B_2 t_s} - y + h_d \right) \qquad (5.27)$$

where t is the depth of the sheet pile, and

$$B_1 = \frac{1}{2t_s} (\sqrt{t_s^2 + b_1^2} - \sqrt{t_s^2 + b_2^2}) \qquad (5.28a)$$

and

$$B_2 = \frac{1}{2t_s} (\sqrt{t_s^2 + b_1^2} + \sqrt{t_s^2 + b_2^2}) \qquad (5.28b)$$

It has been found in this solution that the uplift decreases when the sheet piles are near the upstream (the same conclusion given in earlier discussions). When the sheet piles are located near the center of the base, the uplift pressures are found independent of the pile depths t_s. This case is purely academic and cannot be used in practice.

Soil of finite depth with no embedment and one row of sheet piles This problem (Fig. 5.10c) was solved independently by Muskat in 1936 and Pavlovsky in 1922 (Harr, 1962) by conformal mapping techniques. In these solutions, q/Kh_e values for various ratios $B/2D_q$ and t_s/D_q are found to be almost unaffected by the location of the sheet piles. When the sheet piles are at either end, q/Kh_e is about 95 percent of its value when the sheet piles are located at the center of the base.

From the mathematical analysis it has been found that when the sheet piles are at the heel, not much benefit is gained by increasing the depth t_s of the sheet piles whenever $2D_q/B < \frac{1}{2}$. It is also recommended that $(2t_s/B)_{\max} \simeq 0.5$.

The maximum exit gradient when the sheet piles are at the toe for $t_s/D_q < 0.75$ is

$$(i_{g,e})_{\max} = \frac{h_e}{\pi t_s} \qquad (5.29)$$

The relationship between q/Kh_e and t_s/D_q for various values of $B/2D_q$ was given in a graphic form by Muskat (1937) based on rigorous mathematical analysis. From a practical point of view, the following empirical formula has been found to be satisfactory:

$$\frac{q}{Kh_e} = \frac{D_q}{t_s}\left(0.15 - 0.06\frac{B}{D_q}\right) + 0.23 \tag{5.30a}$$

or

$$\frac{t_s}{D_q} = \frac{0.15 - 0.06\dfrac{B}{D_q}}{\dfrac{q}{Kh_e} - 0.23} \tag{5.30b}$$

The preceding formula agrees quite well with the rigorous solution given by Muskat (1937) for a range of t_s/D_q between 0.4 and 0.6; the maximum error in q/Kh_e is about ± 0.02. If a maximum error of ± 0.04 is permissible, Eq. (5.30) can be used for a wider range of t_s/D_q between 0.3 and 0.7.

Summary of Recommended Design Procedures for Solid Dams or Weirs

1. Use Lane's method (Table 5.1) to find a preliminary length for the floor of the dam (assuming no sheet piles).

2. Find the depth of embedment from soil investigations (geotechnical engineering).

3. Select a depth for the sheet piles t_s between $0.3\ D_q$ and $0.5\ D_q$.

4. Place the sheet piles near the upstream.

5. Sketch a flow net and find q/Kh_e.

6. Apply Eq. (5.30b) to check the selected value of t_s.

7. Sketch another flow net if there is a big difference in the t_s values.

8. Determine the exit gradient and the factor of safety against heaving on the basis of the drawn flow net.

9. Repeat steps 5 through 8 if a downstream row of sheet piles is required. The requirement for additional sheet piles depends on the results of step 8.

10. Find the water-pressure diagram at the base. Compute the total uplift and find its line of action.

11. Complete the stability analysis (safety against sliding and overturning, downwrd forces against the uplift force, and structural strength at critical sections).

The tentative design should be changed whenever any safety requirement is not satisfied. This is usually performed by one or more of the following methods: providing a filter at the downstream exit, providing a toe, adding a shorter row of sheet piles at the downstream, increasing the floor length, and/or increasing the floor depth. If the design is very conservative, that is, uneconomical, the length of the weir should be reduced.

5.5 Seepage through Earth Dams

Steady-state seepage through earth dams is governed by the Laplace equation [Eq. (5.3)], and the flow is characterized as unconfined. The main problem with any solution lies in determining the free surface. Special mathematical techniques, such as the conformal mapping and velocity hodographs, are already available for some cases (Muskat, 1937; Polubarinova-Kochina, 1962; DeWiest, 1969; Bear and Dagan, 1964; Bear, 1979). Mathematical approximations, including mainly Dupuit's assumptions (explained later), are also available (Casagrande, 1937). Several other techniques have been used to render simple solutions (Kashef, 1965). Graphic sketching of flow nets (Fig. 5.3) is a possible solution, but this takes more time and talent as compared with its use in confined flow problems.

The most efficient approach is to obtain computerized numerical solutions using the finite-difference or finite-element techniques. The number of iterations in these techniques is very much reduced if the initial assumptions are close to the final solution. Simple solution methods are discussed in this section, emphasizing the main characteristics of the flow. These simple solutions may be considered final in certain practical situations; otherwise they can be used as initial conditions in the computerized methods to save computer time.

Dupuit's Approximations

Consider an unconfined aquifer in which the flow is uniform (Fig. 3.15). The water table ab is almost parallel to the impervious boundary. In fact, the water table should be slightly curved (such as ab' in the same figure). The water table becomes more deflected near exit surfaces (such as the drain shown in Fig. 5.11) or in proximity to water wells. Diagram $cc'bb'$ in Fig. 5.11 represents the distribution of total head across a vertical section cc' of depth D_x and can be plotted from the flow net. It has been found that curve bb' is almost a parabola (Kashef, 1965b) that is orthogonal to the impervious boundary at b', because the vertical velocity component at b' is zero. The slope of the tangent to this parabola at b gives the vertical component of the gradient i_g at a point c on the water table:

$$i_g = \frac{dh}{ds} \qquad i_{gx} = i_g \cos \delta \qquad \text{and} \qquad i_{gy} = i_g \sin \delta$$

Since $dh/ds = \sin \delta$ (because $\Delta h = \Delta y$ along the water table), then $i_{gy} = \sin^2 \delta$, $i_{gx} = \sin \delta \cos \delta$, and $i_g = \sin \delta$ (Fig. 5.11a). The total head varies

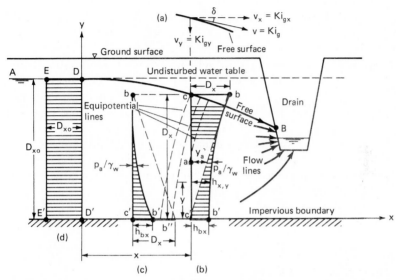

Figure 5.11   Distribution of total and pressure heads across vertical sections in an unconfined flow system. (a) Components of velocity v; (b) total-head diagram across cc', distance $x = x$; (c) pressure-head diagram across cc', distance $x = x$; (d) total-head diagram across DD', distance $x = 0$.

from a maximum value D_x at c to a minimum value h_{bx} at c', representing the base pressure head (Fig. 5.11b). The triangle cbc' in Fig. 5.11b represents the position-head diagram across cc', and $bc'b'$ represents the distribution of the pressure heads p/γ_w across the same section. At any point a on section cc', the total head $h_a = y_a + p_a/\gamma_w$. Plotting the pressure-head diagram with a vertical datum cc' (Fig. 5.11c), the horizontal ordinates indicate the *magnitudes* p/γ_w. If a section DD' is located in a region where the velocities are very small, the water table would be almost horizontal (Fig. 5.11d), and the pressure-head diagram would be represented by a triangular distribution $DD'E'$ with a base pressure head that is D_{xo}, i.e., hydrostatic pressure. The total heads across section DD' are all the same and equal to D_{xo}.

From the properties of the parabola, the area of the total-head diagram at section cc' (Fig. 5.11b) is equal to $\frac{1}{3}D_x^2 + \frac{2}{3}h_{bx}D_x$, and the area of the position-head diagram is $\frac{1}{2}D_x^2$. Therefore, the area P_x of the pressure-head diagram is (Kashef, 1965b)

$$P_x = \tfrac{2}{3}h_{bx}D_x - \tfrac{1}{6}D_x^2(\tfrac{1}{2} - \tfrac{1}{3}\sin^2\delta) \tag{5.31}$$

and

$$h_{x,y} = h_{bx} + \frac{(D_x - h_{bx})y^2}{D_x^2} \tag{5.32}$$

One of the common sets of assumptions introduced in unconfined flow systems are known as *Dupuit's assumptions*, and they suggest that the gradient at the top point in a vertical section through a saturated medium be given by $i_{gx} \simeq dh/dx$ rather than by $dh/ds = i_{gs}$. Thus $i_{gs} \simeq \tan\delta$ at point c (Fig. 5.11a and b) rather than $\sin\delta$. Moreover, the same assumptions (Dupuit, 1863) suggest that this gradient i_{gx} is the same at any other point along that vertical section. Dupuit's assumptions thus neglect the existence of the vertical velocity components and give an average horizontal gradient at any vertical section for an assumed horizontal flow. The resulting total-head diagram at any section would thus be uniform, yielding a hydrostatic pressure-head diagram. Dupuit's assumptions cannot be used unless the deflection in the water-table curve is very small; i.e., the velocities are also very small. The diagrams corresponding to the real flow would be similar to those in Fig. 5.11b and c.

Dupuit's assumptions have been extensively used in two-dimensional gravity flow systems, including gravity wells. They have been generalized (Muskat, 1937) as follows:

$$\frac{\delta^2 \tilde{h}^2}{\delta x^2} + \frac{\delta^2 \tilde{h}^2}{\delta y^2} = 0 \tag{5.33}$$

where $\tilde{h}$ is the depth of the saturated zone if Dupuit's assumptions are used. Various methods have been used to solve Eq. (5.33), including its linearization. However, the present extensive use of numerical analysis provides direct solutions to the Laplace equation [Eq. (5.3)] without the need to introduce Dupuit's assumptions or the generalized form given by Eq. (5.33). Seepage through various types of earth dams is given in the following section without introducing Dupuit's assumptions. The results are very close to those of other more rigorous approaches (Polubarinova-Kochina, 1962). In cases where gradients are small, Dupuit's assumptions are usually used.

Rectangular Cores of Dams

Figure 5.12a represents a cross section of a rock-fill dam (or embankment) with a vertical soil core. The rocks are used for stability, but they do not cause any appreciable resistance to flow. The water surface bc

(Fig. 5.12*a*) is almost horizontal, coinciding with the upstream water level *ab*. Through the soil core, the free surface follows curve *ce*. The discharge point *e* should always be above the tailwater level (which coincides with plane *c'e'* in Fig. 5.12*a*). The analysis of seepage in the rectangular earth section shown in Fig. 5.13*a* should theoretically be the same as for the core of the dam shown in Fig. 5.12*b*. The flow domain is bounded by the free surface *ce* (neglecting capillary action), the outflow surface *ee'*, the impervious boundary *c'e'*, and the upstream side *cc'*. The boundary conditions can be described for the section shown in Fig. 5.12*a* as follows:

cc' = an equipotential line

ce = the free surface, which is a stream line as well as an isobar of intensity zero (atmospheric pressure)

ee' = the outflow surface, which is an isobar, but neither a stream line nor an equipotential line

$c'e'$ = a horizontal stream line

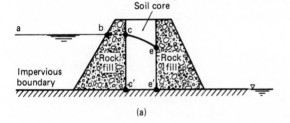

(a)

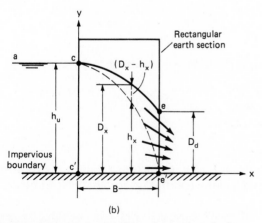

(b)

Figure 5.12 Seepage through rectangular sections. *(a)* Earth core in a rock-fill dam; *(b)* rectangular earth section.

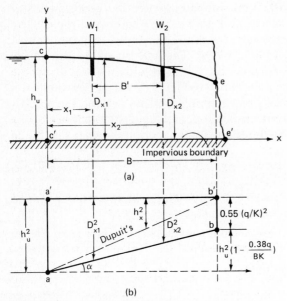

Figure 5.13 Unconfined flow between two sections. (a) Free surface; (b) linear distribution of D_x^2 within the flow zone.

The free surface ce is normal to the upstream side cc' at the point of entrance c and is tangential to the downstream face at e (Casagrande, 1937). At the discharge face ee', the flow lines intersect at angles α, varying gradually downward from zero at e to 90° at e'.

Introducing Dupuit's assumptions, then

$$q = K\tilde{h}\frac{d\tilde{h}}{dx}$$

By integration, the following equations are obtained:

$$q = K\frac{h_u^2 - \tilde{h}_x^2}{2x} \quad \text{(approximate)} \tag{5.34}$$

and

$$q = K\frac{h_u^2}{2B} \quad \text{(exact)}° \tag{5.35}$$

Equation (5.34) gives the values of $\tilde{h}_x$ at various x values in order to trace what is known as *Dupuit's curve* (ce' in Fig. 5.12b). This indicates

---

° Although this equation is based on Dupuit's assumptions, it has been proved by Charney that it is rigorously correct if the vertical velocity components are considered (Polubarinova-Kochina, 1962).

that the flow domain has no outlet when $\tilde{h}_x = 0$ at $x = B$. Dupuit's assumptions do not lead therefore to a determination of the true free surface ce. However, Eq. (5.35) gives the exact value q even though it is based on Dupuit's assumptions. Other methods have been used to find the true free surface ce. The height D_d was found to be (Polubarinova-Kochina, 1962)

$$D_d = 0.742 \frac{q}{K} = \sqrt{0.55} \frac{q}{K} \qquad \text{(exact)} \qquad (5.36)$$

Although the $\tilde{h}_x$ value determined by Dupuit's formula [Eq. (5.34)] does not represent the true height D_x of the free surface at x, it has a physical significance (Kashef, 1965b) expressed by

$$\tilde{h}_x = \sqrt{2P_x} \qquad (5.37)$$

where P_x is the area of the pressure-head diagram [Eq. (5.31)]. The true free surface is approximately determined by (Kashef, 1965b)

$$D_x \simeq h_u \sqrt{1 - \frac{x}{B}\left(1 - \frac{1}{7.27m^2}\right)} \qquad (5.38)$$

where $m = B/h_u$.

Equation (5.38) is slightly modified from the original reference, where the factor of m^2 was 8 rather than 7.27. The same equation satisfies the boundary conditions; at $x = 0$, $D_x = h_u$, and at $x = B$, D_x gives the value of D_d expressed by the exact formula given by Eq. (5.36).

The base pressure head h_{bx} is given by (Kashef, 1977)

$$h_{bx} = \left(1 - \frac{x}{B}\right)\left(\tfrac{3}{4} \frac{h_u^2}{D_x} + \tfrac{1}{4}D_x\right) \qquad (5.39)$$

The total head $h_{x,y}$ at any point (x,y) is determined from Eq. (5.32) by substituting the values of D_x and h_{bx} given, respectively, by Eqs. (5.38) and (5.39).

Equation (5.38) can be written in a dimensionless form in terms of q/K given by Eq. (5.35) as follows:

$$\frac{D_x}{h_u} = \sqrt{1 - \frac{x}{B}\left(1 - \frac{0.38q}{BK}\right)} \qquad (5.40)$$

Applications in extensive unconfined aquifers In practice, the same formulas can be applied for uniform flow in unconfined aquifers if B is known (Fig. 5.13a). Hydraulically, the vertical section cc' corresponds to the section at which a groundwater divide occurs or at which the darcian velocity in the soil is almost negligible. Thus in unconfined systems of natural flow, the same analysis for rectangular cores of earth dams can be used. The water table is plotted using Eq. (5.38), once B and h_u are known.

If two bore holes W_1 and W_2 distance B' apart are made to record D_{x1} and D_{x2} from the field, then with reference to Eq. (5.40) and Fig. 5.13a,

$$\frac{D_{x1}^2 - D_{x2}^2}{B'} = \frac{h_u^2}{B}\left(1 - \frac{0.38q}{BK}\right) = C_q \qquad (5.41)$$

where C_q is a constant. Thus line ab representing the ordinates of D_x^2 along B (Fig. 5.13b) should be a straight line. At the exit ee', the D_x^2 diagram has the ordinate $bb' = D_d^2 = 0.55(q/K)^2$ [Eq. (5.36)]. The distances $B - x_1$ and $B - x_2$ of the bore holes from the discharge face, which may be a cut, a drain, a stream, etc., are already known. Then from the geometry of the D_x^2 diagram (Fig. 5.13b),

$$C_q = \frac{D_{x1}^2 - D_{x2}^2}{B'} = \frac{D_{x1}^2 - 0.55(q/K)^2}{(B - x_1)} = \frac{D_{x2}^2 - 0.55(q/K)^2}{(B - x_2)} \qquad (5.42)$$

and q/K is calculated from Eq. (5.42).

Combining Eqs. (5.35) and (5.41),

$$C_q = \frac{2q}{K} - \frac{0.76}{B}\left(\frac{q}{K}\right)^2 \qquad (5.43)$$

Then B is determined from Eq. (5.43). Using Eq. (5.35), h_u can be evaluated. Once h_u, B, and q/K are known, the water-table curve ce (Fig. 5.13a) can be plotted using Eq. (5.40).

The preceding procedure is simple but accurate. It does not necessitate the separate determination of K or q or looking for the actual values of h_u and B. Current practice is to use Darcy's law to determine $\dot{q}$ if K has been determined by some field procedure (see Chaps. 6 and 8). Thus

$$q \simeq K\left(\frac{D_{x1} + D_{x2}}{2}\right)\left(\frac{D_{x1} - D_{x2}}{B'}\right) = K\frac{D_{x1}^2 - D_{x2}^2}{2B'} = \tfrac{1}{2}KC_q$$

or $C_q = 2q/K$. Comparing this result with Eq. (5.43), C_q is given only by the first term of Eq. (5.43). Therefore, Eq. (5.43) is recommended.

Earth Dams with Cores of Trapezoidal Sections

A trapezoidal section of an earth dam (Fig. 5.14a) is in fact an earth dam or enbankment with a vertical upstream side. In practice, the upstream side should be supported by a triangular rock-fill wedge through which head dissipation is neglected. If there is a sloping upstream, the section of

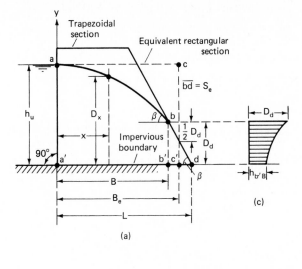

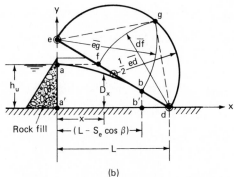

Figure 5.14 Seepage through a trapezoidal earth core of a dam. *(a)* Section of dam; *(b)* graphic construction to locate exit point *b*; *(c)* total-head diagram across section *bb'*.

the dam is approximately treated as a trapezoidal section, but the entrance point *a* (Fig. 5.14*a*) is corrected by moving it to the left, as explained later. This approach simplifies the analysis.

When the downstream slope angle β is small, the flow lines, including the free surface, is almost flat, justifying the use of Dupuit's assumptions. The first published solutions for the free surface and rate of flow through such sections when $\beta \leqslant 30°$ are attributed to Schaffernack and Iterson (Casagrande, 1937). They applied Dupuit's assumptions, but their solution was improved by introducing a boundary condition that recognized the existence of the discharge face S_e. Thus with reference to Fig. 5.14a,

$$q = KS_e \sin \beta \tan \beta \tag{5.44}$$

also,

$$q = -KD_x \frac{dD_x}{dx} \tag{5.45}$$

or

$$2q \, dx = -K \, dD_x^2$$

By integration,

$$2qx + KD_x^2 = \text{constant}$$

The constant is determined by introducing the condition that when $x = 0$, $D_x = h_u$; then the constant is equal to Kh_u^2, and

$$2qx + KD_x^2 = Kh_u^2$$

Substituting the value of q given in Eq. (5.44) in the preceding equation, we get

$$2(S_e \sin \beta \tan \beta)x + KD_x^2 = Kh_u^2 \tag{5.46}$$

When $x = L - S_e \sin \beta$, $D_x = S_e \sin \beta$; thus

$$S_e^2 - \frac{2S_e L}{\cos \beta} + \frac{h_u^2}{\sin^2 \beta} = 0 \tag{5.47}$$

Solving Eq. (5.47) for S_e, we get

$$S_e = \frac{L}{\cos \beta} \pm \sqrt{\frac{L^2}{\cos^2 \beta} - \frac{h_u^2}{\sin^2 \beta}}$$

The plus sign in this equation would yield an imaginary answer; therefore,

$$S_e = \frac{L}{\cos \beta} - \sqrt{\frac{L^2}{\cos^2 \beta} - \frac{h_u^2}{\sin^2 \beta}} \tag{5.48}$$

Thus the rate of flow q [Eq. (5.44)] can be computed in terms of the value of S_e determined by Eq. (5.48). The magnitude of S_e locates the exit point b (Fig. 5.14a). The free surface can thus be drawn by constructing a parabola that is tangential to the horizontal at point a and tangential to the downstream slope at point b.

The exit point b can be determined by a graphic construction (Casagrande, 1937) using Eq. (5.48). The upstream vertical plane is extended upward to intersect the extension of the downstream slope at point e (Fig. 5.14b), and a semicircle with $\overline{ed}$ as the diameter is constructed. A horizontal line is drawn through point a to intersect $\overline{ed}$ at f. With point d as the center, an arc of a circle with a radius $\overline{df}$ is drawn to intersect the semicircle at point g. With e as the center, an arc of a circle with radius $\overline{eg}$ is drawn to intersect the downstream slope at point b, which is the exit point. The proof is as follows:

$$S_e = \overline{bd} = \overline{de} - \overline{eb} = \overline{de} - \overline{eg} = \overline{de} - \sqrt{\overline{de}^2 - \overline{gd}^2}$$

$$= \overline{de} - \sqrt{\overline{de}^2 - \overline{fd}^2} = \frac{L}{\cos \beta} - \sqrt{\frac{L^2}{\cos^2 \beta} - \frac{h_u^2}{\sin^2 \beta}}$$

When β exceeds $30°$, the solution cannot be valid. Leo Casagrande modified Eq. (5.45) for β values between $30°$ and $60°$ by introducing assumptions similar to Dupuit's assumptions but considered the constant gradient across a vertical section as dD_x/ds rather than as dD_x/dx (Casagrande, 1937). It is obvious that these assumptions are not realistic; implicitly, all flow lines are assumed to intersect the impervious base, which is a flow line.

Rigorous mathematical solutions (Polubarinova-Kochina, 1962) for these sections were simplified for practical use (Kashef, 1965b) without any appreciable difference in the results. The known parameters are h_u, L, downstream angle β, and the average hydraulic conductivity K of the body of the dam, while the unknowns are the rate of flow q through the body of the dam, the location of the free surface curve ab, and the flow configuration. Once the exit point b is located, B, D_d (Fig. 5.14b), and accordingly the outflow surface bd are known. The following equations give the solution in this case (Kashef, 1965b):

$$D_d = L \tan \beta - \sqrt{L^2 \tan^2 \beta - [h_u^2/(1 - \tfrac{2}{3} \sin^2 \beta)]} \qquad (5.49)$$

$$B = L - D_d \cot \beta = \sqrt{L^2 - [h_u^2 \cot^2 \beta/(1 - \tfrac{2}{3} \sin^2 \beta)]} \qquad (5.50)$$

$$\frac{q}{K} = \frac{h_u^2 - D_d^2(1 - \tfrac{2}{3} \sin^2 \beta)}{2B} \qquad (5.51)$$

It has been found that q/K for a trapezoidal section is the same as that for an equivalent rectangular section (Kashef, 1965b) with a breadth B_e (Fig. 5.14a); thus

$$B_e = L - \tfrac{1}{2}D_d \cot \beta = \tfrac{1}{2}(L + B)$$
or
$$B_e = B + \tfrac{1}{2}D_d \cot \beta \qquad (5.52)$$

B_e is determined by substituting the values of D_d and B [Eqs. (5.49) and (5.50)] in Eq. (5.52); thus

$$B_e = \tfrac{1}{2}L + \tfrac{1}{2}\sqrt{L^2 - \frac{h_u^2 \cot^2 \beta}{(1 - \tfrac{2}{3} \sin^2 \beta)}} \qquad (5.53)$$

Thus B_e may be determined whenever B and D_d are calculated [Eq. (5.52)], or it can be determined from Eq. (5.53) from the initially known parameters L in h_u and β. The total-head diagram across the vertical section bb' is shown in Fig. 5.14c. The area P_b of the pressure-head diagram across bb' is determined from Eq. (5.31) by substituting $D_x = D_d$

and $\delta = \beta$; thus

$$P_b = D_d^2(\tfrac{1}{2} - \tfrac{1}{3}\sin^2\beta) \tag{5.54}$$

Also

$$P_a = \frac{h_u^2}{2} \tag{5.55}$$

Whether the equivalent rectangular section or the original section are used, q/K has the following forms:

$$\frac{q}{K} = \frac{P_a - P_b}{B} = \frac{P_a}{B_e} = \frac{P_b}{\tfrac{1}{2}D_d\cot\beta}$$

or

$$\frac{q}{K} = \frac{h_u^2}{2B_e} = \frac{D_d(1 - \tfrac{2}{3}\sin^2\beta)}{\cot\beta} \tag{5.56}$$

Also

$$\frac{h_u^2 K}{q} = L + \sqrt{L^2 - \frac{h_u^2\cot^2\beta}{(1 - \tfrac{2}{3}\sin^2\beta)}} \tag{5.57}$$

Another formula for B_e in terms of D_u and D_d can be determined from Eq. (5.56) as

$$B_e = \frac{h_u^2\cot\beta}{2D_d(1 - \tfrac{2}{3}\sin^2\beta)} \tag{5.56a}$$

The value of D_d depends on L, β, and h_u, and it can be written in terms of q/K and β [Eq. (5.56)].

The free surface is normal to the vertical entrance aa' at a and is tangential to the downstream slope at b (Casagrande, 1937). The exit condition has been used in formulating Eq. (5.54). The base pressure $h_{bx} = h_{bB}$ at $x = B$ is therefore given by

$$h_{bB} = D_d(1 - \tfrac{1}{2}\sin^2\beta) \tag{5.58}$$

The free surface is traced by assuming the equivalent rectangular section $acc'a'$ (Fig. 5.14a) and calculating D_x from Eq. (5.38) for $0 \leqslant x \leqslant B$. If there is any difference in the D_x values following this approach and other rigorous solutions (Polubarinova-Kochina, 1962), the maximum error occurs at the last section bb' (Fig. 5.14a) distance B from the origin. A comparative study (Kashef, 1965b) indicated that the maximum error does not exceed 2.9 percent for values of $L/h_u < 2$ and 1.6 percent for values of $L/h_u > 2$ and for a wide range of β from $10°$ to $70°$. The base pressure heads h_{bx} are then calculated from Eq. (5.39) and the total head $h_{x,y}$ from Eq. (5.32). Equations (5.32), (5.38), and (5.39) should be used only for the equivalent rectangular section within region $0 \leqslant x \leqslant B$. The values of h_{bx} and $h_{x,y}$ within the triangular section $bb'd$ may be determined from a flow-net sketch or by means of numerical analysis if such refinement is required.

Trapezoidal sections were partially analyzed by rigorous mathematical solutions using various approaches, including the velocity hodograph method. Some of the results were reduced to simple forms by the same authors. For example (Polubarinova-Kochina, 1962), when β is small,

$$\frac{q}{K} = \frac{h_u^2}{L + \sqrt{L^2 - h_u^2 \cot^2 \beta}} \qquad \text{(Mikhailov and Pavlovsky)} \quad (5.59)$$

When β approaches $\pi/2$,

$$\frac{q}{K} = \tan \beta \left[h_u + (L \tan \beta \, h_u) \ln \left(\frac{L}{L - h_u \cot \beta} \right) \right]$$

$$\text{(Meletchenko and Mikhailov)} \quad (5.60)$$

Pavlovsky recommended Eq. (5.59) for use in all isotropic trapezoidal types irrespective of the β value, and the same equation was recommended by Polubarinova-Kochina (1962) on the basis of field records on some real dams. Other rigorous solutions for D_d were simplified by Mikhailov, resulting in an error of less than 3 percent:

$$\frac{KD_d}{q} = \tfrac{1}{4}(\lambda \cot B + 6 - \lambda) \qquad (5.61)$$

where λ is a constant that has only two values: either 4 when $\cot \beta > 1$ or 3 when $\cot \beta < 1$. When $\beta = \pi/2$, $\cot \beta = 0 < 1$ and then $\lambda = 3$; it follows that $D_d = 0.75q/K$, which is almost the same as that given by Eq. (5.36).

Earth Dams with Horizontal Filter Blankets

An earth dam section $abcd$ is shown in Fig. 5.15a with a horizontal filter blanket along the base starting from O and extending beyond point d to drain the seepage water. It this blanket does not exist, the free surface as well as all other flow lines emerges at the downstream slope cd (as in the case of trapezoidal sections). Because the horizontal blanket is in contact with atmospheric pressure, it diverts all flow lines within the body of the dam away from the downstream surface, increasing its stability and preventing erosion along the downstream face. The pressures along the outflow surface OB are zero. The downstream angle in this case between the bottom impervious boundary at O and the discharge surface OB is $\beta = \pi$. Kozeny (Casagrande, 1937) solved this problem by conformal mapping techniques. Although the analysis and techniques of conformal mapping are outside the scope of this book, Kozeny's solution is explained here for the following reasons:

1. It is one of the simplest methods of conformal mapping, and this gives the reader a rough idea of the basic principles involved.

2. The results of the solution give useful general equations that may be used by practitioners without going into the details of the solution.

3. The solution's application in practice entails some approximations, although it is rigorous. This illustrates the point that mathematical solutions are not always perfect.

4. The results have been used to find some approximate solutions (Casagrande, 1937) for dams with $\pi/3 < \beta < \pi$ (discussed later).

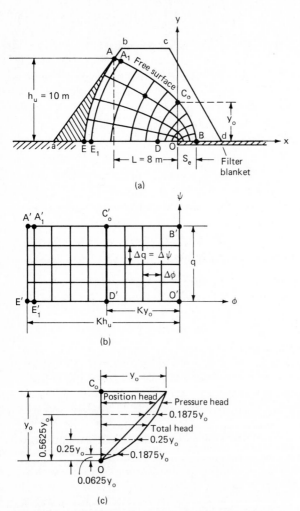

Figure 5.15 Seepage through earth dam with a horizontal filter blanket, Kozeny's solution. *(a)* $\bar{z}$ plane; *(b)* $\bar{w}$ plane; *(c)* distribution of various heads across section OC_o.

In this method, two complex planes $\bar{z}$ (Fig. 5.15a) and $\bar{w}$ (Fig. 5.15b) are defined as follows:

$$\bar{z} = x + iy \quad \text{and} \quad \bar{w} = \phi + i\psi \quad (5.62)$$

where $\qquad i = \sqrt{-1.0} \quad \text{(imaginary)}$

The $\bar{z}$ plane includes the true section, and the $\bar{w}$ plane represents the relationship between ϕ and ψ as related to the true section. The solution requires that a square flow net in the $\bar{w}$ plane (real squares) correspond to the final flow net in the $\bar{z}$ plane, consisting of curvilinear squares. In order to transform the exact squares drawn on the $\bar{w}$ plane to the real section ($\bar{z}$ plane), there is a mapping function between both planes. Determination of this mapping function is complex and is the most important phase of the method. Sometimes the method requires additional auxiliary planes to finalize the solution (Polubarinova-Kochina, 1962). Kozeny found that the mapping function of this problem is expressed by

$$\bar{z} = C_i \bar{w}^2 \quad (5.63)$$

where C_i is a constant.

From Eqs. (5.62) and (5.63),

$$x + iy = C_i(\phi + i\psi)^2$$
$$= C_i(\phi^2 - \psi^2) + 2i\phi\psi C_i \quad (5.64)$$

The real term on the left-hand side of the equation is equal to the real terms on the right-hand side. Thus the imaginary terms (including $i = \sqrt{-1.0}$) are equal; therefore,

$$x = C_i(\phi^2 - \psi^2) \quad (5.65a)$$

$$y = 2C_i\phi\psi \quad (5.65b)$$

or

$$x = C_i\left(\phi^2 - \frac{y^2}{4C_i^2\phi^2}\right) \quad (5.65c)$$

Equation (5.65c) gives the relations between x, y, ϕ, and ψ. In order to find the constant C_i, it is noted that the free surface ACB (Fig. 5.15a) has the conditions $\psi = q$ and $\phi = -Ky$. Substituting these values in Eq. (5.65b), the C_i value is determined:

$$y = -2C_iKyq$$

$$(5.66)$$

or

$$C_i = -\frac{1}{2Kq}$$

Substituting the value of C_i in Eq. (5.65a and b),

$$x = -\frac{1}{2Kq}(\phi^2 - \psi^2) \quad (5.67a)$$

and

$$y = -\frac{1}{Kq}\phi\psi \quad (5.67b)$$

The equation of the free surface AC_oB is thus determined from Eq. (5.67a) by substituting $\psi = q$ and $\phi = -Ky$:

$$x = -\frac{Ky^2}{2q} + \frac{q}{2K} \qquad (5.68)$$

which is a parabolic equation.

Substituting the coordinates x, y of the entrance point $A = (-L, h_u)$ in Eq. (5.68), the magnitude of q is obtained in terms of the known values K, L, and h_u:

$$-L = -\frac{Kh_u^2}{2q} + \frac{q}{2K} \qquad \text{or} \qquad q^2 + 2KLq - q^2h_u^2 = 0$$

Thus $\qquad q = K(-L + \sqrt{L^2 + h_u^2}) \qquad (5.69)$

In order to determine the length S_e of the outflow surface OB (Fig. 5.15a), $x = S_e$ and $y = 0$ are substituted in Eq. (5.68); thus

$$S_e = \frac{q}{2K} \qquad (5.70)$$

The free surface intersects the y axis at C_o, which has the coordinates $x = 0$ and $y = y_o$; then from Eq. (5.68),

$$y_o = \frac{q}{K} \qquad (5.71)$$

From Eqs. (5.69), (5.70), and (5.71), it follows that

$$\frac{q}{K} = y_o = 2S_e = -L + \sqrt{L^2 + h_u^2} \qquad (5.72)$$

Equation (5.68) of the free surface thus may be written in terms of S_e and y_o as follows:

$$x = \tfrac{1}{2}\left(y_o - \frac{y^2}{y_o}\right) = S_e - \frac{y^2}{4S_e} \qquad (5.73)$$

In order to plot the flow net (Fig. 5.15a), it should be noted that the ϕ and ψ lines are represented by vertical and horizontal lines, respectively, on the $\bar{w}$ plane (Fig. 5.15b). If K, h_u, and the magnitude of q [Eq. (5.69)] are known, points O, B, C_o, A, E, and D in Fig. 5.15a correspond to points O', B', C_o', A', E', and D' on the $\bar{w}$ plane (Fig. 5.15b). If a value for n_F, the number of flow channels ($n_F = 4$ in Fig. 5.15a and b), is selected, then the equipotential drop times K on the $\bar{w}$ plane has the same magnitude as q/n_F (note that ϕ and ψ have the same units). Then the square flow net drawn on the $\bar{w}$ plane can be mapped on the $\bar{z}$ plane as illustrated by the following example.

EXAMPLE 5.3 Consider that the earth dam shown in Fig. 5.15a has the following data: $h_u = 10$ m, $L = 8.0$ m, and $K = 10^{-4}$ m/s. Plot the flow net.

Solution From Eq. (5.69),

$$q = 10^{-4}(-8 + \sqrt{61 + 100}) = 4.8 \times 10^{-4} \text{ m}^2/\text{s}$$

Therefore, $S_e = 2.4$ m and $y_o = 4.8$ m.

If four flow channels are selected ($n_F = 4$), then

$$\Delta\psi = \Delta q = \frac{q}{4} = \frac{4.8 \times 10^{-4}}{4} = 1.2 \times 10^{-4} \text{ m}^2/\text{s}$$

Since $\Delta\phi = \Delta\psi$ (Fig. 5.15b), then the equipotential drop is

$$\Delta h = \frac{\Delta\phi}{K} = \frac{1.2 \times 10^{-4}}{10^{-4}} = 1.2 \text{ m}$$

But $\Delta h = h_u/n_E$, where n_E is the number of equipotential drops; therefore,

$$n_E = \frac{h_u}{\Delta h} = \frac{10}{1.2} = 8\tfrac{1}{3}$$

In other words, four curvilinear rectangles in the true section exist between equipotential lines $A_1 E_1$ and AE (Fig. 5.15a) that correspond to the vertical lines $A'E_1'$, and $A'E'$ in the $\overline{w}$ plane. The distance $A'A_1'$ equals

$$E'E_1' = \tfrac{1}{3}\Delta\phi = \frac{1.2 \times 10^{-4}}{3} = 4 \times 10^{-5} \text{ m}^2/\text{s}$$

The flow net in the $\overline{w}$ plane is now already determined. In order to plot it back on the real dam ($\overline{z}$ plane), the coordinates of the points of intersection (corners of the squares) have to be found. For example, point F' in the $\overline{w}$ plane corresponds to point F in the $\overline{z}$ plane.

At F': $\psi = \tfrac{3}{4}q = 3.6 \times 10^{-4}$ m²/s (or $= 3\Delta\psi$)

$\phi = -6\Delta\phi = 6 \times 1.2 \times 10^{-4} = -7.2 \times 10^{-4}$ m²/s

Substituting these values of ψ and ϕ in Eq. (5.67a and b), the coordinates x and y of point F in the true section are determined:

$$x = \frac{-(7.2 \times 10^{-4})^2 - (3.6 \times 10^{-4})^2}{2 \times 10^{-4} \times 4.8 \times 10^{-4}} = -4.05 \text{ m}$$

$$y = -\frac{(7.2 \times 10^{-4})(3.6 \times 10^{-4})}{10^{-4} \times 4.8 \times 10^{-4}} = +5.40 \text{ m}$$

The procedure is repeated for all points of intersection, and the flow net is completed. The free surface can be traced without drawing the entire flow net by means of Eq. (5.68). The equation for any other stream line can be determined from Eq. (5.67a and b) as

$$x = -\frac{1}{2Kq}\left(\frac{K^2 q^2 y^2}{\psi^2} - \psi^2\right) \tag{5.74}$$

A specific stream line such as that on which point F lies has an intensity of $\psi = \tfrac{3}{4}q = \tfrac{3}{4} \times 4.8 \times 10^{-4} = 3.6 \times 10^{-4}$ m²/s, and its equation is obtained by substituting this value in Eq. (5.74).

The equation for the equipotential lines is determined from Eq. (5.67a and b) as

$$x = -\frac{1}{2Kq}\left(\phi^2 - \frac{K^2q^2y^2}{\phi^2}\right) \tag{5.75}$$

For a specific equipotential line, the intensity ϕ is substituted in Eq. (5.75). For example, the equipotential line on which point F lies has a value equal to $-6\Delta\phi = -6 \times 1.2 \times 10^{-4} = -7.2 \times 10^{-4}$.

Discussion of Kozeny's Solution

1. Kozeny's solution gives two systems of confocal parabolas with a common focus at O (Fig. 5.15a).

2. The solution is not valid unless the upstream slope Aa is parabolic (AE in Fig. 5.15a). The solution may, however, be considered practically feasible if the zone AEa is neglected or is considered as if it consisted of a material offering no resistance to the flow (Taylor, 1948), such as a rock fill or a filter.

3. The slope of the free surface at point $C_o = (o, y_o)$ makes an angle $45°$ to horizontal. The slope of any other stream line may be determined from Eq. (5.74) as

$$\frac{dy}{dx} = -\frac{\psi^2}{yKq} \tag{5.76}$$

Therefore, at point $C_o = (0, y_o)$

$$\frac{dy}{dx} = -\frac{q^2}{y_oKq} = -1.0$$

4. The number of equipotential drops is equal to the number of flow channels between curves C_oB and C_oD (Fig. 5.15a), as shown in the corresponding square $C_o'B'O'D'$ in the $\overline{w}$ plane. It should also be noted that the exit surface OB is an equipotential line of intensity $\phi = 0$ (atmospheric). If the portion of the flow net bounded by the equipotential lines C_oD and OB is considered, the rate of flow $q = Ky_o(n_F/n_E) = Ky_o$, which is the same result as that given by Eq. (5.71).

5. The distribution of the total heads and pressure heads may be obtained at any vertical section by substituting the x coordinate of the vertical section in Eq. (5.75). In the preceding example, the distributions of the total heads and pressure heads are obtained along the vertical section OC_o (Fig. 5.15c) by substituting $x = 0$ in Eq. (5.75). The values of $y = \phi^2/Kq$ were obtained at the points of intersection of the flow lines and equipotential lines along OC_o in terms of $y_o = q/K$.

6. The distribution of the base pressure heads h_{b_x} between h_u at E and

zero at 0 along EO (values of x are negative) may be obtained by substituting $y = 0$ in Eq. (5.75); therefore,

$$h_{bx} = \sqrt{2y_o x} \qquad (5.77)$$

Overhanging Earth Dams

These types of earth dams (Fig. 5.16) have a downstream angle greater than 90°. In principle they serve the same objective as dams with horizontal filter blankets because the flow lines are diverted away from the downstream slope ef. The filter in this case is a rock-fill toe $dff'd'$. The free surface ab is normal to the vertical upstream side at a and tangential to the vertical at the exit point b (Casagrande, 1937). The outflow surface bd makes an angle $\pi > \beta > \pi/2$ with the horizontal impervious bed ad.

Rigorous solutions for this type are given elsewhere (Polubarinova-Kochina, 1962). One of the common approximate solutions is based on the flow-net graphic method (Casagrande, 1937) as compared with Kozeny's solution (explained later). This case is intermediate between the cases with $\beta = \pi/2$ and $\beta = \pi$, for which we have exact solutions. By simple interpolation, the rate of flow q and the discharge faces are determined as follows.

Considering an equivalent rectangular earth dam with width B_e, the rate of flow for dams with $\beta = \pi/2$ and dams with $\beta = \pi$ may be, respectively, expressed as follows:

$\beta = \dfrac{\pi}{2}$: $\dfrac{q}{K} = \dfrac{h_u^2}{2B_e} = \dfrac{S_e}{0.742}$ [from Eq. (5.36)]

$\beta = \pi$: $\dfrac{q}{K} = \dfrac{h_u^2}{2B_e} = \dfrac{S_e}{0.5}$ [from Eq. (5.70)]

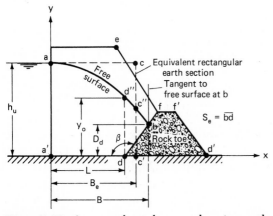

Figure 5.16 Seepage through an overhanging earth dam ($\beta > 90°$).

Therefore,

$$S_e B_e = 0.371 h_u^2 \quad \text{for} \quad \beta = \frac{\pi}{2}$$

and

$$S_e B_e = 0.25 h_u^2 \quad \text{for} \quad \beta = \pi$$

By interpolation,

$$S_e B_e = \left(0.492 - 0.242 \frac{\beta}{\pi}\right) h_u^2 \quad \text{for any value } \pi \geqslant \beta \geqslant \frac{\pi}{2} \quad (5.78)$$

Also, when $\beta = \pi/2$, $B_e = L$, and when $\beta = \pi$,

$$\frac{q}{K} = y_o = \frac{h_u^2}{2B_e} = -L + \sqrt{L^2 + h_u^2} \quad \text{or} \quad B_e = \frac{\frac{1}{2}h_u^2}{-L + \sqrt{L^2 + h_u^2}}$$

By interpolation for any value of $\pi \geqslant \beta \geqslant (\pi/2)$,

$$B_e = \left(\frac{2\beta}{\pi} - 1\right)\left(\frac{h_u^2}{-L + \sqrt{h_u^2 + L^2}} - L\right) + L \quad (5.79)$$

Calculating B_e from Eq. (5.79) and q/K from $q/K = h_u^2/2B_e$ [Eq. (5.56)], the length of the outflow surface $bd = S_e$ can be computed from Eq. (5.78).

The free surface may be plotted approximately within the width B_e using Eq. (5.38), which is rewritten in the form

$$D_x \approx h_u \sqrt{1 - \frac{x}{B_e}\left(1 - \frac{h_u^2}{7.27B_e^2}\right)} \quad (5.79)$$

Beyond the equivalent rectangular section $acc'a'$ (Fig. 5.16), the small segment of the free surface $c''b$ may be plotted as a parabola between the determinted points c'' and b.

The interpolation may be justified between $\beta = \pi/2$ and $\beta = \pi$ because of the relatively small differences between the two cases in the values of D_x and q/K.

Casagrande's Approach

Casagrande (1937) sketched flow nets for each dams of different downstream slopes, that is, $60° \leqslant \beta < 180°$ (Fig. 5.17). The flow nets were then matched with Kozeny's parabola, which is the free surface when $\beta = 180°$, as shown by dashed lines in Fig. 5.17a, b, and c. In each case, the intersection of Kozeny's parabola with the downstream surface is located and the ratio $\Delta S_e/(S_e + \Delta S_e)$ is evaluated. Figure 5.17d indicates the relationship between this ratio and the downstream angle β. The discharge point of the free surface is then determined from this curve.

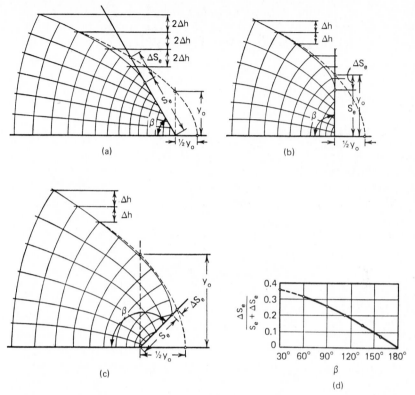

Figure 5.17 Casagrande method for locating exit points for various downstream slopes. (a) $\beta = 60°$, $\Delta S_e/(S_e + \Delta S_e) = 0.32$; (b) $\beta = 90°$, $\Delta S_e/(S_e + \Delta S_e) = 0.26$; (c) $\beta = 135°$, $\Delta S_e/(S_e + \Delta S_e) = 0.14$; (d) the slope of discharge face β versus $\Delta S_e/(S_e + \Delta S_e)$. (A. Casagrande, "Seepage through Dams," J. N. Engl. Water Works Assoc., vol. 51, 1937.)

For example, if $\beta = 60°$, $\Delta S_e/(S_e + \Delta S_e) = 0.32$. By plotting Kozeny's parabola [Eqs. (5.68) and (5.69)], the magnitude $S_e + \Delta S_e$ is measured, and ΔS_e is in this case, equal to $0.32 \times (S_e + \Delta S_e)$. Once the exit point is determined, the flow net can be drawn. It is obvious that the intention of Casagrande (1937) was to facilitate flow-net sketching rather than to find an analytic solution. Casagrande's method or the approximate solutions previously explained also may be used as initial conditions in more refined approaches, such as using computerized numerical solutions.

Casagrande's approach can be checked against the rigorous solution for vertical rectangular earth sections ($\beta = 90°$) given by Eq. (5.36). From Fig. 5.17d, the value of $\Delta S_e/(S_e + \Delta S_e)$ corresponding to $\beta = 90°$ is equal to 0.26. The characteristics of the free surface shown in Fig. 5.17b indicate that the flow-net solution matches Kozeny's solution near the upstream. Thus the rate of flow can be considered the same in both cases because the rate of flow can be determined equally from the portion of the flow net near the upstream. Following this reasoning and applying Eq. (5.71), $q/K = D_d/0.742 = S_e/0.742 = S_e + \Delta S_e$. Therefore, $S_e/(S_e + \Delta S_e) = 0.742$ or $\Delta S_e = 0.3477 S_e$, and the $\Delta S_e/(S_e + \Delta S_e)$ value is 0.258, which is practically the same as that determined from Casagrande's chart (0.26).

Effect of Upstream Slope

The upstream side of an earth dam is usually sloping at an angle $\alpha < (\pi/2)$ if it is not supported (Fig. 5.18b). Most of the previous cases were analyzed on the basis of a vertical upstream surface and the results should be modified according to the following analysis.

It has been proved (Kashef, 1965b) that the rate of flow between any two vertical sections at distances x_1 and x_2 in any state-steady two-dimensional system is given by

$$\frac{q}{K} = \frac{P_{x1} - P_{x2}}{x_2 - x_1} \tag{5.80}$$

where P_x is the area of the pressure-head diagram along a vertical section at x. The equivalent rectangular sections of width B_e in all the cases previously discussed have q/K values consistent with Eq. (5.80), or

$$\frac{q}{K} = \frac{h_u^2}{2B_e} \tag{5.81}$$

In the case of a sloping upstream, the rate of flow q decreases with decreases in α because of increases in the size of the body of the dam. Meanwhile it was shown previously that the total-head distribution across section aa' is as shown in Fig. 5.18a and that (Kashef, 1965b)

$$h_{ba'} = h_u(1 - \tfrac{1}{2} \cos^2 \alpha) \tag{5.82}$$

Moreover, $\qquad P_x$ at $aa' = P_o = h_u^2(\tfrac{1}{2} - \tfrac{1}{3} \cos^2 \alpha) \tag{5.83}$

Thus the rate of flow given by Eq. (5.81) has another value q_{re} that is less than q:

$$\begin{aligned}
\frac{q_{re}}{K} &= \frac{h_u^2(\tfrac{1}{2} - \tfrac{1}{3} \cos^2 \alpha)}{B_e} \\
&= \frac{\tfrac{1}{2} h_u^2}{B_e'} = \frac{\tfrac{1}{2} h_u^2 - h_u^2(\tfrac{1}{2} - \tfrac{1}{3} \cos^2 \alpha)}{\Delta B_e} \\
&= \frac{\tfrac{1}{3}(\cos^2 \alpha) h_u^2}{\Delta B_e}
\end{aligned} \tag{5.84}$$

where $\qquad\qquad\qquad B_e' = B_e + \Delta B_e$

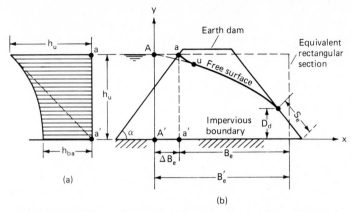

Figure 5.18 Effect of upstream slope on free surface. *(a)* Total-head distribution across section *aa'*; *(b)* initial segment of the free surface.

and ΔB_e is the additional width of the equivalent rectangular section to accommodate the sloping upstream surface (Fig. 5.18*b*), and by use of Eq. (5.84),

$$\Delta B_e = \tfrac{2}{3} B_e' \cos^2 \alpha = B_e \frac{2 \cos^2 \alpha}{3 - 2 \cos^2 \alpha} \qquad (5.85)$$

The reduced flow rate q_{re} may thus be determined from Eq. (5.84) on the basis of the values of B_e, ΔB_e, or B_e' given above. The free surface is normal to the upstream slope at *a*; this results in an inflection point on the curve of the free surface (point *u* in Fig. 5.18*b*). The free surface may still be plotted using Eq. (5.38) by measuring *x* from section *AA'* (Fig. 5.18*b*) and substituting B_e' for *B* and *m* for $B_e'^2/h_u^2$, or

$$D_x \simeq h_u \sqrt{1 - \frac{x}{B_e + \Delta B_e} \left[1 - \frac{h_u^2}{7.27(B_e + \Delta B_e)^2} \right]} \qquad (5.86)$$

After plotting the curve, *Au* (Fig. 5.18) is adjusted to *au* by sketching (Casagrande, 1937) a curve between points *a* and *u*, observing that the curve is normal to the upstream slope at *a*.

Effects of Tailwater Bodies

In the previous analysis it was assumed that each earth dam was constructed on an impervious base with no tailwater (dry downstream). This is the common critical condition assumed in the stability analysis of such dams, coupled with an assumption of the highest possible upstream water level. However, if the dam being considered has a water body downstream (Fig. 5.19), an approximate analysis (Casagrande, 1937) can be made by assuming a fictitious impervious bed ab coinciding with the tailwater level. The previous analysis would be valid for that portion of

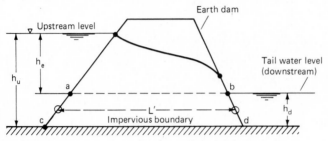

Figure 5.19   Earth dam with downstream water body.

the dam above plane ab, and the upper flow rate q_u could be determined. The rate of flow q_b for the lower portion between ab and the impervious boundary is treated as a confined stratum and Darcy's law [Eq. (3.5)] is used:

$$q_b = h_d K \frac{h_e}{L'} \tag{5.87}$$

where L' is the average length of the flow path and equals $\frac{1}{2}(\overline{ab} + \overline{cd})$ (Fig. 5.19). The total flow $q = q_u + q_b$

The physical meaning of the approximation in this procedure is that a stream line is assumed along ab with no flow crossing it. This is hydraulically equivalent to an impervious boundary along ab. If an exact solution is obtained, stream lines near ab would have very gentle slopes.

Dams Constructed on Soil and Anisotropy Effects

If an earth dam is constructed over a natural soil layer of depth D_q (Fig. 5.20), an additional rate of flow q'_b is added to the rate of flow above ab. Applying Darcy's law [Eq. (3.5)],

$$q'_b = D_q K_q \frac{h_e}{L_s} \tag{5.88}$$

where K_q is the hydraulic conductivity of the soil layer and L_s is as shown in Fig. 5.20. It is obvious that if $h_d = 0$, Eq. (5.88) is still valid.

If the horizontal and vertical hydraulic conductivities are evaluated for earth dams, then the previous methods should be applied to the transformed sections (see Sec. 5.1).

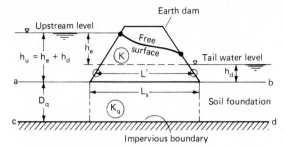

Figure 5.20 Earth dam constructed on a soil layer overlying an impervious boundary.

PROBLEMS AND DISCUSSION QUESTONS

Figures for this section are numbered according to the problem in which they first appear.

5.1   Some geohydrologists claim that in most regions the groundwater table follows the topography. Discuss the validity of this concept with reference to the flow between points a and b in Fig. P-5.1 (sketch not to scale).

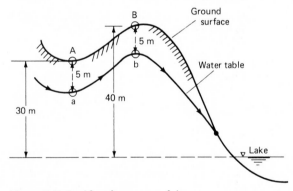

Figure P-5.1   (sketch not to scale)

5.2   A portion of a flow net is shown in Fig. P-5.2 (scale distorted). If $a/b = 1.0$, $\phi_3 - \phi_2 = 2(\phi_2 - \phi_1)$, and $(\psi_1 - \psi_2) = 3(\psi_2 - \psi_3)$, find the ratios a_1/b_1, a_2/b_2, and a_3/b_3.

5.3   A concrete weir (Fig. P-5.3) is to be constructed in a silty sand soil. It is necessary to check the safety of the weir against piping (roofing). Determine the following:
       (a) The safe length of the weir according to Lane's method if the safe weighted creep ratio is 8.5.

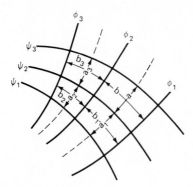

Figure P-5.2   (scale distorted)

(b) The uplift pressures in kilograms per square centimeter at points B and C

(c) The factor of safety against total uplift force assuming a uniform hydraulic gradient. The unit weight of concrete is 24 kg/m³.

(d) Repeat part (c) if the sheet pile is disregarded.

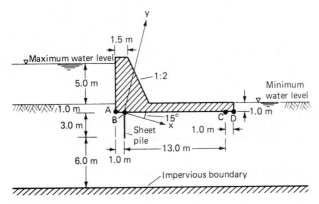

Figure P-5.3

5.4   Construct a flow net for seepage beneath the structure shown in Fig. P-5.3 for each of the following cases:

(a) The flow medium is homogeneous and isotropic and $K = 3 \times 10^{-2}$ cm/s.

(b) The flow medium is homogeneous and anisotropic, $K_x = 6$, and $K_y = 3 \times 10^{-2}$ cm/s. (Two flow nets should be drawn for this case, noting that the origin is at B and the x axis makes an angle 15° with the base of the weir.)

Determine also in each case the rate of seepage and the exit gradient.

5.5   Find the ratio of the total uplift force determined in Problem 5.4b to that determined in Problem 5.3c.

5.6   If the weir shown in Fig. P-5.3 has another sheet pile 0.5 m left of point D and 1.5 m deep, construct a flow net assuming that $K_x = K_y = 3 \times 10^{-2}$ cm/s. Determine the exit gradient and the factor of safety against heaving.

5.7   Figure P-5.7 shows the equipotential lines in a curvilinear square flow net. Copy the figure neatly (correct the lines if necessary) and write down the intensities of the total head of each line with respect to the shown datum. Find also the total, pressure, and position heads at each of the points A, B, C, D, E, and F with respect to the same datum.

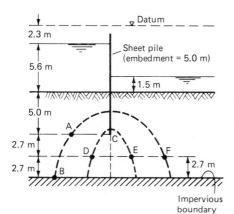

Figure P-5.7

5.8   Complete the flow net in Fig. P-5.7. Select one of the square fields with an inclination of about 45° and compute all the hydraulic forces given in Sec. 5.3 of the text. Draw to scale the corresponding force diagram (similar to that shown in Fig. 5.6d).

5.9   Indicate whether the following surfaces are flow lines, equipotential lines, or isobars:
    (a) Horizontal piezometric surface
    (b) Inclined piezometric surface
    (c) Upstream slope of an earth dam below water level
    (d) Downstream surface of an earth dam between the tailwater level and the exit point of the free surface
    (e) Impervious boundary

5.10   The rate of flow through the rectangular earth section shown in Fig. P-5.10 is given exactly by Eq. (5.35) (when $h_d = 0$). If $h_u = 10.0$ m, $B = 7.5$ m, and $n_F = 8$, find the number of equipotential drops n_E between the equipotential lines ad and bc. Find also the pressure heads at points b, b', c, and e.

5.11   Plot the free surface in Fig. P-5.10. On the same drawing, plot a parabola passing by point b and tangential to the horizontal at a. What is the maximum vertical difference between the two curves?

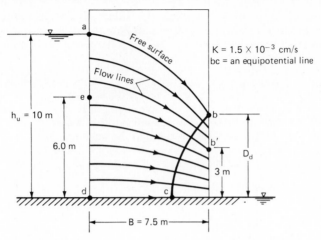

Figure P-5.10

5.12   Find analytically the area of the pressure-head diagram at a vertical section 4.0 m distant from point d in the rectangular earth section shown in Fig. P-5.10. Plot this diagram, indicating the magnitude of the ordinates at the top, bottom, and middle points.

5.13   A drain is constructed at a distance B from a lake (Fig. P-5.13). Field records indicated that the water-table level is 6.10 m above the impervious boundary at station A, 20 m from the lake. If $K = 1 \times 10^{-2}$ cm/s, find the rate of flow toward the drain, the distance B, and the location of the exit point in the drain. Determine also the rate of flow if Dupuit's assumptions are used, and plot Dupuit's curve throughout length B. Plot also the D_x^2 diagram (as shown in Fig. 5.14b). Indicate the ordinates of the plots at distance 5, 10, 15, 20, 25, and B m.

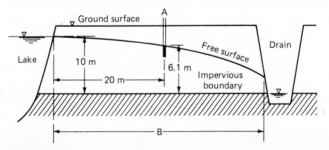

Figure P-5.13

5.14   If h_u and L in Example 5.3 (Fig. 5.15) were, respectively, changed to 12 and 16 m, construct the corresponding flow net. Find also the rate of flow if $K = 10^{-4}$ cm/s.

5.15 A flow net obtained by Kozeny's method is shown in Fig. P-5.15 (sketch not to scale). If $K = 5 \times 10^{-3}$ cm/s and the flow net consists of curvilinear rectangles, find the following:
   (a) Rate of flow
   (b) Length S_e of the outflow surface
   (c) Vertical component of the darcian velocity at point a
   (d) Constant ratio $\Delta S/\Delta n$ of the fields shown in Fig. P-5.15

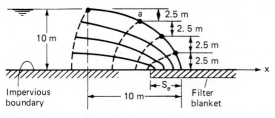

Figure P-5.15 (sketch not to scale)

5.16 With reference to Fig. P-5.16, write down the mathematical expressions describing all boundary conditions and explain their physical meanings. Find also the rate of flow, and plot the free surface using *all* possible approaches explained in the text (including Schaffernack and Iterson's method).

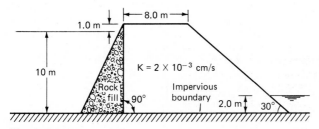

Figure P-5.16

5.17 Plot the free surface and determine the rate of flow through the earth dam shown in Fig. P-5.17 using *all* possible approaches explained in the text.

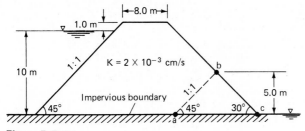

Figure P-5.17

5.18 Using the approximate results of the free-surface location in Problem 5.17, plot a flow net for the same section shown in Fig. P-5.17. Find the rate of flow on the basis of the drawn flow net.

5.19 If the downstream segment *abc* of the dam shown in Fig. P-5.17 were constructed as a rock-fill toe, plot the fre surface and find the rate of flow. Determine also S_e using Casagrande's approach.

5.20 Using the results obtained from Problem 5.16, find the *total* rate of flow of the dam shown in Fig. P.5.20.

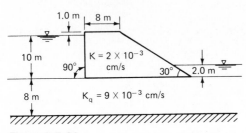

Figure P-5.20

REFERENCES

Bear, J.: *Hydraulics of Groundwater*, McGraw-Hill, New York, 1979.

Bear, J., and G. Dagan: "Some Exact Solutions of Interface Problems by Means of the Hodograph Method," *J. Geophys. Res.*, vol. 69, no. 8, 1964, p. 1563.

Bligh, W. G.: "Dams, Barrages, and Weirs on Porous Foundations," *Eng. News*, vol. 64, 1910, pp. 708–710.

Boothroyd, A. R., E. C. Cherry, and R. Makar: "An Electrolytic Tank for the Measurement of Steady-State Response, Transient Response, and Allied Properties of Networks," *Proc. Inst. Electr. Eng. (Lond.)*, vol. 96, pt. 1, 1949, pp. 163–177.

Cahill, J. M.: "Hydraulic Sand-Model Study of the Cyclic Flow of Salt Water in a Coastal Aquifer," U.S. Geological Survey Professional Paper 575-B, Washington, 1967, pp. 240–244.

Casagrande, A.: Discussion of "Security from Under-Seepage Masonry Dams on Earth Foundations," *Trans. Am. Soc. Civil Eng.*, vol. 100, 1935, pp. 1289–1294.

Casagrande, A.: "Seepage through Dams," *J. N. Engl. Water Works Assoc.*, vol. 51, 1937, pp. 131–172.

Cedergren, H. R.: *Seepage, Drainage and Flow Nets*, 2d ed., Wiley, New York, 1977.

Debrine, B. E.: "Electrolytic Model Study for Collector Wells under River Beds," *Water Resources Res.*, vol. 6, 1970, pp. 971–978.

DeWiest, R. J. M. (ed.): *Flow through Porous Media*, Academic Press, New York, 1969.

Dupuit, J.: *Etudes théroiques et pratiques sur le mouvement des eaux dans les canaux découverts et à travers les terrains perméable*, 2d ed., Dunod, Paris, 1863.

Hague, B.: "Distribution of Electric and Magnetic Fields," *Electrician*, vol. 102, 1929, p. 186.

Hansen, V. E.: "Complicated Well Problems Solved by the Membrane Analogy," *Trans. Am. Geophys. Union*, vol. 33, 1952, pp. 912–916.

Harr, M. E.: *Groundwater and Seepage*, McGraw-Hill, New York, 1962.

Kashef, A. I.: "Numerical Solutions of Steady State and Transient Flow Problems," Ph.D. thesis, Purdue University, West Lafayette, Ind., 1951.

Kashef, A. I.: "Mathematical Developments in Transient Groundwater Hydraulics Using a Cylindrical Coordinate System," *Proceedings of the Symposium on Transient Ground-Water Hydraulics*, Colorado State University, Fort Collins, Colo., 1963, pp. 27–32.

Kashef, A. I.: "Exact Free Surface of Gravity Wells," *Proc. Am. Soc. Civ. Eng.*, vol. 91, no. HY4, 1965a, pp. 167–184.

Kashef, A. I.: "Seepage through Earth Dams," *J. Geophys. Res.*, vol. 70, no. 24, 1965b, pp. 6121–6128.

Kashef, A. I.: *Critical Reviews in Environmental Control*, vol. 7: *Management and Control of Salt-Water Intrustion in Coastal Aquifers*, CRC Press, Cleveland 1977, 217–275.

Kashef, A. I., and M. M. Safar: "Comparative Study of Fresh-Salt Water Interface Using Finite Elements and Simple Approaches," *Water Res. Bull.*, vol. 11, no. 4, 1975, pp. 651–665.

Lane, E. W.: "Security from Under-Seepage Masonry Dams on Earth Foundations," *Trans. Am. Soc. Civil Eng.*, vol. 100, 1935, pp. 1235–1351.

Maasland, D. E. L., and M. W. Bittinger (eds.): *Proceedings of the Symposium on Transient Ground Water Hydraulics*, Colorado State Univ., Fort Collins, Colo., 1965.

Muskat, M.: *The Flow of Homogeneous Fluids through Porous Media*, McGraw-Hill, New York, 1937; 2d printing, 1946, by J. W. Edwards, Ann Arbor, Mich.

Polubarinova-Kochina, P. Y.: *Theory of Groundwater Movement*, translated by R. J. M. DeWiest from the 1952 Russian edition, Princeton University Press, Princeton, N.J., 1962.

Prickett, T. A.: "Ground-Water Computer Models — State of the Art, *Ground Water*, vol. 17, 1979, pp. 167–173.

Richardson, L. F.: "A Freehand Graphical Way of Determining Stream Lines and Equipotentials," *Philosophical Mag. Ser. VI (Lond.)*, vol. 15, 1908, pp. 237–250.

Rushton, K. R., and S. C. Redshaw: *Seepage and Groundwater Flow, Numerical Analysis by Analog and Digital Methods*, Wiley, New York, 1979.

Southwell, R. V.: *Relaxation Methods in Engineering Science*, Oxford University Press, London, 1940.

Southwell, R. V.: *Relaxation Methods in Theoretical Physics*, Oxford University Press, London, 1946.

Taylor, D. W.: *Fundamentals of Soil Mechanics*, Wiley, New York, 1948.

Terzaghi, K., and R. B. Peck: *Soil Mechanics in Engineering Practice*, 2d ed., Wiley, New York, 1967.

Todd, D. K.: *Groundwater Hydrology*, 2d ed., Wiley, New York, 1980.

Yang, S. T.: "Seepage toward a Well Analyzed by the Relaxation Method," Ph.D. thesis, Harvard University, Cambridge, Mass., 1949.

Zienkiewicz, O.: The Finite Element Method in Engineering Science, McGraw-Hill, New York, 1971.

6

HYDROLOGIC PARAMETERS AND RELATED EQUATIONS

Derivations of the mathematical equations that govern different ground-water flow systems and their solutions are covered in several textbooks and scientific papers. In this chapter emphasis is placed on the following:

1. The assumptions and simplifications made in the derivation and solution of these equations

2. The physical significance of the various coefficients included in these equations

3. The practical conditions under which these equations can be used

Once these fundamentals are comprehended, the professional or re-searcher can select the most suitable approach to deal with the problem at hand or come up with an innovative approach.

At present, simple and complex groundwater flow problems are usually solved by computers using numerical analysis (see Chap. 5). The governing equations and boundary conditions must be carefully under-stood in order to properly use these computerized solutions. The choice of a method, such as the finite-difference or finite-element method or any other approach, depends on the type of problem to be solved. Nu-merical analysis has been used successfully in solving complex equations that were not in the past amenable to closed-form mathematical solutions without approximation. For example, under steady-state conditions, the

Laplace equation [Eq. (5.3)] should be used as a basis for the selected numerical method rather than an equation with assumptions such as Eq. (5.33), which is based on Dupuit's assumptions. In certain numerical solutions, it may be possible to eliminate some of the theoretical assumptions implied in the basic equations. Numerical analysis can, for example, theoretically accommodate the changes in hydraulic conductivity within a flow medium rather than simply use an average value throughout the medium; however, it is not feasible to determine these changes in the field.

Most of the general mathematical equations include certain coefficients that are assumed constant with respect to time and space. These coefficients are analyzed to check the validity of such assumptions.

6.1 General Hydrodynamic Equations

With respect to three-dimensional systems with cartesian coordinates x, y, and z, the general equation for groundwater flow in homogeneous and isotropic media under transient conditions of flow is written as follows:

$$\frac{\delta^2\phi}{\delta x^2} + \frac{\delta^2\phi}{\delta y^2} + \frac{\delta^2\phi}{\delta z^2} = \frac{n\gamma_w}{K}\left(\beta_w + \frac{m_v}{n}\right)\frac{\delta\phi}{\delta t} = \frac{S}{T}\frac{\delta\phi}{\delta t} \tag{6.1}$$

where ϕ = velocity potential at point x, y, z, which is equivalent to $-Kh$ + constant

γ_w = unit weight of water

t = time

K = average hydraulic conductivity of the flow medium

n = *total* porosity of soil

β_w = water compressibility ($\approx 4.69 \times 10^{-8}$ cm^2/g), which is the reciprocal of its bulk modulus E_w ($\approx 20,000$ kg/cm^2)

m_v = vertical compressibility of soil skeleton

S = coefficient of storage

T = coefficient of transmissivity, which is equivalent to KD

x, y = axes in horizontal planes

z = vertical axis

$$\frac{S}{T} = \frac{n\gamma_w}{K}\left(\beta_w + \frac{m_v}{n}\right) = \frac{S_s}{K} \approx \frac{1}{v} \approx \frac{1}{c_v} \tag{6.2}$$

S_s = specific storage

v = soil diffusivity, which is equivalent to $T/S = K/S_s \approx c_v$

c_v = coefficient of consolidation of the soil, which is equivalent to $K/\gamma_w m_v$ [Eq. (3.19)]

$$S = n\gamma_w\left(\beta_w + \frac{m_v}{n}\right)D_q \tag{6.3}$$

where D_q is the total thickness of the artesian aquifer or average saturated depth of an unconfined aquifer $(\overline{D}_q)$.

$$S_s = S/D_q \simeq \gamma_w m_v$$

The coefficients S, S_s, m_v, β_w, c_v, and T are discussed in greater detail
later in this chapter.

In cylindrical coordinate systems (r, θ, z), Eq. (6.1) becomes

$$\frac{1}{r}\frac{\delta}{\delta r}\left(r\frac{\delta\phi}{\delta r}\right) + \frac{1}{r^2}\frac{\delta^2\phi}{\delta\theta^2} + \frac{\delta\phi}{\delta z^2} = \frac{S}{T}\frac{\delta\phi}{\delta t} \tag{6.4}$$

where $r =$ radius $= \sqrt{x^2 + y^2}$
$\theta = \tan^{-1}(y/x)$

Equations (6.1) and (6.4) do not have simple closed-form solutions.
Equation (6.1) is usually applied to two-dimensional cases in which one
of the three axes x, y, or z is eliminated. These two-dimensional cases are
general approximations of the actual three-dimensional case. For example, if seepage in a unit length of an earth dam is analyzed, the flow
pattern in a certain section is assumed to be similar to that in any other
parallel section, except of course in the end abutments, where the flow
becomes three-dimensional. Designers usually rely on experience, testing, and sound judgment in analyzing these end sections. Equation (6.4)
governs the radial flow toward a well. If axisymmetric flow is assumed,
which is a common simplification, the circumferential velocities $\delta\phi/\delta\theta$
are eliminated. This is achieved by assuming a vertical well with horizontal aquifer boundaries and by assuming that the initial water table or
piezometric surface is horizontal. If one or all of these surfaces are somewhat sloping, the velocity $\delta\phi/\delta\theta$ can still be neglected as compared with
the magnitudes of the velocities $\delta\phi/\delta x$ and $\delta\phi/\delta y$.

In two-dimensional systems, Eqs. (6.1) and (6.4) reduce to

$$\frac{\delta^2\phi}{\delta x^2} + \frac{\delta^2\phi}{\delta y^2} = \frac{S}{T}\frac{\delta\phi}{\delta t} \tag{6.5}$$

and

$$\frac{1}{r}\frac{\delta}{\delta r}\left(r\frac{\delta\phi}{\delta r}\right) + \frac{\delta^2\phi}{\delta y^2} = \frac{S}{T}\frac{\delta\phi}{\delta t} \tag{6.6}$$

Equations (6.1), (6.4), (6.5), and (6.6) are written in terms of the
velocity potential ϕ. They may also be written in terms of the total head h
or the drawdown s_d, which is equal to $(h_e - h)$, where h_e is the initial
height of the water table or the piezometric surface above the lower
impervious base. In two-dimensional equations [for example, Eqs. (6.5)
and (6.6)], the vertical axis is commonly considered y rather than z.

Equation (6.5) describes two-dimensional flow, such as seepage
through dams and toward subdrains (buried pipes used in agriculture

and highways for water drainage). Equation (6.6) is generally used in gravity-flow wells (water-table wells) after certain adjustments for the proper value of S.

Under steady-state conditions, $\delta\phi/\delta t = 0$; then by using total head h rather than velocity potential ϕ,

$$\frac{\delta^2 h}{\delta x^2} + \frac{\delta^2 h}{\delta y^2} = 0 \qquad \text{(Laplace equation)} \qquad (5.3)$$

and

$$\frac{1}{r}\frac{\delta}{\delta r}\left(r\frac{\delta h}{\delta r}\right) + \frac{\delta^2 h}{\delta y^2} = 0 \qquad (6.7)$$

If flow is vertical, Eq. (6.5) reduces (after replacing ϕ by h) to

$$\frac{\delta^2 h}{\delta y^2} = \frac{S}{T}\frac{\delta h}{\delta t} \qquad (6.8)$$

and from Eq. (6.2), we get

$$\frac{\delta^2 h}{\delta y^2} \approx \frac{1}{c_v}\frac{\delta h}{\delta t} \qquad (6.9)$$

Equation (6.9) is the well-known one-dimensional vertical consolidation equation (Sec. 6.2). It is time-dependent, and its solution gives the rate of settlement under a constant or variable downward effective pressure produced by a structure or by well pumping.

A unidirectional groundwater flow system in a designated direction under steady-state conditions is governed by

$$\frac{\delta^2 h}{\delta x^2} = 0 \qquad \text{and} \qquad \frac{\delta^2 h}{\delta y^2} = 0 \qquad (6.10)$$

The flow toward artesian wells under transient conditions is governed by

$$\frac{1}{r}\frac{\delta}{\delta r}\left(r\frac{\delta h}{\delta r}\right) = \frac{\delta^2 h}{\delta r^2} + \frac{1}{r}\frac{\delta h}{\delta r} = \frac{S}{T}\frac{\delta h}{\delta t} \qquad (6.11)$$

or

$$\frac{1}{r}\frac{\delta}{\delta r}\left(r\frac{\delta s_d}{\delta r}\right) = \frac{\delta^2 s_d}{\delta r^2} + \frac{1}{r}\frac{\delta s_d}{\delta r} = \frac{S}{T}\frac{\delta s_d}{\delta t} \qquad (6.11a)$$

The solution gives the drawdown s_d at any radius r and time t since the start of pumping at a constant rate of flow Q.

Steady-state flow toward artesian wells is governed by

$$\frac{1}{r}\frac{\delta}{\delta r}\left(r\frac{\delta s_d}{\delta r}\right) = \frac{\delta^2 s_d}{\delta r^2} + \frac{1}{r}\frac{\delta s_d}{\delta r} = 0 \qquad (6.12)$$

It should be realized that Eqs. (6.4) through (6.12) are all based on Eq. (6.1) by either transforming it to cylindrical coordinates or by introducing $\delta\phi/\delta t = \delta h/\delta t = \delta s_d/\delta t = 0$ when the flow is steady. The assumptions made in deriving Eq. (6.1) thus provides insight into the limitations imposed on using Eqs. (6.4) through (6.12); these are explained in the following paragraphs.

In deriving Eq. (6.1), a small element $[(\Delta x)(\Delta y)(\Delta z)]$ is considered (see, for example, Jacob, 1950). The continuity condition is based on the natural law of conservation of matter; that is, the net rate of flow entering and leaving the prism is equal to the water released from or acquired by storage. The release of water from storage is produced by the decrease in pore-water pressures and consequently the increase in effective pressures (Sec. 3.9). Thus a prism $[(\Delta x)(\Delta y)(\Delta z)]$ is compressed when its interior effective pressure increases. This compressibility leads to a decrease in the volume of voids, thus releasing water out of the prism. Also, the decrease in water pressure allows the volume of water occupying the pores of the prism to expand, thus releasing an additional amount of water. *Theoretically speaking, a reverse action of the same magnitude would take place as a result of an increase in pore-water pressure by water recharge or cessation of well pumping. This assumption, accepted by some groundwater hydrologists, is incorrect according to laboratory consolidation tests* (Secs. 3.4 and 6.2). Pressure release and further reapplication of pressure leads to hysteresis loops; total recovery of water levels or pressures cannot therefore take place (Sec. 6.2).

The flow medium is assumed homogeneous and isotropic in deriving Eq. (6.1). However, in limited cases of anisotropy, where the hydraulic conductivities at any point in the three directions x, y, or z are, respectively, K_x, K_y, or K_z and are the same at any other point in the medium, Eq. (6.1) changes to

$$K_x \frac{\delta^2 s_d}{\delta x^2} + K_y \frac{\delta^2 s_d}{\delta y^2} + K_z \frac{\delta^2 s_d}{\delta z^2} = S_s \frac{\delta s_d}{\delta t} \qquad (6.13)$$

This equation can be transformed to a cylindrical coordinate system or to any other special form [see, e.g., Eqs. (5.8) and (5.9)].

In deriving Eq. (6.1), Darcy's law is assumed to be valid; this implies that the flow medium is saturated and that flow is laminar. In addition, lateral deformation of the soil skeleton is neglected, assuming that total deformation occurs only in a vertical direction. This deformation of solids is in fact due to the decrease in pore-space size rather than to the elasticity of the solid grains. In groundwater hydrology, the effective stresses produced by the change in pore-water pressure are relatively small. In sands, for example, testing showed (Terzaghi and Peck, 1967) that the sand grains start to be crushed at an effective pressure of about 100 kg/cm² (this pressure corresponds to a water pressure head of about 1000 m). Therefore, the elasticity of the individual soil grains is neglected.

The change in pore-water pressure is assumed to be numerically equal to the change in effective stress in deriving Eq. (6.1). This is true only when the total stress is constant, such as in confined aquifers. Equations (6.1) through (6.6) should therefore be restricted to confined aquifers with constant depths. These equations are also still used for gravity wells after introducing certain adjustments (Chap. 8).

The effects of evaporation, recharge, and leakage are not included in Eq. (6.1). These effects are discussed further in Sec. 6.4.

6.2 Compressibility of Soils

The theoretical, experimental, and practical properties of the compressibility of soils (especially completely saturated clay layers) was investigated about 50 years ago by Terzaghi (1943), and his theory is still used. A consolidation test is conducted on an undisturbed soil trimmed to fit a solid steel ring in the apparatus. The clay sample is confined between two porous disks to simulate a clay layer in the field overlain and underlain by sand layers. Under pressure, the water drains out of the sample into these porous disks as a result of the compressibility of the sample. The test results can also be applied to other conditions in which the clay layer is overlain by a sand layer but underlain by a practically impervious material.

The pressure $\bar{\sigma}$ is applied to the sample in steps. After each increment of load, the deformation is measured over time until it practically ceases (in normal clays, one loading may take about a day). The results are given in a curve similar to that shown in Fig. 6.1 (Peck et al., 1974). Most of the upper portion of this curve resembles the theoretical curve shown in Fig. 6.2, which is a grahic solution to Eq. (6.9). The ordinates of the curve give the degree of consolidation U_c in percent, which may be regarded as approximately the amount of settlement that has occurred after a known time t compared with the ultimate settlement S_u that takes place after time $t = \infty$, at which the compressibility ceases under a certain pressure. The ultimate settlement S_u can easily be computed as explained later. The exact nature of U_c is explained in the geotechnical literature (see, for example, Taylor, 1948). The abscissa of Fig. 6.2 is given in terms of the logarithm of a dimensionless time factor T_v defined as follows:

$$T_v = \frac{c_v}{H_p^2} t \tag{6.14}$$

where t = real time, min
H_p = maximum flow path of the drained water
c_v = the coefficient of consolidation determined from the test [see Eq. (3.19)], m²/min

H_p has only two values: $H_p = D_a$ when the clay layer is bounded by a sand layer at one side and drainage is prevented at the other; otherwise $H_p = \frac{1}{2}D_a$ when drainage is possible at both the top and bottom surfaces of the clay layer.

The lower limb of the experimental curve shown in Fig. 6.1 dips downward, indicating a deviation from theory where the lower portion of the curve (Fig. 6.2) is asymptotic to the horizontal. This is due to what is known as *secondary consolidation* (Peck et al., 1974), which is not considered in deriving Eq. (6.9) and generally is due to plastic deformation. Comparing the theoretical curve (Fig. 6.2) with the upper portion of the experimental curve, c_v and K can be computed on the basis of a factor m_v known as the *coefficient of compressibility*, which is defined as follows:

$$m_v = -\frac{dn}{d\bar{\sigma}} \tag{6.15}$$

where n is the porosity of clay. The minus sign indicates that n decreases when $\bar{\sigma}$ increases, and

$$c_v = \frac{K}{\gamma_w m_v} \tag{3.19}$$

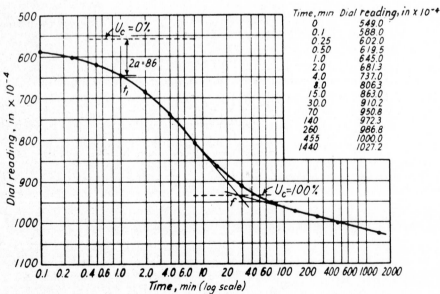

Figure 6.1 Rate of consolidation determined by testing. (*R. B. Peck et al., Foundation Engineering, 2d ed., Wiley, New York, 1974.*)

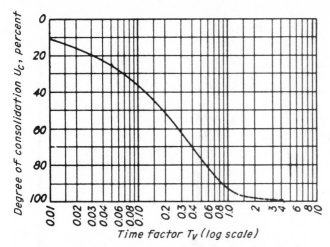

Figure 6.2 Theoretical relationship between degree of consolidation U_c and time factor T_v. (R. B. Peck et al., *Foundation Engineering*, 2d ed., Wiley, New York, 1974.)

The test is repeated on the same sample adding successive increments of loads (about six or more). Curves similar to that shown in Fig. 6.1 are obtained. In each test under a load increment, c_v, m_v, and K are computed. Although these values are *assumed constants*, they vary with pressure increments because of the decrease in n and hence K. The average values of c_v and K determined from the test should correspond to the range of test pressures equivalent to those in the field.

The vertical compressibility of a clay layer reduces the pore volume, thus decreasing K under high pressures. In addition, the porosity decreases, affecting m_v [Eq. (6.15)]. Therefore, under actual conditions, m_v and K vary with pressure and time. The coefficient of consolidation c_v should therefore also vary with time. The assumption that c_v is a constant is not realistic unless the rates of decrease in both m_v and K are such that the ratio K/m_v remains approximately constant under all loads.

For each increment of load, the test curve (Fig. 6.1) indicates that the deformation practically ceases at a final void ratio e that corresponds to the applied load. After conducting six or more tests, a synthetic curve K_u giving the relationship between stress and the corresponding final void ratio e can be plotted as shown in Fig. 6.3. If at a certain pressure $\overline{\sigma}'_o$ (Fig. 6.4) the load is gradually decreased, a rebound curve mm' results. If loading is resumed, a reloading curve $m'm$ results (slightly above the rebound curve). The same trend takes place at any other pressure $\overline{\sigma}''_o$.

The unloading and reloading characteristics in the experiments indicate that the preloading field conditions should be taken into account. In

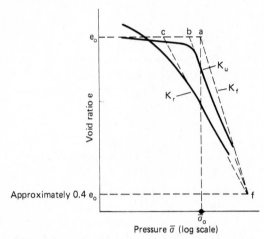

Figure 6.3   Typical *e* versus log $\bar{\sigma}$ curves for undisturbed *(K_μ)* and remolded *(K_r)* samples of normally loaded clay of low sensitivity. *(R. B. Peck et
al., Foundation Engineering, 2d ed., Wiley, New
York, 1974.)*

the field, *preconsolidation pressures* mean that the soil layers were subjected to pressures at some time in the past that exceeded the pressures
at the present time. Preconsolidation of soil may be due to the removal of
loads that existed in the past, such as demolition of structures, thawing of
geologic glaciers, removal of soil by scour action, and even the effects of
soil desiccation by heat. If the present pressure is the maximum that the
clay layer has even been subjected to, the clay is known as *normally
loaded.* The magnitudes of preconsolidation pressures are determined

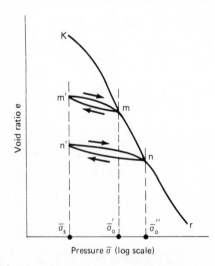

Figure 6.4   Typical *e* versus
log $\bar{\sigma}$ curve for clay remolded
near liquid limit. *(R. B. Peck et
al., Foundation Engineering, 2d
ed., Wiley, New York, 1974.)*

from geologic investigations as well as from study of the results of consolidation tests (Terzaghi and Peck, 1967).

The field curve K_f, the undisturbed curve K_u, and the remolded curve K_r (which is similiar to K_u but obtained from a consolidation test on a sample remolded at the same water content as that of the undisturbed sample) are shown in Fig. 6.3 relative to each other. They all meet at a void ratio $e \simeq 0.4e_o$, where e_o is the natural (in situ) void ratio of the undisturbed soil (Terzaghi and Peck, 1967). On this basis, the field curve K_f is obtained. The major portions of the preceding curves are straight lines, and thus K_f should also be a straight line between point f and point a that corresponds to the existing undisturbed condition with coordinates e_o and $\bar{\sigma}_o$. Point f is located from either the K_u or the K_r curves, which are the result of consolidation tests. Point a is located on the plot in terms of e_o and $\bar{\sigma}_o$. The pressure $\bar{\sigma}_o$ is the effective pressure in the field at the point from which the undisturbed sample was obtained and is calculated from the weights of overburden pressures (Sec. 3.9).

The slope of the K_f curve (Fig. 6.3) is known as the *compression index* C_{co} (dimensionless), and

$$C_{co} = \frac{\Delta e}{\Delta(\log \bar{\sigma})} \tag{6.16}$$

or $\qquad C_{co} = \Delta e$ corresponding to one log cycle of $\bar{\sigma}$

C_{co} ranges usually between 0.1 and 1.0 for clays. In a remolded state, C'_{co} corresponds to the remolded curve K_r and is given approximately by (Terzaghi and Peck, 1967) $C'_{co} = 0.007(w_L - 10)$, and for undisturbed clays, C_{co} is given by $C_{co} = 1.3C'_{co}$; then

$$C_{co} \simeq 0.009(w_L - 10) \tag{6.17}$$

where w_L is the liquid limit of the clay (Sec. 1.3).

If the initial void ratio is e_o and the initial overburden pressure is $\bar{\sigma}_o$, then the total pressure is $\bar{\sigma}_o + \Delta\bar{\sigma}_o$, where $\Delta\bar{\sigma}_o$ is the added pressure increment corresponding to a structural load or to an increase in effective pressure due to pumping (Sec. 8.11). The ultimate settlement S_u due to $\Delta\bar{\sigma}_o$ is determined as follows:

$$S_u = \Delta n D_a = \frac{e_o}{1 + e_o} D_a = \frac{C_{co}}{1 + e_o} D_a \log \frac{\bar{\sigma}_o + \Delta\bar{\sigma}_o}{\bar{\sigma}_o} \tag{6.18}$$

Once the total settlement is computed, U_c is determined at any time $t = T_v H_p^2 / c_v$ [Eq. (6.14)], and the rate of settlement is determined on the basis of the theoretical curve (Fig. 6.2).

As previously mentioned, the added effective pressure increment $\Delta\bar{\sigma}_o$ may be produced by a structural building or by reducing water pressures by pumping. Consequently, land subsidence (Sec. 8.11) takes place when the effective stresses are increased due to pumping or other causes.

In the case of land subsidence due to pumping, the reduction in water pressure and, consequently, the increase in $\Delta\bar{\sigma}$ is continuous over time, whereas in structural engineering, $\Delta\bar{\sigma}$ has only one value (the stress due to a structure). The determination of c_v and K is best illustrated by the following example.

EXAMPLE 6.1[*]   The time-dial reading curve shown in Fig. 6.1 was obtained from a consolidation test on a soft glacial clay (liquid limit $w_L = 43$ percent, plastic limit $w_p = 21$ percent, and natural water content $w_n = 39$ percent) when the pressure was increased from 1.66 to 3.33 kg/cm². The void ratio after 100 percent consolidation of the sample under 1.66 kg/cm² was 0.945 and that under 3.33 kg/cm² was 0.812. The dial was set at zero at the beginning of the test. If the initial height of the sample was 0.75 in (1.9 cm) and drainage was permitted at both faces of the sample, compute the hydraulic conductivity K corresponding to the stated increment of pressure.

Solution   To determine dial reading R_d at 0 percent (corresponding to $U_c = 0$ percent of the theoretical curve; Fig. 6.2), select *any* time t_1 near the upper portion of the curve. The difference in dial readings corresponding to t_1 and $\frac{1}{4}t_1$ is designated as a (Fig. 6.5). The $U_c = 0$ percent line that corresponds to dial reading R_d is determined by scaling a vertical distance a above point d (Fig. 6.5) or $2a$ above point c. Selection of time t_1 is arbitrary on condition that it should correspond to a deformation less than $U_c = 50$ percent (this is based on the characteristics of the upper portion of the theoretical curve; Fig. 6.2).

Select $t_1 = 1.0$ min (Fig. 6.1). Thus when $t_1 = 1.0$ min, R_d[†] $= 645$ and when $t_1/4 = 0.025$ min, $R_d = 604$; therefore, $a = 645 - 602 = 43$, and $2a = 86$. Consequently, $U_c = 0$ percent takes place at $R_d = 645 - 86 = 559$, and $U_c = 100$ percent takes place at $R_d = 936$ (this is determined graphically from Fig. 6.1 by locating the point of intersection f of the tangent at the point of inflection of the experimental curve with the extension of the lower straight-line portion of the curve). Then at $U_c = 50$ percent, $R_d = 559 + (936 - 559)/2 = 748$, and $t = 4.5$ min $= 270$ s.

The thickness of the sample is $\frac{1}{2}(0.7500 - 0.0748) = 0.3376$ in (after it was compressed under the load increment). And since $c_v = T_v H_p^2/t$ [Eq. (6.14)] and $T_v = 0.197$ (from Fig. 6.2 corresponding to $U_c = 50$ percent) (note that any other U may be used — preferably 50 percent or less — and the appropriate T_v is read from Fig. 6.2), then

$$c_v = \frac{0.197}{270}(0.3376 \times 2.54)^2 = 5.37 \times 10^{-4} \text{ cm}^2/\text{s}$$

$$m_v = \frac{\Delta n}{\Delta\bar{\sigma}} = \frac{\Delta e}{(1 + e_o)\Delta\bar{\sigma}} = \frac{0.945 - 0.812}{3330 - 1660} \times \frac{1}{(1 + 0.945)}$$

$$= 4.09 \times 10^{-5} \text{ cm}^2/\text{g}$$

and $K = c_v \gamma_w m_v = (5.37 \times 10^{-4}) \times 1 \times (4.09 \times 10^{-5}) = 2.20 \times 10^{-8} \text{ cm/s}$

---

[*] Example 6.1 is based on Example 2, page 74, in Peck et al. (1974).
[†] All R_d values in this example are given in inches $\times 10^{-4}$ (Peck et al., 1974).

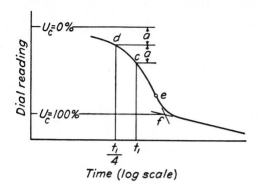

Figure 6.5 Graphic construction for determining deformations corresponding to $U_c = 0$ percent and $U_c = 100$ percent in consolidation test. (*R. B. Peck et al., Foundation Engineering, 2d ed., Wiley, New York, 1974.*)

The preceding summary does not include other consolidation theories, such as those for two- and three-dimensional consolidation (Biot, 1941) or consolidation of anisotropic soils (Biot, 1955). In addition, $e - \log \bar{\sigma}$ curves for preconsolidated soil are not discussed (Terzaghi and Peck, 1967). These occur in groundwater hydrology as a result of, for example, subsequent pumping and shutdowns of water wells. Moreover, the summary does not include the details of testing and the consolidation characteristics of other soil types, such as organic, sensitive, residual, and collapsible soils. These details are found in the literature on soil mechanics (geotechnical engineering) and site investigations for engineering foundations (see, e.g., Terzaghi and Peck, 1967; Bowles, 1978; American Society for Testing and Materials, 1982; and the references therein).

Probably 90 percent of the total amount of groundwater used in the United States is withdrawn from sands and gravels (Walton, 1970). The consolidation characteristics of these cohesionless soils (Fig. 6.6) have some resemblance to those of preloaded clays, but with small values of C_{co} (Peck et al., 1974). The compressibility of loose sands is very excessive as compared with that of dense sands. Under very high pressures, the void ratio or porosity of loose sands may not be reduced to that of *natural* dense sands of the same gradation. The existence of small percentages of plate-shaped particles (such as mica or clay) can greatly increase the compressibility (Terzaghi and Peck, 1967). Well-graded sands are less compressible than uniform or gap-graded sand. Sandy soils with rounded particles of sand are usually less compressible than other soils of comparable gradation. From Fig. 6.6, $C_{co} \simeq 0.17$ for loose sands and about 0.05 for dense sands. Land subsidence above sandy aquifers occurs mainly as a result of the compressibility of the aquitards above or below the aquifers. The aquifers subside only a small fraction compared with the aquicludes or aquitards.

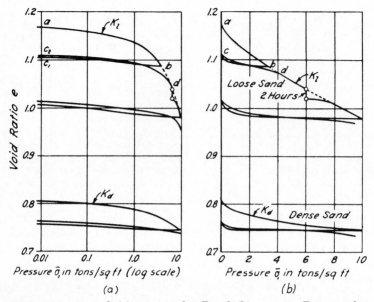

Figure 6.6 Typical *(a)* e versus log $\bar{\sigma}$ and *(b)* e versus $\bar{\sigma}$ curves for initially loose (K_e) and initially dense (K_d) laterally confined sand (10 tons per square foot is equivalent to 9.765 kg/cm²). *(R. B. Peck et al., Foundation Engineering, 2d ed., Wiley, New York, 1974.)*

6.3 Storage Coefficient, Specific Yield, Soil Diffusivity, and Transmissivity

Storage Coefficient

The coefficient of storge S is given in Eqs. (6.1) through (6.6) and in Eqs. (6.8) and (6.11). In groundwater hydraulics, S is assumed constant and is normally determined by field tests (Chap. 8). S is given analytically as follows (Jacob, 1950):

$$S = n\gamma_w \left(\beta_w + \frac{m_v}{n} \right) D_q \tag{6.19}$$

The term β_w in Eq. (6.19) is the water compressibility or the reciprocal of its bulk modulus E_w and can be expressed as

$$\beta_w = \frac{dp/\rho_o}{dp} \simeq \frac{d\gamma_w/\gamma_w}{dp} = \frac{-d(\Delta V_w)/\Delta V_w}{dp} = \frac{dn/n}{dp} \tag{6.20}$$

where ΔV_w = the volume of water in an elemental volume V
ρ = the water mass density
ρ_o = a reference unit mass of water conveniently taken at atmospheric pressure

The average value of β is 4.69×10^{-8} cm²/g (3.3×10^{-6} in²/lb), and therefore,

$$\frac{1}{\beta_w} = E_w \approx 2 \times 10^7 \text{ g/cm}^2 = 47.8 \text{ m}^2/\text{kN} \approx 20,000 \text{ kg/cm}^2$$

Other symbols D_q, n, γ_w, and p have already been defined. The value of m_v may be determined experimentally as explained in the previous section. The value of the porosity n may be indirectly computed from the volume-weight relationships for undisturbed samples (Sec. 1.3). However, in sands and other granular materials, undisturbed samples are difficult and expensive to obtain. Equation (6.19) would then serve only to clarify the physical nature of S rather than to compute it.

From Eqs. (6.15), (3.19), and (6.19), we have

$$S = n\gamma_w\beta D_q + \gamma_w m_v D_q = \frac{d\gamma_w}{dp}nD_q - \gamma_w\frac{dn}{d\bar{\sigma}}D_q$$

Considering a vertical prism of height D (the entire thickness of the aquifer) and of a unit cross section in a confined aquifer (Fig. 6.7a), then $D_q = V$ numerically, $n = V_v/V = V_w/V$ (in a saturated medium), and $dn = dV_w/V$. Since $dp = -d\bar{\sigma}$, then the preceding equation reduces to

$$S = \frac{V_w d\gamma_w + \gamma_w dV_w}{dp} = \frac{V_w d\gamma_w + \gamma_w dV_w}{\gamma_w dh} = \frac{d(V_w\gamma_w)}{\gamma_w dh}$$

$$= \frac{dV_w}{dh}$$

(6.21)

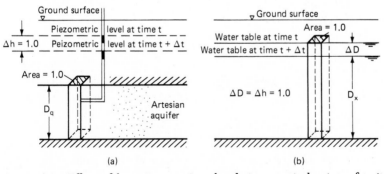

(a)   (b)

Figure 6.7 Effect of lowering pressure heads in a vertical prism of unit cross-sectional area. *(a)* Artesian aquifer; *(b)* unconfined aquifer.

For a unit value of dh, $S = dV_w$, which is the change in the volume of water in the pores of the prism under consideration owing to a change of unit total head. Thus S is defined as "the volume of water released from storage, from a soil prism of a unit cross-sectional area and a depth equal to the entire depth D_q of the aquifer, due to a unit change in the hydraulic head" (Jacob, 1950). The volume of water released from storage consists of two components: water expansion due to the compressibility characteristics of water and water released due to the decrease in pore sizes resulting from the compressibility of the soil skeleton. The first portion has a percentage of $D_q n \gamma_w \beta_w / S$ with respect to the total amount S [Eq. (6.19)], while the second portion has the balance of $D_q \gamma_w m_v / S$. The ratio of the volume of water released from storage due to water expansion to that released due to the compressibility of the soil skeleton is then equal to $n\beta_w / m_v$. Considering that the range of m_v for cohesionless materials between 1×10^{-5} and 5×10^{-7} cm^2/g and that the porosity n varies from 20 to 40 percent, $n\beta / m_v$ therefore has a range between a maximum of 0.04 and a minimum of 0.001 ($\beta_w \simeq 5 \times 10^{-8}$ cm^2/g). Depending on the characteristics of the aquifer, the percent ratio $(n \times 100\beta_w)/m_v$ would thus vary between 0.1 and 4 percent.

The coefficient of specific storage S_s is defined as

$$S_s = \frac{S}{D_q} = n\gamma_w \left(\beta + \frac{m_v}{n} \right) \tag{6.22}$$

The S_s unit is cm^{-1}. This coefficient may be useful in conducting accurate analyses where m_v and n vary with depth and time. Although the use of S_s as a parameter is uncommon, it is more logical to use than S, which depends on the depth of aquifer D_q. In comparing two aquifers of the same soil characteristics but with different depths, S_s would be constant for both, while S would vary.

The range of S values in confined aquifers varies between 1×10^{-5} and 1×10^{-3} (Walton, 1970) for all soils. In productive aquifers, the range varies between 5×10^{-5} and 1×10^{-2}.

Specific Yield

In unconfined aquifers, a drawdown produces water drainage by gravity through height ΔD (Fig. 6.7b) and water release from storage from within height D_x. It has already been shown that wells, springs, drains, and seeps cannot withdraw all the water existing in the soil pores (Sec. 1.3). Water is partially retained around the particles by the forces of molecular attraction between the solid surfaces of these particles and the water molecules (adhesion) and between the water molecules close to these solid surfaces (cohesion). The ratio of the volume of water retained in a soil mass to the total volume of this mass is known as the *specific retention.* The difference between the porosity n and the specific retention after drainage by gravity is completed is known as the *specific yield* S_y. The water yield by gravity produced by a certain drawdown is not instantaneous. The yield capacity therefore increases gradually with time from zero to its maximum value S_y during the drainage process. The intermediate stages of water yield indicate that the intermediate delayed yield S_{ya} is less than S_y, which is also known as the *gravity yield* (Rasmussen and Andreasen, 1959). In practice, S_{ya} is used as the ultimate yielding capacity S_y, which is assumed constant. Its value is determined by field tests (Chap. 8) and, in fact, depends on the period of the test. The portion of the prism below the lowering water table after its drawdown (Fig. 6.7b) is also affected by the drop in head Δh in the same manner as in confined aquifers. Therefore, the total amount of water released from storage in unconfined aquifers after a certain time corresponds to a storage coefficient $S_{t,y}$ in which

$$S_{t,y} \simeq S_{ya} + S$$
$$S_{t,y} = S_y + \gamma_w nD\left(\beta + \frac{m_v}{n}\right) \simeq S_y \tag{6.23}$$

In unconfined aquifers, the value of S_y varies between 0.02 and 0.30, and therefore, S in Eq. (6.23) is negligible as compared with S_y ($\simeq S_{ya}$). It is sufficient to use S_y instead of $S_{t,y}$. If the field tests conducted in unconfined aquifers are reliable (Chap. 8), the determined value of S_y represents $S_{t,y}$ [including S in Eq. (6.23)] under the conditions of testing.

The equation governing the flow toward gravity wells in unconfined aquifers is similar to Eq. (6.6), replacing S by $S_{t,y} \simeq S_y$, and in terms of the drawdown s_d is given by

$$\frac{1}{r}\frac{\delta}{\delta r}\left(r\frac{\delta s_d}{\delta r}\right) + \frac{\delta^2 s_d}{\delta y^2} \simeq \frac{S_y}{T_c}\frac{\delta s_d}{\delta t} \tag{6.24}$$

where $T_c = K\overline{D}_q$. It should be noted that $\overline{D}_q$ in Eq. (6.24) is not constant.

If Eq. (6.1) is derived on the basis of analyzing a vertical element with a height equal to the depth of the *saturated zone* $\tilde{h}$, then by introducing Dupuit's assumptions, the following equation is obtained (Muskat, 1937):

$$\frac{\delta}{\delta x}\left(\tilde{h}\frac{\delta\tilde{h}}{\delta x}\right) + \frac{\delta}{\delta y}\left(\tilde{h}\frac{\delta\tilde{h}}{\delta y}\right) \simeq \frac{S_y}{K}\frac{\delta\tilde{h}}{\delta t} \tag{6.25}$$

This equation is nonlinear and known as the *Boussinesq equation* or the *Dupuit-Forchheimer equation*. When the spatial variation of $\tilde{h}$ remains small, Eq. (6.25) may be linearized in the following form:

$$\frac{\delta^2\tilde{h}}{\delta x^2} + \frac{\delta^2\tilde{h}}{\delta y^2} \simeq \frac{S_y}{K\overline{D}_q}\frac{\delta\tilde{h}}{\delta t} = \frac{S_y}{T_c}\frac{\delta\tilde{h}}{\delta t} \tag{6.26}$$

where T_c is the corrected value of T used in unconfined aquifers, or $T_c = K\overline{D}_q$, where $\overline{D}_q$ is the average thickness of the saturated zone. It is common to correct T determined from pumping tests in unconfined aquifers to T_c (Lohman, 1972).

Boulton treated gravity-flow wells in a somewhat different fashion, considering the delayed yield (Boulton, 1954a, 1954b, 1963, 1964). At the early time of pumping, the gravity well performance is similar to that of a leaky artesian well and $S_{t,y}$ is equal to S. At a later time when gravity drainage is completed, $S_{t,y}$ becomes equal to S_y (Sec. 8.10).

Soil Diffusivity and Transmissivity

In the geotechnical engineering field, Eq. (6.9) is extensively used in the analysis of settlement. It was derived in a manner similar to Eq. (6.6). Comparing Eqs. (6.8) and (6.9), we get

$$\frac{K}{\gamma_w m_v} = c_v \simeq \frac{T}{S} = v \tag{6.27}$$

Since c_v is determined experimentally on a small specimen (Sec. 6.2), the water compressibility β is practically negligible. Therefore, $c_v \simeq T/S = v$, where v is the soil diffusivity which combines the parameters T and S whether β_w is neglected or not.

The transmissivity T, or T_c, is the product of K and D_q, or $\overline{D}_q$. In practice, D_q varies in different locations, and the impervious boundary is not usually well defined because of the existence of transitional deposits of soils and rocks. Instead of finding D_q and K separately, Theis (1935) recommended the use of the coefficient of transmissivity T, that is,

$$T = KD_q \quad \text{(confined aquifers)} \tag{6.28}$$

and

$$T_c = K\overline{D}_q \quad \text{(unconfined aquifers)} \tag{6.29}$$

The coefficient T is normally determined directly from field tests (Chap. 8). This eliminates complex and expensive investigations to find D_q (or $\overline{D}_q$) and K separately.

6.4 Leakage and Effects of Rainfall and Evaporation

The layout shown in Fig. 6.8 is a hypothetical condition under equilibrium where the water table AB of the unconfined aquifer and the piezometric surfaces CD and EF of, respectively, the artesian aquifer and the aquiclude are all horizontal. It is assumed that no natural or artificial activities, such as water withdrawal or recharge, are taking place. It is also assumed that the water table AB coincides with the piezometric surface of the aquitard. The hydraulic conductivities K_a, K, and K'_a and thicknesses D_a, D_q and D'_a for the three formations below the unconfined aquifer are indicated in Fig. 6.8.

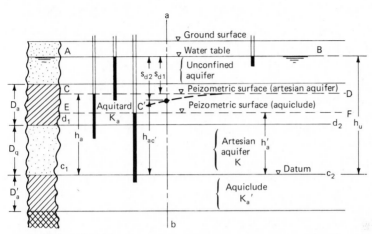

Figure 6.8 Schematic drawing to illustrate the mechanics of leakage.

It is always assumed that an aquiclude is impervious and that no water percolates across its boundary ($c_1 c_2$ in Fig. 6.8). Practically, aquicludes are porous structures that may be jointed or fractured, but their hydraulic conductivities are very small compared with other formations. The voids of these aquicludes are connected with those of the overlying artesian aquifer. The amount of water passing across $c_1 c_2$ is negligible; measurable amounts of water occur only after a relatively very long period of time during which the water levels and piezometric surfaces change their locations. Assuming that the water levels and piezometric surfaces are in equilibrium for a reasonable period of time, the occurrence of leakage can be illustrated.

All nonmetallic structures are porous (Chap. 2), and their voids are interconnected. Little amounts of water cross what are known as impervious boundaries, such as $d_1 d_2$ and $c_1 c_2$ of the artesian aquifers shown in Fig. 6.8. At any of these boundaries the movement may be upward or downward depending on the relative positions of AB, CD, and EF. Vertical seepage across the boundaries is known as *leakage*, and its magnitude can be determined by means of Darcy's law [Eq. (3.5)].

With reference to Fig. 6.8 and the indicated total heads h_u, h_a, and h'_a, leakage rate across $d_1 d_2$ is

$$v_a \downarrow = K_a \frac{h_u - h_a}{D_a} = K_a \frac{s_{d1}}{D_a} = \frac{s_{d1}}{R_a} \qquad \text{(downward)} \qquad (6.30)$$

where $R_a = D_a/K_a$ is known as the *hydraulic resistance* (in s^{-1}). Leakage across $c_1 c_2$ is

$$v_a' \downarrow = K_a' \frac{h_a - h_a'}{D_a'} \qquad \text{(downward)} \qquad (6.31)$$

Here v_a and v_a' have velocity dimensions. For example, if $K_a = 1 \times 10^{-3}$ cm/s and $K_a' = 1 \times 10^{-8}$ cm/s, v_a would be 100,000 times v_o' at the same gradients. Owing to the fact that v_a is already very small, v_o' is usually neglected, although it does exist (see also Chap. 9).

In the preceding analysis, water is assumed to leak at a steady state without affecting the existing water pressures. The aquitard then acts as a transmitting medium without suffering any compressibility due to the changes in water pressures within its boundaries. This is known as leakage without releasing water from storage from within the aquitard. Over a long period of time, large changes in hydraulic heads result in appreciable effective stresses, which in turn result in volume changes in the pores leading to water release from storage. Leakage under these conditions has two components: water flow due to head changes and water flow produced by the compressibility of the acquitard.

If there is a decline in the level CD at a certain vertical section ab due to such factors as pumping water out of the aquifer, then h_a reduces to $h_{ac'}$, the drawdown s_{d1} increases to s_{d2} (Fig. 6.8), and v_a changes to a larger value, or

$$v_a = K_a \frac{s_{d1}}{D_a} = \frac{s_{d1}}{R_a} \qquad (6.32)$$

A factor known as the *leakage factor* B_k (in centimeters) is introduced:

$$B_k = \sqrt{R_a K D_q} = \sqrt{\frac{K D_a D_q}{K_a}} = \sqrt{\frac{T D_a}{K_a}} \qquad (6.33)$$

Several procedures have been developed to include leakage (Chap. 8) with or without accounting for the water released from storage from the confining layer (aquitard). The general governing equation for flow toward a fully penetrating well in a leaky artesian aquifer without water released from storage within the aquitard (Jacob, 1946) is as follows:

$$\frac{\delta^2 s_d}{\delta r^2} + \frac{1}{r} \frac{\delta s_d}{\delta r} - \frac{s_d}{B_k^2} = \frac{S}{T} \frac{\delta s_d}{\delta t} \qquad (6.34)$$

When leakage and water pressures stabilize and reach the steady state, Eq. (6.34) reduces to

$$\frac{\delta^2 s_d}{\delta r^2} + \frac{1}{r} \frac{\delta s_d}{\delta r} - \frac{s_d}{B_k^2} = 0 \qquad (6.35)$$

The flow of water toward a fully penetrating well in a leaky artesian aquifer considering water released from storage within the aquitard is given by (Hantush, 1964; Walton, 1970):

$$\frac{\delta^2 s_d}{\delta r^2} + \frac{1}{r}\frac{\delta s_d}{\delta r} + \frac{K_a}{T}\frac{\delta s_{da}}{\delta y} = \frac{S}{T}\frac{\delta s_d}{\delta t} \tag{6.36}$$

where s_{da} is the drawdown in the aquitard. In Eqs. (6.34), (6.35), and (6.36), it is assumed that the water table (or piezometric surface) in the aquitard coincides initially with that of the main artesian aquifer.

Evaporation lowers the water table in unconfined aquifers. The same effect is caused by evapotranspiration in plants, especially when the water table is shallow and close to the ground surface. If a constant rate W_r (in centimeters per second) is estimated for rainfall, evaporation, or evapotranspiration, then in the general equations governing flow in unconfined aquifers [Eqs. (6.24) and (6.26)], a term $\pm W_r(D_q/K)$ should be added to the left-hand side. In the Boussinesq equation [Eq. (6.25)], the added term should be $\pm W_r/K$. The positive sign corresponds to vertical accretion, such as rainfall, whereas the negative sign corresponds to losses, such as evaporation or evapotranspiration.

6.5 Determination of Aquifer Parameters

Idealization of Aquifers

Confined aquifers are usually idealized by assuming that their boundaries aa and bb as well as the initial undisturbed piezometric surfaces are horizontal (Fig. 6.9), and accordingly, the thickness of the aquifer D_q is also assumed to be uniform. Such idealization is necessary in order to simplify the analysis of the flow systems, especially those produced by artificial means such as well pumping. If cc is assumed horizontal, the

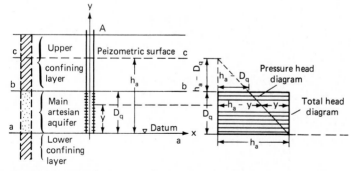

Figure 6.9 Total and pressure heads in idealized confined aquifer.

water is presumably stagnant. This assumption can be justified for two reasons:

1. The natural velocities in aquifers are usually very small, producing negligible slopes in the piezometric surfaces.

2. Most of artificial effects such as well pumping are usually concentrated in relatively small regions within the large extensive aquifers. Within these small regions, the sloping surfaces can reasonably be approximated as horizontal.

An observation well installed at A (Fig. 6.9) would allow the water to rise to level cc. In its simplest form, an observation well is an open-bottom pipe that may be perforated or screened throughout a portion of its length which is drilled into an aquifer. Generally, if the observation well has an open bottom only, the water level corresponds to the pressure head at that bottom. In perforated or screened observation wells, water levels correspond to the average pressure heads through the vertical distance covering the perforations. In idealized confined aquifers, the heads are the same at any vertical section. The total- and pressure-head diagrams are linear, as shown on the right-hand side of Fig. 6.9. Although it is common to assume that the soil of an aquifer is homogeneous and isotropic, simple cases of anisotropy can be analyzed by adjusting the general equations (see Sec. 5.1). The flow in heterogeneous formations with interbedded pockets or layers of different materials can be theoretically analyzed by numerical methods. However, determination of the size and pattern of these pockets and layers and evaluation of their changing permeabilities (in location and direction) would defy any reasonable practical means. The expenses would be exorbitant, and the errors in obtaining such sporadic records cannot be satisfactorily predicted or controlled. The location of test borings and the selection of test samples do not follow rigid scientific rules, and the true picture of the natural deposits of an aquifer would be difficult to determine. The idealization of field and hydrologic conditions is compensated for by including in the mathematical equations such parameters as S and T, which are determined from field tests rather than by theory.

Determination of the Hydraulic Conductivity from Piezometric Maps

Confined aquifers If water is flowing naturally in an aquifer under a steady-state condition, the surface cc (Fig. 6.10) should have a constant slope. By means of two observation wells A_1 and A_2 installed along the flow direction and spaced at a distance B, the darcian velocity v and the

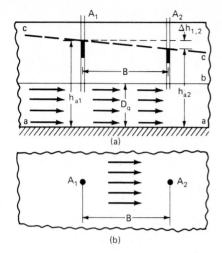

Figure 6.10 Field determination of groundwater velocity in confined aquifers when flow direction is known. *(a)* Section; *(b)* plan view.

rate of flow q can be determined from the recorded $\Delta h_{1,2} = h_{a1} - h_{a2}$ in terms of the hydraulic conductivity K and the depth of aquifer D_q:

$$v = Ki_g = K\frac{\Delta h_{1,2}}{B} \qquad (6.37)$$

or

$$K = \frac{Bv}{\Delta h_{1,2}} \qquad (6.37a)$$

and

$$q = vD_q = KD_q\frac{\Delta h_{1,2}}{B} = T\frac{\Delta h_{1,2}}{B} \qquad (6.38)$$

The pressure- and total-head diagrams at A_1 and A_2 within the aquifer are similar to those shown in Fig. 6.9 except that $h_a \equiv h_{a1}$ at A_1 and $h_a \equiv h_{a2}$ at A_2. The ordinates of these diagrams at A_2 are less than their equivalents at A_1 by the amount $\Delta h_{1,2}$, which is the head loss between stations A_1 and A_2.

In practice, the direction of flow is not known. Instead of the two observation wells A_1 and A_2 (Fig. 6.10), at least three observation wells A_1, A_2, and A_3 are installed to form a triangle $A_1A_2A_3$ (Fig. 6.10). The direction of the flow is then determined as illustrated by the following example.

EXAMPLE 6.2 With reference to Fig. 6.11, find the direction of natural flow if the recorded field drawdowns s_d were 5, 20, and 40 cm at A_1, A_2, and A_3, respectively.

Solution The equipotential lines can be constructed as follows: A certain equipotential drop such as, say, 5 cms is selected. Then by interpolation and/or extrapolation, find the points along the sides of the triangle $A_1A_2A_3$

and/or their extensions that have drawdowns with drops of 5 cm, that is, 5, 10, 15, 20, 25, 30, 35, and 40 are found. The points of equal drawdown intensities are joined by straight lines. Parallel equipotential lines can then be drawn. The line normal to the equipotential lines gives the direction of flow.

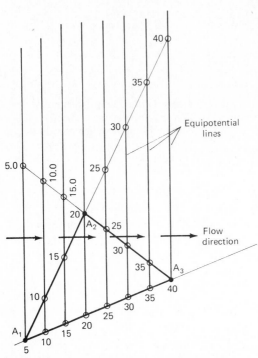

Figure 6.11 Field determination of the direction and velocity of groundwater by means of three observation wells (Example 6.2). Numbers are drawdowns in centimeters.

This method is based on the principles of flow nets (Sec. 5.1). Because the limited area $A_1A_2A_3$ is small compared with the extensive regional aquifer, the portion of the flow net covering this area is assumed to consist of real squares rather than curvilinear squares. It is also assumed in the preceding example that flow is uniform and steady. The gradient i_g is constant and is obtained by dividing the equipotential drop (5 cm) by the normal distance between any two consecutive equipotential lines (scaled from the plot). Equations (6.37) and (6.38) can be used to find v and q if K is known. However, if K is unknown, v has to be measured in the field (Chap. 4), and then Eq. (6.37a) can be applied.

In nature, flow is neither steady nor uniform because it is subjected to natural and artificial changes, such as recharging of an artesian aquifer by infiltration of rainfall through the outcrop zone or by tapping of the aquifers by water wells. The groundwater velocities increase around such activities as well as in the vicinity of natural depressed features such as gullies or streams. The hydraulic conductivity, the depth of the aquifer, and the soil structure also differ in nature at various locations. Recording the water pressures by means of sporadic observation wells in extensive aquifers would lead to distorted equipotential lines affected by these activities (see Fig. 3.14). The flow lines are then traced orthogonal to the equipotential lines to construct a flow net consisting of curvilinear squares (Lohman, 1972). The coefficient of transmissivity T is estimated as shown in the following example.

EXAMPLE 6.3 The average actual discharge by pumping out the aquifer shown in Fig. 3.14 was estimated in 1945 as 1×10^6 ft^3 per day ($= 0.328$ m^3/s) between equipotential lines of intensities 30 ft (9.144 m) and 60 ft (18.288 m), where $n_E = 3$, $n_F = 15$, $h_e = 60 - 30 = 30$ ft $= 9.144$ m. The rate of flow, therefore, through the entire depth of the aquifer is

$$Q = D_q K h_e \frac{n_F}{n_E} \quad \text{(based on Sec. 5.1, where } Q = q D_q)$$

or
$$Q = T h_e \frac{n_F}{n_E}$$

Then $\quad 0.328 = T \times 9.144 \times \dfrac{15}{3}$

or $\quad\quad T = 7.2 \times 10^{-3}$ m^2/s $(= 6700$ ft^2 per day)

Although this method is acceptable in practice, one should be aware of the approximations involved. The pumping cannot be continuous and uninterrupted for years except under uncommon situations. Therefore, the flow due to pumping is normally transient because of these interrupted operations. This results in a piezometric map different than that shown in Fig. 3.14 depending on the time over which the records were taken. The flow configuration also varies in different seasons. The flow net is not a solution for transient conditions or even steady-state conditions in heterogeneous and anisotropic media. The small differences in the records on a regional scale allow this kind of approximation. It is recommended, however, that the T value be checked by other means, such as pumping tests (Chap. 8).

Unconfined aquifers In unconfined aquifers, similar idealizations are also introduced. The undisturbed water table cc and the impervious boundary aa (Fig. 6.12), are assumed horizontal. The initial total head h_u and pressure head $(p/\gamma_w)_{x,y}$ are shown in Fig. 6.12a at distance x_1 from the origin. Any solid observation well with an open bottom indicates a level of water at a height h_u above the impervious boundary aa irrespective of the depth of that observation well. If there is a natural flow, the water table cc deflects to bb. Any two shallow observation wells at distances x_1 and x_2 record water levels close to the levels of D_{x1} and D_{x2} of the water table. The total- and pressure-head diagrams at x_1 are shown in Fig. 6.12b and d. Lines de and fg of these diagrams are curves rather than

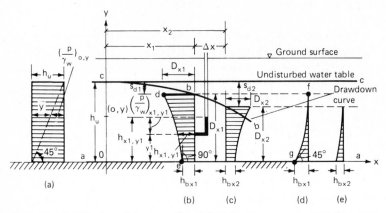

Figure 6.12 Total- and pressure-head distribution across vertical sections in unconfined aquifers. (a) Total-head diagram corresponding to undisturbed water table; (b) and (c) total-head diagrams at, respectively, x_1 and x_2 corresponding to deflected water table; (d) and (e) pressure-head diagrams at, respectively, x_1 and x_2 corresponding to deflected water table.

straight lines because the equipotential lines within the flow medium are deformed (Sec. 5.1). The tangent at point d to the curve de in Fig. 6.12b has a maximum slope angle with the vertical indicating a maximum vertical velocity component along the vertical section at x_1. The tangent at point e is vertical because the vertical velocity component at e is zero (bottom point of section x_1). If the pressure-head ordinates are plotted across a vertical reference line such as that shown in Fig. 6.12d, the slope at point g makes an angle $\pi/4 = 45°$. An observation well at x_1 with its open bottom at a distance y_1 above the base records only the total head $h_{(x1,y1)} = D_{x1} - s_{(x1,y1)}$ or the pressure head $(p/\gamma_w)_{(x1,y1)}$ above the bottom level of the observation well. The same discussion applies to the vertical section at x_2, where the total- and pressure-head diagrams are similar to those at x_1 but with ordinates of different magnitudes.

Unlike artesian aquifers, the different elevations of the bottoms of observation wells in unconfined aquifers lead to different records of water levels at the same location. Since groundwater velocities are small, piezometric maps such as that shown in Fig. 3.14 are acceptable for use in unconfined aquifers. The flow direction is also determined by a construction similar to that shown in Fig. 6.11. The estimates of q and T are made in a similar way as that for confined aquifers.

A piezometric map in an unconfined aquifer represents a contour map of the *upper surface* of the water table rather than the water pressures within the interior of the soil medium. If the direction of flow is known, it has been shown in Sec. 5.5 that Eqs. (5.42) and (5.43) are more accurate for evaluating q/K or K in unconfined aquifers than using Eqs. (6.37) and (6.38) (see Sec. 5.5 and Fig. 5.13).

In applying piezometric maps, it is implicitly assumed that the coefficient of storage $S = 0$ in confined aquifers and S_y or $S_{ya} = 0$ in unconfined aquifers. The piezometric maps are therefore used to find K or T rather than S. These maps are also useful in finding the directions of groundwater flow and in delineating the zones of discharge and recharge.

Field records used in constructing the piezometric maps should be adjusted before their use in order to eliminate such environmental effects as those produced by tides and changes in barometric pressures, as explained in the following.

The records should also be adjusted in such a way that all levels correspond to the same time period in a certain season.

Rainfall Effects

After a rainfall spell, the water level in an observation well in shallow unconfined aquifers rises instantaneously. This may be due to one or more of the following three reasons:

1. The infiltrating water around the casing may leak into the well through its perforations unless the well is properly sealed.

2. Direct rainfall may fall into the well unless it is capped.

3. The pressure of the entrapped air within the zone of aeration may increase (Todd, 1980) because the rainwater seals the pores near the ground surface and the infilterating water compresses the underlying air (Fig. 6.13).

Usually, observation wells are capped and sealed near the top, and the perforations (if there are any) are located near their lower ends. Thus the most probable cause for the rise of water levels as a result of a rainfall would be the entrapped air. The records should be adjusted by the use of the following equation (Todd, 1980):

$$\Delta h = \frac{d_m}{d_v - d_m} \frac{p_a}{\gamma_w} \tag{6.39}$$

where Δh = the water level rise above the water table due to air en-
trappment

d_m = the depth of the saturated zone of rain infiltration (Fig. 6.13)

p_a = atmospheric pressure $\simeq 1000\gamma_w$ (γ_w = 1 g/cm^3)

d_v = the vertical distance between the ground surface and the water table, which is the depth of the vadose zone

After a short period of time, Δh dissipates as a result of the escape of air.

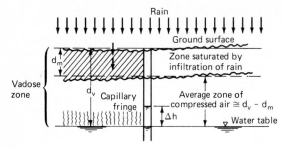

Figure 6.13 Water-level rise in an observation well after a rain spell.

Evaluation of the Coefficient of Storage from Changes in Effective Stresses

Water levels in observation wells drilled in artesian aquifers are affected by changes in the stages of streams, lakes, and oceans and by many other factors, such as atmospheric pressure changes (Jacob, 1950), earth tides (Robinson, 1939), passing trains (Jacob, 1939) or trucks, gusty winds (Heath and Trainer, 1968), and vibrations due to earthquakes (DaCosta, 1964) or explosions of different types (Ineson, 1963). All these factors have one thing in common: changing the effective stresses.

In unconfined aquifers, atmospheric pressure changes are transmitted directly to the water table. This means that the water level in an observation well could coincide with the water table. Air entrapped below the water table changes the recorded water levels in observation wells in the same manner as in artesian aquifers but to a smaller degree (Todd, 1980). The air entrapped between the saturated zone due to infiltration of rainfall and the natural water table lead to a rise in the water levels in observation wells for a short time [Eq. (6.39) and Fig. 6.13].

Fluctuations caused by earthquakes, gusty winds, passing trains or trucks, and earth tides are of scientific interest only because the changes are insignificant compared with those produced by changes in barometric pressure and ocean tides (Heath and Trainer, 1968; Carr and Van Der Kamp, 1969). It is frequently necessary to remove these fluctuations from water-level hydrographs before using the latter in groundwater-resource management. Practical techniques for removing all the effects produced by changes in effective stresses are available elsewhere (Heath and Trainer, 1968).

Tidal effects   If H_{ot} is the stage of the tide, the total stress σ_t at point m (Fig. 6.14) on the ocean front is increased by $\gamma_{st}H_{ot}$, where γ_{st} is the unit weight of salt water. Since $d\sigma_t = dp + d\bar{\sigma}$ (Sec. 3.9), then

$$\gamma_{st}dH_{ot} = dp + d\bar{\sigma} \tag{6.40}$$

The rise in water level dh in an observation well installed very close to the shore is equal to

$$dh = \frac{dp}{\gamma_w} \tag{6.41}$$

where γ_w is the unit weight of fresh water.

Tidal efficiency (TE) is defined as dh/dH_{ot}, or

$$\text{TE} = \frac{dh}{dH_{ot}} = \frac{dp}{\gamma_w dH_{ot}} = \frac{dp/d\bar{\sigma}}{[1 + (dp/d\bar{\sigma})]\gamma_w/\gamma_{st}} \tag{6.42}$$

In practice, the usual approximate average value of TE has a range between 30 and 80 percent (Walton, 1970).

From Eqs. (6.15) and (6.20), we get

$$\frac{dp}{d\bar{\sigma}} = \frac{m_v}{n\beta_w}$$

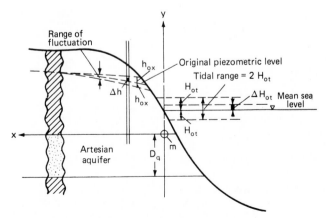

Figure 6.14   Tidal effects on water levels in a piezometer.

Substituting this relationship in Eq. (6.42), we get

$$\text{TE} = \frac{dh}{dH_{ot}} = \frac{(m_v/n\beta_w)\gamma_{st}}{[1 + (m_v/n\beta_w)]\gamma_w} \tag{6.43}$$

Since $\gamma_s/\gamma_w \approx 1.025 \approx 1.0$ (see Chap. 9), then

$$\text{TE} = \frac{dh}{dH_{ot}} \approx \frac{m_v/n\beta_w}{1 + (m_v/n\beta_w)} \tag{6.43a}$$

From this equation, $1 + (m_v/n\beta_w) = 1/(1 - \text{TE})$. Substituting this value in Eq. (6.3), we get the coefficient of storage:

$$S = D_q n\gamma_w\beta_w\left(\frac{1}{1 - \text{TE}}\right) \tag{6.44}$$

EXAMPLE 6.4 A uniform tide exists at the boundary of an artesian aquifer with a depth of 100 m. The water level in an observation well fluctuates simultaneously with an amplitude 40 percent of that of the tide. Evaluate m_v and S if the porosity of the aquifer material is 30 percent.

Solution Substitute $n = 0.30$, TE $= 0.40$, and $\beta_w = 5 \times 10^{-8}$ cm^2/g in Eq. (6.43a); then $m_v = 1.0 \times 10^{-8}$ cm^2/g. Applying Eq. (6.44),

$$S = 0.30 \times 1 \times 10,000 \times 5 \times 10^{-8}\left(\frac{1}{1 - 0.40}\right) = 2.5 \times 10^{-3}$$

[Note that if γ_{st}/γ_w is substituted in Eq. (6.43) as equal to 1.025 rather than 1.0, then $S = 2.2 \times 10^{-3}$.]

In the preceding example the tidal efficiency was recorded in a well near the shore. If the ocean level varies with a simple harmonic motion, sinusoidal fluctuations in water level in the observation wells occur in response to the tidal changes. Thus if

$$dH_{ot} = H_{ot} \sin \frac{2\pi}{t_{ot}} t \tag{6.45}$$

the net rise or fall dh in the piezometric surface is found to be (Todd, 1980)

$$dh = H_{ot} \exp\left(-x\sqrt{\frac{\pi S}{t_{ot}T}}\right) \sin\left(\frac{2\pi t}{t_{ot}} - x\sqrt{\frac{\pi S}{t_{ot}T}}\right) \tag{6.46}$$

where $H_o =$ the amplitude or half range of the tide (Fig. 6.14)
$dH_o =$ stage of tide above or below the mean sea level
$t_{ot} =$ tidal period
$t =$ time through which dH_{ot} occurs
S and $T =$ respectively the coefficients of storage and transmissivity
$x =$ distance of observation well from shore

The amplitude h_{oxt} of water-level fluctuations in an observation well distance x from the shore is given by

$$h_{oxt} = H_{ot} \exp\left(-x\sqrt{\frac{\pi S}{t_{ot}T}}\right) \tag{6.47}$$

Equation (6.46) or (6.47) may be used to evaluate $S/T = 1/v$ if the field records dh (or h_{oxt}), x, t_{ot}, and t are available.

The time lag t_L between the change of the stage of tide and the response in an observation well is given by

$$t_L = x\sqrt{\frac{t_{ot}S}{4\pi T}} \tag{6.48}$$

In unconfined aquifers, Eqs. (6.45) through (6.48) are approximately applicable.

Changes in barometric pressure A reverse effect occurs from changes in barometric pressure compared with that produced by tidal effects; an increase in barometric pressure produces a decline in the water level in an observation well and vice versa. Since the change in atmospheric pressure p_a produces a change in the total stress σ_t, then dp_a should be equal to the sum of the changes $d\bar{\sigma}$ and dp, or

$$dp_a = dp + d\bar{\sigma} \tag{6.49}$$

The water level is lowered by dh as a result of the increase dp_a in atmospheric pressure; then $\gamma_w dh = dp - dp_a = -d\bar{\sigma}$, or

$$\frac{\gamma_w dh}{dp_a} = \frac{-d\bar{\sigma}}{dp + d\bar{\sigma}} = \frac{-1}{1 + (dp/d\bar{\sigma})}$$
$$= \text{barometric efficiency (BE)} \qquad \text{(Jacob, 1950)}$$

However, $dp/d\bar{\sigma} = m_v/n\beta_w$ [Eqs. (6.15) and (6.20)], so

$$\text{BE} = \frac{\gamma_w dh}{dp_a} = \frac{-1}{1 + (m_v/n\beta_w)} \tag{6.50}$$

Comparing Eqs. (6.43a) and (6.50), it is obvious that BE + TE is numerically equal to 1.0, and the coefficient of storage S may be written as [see Eq. (6.44)]

$$S = D_q n \gamma_w \beta_w \frac{1}{\text{BE}} \tag{6.51}$$

In the case shown in Fig. 6.15, BE was found to be 0.52 based on field records. Once BE is evaluated, S is computed from Eq. (6.51) and m_v can be computed from Eq. (6.50). BE ranges in most cases between 20 and 70 percent.

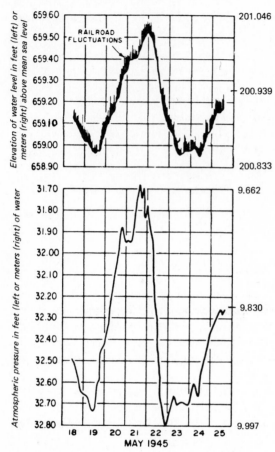

Figure 6.15 Effects of atmospheric pressure and moving trains on the water level in a well at Savoy, Illinois. *(Top)* Water levels in well at Savoy, Illinois; *(bottom)* atmospheric pressure at Urbana, Illinois, BE $= \Delta w/\Delta B_p \times 100 = 0.52/1.02 \times 100 = 52$ percent. *(W. J. Roberts and H. E. Romine, "Effects of Train Loading on the Water Level in a Deep Glacial-Drift Well in Central Illinois," Trans. Am. Geophys. Union, vol. 28, no. 6, 1947, p. 914.)*

Borehole Methods

The most accurate, reliable, and commonly used method of determining aquifer characteristics is by controlled aquifer pumping tests (U.S. Bureau of Reclamation, 1977). The methods are explained in Chap. 8. They are usually expensive; a reasonably sophisticated test may cost from $2000 to $10,000, not including the cost of the well (U.S. Bureau of Reclamation, 1977). The number of tests depends on the size of the area, the degree of heterogeneity of the aquifer or aquifers involved, the known (or determined) boundary conditions, and the type of project.

In almost all major water-resources projects and other engineering projects, exploratory boreholes are made to investigate the soil properties. These boreholes are used to evaluate approximately the hydraulic conductivity in their vicinity from analyzing the records of rate of rise or drop in the water levels. There are scores of different methods available (Bouwer, 1978; U.S. Bureau of Reclamations, 1963 and 1977; NAVFAC, 1982); few are illustrated in the following section. The main advantages of these methods are

1. Pumping is not needed.

2. Observation wells are not required.

3. The period of the test is relatively short.

4. Boreholes are already available for soil investigations.

5. The results are not affected by disturbances in their vicinity, such as interference from other wells.

The disadvantages of the borehole methods are that K is a crude estimate representing only the average value in the immediate vicinity of the borehole and S cannot generally be evaluated. In pumping tests (Chap. 8), the values of T and S represent the weighted averages of relatively larger zones within the influence of pumping.

Slug tests   A pipe is drilled for a distance d_p below the water table (Fig. 6.16). The upper portion is solid and the lower is perforated

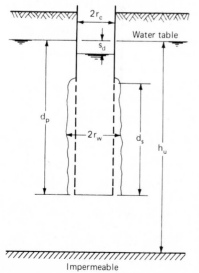

Figure 6.16   Slug test to determine hydraulic conductivity. (*H. Bouwer, Groundwater Hydrology, McGraw-Hill, New York, 1978.*)

through a depth d_s. The water enters through a perforated portion surrounded by a gravel pack or a screened length that should have slot (or hole) sizes designed according to the type of surrounding soil (Sec. 3.10).

Bouwer and Rice (Bouwer, 1978) gave the following formula based on records taken after the sudden removal of a slug of water:

$$K = \frac{r_c^2}{2Ntd_s} \ln \frac{h_o}{h_t} \qquad (6.52)$$

where d_s = height of perforated, screened, or uncased section
$h_0 = h$ at time zero
$h_t = h$ at time t
t = time since h_0 was recorded
r_c = radius of upper solid portion of the pipe

$$N = \frac{1.1}{\ln (d_p/r_w)} + \frac{C_1 + C_2 \ln [(h_u - d_p)/r_w]}{d_s/r_w} \qquad (6.53)$$

where r_w is the radius between borehole center and undisturbed soil $(r_c +$ thickness of gravel pack). h_u, d_p, and d_s are as shown in Fig. 6.16; C_1 and C_2 are dimensionless values determined from the chart given in Fig. 6.17 with respect to the ratio d_s/r_w. N in Eq. (6.53) is equal to

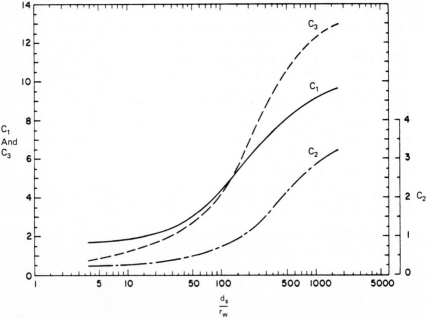

Figure 6.17 Curves required in a slug test. (*H. Bouwer, Groundwater Hydrology, McGraw-Hill, New York, 1978.*)

$1/\ln (R_e/r_w)$, where R_e is the radius of influence in Thiem's equation (Chap. 8).

The value of $\ln [(h_u - d_s)/r_w]$ in Eq. (6.53) has a maximum value of 6 even if the impermeable boundary is very deep ($h_u = \infty$). If the observation well completely penetrates the aquifer, the N value is (Bouwer, 1978)

$$N = \frac{1.1}{\ln (d_p/r_w)} + \frac{C_3}{d_s/r_w} \qquad (\text{when } h_u = d_p) \qquad (6.54)$$

where C_3 is a dimensionless parameter determined also from Fig. 6.17.

Several readings for h_t at different times should be taken. If these values are plotted as log h_t versus t, a straight line is obtained in accordance with Eq. (6.52). The value $1/t \ln (h_0/h_t)$ is preferably determined from the graphic plot. If we select any two points on the straight line h_{t1} and h_{t2} at times t_1 and t_2, then from Eq. (6.52), we get

$$K = \frac{2.3r_c^2 \log (h_{t1}/h_{t2})}{2Nd_s(t_2 - t_1)} \qquad (6.55)$$

The K value determined in this test represents the average horizontal hydraulic conductivity of the soil within a cylindrical zone of a height slightly larger than d_s and a radius of about r_w exp N (Bouwer, 1978).

The values of r_w in Eqs. (6.53) and (6.54) as well as in Fig. 6.17 are very approximate if a gravel pack is made. In the case of a screened length d_s in direct contact with the natural soil, r_w is taken as equal to r_c.

The methods include approximations other than the estimation of r_w. Equation (6.55) was based on Thiem's equation (Chap. 8), which is a steady-state equation. Also it was assumed that drawdown of the water table after water removal around the pipe is negligible. Usually the amount of water removed is relatively small, and the recovery takes less than a minute or a little bit longer. If several recordings are obtained within this short time, the water-level measurements should be made by sensitive devices, such as pressure transducers and fast strip-chart recorders or x-versus-y plotters to record the transducer output (Bouwer, 1978). For example, if a slug of water is removed by a bailer with an inside diameter of 3 in (7.62 cm) and a length of 24 in (60.96 cm), the volume of water removed is 1298.8 cm³. If $2r_c = 6$ in (15.24 cm), the water level in the pipe drops 7.12 cm, and several recordings should be taken while the water level recovers above this small drop. The time $t_{90\%}$ required for 90 percent recovery is given by the following equation (Bouwer, 1978):

$$t_{90\%} = 0.0527 \frac{r_c}{NKd_s} \qquad (6.56)$$

Bailer method   This method was developed by Skibitzke (Lohman, 1972). The coefficient of transmissivity T is determined from the recovery of water level in an artesian well for which the effective radius r_w is already known after bailing out a certain water volume V_w. If s'_d is the residual drawdown at time t measured since bailing stopped, then

$$T \simeq \frac{V_w}{4\pi s'_d t} \tag{6.57}$$

Equation (6.57) requires single values of V_w and s'_d corresponding to relatively large times t. The method is recommended only when the piezometric surface is near the ground surface.

The bailing may consist of any number of cycles. If the residual drawdown s'_d is recorded after n bailing cycles, the following equation applies:

$$T = \frac{1}{4\pi s'_d} \left(\frac{V_{w1}}{t_1} + \frac{V_{w2}}{t_2} + \frac{V_{w3}}{t_3} + \cdots + \frac{V_{wn}}{t_n} \right) \tag{6.58}$$

where t_n is the time elapsed between the time when the nth bailer of volume V_{wn} was removed and the time of observation of s'_d, which is a single value in this equation.

If $V_{w1} = V_{w2} = V_{w3} = \cdots = V_{wn}$, then

$$T = \frac{V_w}{4\pi s'_d} \left(\frac{1}{t_1} + \frac{1}{t_2} + \frac{1}{t_3} + \cdots + \frac{1}{t_n} \right) \tag{6.59}$$

V_w and s'_d are single values in Eq. (6.59).

Rate-of-water-drop method   The drilled hole in this method (Zanger, 1953) is cased from the ground surface to the top of the zone to be tested and extends without support for a suitable depth t_b below the casing (Fig. 6.18b). Water is added in the casing to a level above the water table and then allowed to drop. If a drop Δh between levels a and b is recorded through a time interval Δt, then

$$K = \frac{\pi r_c^2 (\Delta h / \Delta t)}{C_k r_b h_m} \tag{6.60}$$

where r_c = inside radius of the casing
$\quad r_b$ = mean radius of the uncased hole
$\quad h_m$ = mean distance between the levels a and b (Fig. 6.18b) and the water table
$\quad C_K$ = a dimensionless coefficient determined from charts given in Fig. 6.18a and c.

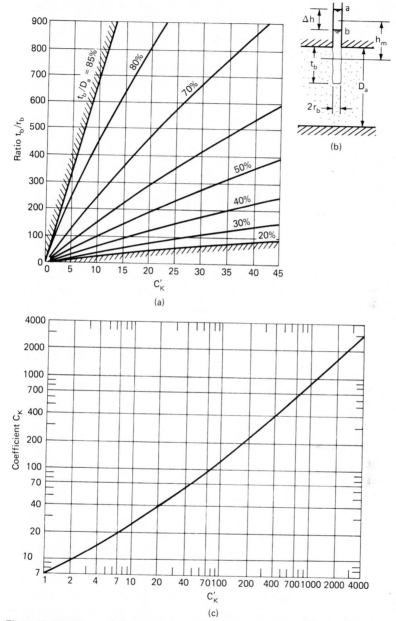

Figure 6.18 Permeability test by the rate-of-water-drop method. *(a)* Relationship between t_b/r_b and C_k' for various values of t_b/D_a; *(b)* layout of the test; *(c)* relationship between C_k' and C_k. (*After C. N. Zanger, Theory and Problems of Water Percolation, U.S. Bureau of Reclamation, Engineering Monograph No. 8, Washington, 1953.*)

The chart in Fig. 6.18a is entered at the point corresponding to the ordinate t_b/r_b and one moves horizontally to intersect one of the family of curves corresponding to t_b/D_a (the degree of penetration). The value of C'_k corresponding to that point of intersection is read. Then from Fig. 6.18c, the coefficient C_k is determined. For example, if $t_b/r_b = 200$ and $t_b/D_a = 40$ percent, C'_k would be 35. The coefficient C_k is then determined from Fig. 6.18c (about 60). This method is not recommended when $t_b/D_a > 85$ percent. When $t_b/D_a < 20$ percent, $C'_k = t_b/r_b$ is introduced and C_k is found directly from Fig. 6.18c.

Several other methods have been developed to find K from available boreholes or by digging auger holes in shallow aquifers, such as those explained by Boast and Kirkham (1971), the U.S Bureau of Reclamation (1977), and the Department of the Navy (Naval Facilities Engineering Command, 1982).

Rough Estimates of *K* and *S*

In preliminary investigations, the order of magnitude of the aquifer parameters may be needed before any decision is made in planning the field tests. Experience and judgment are necessary to evaluate S and K in a crude manner. The following approximations are used to make such rough estimates.

It has been shown that the coefficient of storage S varies between 10^{-5} and 10^{-3} in most productive aquifers (Walton, 1970; and Sec. 6.3). Thus a rule of thumb is to assume a constant specific storage $S_s = S/D_q$ as equal to 3×10^{-7} m^{-1} (Lohman, 1972). The coefficient S for a range of D between 30 and 300 m thus gives essentially the practical range of S $(30 \times 3 \times 10^{-7} = 10^{-5}$ and $300 \times 3 \times 10^{-7} = 10^{-3})$.

The specific yield S_y for unconfined aquifers ranges from about 0.1 to 0.3. Therefore, a crude estimate in any situation is to assume that $S_y = 0.2$, which is the average between the upper and lower limits of the practical range. A better evaluation might be determined either by testing using the neutron moisture probes (Meyer, 1962) or, of course, by conducting pumping tests (Chap. 8).

Hydraulic conductivity K has been evaluated by various investigators on the basis of the classification of aquifer materials. Table 3.1 and Fig. 3.9 list these values.

The methods explained in this section for the determination of aquifer parameters should be used only on small jobs or whenever preliminary values are needed to plan a large project. Chapter 8 includes most of the updated techniques for finding accurate values of these parameters. Cost plays a big role in selection of the method.

PROBLEMS AND DISCUSSION QUESTIONS

Figures for this section are numbered according to the problem in which they first appear.

6.1 (a) Prove that the soil diffusivity is approximately equal to the coefficient of consolidation.

(b) State *all* the assumptions made in the derivation of Eq. (6.1), and explain their effects on use of the equation.

6.2 Find the $U_c = 0$ percent in Example 6.1 corresponding to the following values of t_1:

(a) 0.4 min

(b) 2.0 min

(c) 10 min

Explain why there is an excessive discrepancy as compared with $t_1 = 1.0$ min when $t_1 = 10$ min is selected.

6.3 An artesian aquifer 100 m deep has a porosity of 30.5 percent and a coefficient of compressibility of 3×10^{-6} cm²/g. If the piezometric surface drops 2.5 m, determine the volume of water released from storage from a vertical prism with a cross-sectional area of 1×1 m and a depth that is the same as the aquifer. Assume that the aquifer receives no recharge.

6.4 Using the data from Problem 6.3, determine the volume of water released from storage due to water expansion only within an area of 1 km². Assume that the drop in head is the same within this area.

6.5 An unconfined aquifer is separated from an artesian aquifer by an aquitard with an averge thickness of 10 m (Fig. P-6.5). A tightly cased observation well is installed in the artesian aquifer. Its bottom is about 3.0 m below the bottom of the aquitard. The water level is recorded in this well at 13.0 m below the ground surface. If the water-table level in the unconfined

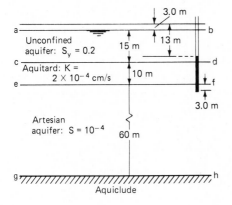

Figure P-6.5

aquifer is 3.0 m below the ground surface, determine the steady-state leakage through the aquitard within an area of 1 km² and the hydraulic resistance. Assume that the aquitard is incompressible.

6.6 During a long dry season, the water table in Fig. P-6.5 dropped 10 m (average). The water was depleted due to evaporation and water withdrawal by pumping. Assuming that the same conditions prevail within 1 km², determine the amount of water loss due to evaporation and water withdrawal.

6.7 Assuming that the soil material in both the unconfined and artesian aquifers shown in Fig. P-6.5 has the same properties, compute the amount of water released from storage in the lower 5.0 m of the unconfined aquifer over an area of 1 km² during the process of water-table drawdown from its original level to its final level at 13.0 m below the ground surface. What is the percent of this amount of water compared with that computed in Problem 6.6?

6.8 If the hydraulic conductivity of the artesian aquifer shown in Fig. P-6.5 is 6×10^{-3} cm/s, find its coefficient of transmissivity in square meters per day. Also compute the leakage factor in centimeters.

6.9 Rework Example 6.2 if the drawdowns Δs_d are 42.5, 18.2, and 6.5 at A_1, A_2, A_3, respectively.

6.10 The water level in an observation well in an unconfined aquifer rises 3.33 m after a rainfall. If the depth of the vadose zone is 7.7 m, find the depth of the zone of compressed air above the water table.

6.11 A uniform tide exists at the boundary of an artesian aquifer. The water level in a cased observation well fluctuates simultaneously with an amplitude of 44 percent of the tide. If the coefficients of compressibility and storage of the aquifer are, respectively, 8×10^{-7} cm²/g and 2×10^{-3}, determine the coefficient of specific storage of the aquifer. Assume that the porosity of the aquifer is 30 percent.

6.12 Find the barometric efficiency in the preceding example.

6.13 A slug test is conducted using a bailer in an unconfined aquifer with a depth of 13.52 m. The water table is 1.52 m below the ground surface. With reference to Fig. 6.16, the following data are available: $r_c = 7.62$ cm, $d_p = 6.1$ m, $r_w = 12$ cm, and $d_s = 4.572$ m. The bailer removes 1298.8 cm³ of water. The water drop was 6 cm when the recorded time was set at zero. After 35 seconds, the water drop was recorded as 2 cm. Find the average horizontal hydraulic conductivity of the aquifer.

6.14 Find the time corresponding to 90 percent recovery in Problem 6.13.

6.15 Construct a semilog plot for the logarithm of recovery versus t using the information and results of Problem 6.13. Determine from this plot the approximate recovery levels at times 2, 5, 10, 15, and 20 s.

6.16 Draw to scale the layout of the slug test performed in Problem 6.13, and show on the drawing the limits of the zone for which the horizontal hydraulic conductivity was calculated.

6.17   The rate-of-water-drop method (Zanger, 1953) was used to determine the hydraulic conductivity of the soil below the water table. With reference to Fig. 6.18b, the following data are available: $r_c = r_b = 5$ cm, $t_b = 10$ m, $D_a = 25$ m, and the water table is 2.5 m below the ground surface. Water is added to the casing and allowed to drop. When recording starts, the water level in the casing is 4.5 m above the water-table level. After 3 minutes and 38 seconds, the water drops to 3.5 m above the water table. Determine the hydraulic conductivity of the aquifer.

REFERENCES

American Society for Testing and Materials: *Annual Book of ASTM Standards, Soil and Rock, Building Stones*, pt. 19, Philadelphia, 1982.

Biot, M. A.: "General Theory of Three-Dimensional Consolidation," *J. Appl. Phys.*, vol. 12, no. 2, 1941, pp. 155–164.

Biot, M. A.: "Theory of Elasticity and Consolidation for a Porous Anisotropic Solid," *J. Appl. Phys.*, vol. 26, no. 2, 1955, pp. 182–185.

Boast, C. W., and Kirkham, D.: "Auger Hole Seepage Theory," *Soil Sci. Soc. Am. Proc.*, vol. 35, 1971, pp. 365–373.

Boulton, N. S.: *Unsteady Radial Flow to a Pumped Well Allowing for Delayed Yield from Storage*, International Association of Scientific Hydrology Publication no. 37, 1954a, pp. 472–477.

Boulton, N. S.: "The Drawdown of the Water Table under Nonsteady Conditions Near a Pumped Well in an Unconfined Formation," *Proc. Inst. Civil Eng. (Lond.)*, vol. 3, 1954b, pp. 564–579.

Boulton, N. S.: "Analysis of Data from Nonequilibrium Pumping Tests Allowing for Delayed Yield from Storage," *Proc. Inst. Civil Eng. (Lond.)*, vol. 26, 1963, pp. 469–482.

Boulton, N. S.: Closure discussion on Boulton, 1963, and discussions by R. W. Stallman, W. C. Walton, and J. Ineson, *Proc. Inst. Civil Eng. (Lond.)*, vol. 28, 1964, pp. 603–610.

Bouwer, H.: *Groundwater Hydrology*, McGraw-Hill, New York, 1978.

Bowles, J. E.: *Engineering Properties of Soils and Their Measurement*, 2d ed., McGraw-Hill, New York, 1978.

Carr, P. A., and Van Der Kamp, G. S.: "Determining Aquifer Characteristics by the Tidal method," *Water Resources Res.*, vol. 5, 1969, pp. 1023–1031.

DaCosta, J. A.: "Effect of Hegben Lake Earthquake on Water Levels in Wells in the United States," U.S. Geological Survey Professional Paper 435, Washington, 1964, pp. 167–178.

Hantush, M. S.: "Hydraulics of Wells," in V. T. Chow (ed.), *Advances in Hydroscience*, vol. 1, Academic Press, New York, 1964, pp. 281–432.

Heath, R. C., and F. W. Trainer: *Introduction to Groundwater Hydrology*, Wiley, New York, 1968.

Ineson, J.: "Form of Groundwater Fluctuations due to Nuclear Explosions," *Nature*, vol. 198, 1963, pp. 22–23.

Jacob, C. E.: "Fluctuations in Artesian Pressure Produced by Passing Railroad-Trains as Shown in a Well on Long Island, New York," *Trans. Am. Geophys. Union*, vol. 20, 1939, pp. 666–674.

Jacob, C. E.: "Radial Flow in Leaky Artesian Aquifers," *Trans. Am. Geophys. Union*, vol. 27, no. II, 1946, pp. 198–208.

Jacob, C. E.: "Flow of Ground Water," in H. Rouse (ed.), *Engineering Hydraulics*, Wiley, New York, 1950, chap. 5, pp. 321–386.

Lohman, S. W.: "Groundwater Hydraulics," U.S. Geological Survey Professional Paper 708, Washington, 1972.

Meyer, W. R.: "Use of Neutron-Moisture Probe to Determine the Storage Coefficient of an Unconfined Aquifer," U.S. Geological Survey Professional Paper 450E, Washington, 1962, pp. E174–E176.

Muskat, M.: *The Flow of Homogeneous Fluids through Porous Media,* McGraw-Hill, New York; second printing 1946 by J. W. Edwards, Ann Arbor, Mich.

Naval Facilities Engineering Command: "Soil Mechanics," Design Manual 7.1, Department of the Navy, NAVFAC, Alexandria, Va., 1982.

Peck, R. B., W. E. Hanson, and T. H. Thornburn: *Foundation Engineering,* 2d ed., Wiley, New York, 1974.

Rasmussen, W. C., and G. E. Andreasen: "Hydrologic Budget of the Beaverdam Creek Basin, Maryland," U.S Geological Survey Water Supply Paper 1472, Washington, 1959.

Roberts, W. J., and H. E. Romine: "Effects of Train Loading on the Water Level in a Deep Glacial-Drift Well in Central Illinois," *Trans. Am. Geophys. Union,* vol. 28, no. 6, 1947, pp. 912–917.

Robinson, T. W.: "Earth-Tides Shown by Fluctuations of Water-Levels in Wells in New Mexico and Iowa," *Trans. Am. Geophys. Union,* vol. 20, pt. IV, 1939, pp. 656–666.

Taylor, D. W.: *Fundamentals of Soil Mechanics,* Wiley, New York, 1948.

Terzaghi, K.: *Theoretical Soil Mechanics,* Wiley, New York, 1943.

Terzaghi, K., and R. B. Peck: *Soil Mechanics in Engineering Practice,* 2d ed., Wiley, New York, 1967.

Theis, C. V.: "The Relation between the Lowering of the Piezometric Surface and the Rate and Duration of Discharge of a Well Using Groundwater Storage," *Trans. Am. Geophys. Union,* vol. 16, 1935, pp. 519–524.

Todd, D. K.: *Ground Water Hydrology,* 2d ed., Wiley, New York, 1980.

U.S. Bureau of Reclamation: *Earth Manual,* U.S. Department of Interior, Washington, 1963.

U.S. Bureau of Reclamation: *Ground Water Manual,* U.S. Department of Interior, Washington, 1977.

Walton, W. C.: *Groundwater Resource Evaluation,* McGraw-Hill, New York, 1970.

Zangar, C. N.: *Theory and Problems of Water Percolation,* Bureau of Reclamation, Engineering Monograph No. 8, Washington, 1953.

CHAPTER 7

MANAGEMENT OF GROUNDWATER

The ideal purpose of groundwater management in a basin is to develop the maximum possible groundwater to satisfy the requirements of all users within the basin and to meet specific predetermined conditions, such as the level of water quality, the costs of development and operation, and certain legal, social, and political constraints. Management procedures should be adequate to avoid present and future potential detrimental effects, such as excessive water depletion, deterioration of water quality (Chap. 4), and land subsidence due to excessive pumping (Chap. 8).

Many current groundwater problems could have been avoided if good planning had been practiced in the past. The limited water demand of past generations was an important reason that subsequent undesirable effects were overlooked or underestimated (Kashef, 1971). The continuous increase in population and the recent increase in water demands due to the continuously rising standard of living created a new situation that necessitated proper planning and sound management of groundwater resources. Aquifers can no longer be looked on as everlasting sources of abundant water of good quality. Increases in urban wastes and expansion of industry and agriculture lead to a deterioration in the quality of groundwater. In coastal aquifers, increases in water demand have resulted in inland progress of saltwater intrusion. All these problems and others dictated the urgent need for groundwater management. Optimum economic water management requires an integrated approach to the management of both surface waters and groundwaters.

Too often comprehensive groundwater management is only an afterthought, considered when water shortages or other detrimental effects have occurred (Posey, 1984). In the past, development of groundwater in a basin did not usually follow an orderly managerial procedure. The pattern, depth, and capacity of individual wells in a basin were initially constructed in an unspecified manner. Once they proved to have adequate yield, competitors were encouraged to construct more and more wells without regard to the consequences.

The management program should be carried out by an appropriate organization that has the ability and the power to implement the program and its related policies. Such an organization is typically a government agency or group of agencies or water-management districts chartered and supervised by the local or state governments (Posey, 1984). Generally, the functions of groundwater management are as follows:

1. *Regulation of water consumption.* Water consumption can be regulated either directly by allocation or indirectly by a fee or tax on consumption. The objective of this function is to maintain the aquifer yield at a satisfactory level and to prevent the mining of the aquifer when water withdrawals through a specified period of time exceed the aquifer recharge during the same period.

2. *Augmentation of water supply.* Several methods are used to increase the water supply, such as artificial recharge, relocation of wells, or importing water.

3. *Aquifer restoration.* Certain measures should be taken to restore the integrity of the aquifer against pollution and excessive withdrawal. The latter effect would deplete the groundwater levels or piezometric heads, which might require deepening the wells or increasing pump lifts. In coastal aquifers, excessive withdrawal could lead to the expansion of saltwater encroachment (Chap. 9).

As explained in Chap. 4, the management program should be conducted on a regional basis by defining the boundaries of the groundwater basin. In humid areas, generally very few groundwater basins can be considered overdeveloped. However, the development of groundwater without following well-investigated plans creates problems because of the erratic nature of groundwater withdrawal in space and over time. The limitations in groundwater supplies require proper groundwater management. In some arid and semiarid regions where groundwater is the major source of water supply, the basin has to be managed in order to guarantee a perennial water supply. Unfortunately, usually attention is concentrated only on problem areas; other areas are completely ignored until problems surface. This approach of remedial action on a one-to-one basis should be abandoned, and proper management of the entire basin should be sought.

Heavy water withdrawal in a specific area in a groundwater basin from closely spaced wells can create shortages in other areas in the basin and other detrimental effects. If the number, pattern, and capacities of the wells are properly designed, the same total amount of water can frequently be obtained without creating the problems produced by the localized heavy withdrawals. This is one phase of groundwater management in which it can be shown that problems of water shortage and water withdrawals are not due to the lack of water, but to the lack of its proper distribution.

Management plans would be greatly enhanced if they were based on the conjunctive use of surface waters and groundwaters. Generally, in all areas, whether arid, semiarid, or humid, appreciable quantities of surface waters are either wasted to the sea or rejected from infiltration to reach the groundwater. Such waters can be used to augment the groundwater for future use by means of artificial recharge of basin aquifers.

7.1 Technical Procedures of Basin Management

The major technical procedures for adequate groundwater resources management in a well-defined basin are explained in this section. The legal, social, economic, and political aspects are outside the scope of this textbook. The general sequence of procedures is summarized as follows:

1. Evaluation of water needs (present and future) in the basin (see Chap. 4).

2. Data collection and field work.

3. Evaluation of the basin yields and related investigations.

4. Feasibility studies for satisfying the desirable water demands.

5. Planning a comprehensive design for water withdrawal and, if needed, for water recharge. Various plans should be made; the decision making as to the selection of an appropriate plan should be based on the legal, economic, social, and political aspects, as well as experience and proper judgment. Any plan should give in detail its impact on such aspects as changes in the original groundwater configuration, the probable zones of future water shortages, the side effects of water withdrawal,

possible changes in social and economical activities in the region, and possible changes in the quality of the groundwater. The existing and previous activities within the basin should be recognized in any of these plans.

6. Details of implementing the selected plans should be specified. These details should include the methods of engineering construction and the operations procedures and costs.

This sequence of steps may vary depending on the nature of the plan and various local constraints (legal, social, economic, political, etc.).

It is apparent that a comprehensive groundwater management plan is rather complex and costly. It encompasses not only groundwater hydrology, but also other related fields cited earlier in this book. Therefore, it is almost impossible to give fixed rules for proper management that can be applied to all cases. Accordingly, only a summary of the most common aspects are discussed: planning groundwater investigations, the water budget, basin yield, and groundwater recharge. Analytical methods are limited to two cases: (1) seepage from streams and (2) infiltration from recharge basins. The reader should consult the cited references for further studies of these topics and others related to the field of groundwater management.

7.2 Planning Groundwater Investigations

Groundwater management in a certain basin consists of the evaluation of individual projects within the basin as well as their interactions with each other. The main goal is to regulate water usage, restore the aquifer, and augment the water supply whenever necessary. Examples of such individual projects are

1. Planning a large well field for irrigation or industrial or water supply purposes

2. Designing drainage systems for agricultural land or urban areas or other engineering structures such as highways and airports

3. Locating and designing groundwater recharge facilities and their relations with groundwater storage

4. Estimating water losses from surface-water reservoirs or water streams and their effects on groundwater supply and/or their adverse impact on agricultural land

5. Planning conjunctive surface-water and groundwater utilization

The success of any of these projects depends on the experience, judgment, and technical abilities of the specialists in charge. It also depends on the quality and frequency of collected data and experimental tests. Each of these projects and any other project may present unique problems and require different concepts, data, approaches, funding, and time considerations.

A basic data checklist suggested as a guide by the U.S. Bureau of Reclamation (known for a short period as the Water and Power Resources Service) is given in the following (Water and Power Resources Service, 1981). This list is used only as a guide to plan field investigations and tests. The data are normally available in published and unpublished reports and records. On many occasions, however, the data have to be completed by additional investigations, especially in areas where published data are lacking.

1. *Maps, cross sections, and fence diagrams:*
   a. Planimetric
   b. Topographic
   c. Geologic
      (1) Structure
      (2) Stratigraphy
      (3) Lithology
   d. Hydrologic
      (1) Location of wells, observation holes, and springs
      (2) Groundwater table and piezometric contours
      (3) Depth to water
      (4) Quality of water
      (5) Recharge, discharge, and contributing area
   e. Vegetative cover
   f. Soils
   g. Aerial photographs
2. *Data on wells, observation holes, and springs:*
   a. Location, depth, diameter, types of well, and logs
   b. Static and pumping water level, hydrographs, yield, specific capacity, quality of water
   c. Present and projected groundwater development and use
   d. Corrosion, incrustation, well interference, and similar operation and maintenance problems
   e. Location, type, geologic setting, and hydrographs of springs
   f. Observation-well networks
   g. Water-sampling sites
3. *Aquifer data:*
   a. Type, such as unconfined, artesian, or perched
   b. Thickness, depth, and formational designation
   c. Boundaries

    d. Transmissivity, storativity, and permeability
    e. Specific retention
    f. Discharge and recharge
    g. Groundwater and surface-water relationships
    h. Aquifer models
4. *Climatic data:*
    a. Precipitation
    b. Temperature
    c. Evapotranspiration
    d. Wind velocities, directions, and intensities
5. *Surface water:*
    a. Use
    b. Quality
    c. Runoff distribution, reservoir capacities, and inflow and out-flow data
    d. Return flows and section gain or loss
    e. Recording stations
6. *Local drilling facilities and practices:*
    a. Sizes and types of drilling rigs locally available
    b. Logging services locally available
    c. Locally used materials, well designs, and drilling practices
    d. State or local rules and regulations

One of the main purposes of field investigations is to prepare a piezometric contour map for the area of the groundwater project similar to that shown in Fig. 3.14. Such maps are fairly accurate for artesian aquifers, but for unconfined aquifers, their accuracy depends on the depth of the observation wells, as was discussed in Chap. 6. Piezometric contour maps are useful for finding the following:

1. The direction and configuration of groundwater flow.

2. Estimates of local hydraulic conductivities when the groundwater velocities are measured in the field (using tracers, for example).

3. Possible relationships between surface water and groundwater.

4. The location of discharge and recharge wells.

5. Delineating the areas where additional observation wells should be installed.

6. Detecting groundwater mounds, which can result from downward seepage of surface water or upward leakage in areas of local recharge (Water and Power Resources Service, 1981). In an ideal aquifer, the hydraulic gradient is maximum at the center of the mound and decreases radially at a declining rate away from the center. In a piezometric contour map, the flow net would then be finer around the center and would change gradually to a coarser flow net.

7. Detecting unexpected combinations of factors. Contour-line spacings are dependent on the hydraulic gradient, the flow rate, aquifer thickness, and hydraulic conductivity. Thus areal changes in contour spacing may indicate changes in aquifer conditions if the aquifer is practically homogeneous. Owing to the heterogeneity of most aquifers, careful interpretation of the contour lines is required in order to investigate all possible combinations of factors.

8. The change in groundwater storage ΔS_g [Eqs. (7.1) and (7.2)], as explained later.

In stratified formations where more than one aquifer exist, individual piezometric maps should be prepared for each individual aquifer.

Since most aquifers have relatively high degrees of hydraulic conductivity, groundwater levels or piezometric heads are measured in cased observation wells, which are simply stand pipes. Most recordings of groundwater levels and pressures are determined by measuring water levels and pressures in these observation wells. Several different devices exist for making such recordings. The chalked-tape method has an accuracy of about 0.5 cm (0.02 ft) and cannot be used for depths of more than 60 m (200 ft). Steel tape with a popper may also be used (Water and Power Resources Service, 1981). In cases of deep water or when the inner surfaces of the casings are wet as a result of water condensation or leakage from pump columns, an electric water-level indicator can be used. This and other sensitive devices and methods, such as the automatic water-stage recorder or air-line method (Bouwer, 1978) for measuring the water level in a producing well, are commercially available. The water levels in a producing well are usually in disturbance; an access hole for probe insertion may be required together with an air line and gauge. Bordon gauges or mercury manometers are usually suitable for flowing artesian wells (Water and Power Resources Service, 1981).

Recently, groundwater hydrologists have become involved in investigations of leakage (Chaps. 3, 6, and 8) and land subsidence due to well pumping (Chap. 8). These problems are associated with aquitards and aquicludes, which have medium to low hydraulic conductivities. Observation wells are not suitable in such situations because of the relatively large time lag necessary for the water to reach its equilibrium level. In such strata of low to medium hydraulic conductivities, piezometers have to be used. Generally, there are three main types of these piezometers: pneumatic, hydraulic, and electric (Terzaghi and Peck, 1967; Water and Power Resources Service, 1981; and references therein). Each type has its advantages and disadvantages. Piezometers are used not only for complex subsurface conditions where different water levels are encountered, but also for observing the influence of pumping tests (Chap. 8) by minimizing lag time.

It was stated earlier that piezometric maps can be used to estimate the average hydraulic conductivity of an aquifer at a certain location. This requires field determination of groundwater velocities. Tracers can be introduced at one location and detected at some other location in order to study the flow directions and rates. The interpretation of field experiments using tracers is not simple (Kaufman and Todd, 1962; Drost et al., 1968). Selection of tracer type depends on the local conditions. An ideal tracer that overcomes all kinds of shortcomings does not exist. In general, tracers are subject to dispersion, diffusion, dilution, and absorption. Some tracers may react with the water, and others may be uneconomical to use. The selection of a tracer should therefore be based on minimizing the number of problems resulting from its use in a certain aquifer under specified field conditions. Possible tracers include water-soluble dyes, soluble chloride and sulfate salts and sugars, strong electrolytes, and radioisotopes (Todd, 1980; Keeley and Scalf, 1969; Wiebenga et al., 1967). Detection of a tracer is based on one or more of the following effects: color, chemical composition, electric conductivity, nuclear radiation, mass spectrography, and flame spectrophotometry. Organic dyes (mostly sodium fluorescein) are detected in very low concentrations, but they are absorbed. Although most radioactive substances are very sensitive tracers and can be measured at very low mass concentrations, they are affected by base exchange and absorption and they may contaminate the aquifer (Gaspar and Oncescu, 1972). Tritium, which is a radioisotope, has the advantages of not being absorbed and of not contaminating water. The use of tracers to study water motion over long distances is uneconomic because of slow natural water velocities.

Naturally occurring radioisotopes such as tritium (hydrogen 3) and carbon 14 have been used for groundwater dating (Wigley, 1975), as well as in evaluation regional groundwater flow (Back et al., 1970; Vogel, 1970).

A reconnaissance survey, which includes gathering and digesting geologic and engineering information and making one or more site inspections, should first be initiated before any other investigation is made. Then a preliminary exploration program should be conducted and should include at least the following (Water and Power Resources Service, 1981):

1. Existing flow patterns and hydraulic boundaries

2. Geometric boundaries of the aquifer, aquicludes, and aquitards

3. Environmental factors that may affect the existing flow patterns

4. Physical and hydraulic properties of the main and contributing aquifers

5. Site investigations of engineering structures, such as pumping stations and pipelines

6. Proper methods of soil exploration (U.S. Bureau of Reclamation, 1965), including some of or all the following: borings, sounding, soil sampling, field tests, geophysical explorations, aerial photogrammetry and remote sensing, pumping tests, and other hydrologic tests

During the boring operation and soil sampling, some field records are obtained, such as water levels and penetration resistance; the latter is used in engineering projects rather than in hydrologic studies. All geophysical methods have their limitations and should be complemented by boring programs.

7.3 Water Budget of Groundwater Basins

Under normal conditions, the optimum development and management of groundwater are based on equating the average water replenishment (recharge) of the basin with the average water withdrawal from the basin (discharge). The wasted natural discharge should be salvaged in order to keep the required balance. Discharges into seas or oceans are real waste; however, discharges into another basin are a gain to that basin unless it is eventually wasted. Interbasin water transfer may be planned in terms of imported or exported water depending on which basin is managed.

Recharge into or discharge out of a basin may be artificial and/or natural. Natural recharge generally results from deep percolation, precipitation, seepage from streams and lakes, and subsurface underflow. Artificial recharge can be planned by such methods as water spreading and recharge wells or it may be incidental (not originally planned for recharge purposes), such as the deep percolation of irrigation return flow or seepage from surface-water reservoirs. Although irrigation water is a good source of artificial recharge, whether it is planned or not, it may produce a high increase in water-table levels leading to waterlogging of the soil and eventual reductions in crop yield. If the water is higher than a level 1.5 to 2.0 m below the root zone, the land should be drained (Bouwer, 1978).

Examples of natural discharge are groundwater seepage to streams, flow out of springs and flowing artesian wells, subsurface underflow, and evapotranspiration. Artificial discharge is usually produced by pumping wells or infiltration galleries or subdrains.

Since the data obtained to conduct management programs are not void of inaccuracies and the general analyses are based on inexact sciences (such as hydrology), the hydrologic-balance equation cannot be achieved in most individual years in specific basins. Furthermore, the storage of groundwater in a basin cannot be maintained at a constant value because of seasonal and yearly variations. The recharge of aquifers in wet years allows the storage of excess water for use in dry years. The water budget is not restricted to balancing the amount of water available within the basin; it may be necessary to augment the water resources by importing water to the basin whenever water demands exceed water discharge. Within the basin itself, water also can be transferred from one location of excess water to another with a shortage.

The quantitative statement of the balance between water gains and losses in a certain basin during a specified period of time is known as the *water budget* (Walton, 1970). Generally, there are two types that should be considered: the *hydrologic budget* and the *groundwater budget*. The former accounts for all surface and subsurface waters entering, leaving, and stored within the groundwater basin; the latter is restricted to the balance of groundwater, whatever the sources might be.

In a hydrologic budget, inflow I is recharge to the basin originating from precipitation P_i, which consists of rain and snow. Outflow O consists of discharge water that has the components of stream flow Q_{st}, evapotranspiration (ET), and subsurface underflow Q_{sub}. Changes in both soil moisture ΔS_m and groundwater storage ΔS_g balance the difference between inflow and outflow (Schicht and Walton, 1961).

In a groundwater budget, inflow I is the groundwater recharge R_g and outflow O consists of the groundwater runoff Q_g (known also as *base flow*), groundwater evapotranspiration $(ET)_g$, and subsurface underflow Q_{sub}. Groundwater storage ΔS_g is the balance between inflow and outflow in a groundwater budget (Schicht and Walton, 1961).

These budgets are written in equation form as follows:

$$\text{Hydrologic budget:} \quad I - O = P_i - (Q_{st} + ET + Q_{sub})$$
$$= \pm(\Delta S_m + \Delta S_g) \tag{7.1}$$

$$\text{Groundwater budget:} \quad I - O = R_g - [Q_g + (ET)_g + Q_{sub}]$$
$$= \pm \Delta S_g \tag{7.2}$$

In analyzing groundwater basins on a regional scale, inflow $I = P_i$ in the hydrologic budget is not appreciably affected by the relatively small recharges from ponds and the like. In addition, outflow O is not affected by the relatively small discharges of domestic and livestock wells (Walton, 1970). Other small elements can generally be eliminated from the hydrologic budget because they are within the magnitude of error made in evaluating the main elements. Stream flow Q_{st} consists of two components: surface runoff and groundwater runoff (or base flow). Surface runoff reaches the streams more rapidly and is discharged from the basin within a few days, while groundwater runoff moves slowly to the streams.

On many occasions after precipitation ceases by 3 to 5 days, stream flow Q_{st} is derived entirely from groundwater runoff Q_g (Walton, 1970).

Groundwater runoff Q_g or base flow is not subject to a wide range of fluctuations and is indicative of the aquifer characteristics within a basin (Johnston, 1971). However, Singh (1968) presented some factors that affect base flow, such as the storm period and bank storage. *Bank storage* is defined as the volume of water stored in the aquifer after a flood (from the seepage of the stream) and that subsequently released after the flood back to the stream (Cooper and Rorabaugh, 1963). Base flow is estimated in different ways. Schicht and Walton (1961) separated surface runoff and base flow by fitting the graphs: (1) rating curves of mean groundwater levels versus runoff for periods of maximum and minimum evapotranspiration, (2) mean groundwater stage versus time for a 1-year period, and (3) a stream-flow hydrograph for 1 year. Another approach is based on measurements of chemical concentrations (Kunkle, 1965; Visocky, 1970; Pinder and Jones, 1969). In many areas of the United States, maximum groundwater runoff occurs during spring and early summer months; the minimum is usually during late summer and fall months.

The rate of evapotranspiration (ET) includes surface and soil evaporation as well as groundwater evapotranspiration $[(ET)_g]$. The value of ET is much greater than $(ET)_g$ unless the water table is shallow and the area is heavily vegetated. The relative values of the various terms in Eq. (7.1) fluctuate excessively during various seasons or during short periods of time. However, if the time period is considered as 1 year, the variations should be practically the same from year to year. It is preferable to apply the water year extending from October through September rather than a calendar year (Todd, 1980). Moreover, during a period of 1 year, the change in soil moisture ΔS_m is negligible and can practically be canceled out in Eq. (7.1). This simplifies the equation and leads to the evaluation of ET indirectly after estimating the values of P_i, Q_{st}, Q_{sub}, and ΔS_g. Practically, it is customary to evaluate P_i and Q_{st}. Subsurface underflow Q_{sub} is estimated from Darcy's law, while ΔS_g is evaluated from the groundwater levels taken at two different periods of time (Walton, 1970). Each time a piezometric contour map is constructed. The overlay of the two maps permits an estimate of the change in groundwater storage ΔS_g if the average coefficient of storage S has been predetermined (see Chaps. 6 and 8). The volume change determined from both maps multiplied by the porosity gives the change in gross storage in dry soils. Apparently, gross storage cannot be used unless the water table in unconfined aquifers was rising during the time span of both piezometric maps.

In the groundwater budget [Eq. (7.2)], the terms Q_{sub} and ΔS_g are essentially equal in magnitude to those given in Eq. (7.1). Groundwater recharge R_g is only a portion of precipitation P_i that reaches the groundwater. Estimates of R_g have been based on measurements of precipitation and groundwater levels (Jacob, 1943 and 1944; Hantush, 1957; Malmberg, 1965; Domenico, 1972). Groundwater recharge R_g can also be determined from Eq. (7.2) after properly evaluating Q_g, $(ET)_g$, Q_{sub}, and ΔS_g. The last two terms are determined in the same manner as in Eq. (7.1).

An approximate estimate of groundwater evapotranspiration can be determined from rating curves of mean groundwater storage versus groundwater runoff; the difference in groundwater runoff between the curve for the period April through November and the curve for the period November through March is the approximate groundwater evapotranspiration (Walton, 1970). Evapotranspiration increases as the depth to groundwater decreases. For the same water-table level, evapotranspiration of deep-rooted vegetation is greater than that of shallow-rooted vegetation.

Another approximate method to estimate the total quantity of groundwater withdrawn by evapotranspiration during a day was suggested by White (1932). His method was based on hourly records of the rates of rise and drop of water levels during a period of 1 day and the value of the specific yield S_y of the soil near the water table. Transpiration and evaporation produce diurnal fluctuations, the magnitude of which depends on the type of vegetation, the season, and the weather. Maximum water-table levels usually occur in midmorning and then drop gradually until early evening, after which they start to rise again.

Phreatophytes consume large amounts of groundwater. In an extensive study made by Gatewood et al. (1950) on an area of 3765 ha during a period of 12 months, it was found that transpiration from these plants amounted to 0.75 m from groundwater and 0.16 m from precipitation.

Evaporation of groundwater increases as the water-table level increases close to the ground surface. Its rate depends also on the soil structure and capillary tension. Field measurements made by White (1932) indicated that when water tables are within 1 m of the ground surface, evaporation is controlled by atmospheric conditions, but below this depth, the rate of evaporation decreases markedly with depth.

In bare soils, evaporation is less than in vegetated soils. The determination of evaporation in bare soil is rather complicated because it is related to various factors (Gardner and Fireman, 1958; Hellwig, 1973; Veihmeyer and Brooks, 1954). For example, when the surface temperature is very high, an upward vapor movement is established in response to a vapor-pressure gradient (Ripple et al., 1972).

In certain cases, the management of a basin may require exporting or importing water. These waters can be derived from surface waters or groundwaters or both. In the budget equations, the rates of these waters should be appropriately added to inflow or outflow. For example, the rate of exported water should be added to outflow in Eq. (7.1). However, in applying Eq. (7.2) of the groundwater budget, only the portion of exported water that was derived from groundwater is added to outflow.

7.4 Basin Yield

In managing a groundwater basin, one of the main goals is to evaluate the maximum annual groundwater yield of the basin that can be withdrawn and used without producing undesirable effects. Investigation of such yields has been accompanied by the introduction of certain terms, such as safe yield, sustained yield, mining yield, and perennial yield, that should be explained. Some of these terms are controversial and have been subjected to long debates. Various concepts related to these terms are explained in the following section.

Safe Yield (or Sustained Yield)

Safe yield was first defined by Lee (1915) as "the limit to the quantity of water which can be withdrawn regularly and permanently without depletion of the storage reserve." It has been observed that *regular* and *permanent* water withdrawal are too vague to be specified. Moreover, the undesirable effects were in the past confined only to the depletion of storage. Many others attempted to clarify the definition of safe yield by specifying a time period for withdrawal and/or expanding on the undesirable effects. The main definitions, other than Lee's, are summarized in the following paragraphs.

Meinzer (1923) defined safe yield as "the rate at which water can be withdrawn from an aquifer for human use without depleting the supply to the extent that withdrawal at this rate is no longer economically feasible." The undesirable effects were concentrated only as water depletion in both Lee's and Meinzer's definition. However, the economic factor was emphasized in the latter.

Conkling (1946) expanded Meinzer's definition by specifying the period of withdrawal as 1 year and stating that the safe yield should not

1. Exceed average annual recharge

2. Lower the water table so that the permissible cost of pumping is exceeded

3. Lower the water table so as to permit intrusion of water of undesirable quality

It should be noted that the quality criterion was added in Conkling's definition as compared with Lee's and Meinzer's definitions.

Banks (1953) added a fourth condition: the protection of existing rights.

Todd (1959a) grouped all the detrimental conditions as "undesired result" and briefly defined the safe yield as "the amount of water which can be withdrawn (from a groundwater basin) annually without producing an undesired result."

As stated by Domenico (1972), "The controversial nature of the concept is clearly demonstrated in 43 pages of discussion of Conkling's (1946) 28-page paper, by no less than 10 authorities." Thomas (1951) and Kazmann (1956) have suggested abandonment of the term because of its indefiniteness.

Methods of determining safe yield are generally based on field-collected data. The Hill method (Domenico, 1972) is based on linear plotting between the average annual change in water levels and the average annual pumpage. Safe yield is then determined as the average annual pumpage corresponding to zero average annual change in water levels. Other methods are also used (Domenico, 1972). These methods indicate that safe yield cannot have a unique value because its value is determined on the basis of historical data during arbitrarily selected time periods. Thus the term is meaningless from a hydrologic standpoint (Todd, 1980).

Modification of the Safe-Yield Concept

The concept of safe yield (sustained yield) has been modified by introducing two types of yield whose values vary with time (American Society of Civil Engineers, 1972): mining yield and perennial yield. Any quantitative determination for these two types is based on determined and/or assumed specified operating conditions. If any of these conditions changes over time, the value of the yield also changes. The safe-yield concept as well as the concepts of mining and perennial yield have one common feature, namely, that the annual yield should not produce undesirable results.

These adverse effects are generally caused by exceeding the annual recharge and in turn lowering water-table levels or decreasing piezometric heads. The progressive reduction of the groundwater reserve results in uneconomic pumping conditions. The lowered water levels increase pumping lifts, which may necessitate deepening wells, lowering pump bowls, and/or installing larger pumps. These changes and repairs are usually costly and time-consuming. They disrupt normal operations, such as shutting down some production wells during changes. Excessive water withdrawal also may interfere with existing water rights

(McClesky, 1972), may lead to land subsidence (Mann, 1960, 1963; Young, 1970; and Chap. 8), may increase the possibility of pumping underlying connate brines or increasing the progress of saltwater intrusion in coastal aquifers (Chap. 9), may lead to degradation of water quality, and may produce negative environmental and social impacts (Domenico, 1972). However, some benefits may be obtained from lowering water levels (Todd, 1980), such as increases in subsurface inflow and groundwater recharge from losing streams, decreases in subsurface outflow and discharge into gaining streams, and reductions in uneconomic evapotranspiration losses. Rising water levels have reverse effects, but they may produce undesirable effects in agricultural land, especially waterlogging, which decreases crop yield.

Again, it should be emphasized that the main difference in the concepts of safe yield and perennial yield introduced by the American Society of Civil Engineers (1972) is that the safe-yield concept is based on a unique value, whereas the perennial-yield concept is based on changing values according to the nature of the groundwater levels within the basin.

When pumping exceeds the perennial supply, one or more of the following corrective measures should be taken (Walton, 1970):

1. Water conservation by preventing waste.

2. Proportionate reduction of pumping capacities or pumping periods in production wells.

3. Prohibition of further development or, whenever economic conditions permit, water importation from areas of abundant water. Such areas may be located in other basins or within the same managed basin (e.g., during the wet season).

4. Recycling and reclaiming used water.

5. Artificial groundwater recharge from surplus stream waters or by other methods, as explained later. The success of these recharge methods depends on the aquifer transmissivity; if it is inadequate, the water levels continue to decline even if the recharge method is adequate.

Mining Yield

When annual withdrawn groundwater exceeds annual replenishment, the excess water should be released from storage (Chaps. 6 and 8). This process lowers water tables and piezometric surfaces to irreversible positions if the withdrawal is continuous. The water yield under these conditions is known as the *mining yield* (Domenico, 1972). In other words, the groundwater resource is treated as any other mineral source and its mining is, in fact, storage depletion.

In the past, geologists estimated the available groundwater reserve in an aquifer on the basis of the total volume of the pores. As indicated in Chap. 6 and elsewhere in this textbook, the water within the pores cannot be fully extracted. Moreover, the techniques used for withdrawal (mainly wells) are not capable of extracting all stored water. For example, in an artesian aquifer with an area A, porosity n, depth D_q, and average coefficient of storage S, the theoretical available volume of water is $A \times D_q \times S < A \times D_q \times n$, but the water wells in such an area cannot withdraw this amount because of the limitations imposed by groundwater velocities, time, configuration of water movement, and withdrawal techniques. Thus there is a difference between mining groundwater and mining a mineral resource. In the latter case, the entire amount of mineral can be taken out, while in the case of groundwater, an appreciable amount of water is left intact. Thus the process of mining groundwater implies a continuous water withdrawal resulting in a continuously depleted water supply, but one that is limited by technical means and economic factors.

Mining yield is of two types (Walton, 1970; American Society of Civil Engineers, 1972):

1. *Maximum mining yield.*   This is the total volume of water in storage that can be feasibly extracted and utilized but cannot be replenished.

2. *Permissive mining yield.*   This is the maximum volume of water in storage that can be economically and legally extracted without producing an undesirable effect. The permissive mining yield is a portion of the maximum mining yield.

Pumping lifts constitute the most outstanding economic factor. It should be noted that there is a time limit (a decade or several decades) for mining a groundwater basin beyond which continuous pumping may not be possible or during which more economic methods become available, such as water importation or desalting.

Water mining is justified by several workers, while others are against it. For example, Sasman and Schicht (1978) stated that water in storage is of no value unless it is used. As explained earlier, large amounts of water in storage cannot be used and the existence of this unusable water cannot be looked on as "of no value." The existence of such deep water in storage, in fact, forms a foundation for shallow waters that can be economically exploited. Experience has shown that there are more drawbacks to mining water than benefits gained (Kashef, 1981). In the author's opinion, groundwater mining is nothing more than the result of a selfish attitude of a present generation that ignores the deleterious effects on future generations. A similar example is present-day water pollution, which knowingly or unknowingly resulted from a complete disregard for other water users (Renshaw, 1963; Kashef, 1971).

In estimating maximum mining yield or permissive yield, all the undesired effects should be taken into consideration, except, of course, the progressive reductions in groundwater resources on which these types of yield are based.

Perennial Yield

The *perennial yield* of a groundwater basin defines the *variable* annual yield that can be continuously withdrawn under specified managerial procedures without producing undesired results. It is measured in a rate unit. Any water withdrawal in excess of the annual perennial rate is known as an *overdraft.* Problem areas of overdraft constitute the main groundwater problem in the United States (Thomas, 1951). The perennial-yield estimate varies from year to year and depends on the patterns of recharge, development, and water use in the basin. These constraints, as well as consideration of undesired results, may limit the perennial yield to a value lesser than the water recharged to the basin that should reach the water table in unconfined aquifers or join the flow in artesian aquifers. The perennial yield is estimated on the basis of the hydrology of the basin and Eq. (7.2) (Mann, 1969).

The perennial yield is further subdivided to two types:

1. Deferred perennial yield
2. Maximum perennial yield

In *deferred perennial yield,* two different pumping rates are applied. The initial rate is larger and produces a rate that exceeds the perennial yield (overdraft) without producing undesirable effects. The second rate is smaller than the first and starts after predetermined water levels are reached. Pumping in the second stage corresponds to the perennial-yield value that should be maintained thereafter. The main reason for such a technique is to increase the storage capacity of the aquifer during the first stage of pumping, which allows a larger perennial yield during the second stage (Peters, 1972).

Maximum perennial yield is determined by considering all possible and available sources, including artificial and natural recharge, water importation, and surface water. This leads to management of the basin as one unit in the most economic way because of the conjunctive use of both groundwater and surface-water sources, which are more costly to operate independently. Increases in perennial yield in unconfined aquifers may be accomplished by relocating wells near recharging sources such as streams to make use of induced infiltration. However, in confined aquifers, the rate of flow within the aquifer governs the perennial yield (Todd, 1980). In large confined aquifers, continuous pumping from storage may last for many years without establishing an equilibrium with basin recharge (Todd, 1980). The perennial yield is subject to change due to the changes in vegetation, purpose of pumping, water-quality standards, power costs, and urbanization.

Usually a reasonably reliable estimate of perennial yield can be made after the occurrence of an overdraft whenever water shortages or other detrimental effects have occurred.

7.5   Recharge to Groundwater

Groundwater in a certain basin is recharged from either surface water within the basin or groundwater percolating from another basin. In both cases, the recharge may be natural or artificial.

Natural Recharge

Natural recharge is usually produced under one or more of the following conditions:

1. *Deep infiltration of precipitation.*   As stated earlier, the amount of infiltrated water varies from one area to another and varies seasonally within the same area. The net amount of infiltration (after excluding evapotranspiration or water that did not reach the groundwater body) is influenced by topography, vegetative cover, type of surface and subsurface soils, the intensity and frequency of rain and snow, and presaturation conditions of the upper soil layers.

2. *Seepage from surface water.*   Surface-water bodies in streams, lakes, or natural water flooding on the land seep downward until they reach groundwater bodies. The amount of such seepage depends on soil characteristics, topography, depth to groundwater, and climatic conditions. This amount is very small if the water table is very high in humid and subhumid areas. In arid regions, the streams may lose all their waters into the aquifer. However, in irrigated areas of arid regions where the water table is very high and the soil is waterlogged due to the low efficiency of irrigation managements, seepage from surface water becomes practically negligible.

3. *Underflow from another basin.*   A certain groundwater basin may be recharged by underflow from a nearby hydraulically connected basin. Also within the same basin an aquifer may be recharged from nearby aquifers. The amount of this recharge depends on several factors, such as the configuration of flow, the extent of hydraulically connected boundaries, and the physical properties of the aquifers.

In unconfined aquifers, recharge areas may cover the entire aquifer or may be limited by the presence of natural or artificial impermeable materials. In artesian aquifers, recharge areas are usually restricted to their outcrops, unless they are hydraulically connected with other aquifers or stream bodies. Recharge water may percolate from an adjacent area to an aquifer and eventually supply the aquifer. The combination of such an adjacent recharge area and that above the aquifer is termed a *contributing area* (Water and Power Resources Service, 1981).

The amount of recharge changes also from season to season. During the season of high temperatures, evapotranspiration and crop water requirements are usually excessive, permitting little rain to infiltrate. When the ground is frozen during winter months, recharge is almost nil. Over a long period of time, the rate of groundwater recharge can be estimated by calculating the difference between infiltration and evapotranspiration (Chap. 4 and Sec. 7.3). Infiltration is the net portion of rainfall that percolates in the ground excluding runoff and evapotranspiration. Deep percolation has also been estimated by field tests (Bouwer, 1978). Detailed techniques for estimating infiltration, surface runoff, evapotranspiration, and deep percolation can be found in many standard textbooks on hydrology and agricultural engineering (see, for example, the references cited by DeWiest, 1965; Walton, 1970; Bouwer, 1978; Todd, 1980 and the references cited in Chaps. 4 and 7 of this book).

Artificial Recharge

The natural supply of groundwater may be augmented by recharging the groundwater basin artificially. Planned artificial recharge is conducted by various methods (International Association of Scientific Hydrology, 1967; Signor et al., 1970; Todd, 1959b; Water Research Association, 1971), such as spreading grounds, infiltration ponds, or recharge wells and galleries. Unforeseen or unintentional recharge is derived from activities planned for other purposes (Water and Power Resources Service, 1981; United Nations, 1975). Examples of these incidental recharges are irrigation application, spreading sewage effluent on the ground, seepage from septic tanks, natural ponds, canals and ditches, and leakage of water mains. Downward leakage due to discharge wells is another phase of incidental artificial recharge that may compensate for part of the withdrawn water. Polluted water may reach the groundwater from polluted surface waters and seepage from septic tanks, landfills and waste-disposal facilities (Bouwer, 1974; Bouwer et al., 1978; Schicht, 1971, Baier and Wesner, 1971). This is not, of course, desirable. Large amounts of incidental recharge derive from water-impounding facilities, such as surface-water reservoirs in the upstream of dams and weirs.

Artificial recharge from natural streams can be enhanced by induced infiltration (Walton, 1963, 1970). Pumping production wells are installed near the stream and discharged continuously until the drawdowns are lowered to permit the stream water to recharge the aquifer. The rates of discharge, recharge, and drawdowns should be investigated to produce optimum recharge conditions (Chap. 8). Induced recharge ensures a continuous water supply and provides water free from organic matters and pathogenic bacteria (Kazmann, 1948; Klaer, 1953). If water is pumped directly from a stream, its temperature is not constant through the years, it has a certain degree of turbidity, and it may have a low water quality. Induced infiltration allows the pumping of a blend of surface water and groundwater free from turbidity and of almost constant temperature. As already mentioned elsewhere, groundwater temperature is almost uniform throughout the year in a certain location and is not affected by surface temperatures, as is surface water. In addition, the mixture of surface water and groundwater obtained from induced recharge has the benefit of decreasing the mineralized nature of groundwater and allowing the withdrawal of a good-quality water free of turbidity. The rate of induced recharge would increase if the water table is maintained below the streambed. It is preferable to locate wells near segments of the stream where the surface-water velocities are relatively high; this eliminates the possibility of silting in these segments and the subsequent obstruction of flow. The mechanics of flow toward wells located near streams is discussed in detail in Chap. 8. Mapping of the produced drawdowns would be useful in estimating the rates of induced recharge over time.

Other than augmenting the groundwater supply, artificial recharge is used to dispose of storm-water runoff and store it for future use. The stored water reduces the costs of pumping and piping. Artificial recharge is also one of the principal means of saltwater control in coastal aquifers (Chap. 9). Injecting hot water from industrial complexes into the ground cools the water for reuse by pumping it again from the ground (see Example 8.4). As previously stated, groundwater temperature is almost uniform throughout the year in a certain location. Therefore, groundwater temperature should be cooler than surface-water temperature during summer and warmer during the winter. This phenomenon has been used practically for energy purposes (Meyer, 1976; Meyer and Todd, 1973; Kazmann, 1978). Employing infiltration basins, artificial recharge is useful in treating sewage effluent or other wastewater. In such cases, the vadose zone and upper layers of the aquifers act as natural filters. Recharge wells are effective in stopping or reducing land subsidence produced by withdrawing large amounts of groundwater (Chap. 8). Artificial recharge provides flexibility in coordinating between surface water and groundwater (conjunctive use) and provides subsurface storage for local and imported water.

The spreading methods of recharge (Muckel, 1959) entail the spreading of water over the ground surface for an extended period of time. These methods consist of various techniques that depend on the physical and economic feasibility of the investigated area of recharge:

1. *Basin method.* This method consists of excavating basins in the ground or surrounding the natural ground surface with dikes or levees. The layout of a basin or a series of basins and methods of their construction and maintenance can be found in Todd (1959*b* and 1980), Task Group on Artificial Ground Water Recharge (1965), Bianchi and Muckel (1970), Richter and Chung (1959), Goss et al. (1973), Nightingale and Bianchi (1973), Bulten et al. (1974), and Bouwer (1978). It is of interest to point out that the hydraulic techniques used in a series of basins (American Society of Civil Engineers, 1972) are similar to those used during the pharaohs' dynasties in Egypt in "the basin irrigation systems" (the last of these basins was abandoned recently, about 1970, after the construction of the High Aswan Dam).

2. *Flooding and irrigation methods.* These methods entail the flooding of a large area of relatively flat land or deliberately using excess irrigation water during the nonirrigating seasons. The operation of these methods requires close observation and regulation of the canals, ditches, and gullies within the spreading area.

3. *Recharge from depressed areas.* This method includes the recharge applied to losing natural stream channels or specially constructed networks of ditches and furrows or pits (Bianchi and Muckel, 1970; Kelly, 1967; McWhorter and Brookman, 1972; Dvoracek and Scott, 1963; Scott and Aron, 1967; Suter and Harmeson, 1960).

4. *Method of water recharge by wells.* Theoretically speaking, recharge wells produce the reverse effects of discharge wells (Chap. 8). The theoretical equations are essentially the same except for changing signs (Chap. 8). Practically, however, the construction may be different (Task Group on Artificial Ground Water Recharge, 1965). It seems that the common practical problems in water-well recharge are (Kashef, 1971) water turbidity, clogging of wells, accumulated air problems, effect of injection and shutdown periods, and study of the effectiveness of the procedure (Sternau, 1967; Rahman et al., 1969; Rebhun and Schwartz, 1968; Harpaz, 1971; Hauser and Lotspeich, 1967; Valliant, 1964; Todd, 1980; Behnke, 1969; Ripley and Saleem, 1973; Berend, 1967, and Bouwer, 1978).

Examples of practical projects of artificial recharge are those on Long Island, New York (Parker et al., 1967; Prill and Aaronson, 1973; Seaburn and Aaronson, 1974; Johnson, 1948; Baffa, 1970; Koch et al., 1973;

Ragone, 1977; Vecchioli, 1972), Sweden (Jansa, 1951, 1954), New Jersey (Barksdale and De Buchananne, 1946; Groot, 1960), and Peoria, Illinois (Suter, 1954).

7.6 Analytical Methods

Generally, the procedures of planning, designing, and operating a basin are based on various analytical methods incorporating the analysis of numerous field data and laboratory tests (including statistical analyses) and the use of different types of models and optimization techniques (Domenico, 1972). All these investigations required for managing a groundwater basin demand skill, experience, imagination, and sound judgment on the part of the groundwater technical specialists involved. The decision making in selecting a plan and the final layout should be considered as unique to the basin involved. No typical procedures can be established for all cases and conditions. This is so in any engineering project.

The preceding sections addressed briefly highlights of the different elements in managing a groundwater basin, hoping to stimulate those who have not as yet given the subject some attention. In the author's opinion, the cited references are more or less comprehensive or at least can be used as the basis for further studies. In addition to the references cited already in each topic discussed, detailed information on coordinated basin management in the United States and elsewhere is given by Beaver and Frankel (1969), Bittinger (1964), Brown et al. (1978), California Department of Water Resources (1968), Conover (1961), Downing et al. (1974), Franke and McClymonds (1972), Hansen (1970), Johnson et al. (1972), Malmberg (1975), Moore and Snyder (1969), Morel-Seytoux (1975), Mundorff et al. (1976), Nace (1960), Orlob and Dendy (1973), Ralston (1973), Rayner (1972), Revelle and Lakshminarayana (1975), Taylor and Luckey (1974), Todd (1980), Walton (1970), and Weschler (1968).

It has already been noted that determination of the elements of the water budget cannot be based on pure analytical methods. Some of these elements belong to fields other than groundwater. For example, studies of precipitation, evapotranspiration, groundwater runoff (base flow in streams), surface runoff, stream flow, and evaporation of surface water have been investigated by hydrologists in the past, long before the advances in groundwater sciences or groundwater-resources management were established. These relatively new fields are characterized by certain complexities; they include, besides the engineering aspects, a complex fabric encompassing social, economic, legal, administrative, and political constraints. Practical engineers are accustomed to closed-form solutions to their routine problems. However, in real-life engineering projects, more than one plan is required, and in each plan, various elements should be thoroughly evaluated. Such an approach is even more important in the management of groundwater basins. Direct solutions may be available for certain elements, such as seepage characteristics from surface water (in ponds, lakes, streams, canals, pits, spreading basin, and irrigation water). Other elements may require more sophisticated approaches; examples of these are perennial yield of a basin (Sec. 7.4), irrigation efficiency, characterization and migration of water pollution, the shapes of the plumes (Chap. 4), and the best locations of wells to activate induced infiltration. Most probably in these cases computerized numerical solutions (based on finite-difference and/or finite-element techniques) would be used and different models would be attempted. Modeling techniques should be supported with reliable data obtained from a well-planned program of field monitoring and laboratory tests. The final and most important stage is to synthesize these elements within the framework of management of the entire basin and come up with an acceptable plan. Such a plan should be flexible in order to adjust the operations whenever necessary, should be able to predict future impacts, and should satisfy the users within the political, economic, legal, social, and administrative constraints. Augmentation of water supply and restoration of the aquifer should always be considered in these plans.

In the following section, analytical solutions for only two of the common elements used in planning management procedures are presented. Other elements are beyond the scope of this book.

Seepage from Streams

The configuration and amount of seepage from ditches, canals, and natural streams within a groundwater basin are required in the evaluation of basin perennial yield and generally in the managing process. Methods of solution have been presented by Muskat (1937) and Polubarinova-Kochina (1962). Infiltration solutions, in general, were given by Philip (1969), Selim and Kirkham (1973), and Zyvoloski et al. (1976). Solutions supplemented with dimensionless charts were also given by Bouwer (1978).

The analytical methods given by Muskat and Polubarinova-Kochina are based on rigorous solutions including inverse methods. The results of some cases are given. All these solutions are based on steady-state flow. The solutions given by Bouwer (1978) are based on resistance-network analogs (R analogs), and he gave partial results in dimensionless charts. Bouwer (1978) also presented some transient solutions.

In this case, approximate solutions can be obtained on the basis of Dupuit's assumptions if the level of the natural groundwater table is known (Bouwer, 1978). In other cases, Dupuit-Forchheimer assumptions are used (Polubarinova-Kochina, 1962). Generally, the flow-net method (Chap. 5) can be used for approximations once the boundary conditions are well defined.

The method given by Polubarinova-Kochina (1962) is an application of a semi-inverse method. The solution was determined before the shape of the stream was known. The latter was determined from the solution. The shapes of a stream with a ratio $B_w/d_c = 4.5$ is shown in Fig. 7.1, where B_w is the width of the top of the stream and d_c is the maximum depth of water at the center of the stream. With respect to the x and y axes shown in Fig. 7.1, the shape of the stream is given by the following equation:

$$\pm x = -\sqrt{d_c^2 - y^2} + \frac{B_w + 2d_c}{\pi} \cos^{-1} \frac{y}{d_c} \qquad (7.3)$$

The aquifer is assumed to extend downward to a highly pervious stratum at infinity.

The rate of seepage q from the perimeter of the stream (per unit length of the stream) is given by

$$q = K(B_w + 2d_c) \qquad (7.4)$$

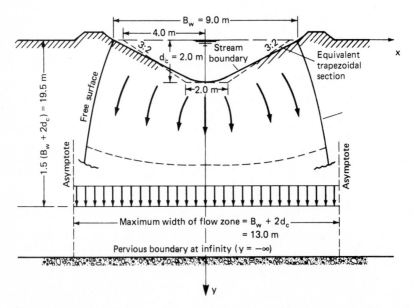

Figure 7.1 Seepage from a stream with a curved perimeter; pervious lower boundary at infinity.

The asymptotes of the two free surfaces at both sides of the stream (tangents at infinity) are at a distance $(B_w + 2d_c)$. In other words, the rate of flow q = area × darcian velocity $v_{y=\infty}$ at the pervious surface (at infinity); thus from Eq. (7.4),

$$q = K(B_w + 2d_c) = (B + 2d_c)v_{y=\infty} \tag{7.5}$$

or

$$v_{y=\infty} = K$$

It has been found in this solution that at a depth $y = 1.5(B_w + 2d_c)$, the distance between the two free surfaces is very close to the distance between the asymptotes. Thus the solution can be seen as giving the seepage from a stream into an aquifer overlying a pervious layer at a finite depth of about $y = 1.5(B_w + 2d_c)$ or greater.

According to Polubarinova-Kochina (1962), the preceding solution was first given by Kozeny but later proved in another way by several others. From a practical point of view, the shape of the stream can be approximated to the shape of an actual stream under investigation by plotting various sections in accordance with Eq. (7.3) for various values of B_w and d_c.

When the lower boundary is impervious (at infinity), as shown in Fig. 7.2, the following equations are obtained (Polubarinova-Kochina, 1962):

$$\pm x = \sqrt{d_c^2 - y^2} + \frac{B_w - 2d_c}{\pi} \cos^{-1}\frac{y}{d_c} \qquad \text{(stream equation)} \tag{7.6}$$

$$q = K(B_w - 2d_c) \tag{7.7}$$

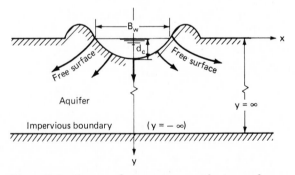

Figure 7.2 Seepage from a stream with a curved perimeter; impervious lower boundary at infinity.

The equation for the free surface (right-hand side free surface) is

$$x = d_c \exp\left(\frac{\pi Ky}{q}\right) + \frac{q}{2K} \tag{7.8}$$

The left-hand side of the free surface (Fig. 7.2) is symmetrical with the right-hand side. In this case, the darcian velocity at the impervious boundary is equal to zero.

A more practical trapezoidal section of a canal was considered, but again with a theoretically pervious boundary at $y = -\infty$ (Polubarinova-Kochina, 1962). The darcian velocity at this boundary is equal to K. The rate of flow q is given by

$$q = K(B_w + A_q d_c) \tag{7.9}$$

B_w and d_c are as indicated in Fig. 7.1, and A_q is a numerical coefficient that varies between 2.25 and 3.25 for most practical trapezoidal sections. Table 7.1 gives the x and y values using the preceding equation in drawing the stream boundary shown in Fig. 7.1 ($B_w/d_c = 4.5$). The aquifer is assumed to extend downward to infinity, where a highly pervious stratum exists.

Infiltration from Recharge Basins

As explained earlier, artificial recharge can be accomplished by the spreading methods either by flooding a basin, ditch and furrow, or natural channel or by irrigation. Another general method is to fill recharge basins with water from nearby surface-water sources either by pumping the water or by diverting streams to the basins. The primary purpose of these methods is to allow water to infiltrate into the soil from a relatively large area and over an extended period of time. High infiltration rates are desirable. However, infiltration rates decrease with time. At the beginning, infiltration rates are small because of dispersion and swelling of the initially dry soil particles. After passing this stage, infiltration rates in-

TABLE 7.1 Plotting the Stream Boundary in Accordance with Eq. (7.3)*

y	$-\sqrt{d_c^2 - y^2}$	$\dfrac{B_w + 2d_c}{\pi} \cos^{-1} \dfrac{y}{d_c}$	x
2.00	0	0	0
1.75	-0.968	2.091	1.123
1.50	-1.323	2.991	1.668
1.25	-1.561	3.706	2.145
1.00	-1.732	4.333	2.601
0.75	-1.854	4.909	3.055
0.50	-1.936	5.454	3.518
0.25	-1.984	5.981	3.997
0.00	-2.000	6.500	4.500

* $B_w = 9.0$ m, $d_c = 2$ m (see Fig. 1.7); note that all figures are given in meters.

crease as a result of the dissolving of entrapped air in the water. Finally, the infiltration rates decrease gradually because of microbial growth (Walton, 1970) unless the soil is treated with organic matter and chemicals. Alternating wet and dry periods in operating the basins allow more water to infiltrate over a certain period of time rather than continuously maintaining the water levels in these basins.

The seepage rates from losing natural streams are given approximately by Eqs. (7.4) and (7.7) depending on the lower boundary of the aquifer. In the ditch or furrow method, the channels are usually shallow, flat-bottomed, and closely spaced. Therefore, the surface water in these channels may be assumed interconnected and the infiltration would be similar to that of a large basin. The analytical methods for such infiltration are given in the following paragraphs.

Infiltration from a recharge basin produces a groundwater mound above the original water table such as that shown diagrammatically in Fig. 7.3. The groundwater mound grows over time, and once the infiltration stops, it decays gradually. The growth and decay of a groundwater mound was analyzed by Bittinger and Trelease (1965), Glover (1964), Hantush (1967), Bianchi and Muckel (1970), Marino (1975a and b), and several others. For engineering problems, it may be preferable to use the methods of Glover (1964) and Hantush (1967).

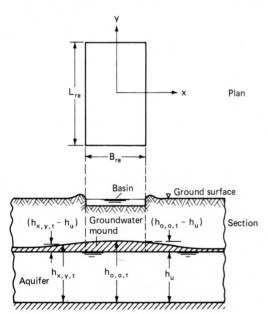

Figure 7.3 Groundwater mounds beneath rectangular recharge basins.

The Glover method (1964) was presented by Bianchi and Muckel (1970) in a dimensionless chart (Fig. 7.4) that gives the rise of the center groundwater mound below a square or rectangular area. Circular areas are treated as squares with the same surface area. Four curves are given in the chart for values of $L_{re}/B_{re} = 1, 2, 4$, and ∞, where L_{re} and B_{re} are, respectively, the length and breadth of the recharge basin in the directions shown in Fig. 7.3. These curves give the relationship between $B_{re}/\sqrt{4Tt/S_y}$ and $(h_{0,0,t} - h_u)/(v_{am}t/S_y)$, where T is the coefficient of transmissivity of the aquifer, S_y is the specific yield, t is the time after recharge starts to reach the water table, v_{am} is the arrival recharge rate per unit area (velocity unit) at the water table of infiltrated water, and $h_{0,0,t}$ and h_u are as indicated in Fig. 7.3. The constant value v_{am} is usually computed indirectly from the quantity of water necessary to maintain a constant water level in the basin.

The curves shown in Fig. 7.4 indicate that when $L_{re}/B_{re} \geq 4$, the mound rises at essentially the same rate as that below an infinitely long recharge basin. The curve corresponding to $L_{re}/B_{re} = \infty$ may be approximately used for recharge of long streams or trenches unless their bottoms are close to the water table, in which case the infiltration may be approximately determined by a flow net assuming the water table as equivalent to an impervious boundary.

Bianchi and Muckel (1970) also gave dimensionless charts for *square*

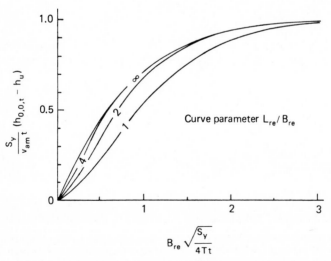

Figure 7.4 Dimensionless graph to determine the rise of center of groundwater mound. *(W. C. Bianchi and D. C. Muckel, "Ground-Water Recharge Hydrology," Agricultural Research Service 41-161, U.S. Department of Agriculture, Washington, 1970.)*

recharge basins, by which the growth of the mound can be obtained below any point (not restricted to the center, as given by Fig. 7.4). The decay in the mound can also be obtained by considering a negative recharge after the cessation of infiltration. For example, if infiltration ceases after time t_1, then the decay after time t' measured after the cessation of infiltration will be determined as the difference between the growth after time $t = t_1 + t'$ and the negative growth throughout time t'.

EXAMPLE 7.1 A recharge basin 50 m wide and 100 m long is used to recharge an unconfined aquifer in which the natural water table is 10 m above the impervious boundary. Determine the rise in the mound below the center of the basin after 15 days from the instant the infiltrated water reached the water table. The infiltration rate $v_{am} = 17$ cm per day, and the coefficients of transmissivity T and specific yield S_y are, respectively, 0.034 m² per minute and 0.12.

Solution Since $t = 15$ days, $v_{am} = 17$ cm per day $= 0.17$ m per day, and $T = 0.034$ m² per minute $= 48.96$ m² per day, then

$$\frac{B_{re}}{\sqrt{4Tt/S_y}} = \frac{50}{\sqrt{4 \times 48.96 \times 15/0.12}} = 0.32$$

Since $L_{re}/B_{re} = 100/50 = 2$, then from Fig. 7.4,

$$\frac{h_{0,0,t} - h_u}{v_{am}t/S_y} \approx 0.2 \quad \text{or} \quad h_{0,0,t} - h_u = 0.2v_{am}t/S_y$$

or the rise in the groundwater mound after 15 days below the center of the basin equals $0.2 \times 0.17 \times 15/0.12 = 4.25$ m.

Hantush (1967) developed equations to predict the growth and decay of groundwater mounds below rectangular recharge basins. The equation for growth is written as

$$h_{x,y,t} - h_u = \frac{v_{am}t}{4S_y} [F(\alpha_1, \beta_1) + F(\alpha_1, \beta_2) + F(\alpha_2, \beta_1) + F(\alpha_2, \beta_2)]$$

$$\alpha_1 = \left(\frac{B_{re}}{2} + x\right)\sqrt{S_y/4Tt} \qquad \beta_1 = \left(\frac{L_{re}}{2} + y\right)\sqrt{S_y/4Tt} \quad (7.10)$$

$$\alpha_2 = \left(\frac{B_{re}}{2} - x\right)\sqrt{S_y/4Tt} \qquad \beta_2 = \left(\frac{L_{re}}{2} - y\right)\sqrt{S_y/4Tt}$$

Other symbols have already been defined. The values of the function $F(\alpha, \beta)$ are tabulated in Table 7.2. In Hantush's method, growth (as well as decay) can be obtained below any point within or outside the boundary of the basin. Equation (7.10) is valid only if $h_{x,y,t} - h_u < 0.5h_u$.

The equation of decay is given by Hantush (1967) as

$$h_{x,y,t} - h_u = Z(x, y, t) - Z(x, y, t - t') \qquad (7.11)$$

TABLE 7.2 Values of the Function $F(\alpha, \beta)$ for Different Values of α and β

α \ β	0.02	0.04	0.06	0.08	0.10	0.14	0.18	0.22	0.26	0.30	0.34	0.38	0.42	0.46	0.50	0.54	0.58	0.62
0.02	0.0041	0.0073	0.0101	0.0125	0.0146	0.0184	0.0216	0.0243	0.0267	0.0288	0.0306	0.0322	0.0337	0.0349	0.0361	0.0371	0.0380	0.0387
0.04	0.0073	0.0135	0.0188	0.0236	0.0278	0.0353	0.0416	0.0470	0.0518	0.0559	0.0596	0.0628	0.0657	0.0683	0.0705	0.0725	0.0743	0.0759
0.06	0.0101	0.0188	0.0266	0.0335	0.0398	0.0509	0.0602	0.0684	0.0754	0.0817	0.0871	0.0920	0.0963	0.1001	0.1035	0.1065	0.1091	0.1115
0.08	0.0125	0.0236	0.0335	0.0425	0.0508	0.0652	0.0776	0.0884	0.0978	0.1060	0.1133	0.1197	0.1254	0.1305	0.1350	0.1389	0.1425	0.1456
0.10	0.0146	0.0278	0.0398	0.0508	0.0608	0.0786	0.0939	0.1072	0.1188	0.1290	0.1381	0.1461	0.1532	0.1595	0.1650	0.1700	0.1744	0.1783
0.14	0.0184	0.0353	0.0509	0.0652	0.0786	0.1025	0.1232	0.1414	0.1573	0.1714	0.1839	0.1949	0.2048	0.2135	0.2212	0.2281	0.2343	0.2397
0.18	0.0216	0.0416	0.0602	0.0776	0.0939	0.1232	0.1490	0.1716	0.1916	0.2094	0.2251	0.2391	0.2515	0.2626	0.2724	0.2812	0.2890	0.2959
0.22	0.0243	0.0470	0.0684	0.0884	0.1072	0.1414	0.1716	0.1984	0.2222	0.2433	0.2621	0.2789	0.2938	0.3071	0.3189	0.3295	0.3389	0.3472
0.26	0.0267	0.0518	0.0754	0.0978	0.1188	0.1573	0.1916	0.2222	0.2494	0.2737	0.2954	0.3147	0.3320	0.3474	0.3612	0.3735	0.3844	0.3941
0.30	0.0288	0.0559	0.0817	0.1060	0.1290	0.1714	0.2094	0.2433	0.2737	0.3009	0.3252	0.3470	0.3665	0.3839	0.3995	0.4134	0.4257	0.4368
0.34	0.0306	0.0596	0.0871	0.1133	0.1381	0.1839	0.2251	0.2621	0.2954	0.3252	0.3520	0.3761	0.3976	0.4169	0.4341	0.4495	0.4633	0.4756
0.38	0.0322	0.0628	0.0920	0.1197	0.1461	0.1949	0.2391	0.2789	0.3147	0.3470	0.3761	0.4022	0.4256	0.4466	0.4654	0.4823	0.4973	0.5108
0.42	0.0337	0.0657	0.0963	0.1254	0.1532	0.2048	0.2515	0.2938	0.3320	0.3665	0.3976	0.4256	0.4508	0.4734	0.4937	0.5119	0.5281	0.5427
0.46	0.0349	0.0683	0.1001	0.1305	0.1595	0.2135	0.2626	0.3071	0.3474	0.3839	0.4169	0.4466	0.4734	0.4975	0.5191	0.5385	0.5559	0.5715
0.50	0.0361	0.0705	0.1035	0.1350	0.1650	0.2212	0.2724	0.3189	0.3612	0.3995	0.4341	0.4654	0.4937	0.5191	0.5420	0.5626	0.5810	0.5975
0.54	0.0371	0.0725	0.1065	0.1389	0.1700	0.2281	0.2812	0.3295	0.3735	0.4134	0.4495	0.4823	0.5119	0.5385	0.5626	0.5842	0.6036	0.6209
0.58	0.0380	0.0743	0.1091	0.1425	0.1744	0.2343	0.2890	0.3389	0.3844	0.4257	0.4633	0.4973	0.5281	0.5559	0.5810	0.6036	0.6238	0.6420
0.62	0.0387	0.0759	0.1115	0.1456	0.1783	0.2397	0.2959	0.3472	0.3941	0.4368	0.4756	0.5108	0.5427	0.5715	0.5975	0.6209	0.6420	0.6609
0.66	0.0394	0.0773	0.1136	0.1484	0.1818	0.2445	0.3020	0.3547	0.4027	0.4466	0.4865	0.5227	0.5556	0.5854	0.6122	0.6364	0.6582	0.6778
0.70	0.0401	0.0785	0.1154	0.1509	0.1849	0.2488	0.3075	0.3612	0.4104	0.4553	0.4962	0.5334	0.5672	0.5977	0.6254	0.6503	0.6728	0.6929
0.74	0.0406	0.0796	0.1171	0.1531	0.1876	0.2526	0.3123	0.3671	0.4172	0.4630	0.5048	0.5429	0.5774	0.6087	0.6371	0.6627	0.6857	0.7064
0.78	0.0411	0.0806	0.1185	0.1550	0.1900	0.2559	0.3166	0.3722	0.4232	0.4699	0.5125	0.5513	0.5865	0.6185	0.6475	0.6736	0.6972	0.7184
0.82	0.0415	0.0814	0.1198	0.1567	0.1921	0.2589	0.3203	0.3768	0.4286	0.4760	0.5192	0.5557	0.5946	0.6272	0.6567	0.6834	0.7074	0.7291
0.86	0.0419	0.0822	0.1209	0.1582	0.1940	0.2615	0.3237	0.3808	0.4333	0.4813	0.5252	0.5653	0.6017	0.6348	0.6648	0.6920	0.7165	0.7386
0.90	0.0422	0.0828	0.1219	0.1595	0.1957	0.2638	0.3266	0.3844	0.4374	0.4860	0.5305	0.5711	0.6080	0.6416	0.6721	0.6996	0.7245	0.7469
0.94	0.0425	0.0834	0.1228	0.1607	0.1971	0.2658	0.3292	0.3875	0.4411	0.4902	0.5351	0.5762	0.6136	0.6476	0.6784	0.7063	0.7316	0.7543
0.98	0.0428	0.0839	0.1236	0.1617	0.1984	0.2676	0.3314	0.3902	0.4442	0.4938	0.5392	0.5807	0.6184	0.6528	0.6840	0.7123	0.7378	0.7608
1.00	0.0429	0.0842	0.1239	0.1622	0.1990	0.2684	0.3324	0.3914	0.4457	0.4955	0.5410	0.5827	0.6206	0.6552	0.6865	0.7150	0.7406	0.7638
1.20	0.0437	0.0858	0.1263	0.1654	0.2030	0.2740	0.3396	0.4001	0.4558	0.5070	0.5540	0.5969	0.6362	0.6719	0.7044	0.7339	0.7605	0.7846
1.40	0.0441	0.0866	0.1275	0.1669	0.2049	0.2767	0.3431	0.4043	0.4608	0.5127	0.5603	0.6039	0.6438	0.6801	0.7132	0.7432	0.7704	0.7949
1.80	0.0444	0.0871	0.1283	0.1680	0.2062	0.2785	0.3454	0.4071	0.4641	0.5165	0.5645	0.6086	0.6489	0.6856	0.7190	0.7494	0.7769	0.8018
2.00	0.0444	0.0871	0.1284	0.1681	0.2064	0.2787	0.3457	0.4075	0.4645	0.5169	0.5651	0.6092	0.6495	0.6863	0.7198	0.7502	0.7778	0.8027
2.20	0.0444	0.0872	0.1284	0.1682	0.2065	0.2788	0.3458	0.4076	0.4646	0.5171	0.5653	0.6094	0.6497	0.6865	0.7200	0.7505	0.7781	0.8030
2.50	0.0444	0.0872	0.1284	0.1682	0.2065	0.2788	0.3458	0.4077	0.4647	0.5172	0.5653	0.6095	0.6498	0.6867	0.7202	0.7506	0.7782	0.8032
3.00	0.0444	0.0872	0.1284	0.1682	0.2065	0.2789	0.3458	0.4077	0.4647	0.5172	0.5654	0.6095	0.6499	0.6867	0.7202	0.7506	0.7782	0.8032

β \ α	3.00	2.50	2.20	2.00	1.80	1.40	1.20	1.00	0.98	0.94	0.90	0.86	0.82	0.78	0.74	0.70	0.66
0.02	0.0444	0.0444	0.0444	0.0444	0.0444	0.0441	0.0437	0.0429	0.0428	0.0425	0.0422	0.0419	0.0415	0.0411	0.0406	0.0401	0.0394
0.04	0.0872	0.0872	0.0872	0.0871	0.0871	0.0866	0.0858	0.0842	0.0839	0.0834	0.0828	0.0822	0.0814	0.0806	0.0796	0.0785	0.0773
0.06	0.1284	0.1284	0.1284	0.1284	0.1283	0.1275	0.1263	0.1239	0.1236	0.1228	0.1219	0.1209	0.1198	0.1185	0.1171	0.1154	0.1136
0.08	0.1682	0.1682	0.1682	0.1681	0.1680	0.1669	0.1654	0.1622	0.1617	0.1607	0.1595	0.1582	0.1567	0.1550	0.1531	0.1509	0.1484
0.10	0.2065	0.2065	0.2065	0.2064	0.2062	0.2049	0.2030	0.1990	0.1984	0.1971	0.1957	0.1940	0.1921	0.1900	0.1876	0.1849	0.1818
0.14	0.2789	0.2788	0.2788	0.2787	0.2785	0.2767	0.2740	0.2684	0.2676	0.2658	0.2638	0.2615	0.2589	0.2559	0.2526	0.2488	0.2445
0.18	0.3458	0.3458	0.3458	0.3457	0.3454	0.3431	0.3396	0.3324	0.3314	0.3292	0.3266	0.3237	0.3203	0.3166	0.3123	0.3075	0.3020
0.22	0.4077	0.4077	0.4076	0.4075	0.4071	0.4043	0.4001	0.3914	0.3902	0.3875	0.3844	0.3808	0.3768	0.3722	0.3671	0.3612	0.3547
0.26	0.4647	0.4647	0.4646	0.4645	0.4641	0.4608	0.4558	0.4457	0.4442	0.4411	0.4374	0.4333	0.4286	0.4232	0.4172	0.4104	0.4027
0.30	0.5172	0.5172	0.5171	0.5169	0.5165	0.5127	0.5070	0.4955	0.4938	0.4902	0.4860	0.4813	0.4760	0.4699	0.4630	0.4553	0.4466
0.34	0.5654	0.5653	0.5653	0.5651	0.5645	0.5603	0.5540	0.5410	0.5392	0.5351	0.5305	0.5252	0.5192	0.5125	0.5048	0.4962	0.4865
0.38	0.6095	0.6095	0.6094	0.6092	0.6086	0.6039	0.5969	0.5827	0.5807	0.5762	0.5711	0.5653	0.5587	0.5513	0.5429	0.5334	0.5227
0.42	0.6499	0.6498	0.6497	0.6495	0.6489	0.6438	0.6362	0.6206	0.6184	0.6136	0.6080	0.6017	0.5946	0.5865	0.5774	0.5672	0.5556
0.46	0.6867	0.6867	0.6865	0.6863	0.6856	0.6801	0.6719	0.6552	0.6528	0.6476	0.6416	0.6348	0.6272	0.6185	0.6087	0.5977	0.5854
0.50	0.7202	0.7202	0.7200	0.7198	0.7190	0.7132	0.7044	0.6865	0.6840	0.6784	0.6721	0.6648	0.6567	0.6475	0.6371	0.6254	0.6122
0.54	0.7506	0.7506	0.7505	0.7502	0.7494	0.7432	0.7339	0.7150	0.7123	0.7063	0.6996	0.6920	0.6834	0.6736	0.6627	0.6503	0.6364
0.58	0.7782	0.7782	0.7781	0.7778	0.7769	0.7704	0.7605	0.7406	0.7378	0.7316	0.7245	0.7165	0.7074	0.6972	0.6857	0.6728	0.6482
0.62	0.8032	0.8032	0.8030	0.8027	0.8018	0.7949	0.7846	0.7638	0.7608	0.7543	0.7469	0.7386	0.7291	0.7184	0.7064	0.6929	0.6778
0.66	0.8257	0.8257	0.8255	0.8252	0.8243	0.8171	0.8064	0.7846	0.7816	0.7748	0.7671	0.7584	0.7486	0.7375	0.7250	0.7110	0.6953
0.70	0.8460	0.8460	0.8458	0.8454	0.8445	0.8370	0.8259	0.8034	0.8002	0.7932	0.7852	0.7762	0.7660	0.7546	0.7417	0.7272	0.7110
0.74	0.8642	0.8642	0.8640	0.8636	0.8627	0.8549	0.8434	0.8201	0.8168	0.8096	0.8014	0.7921	0.7816	0.7698	0.7566	0.7417	0.7250
0.78	0.8805	0.8805	0.8803	0.8799	0.8789	0.8710	0.8591	0.8351	0.8317	0.8243	0.8159	0.8063	0.7956	0.7834	0.7698	0.7546	0.7375
0.82	0.8951	0.8951	0.8949	0.8945	0.8935	0.8853	0.8731	0.8485	0.8450	0.8374	0.8288	0.8190	0.8080	0.7956	0.7816	0.7660	0.7486
0.86	0.9081	0.9081	0.9079	0.9075	0.9065	0.8980	0.8855	0.8604	0.8569	0.8491	0.8402	0.8302	0.8190	0.8063	0.7921	0.7762	0.7584
0.90	0.9197	0.9197	0.9195	0.9191	0.9180	0.9094	0.8966	0.8710	0.8674	0.8594	0.8504	0.8402	0.8288	0.8159	0.8014	0.7852	0.7671
0.94	0.9300	0.9300	0.9298	0.9294	0.9282	0.9195	0.9064	0.8803	0.8767	0.8686	0.8594	0.8491	0.8374	0.8243	0.8096	0.7932	0.7748
0.98	0.9391	0.9391	0.9389	0.9384	0.9373	0.9284	0.9151	0.8886	0.8849	0.8767	0.8674	0.8569	0.8450	0.8317	0.8168	0.8002	0.7816
1.00	0.9433	0.9432	0.9430	0.9426	0.9414	0.9324	0.9191	0.8924	0.8886	0.8803	0.8710	0.8604	0.8485	0.8351	0.8201	0.8034	0.7846
1.20	0.9729	0.9728	0.9726	0.9722	0.9709	0.9614	0.9472	0.9191	0.9151	0.9064	0.8966	0.8855	0.8731	0.8591	0.8434	0.8259	0.8064
1.40	0.9878	0.9878	0.9875	0.9871	0.9858	0.9759	0.9614	0.9324	0.9284	0.9195	0.9094	0.8980	0.8853	0.8710	0.8549	0.8370	0.8171
1.80	0.9980	0.9979	0.9972	0.9972	0.9959	0.9858	0.9709	0.9414	0.9373	0.9282	0.9180	0.9065	0.8935	0.8789	0.8627	0.8445	0.8243
2.00	0.9934	0.9992	0.9990	0.9985	0.9972	0.9871	0.9722	0.9426	0.9384	0.9294	0.9191	0.9075	0.8945	0.8799	0.8636	0.8454	0.8252
2.20	0.9998	0.9997	0.9997	0.9992	0.9985	0.9875	0.9726	0.9430	0.9389	0.9298	0.9195	0.9079	0.8949	0.8803	0.8640	0.8458	0.8255
2.50	1.0000	1.0000	0.9998	0.9990	0.9979	0.9878	0.9728	0.9432	0.9391	0.9300	0.9197	0.9081	0.8951	0.8805	0.8642	0.8460	0.8257
3.00	1.0000	1.0000	0.9998	0.9992	0.9980	0.9878	0.9729	0.9433	0.9391	0.9300	0.9197	0.9081	0.8951	0.8805	0.8642	0.8460	0.8257

SOURCE: M. S. Hantush, "Growth and Decay of Groundwater-Mounds in Response to Uniform Percolation," *Water Resources Res.*, vol. 3, 1967, p. 227.

where t' is the time since water ceased to arrive at the water table, $Z(x, y, t) - Z(x, y, t - t')$ represents the right-hand side of Eq. (7.10) with t and $(t - t')$ as time factors. Time t includes the time since the start of infiltration until the end of the period of time t'.

EXAMPLE 7.2 Using the data from Example 7.1, find the growth in the groundwater mound below the center point $(x, y \equiv 0, 0)$ and point $(x, y) \equiv (20 \text{ m}, 0)$ after 15 days from the time that infiltration water reaches the water table. If infiltration stopped after 15 days, determine the decay in the groundwater mound 6 days later below the center of the basin $(x, y = 0, 0)$.

Solution

$$\frac{S_y}{4Tt} = \sqrt{\frac{0.12}{4 \times 48.96 \times 15}} = 0.00639 \text{ m}^{-1}$$

Below the center $x, y \equiv 0, 0$,

$$\alpha_1 = 0.00639(25) = 0.16 = \alpha_2$$
$$\beta_1 = 0.00639(50) = 0.32 = \beta_2$$

From Table 7.2 (by interpolation)

$$F(\alpha_1, \beta_1) = F(\alpha_1, \beta_2) = F(\alpha_2, \beta_1) = F(\alpha_2, \beta_2) = F(0.16, 0.32) \cong 0.1975$$

Applying Eq. (7.10), we see that

$$h_{0,0,t} - h_u = \frac{0.17 \times 15}{4 \times 0.12} \times 4 \times 0.1975 = 4.197 \text{ m} < 0.5 h_u$$

(Compare with the value 4.25 m determined on the basis of Fig. 7.4 in Example 7.1.) Below point $x, y \equiv 20, 0$,

$$\alpha_1 = 0.00639(25 + 20) = 0.288$$
$$\alpha_2 = 0.00639(25 - 20) = 0.032$$
$$\beta_1 = \beta_2 = 0.00639(50) = 0.320$$

From Table 7.2,

$$F(\alpha_1, \beta_1) = F(0.288, 0.32) \approx 0.31 \cong F(\alpha_1, \beta_2)$$
$$F(\alpha_2, \beta_1) = F(0.032, 0.32) \approx 0.0438 \cong F(\alpha_2, \beta_2)$$

Applying Eq. (7.10), we see that

$$h_{20,0,t} - h_u = \frac{0.17 \times 15}{4 \times 0.12} \times (2 \times 0.31 + 2 \times 0.0438) = 3.76 \text{ m}$$

The decay below the center of the basin 6 days after infiltration stops is determined as follows:

$$t = 15 + 6 = 21 \text{ days}$$
$$t - t' = 21 - 6 = 15 \text{ days}$$

After 21 days

$$\frac{S_y}{4Tt} = \sqrt{\frac{0.12}{4 \times 48.96 \times 21}} = 0.0054 \text{ m}^{-1}$$

$$\alpha_1 = 0.0054(25) = 0.135 = \alpha_2$$
$$\beta_1 = 0.0054(50) = 0.27 = \beta_2$$

From Table 7.2,

$$F(0.135, 0.27) = 0.157$$

With reference to Eq. (7.11),

$$Z(0, 0, 21) = \frac{v_{am}t}{4S_y} \times 4 \times 0.029 = \frac{0.17 \times 21}{4 \times 0.12} \times 4 \times 0.157 = 4.67 \text{ m}$$

and $Z(0, 0, 15) = 4.197$ m

Therefore, $h_{x,y,t} - h_u = 4.67 - 4.197 = 0.473$ m above the original water table, or 6 days after the stop of infiltration, the center of the mound dropped 4.177 m $(= 4.67 - 0.473)$.

Specially developed equations for infinitely long rectangular areas and circular areas also were developed by Hantush (1967). Practically, an infinitely long rectangular area may be looked on as a rectangle with a ratio $L_{re}/B_{re} \geq 4$; a circular area may be treated as a square of an equivalent area to the circle. Solutions for basins above sloping water tables are also available (Baumann, 1965; Bianchi and Muckel, 1970).

PROBLEMS AND DISCUSSION QUESTIONS

7.1   What are the main goals of groundwater management?

7.2   What types of organizations should be responsible for planning and implementing management programs?

7.3   State the major problems that prompted the need for expanding the field of groundwater management.

7.4   Explain the various practical uses of piezometric contour maps.

7.5   Observation wells are customarily used for water elevation records. In some cases, piezometers should be used, and in others, they may be preferred for use. What are these cases?

7.6   What are the major differences between the concepts of safe yield (sustained yield) and perennial yield?

7.7   Does mining yield satisfy the requirements of safe yield? Why?

7.8   Define or explain the following:
(a) Preferred perennial yield   (d) Contributing recharge areas
(b) Maximum perennial yield   (e) Induced recharge
(c) Conjunctive use   (f) Overdraft

7.9   What are the major corrective measures that should be considered when water withdrawal exceeds the perennial supply?

7.10   Explain the difference between maximum and permissive mining yields.

7.11   Give three examples for each of the following:
(a) Natural recharge
(b) Artificial recharge
(c) Incidental water recharges

7.12   Explain the conditions under which irrigation waters are considered as artificial recharge or incidental recharge.

7.13   What practical problems are encountered in the field as a result of recharge wells?

7.14   A water channel has a trapezoidal cross section with side slopes $1:1$. The bed width of the channel is 12 m, and the maximum water depth is 10 m. Assuming that the soil is infinitely deep with a theoretically pervious boundary at infinity and that the average hydraulic conductivity $K = 1 \times 10^{-2}$ cm/s, determine the rate of seepage q from a segment of the channel 1.0 m long.

7.15   If all conditions in Problem 7.14 remain unchanged except that the pervious boundary is at a finite depth of 40 m below the water level, construct by neat sketching a flow net. Find the following on the basis of your drawing:
(a) Rate of seepage q
(b) Darcian velocity v at the pervious boundary
(c) The maximum width between the two free surfaces
Find also the percentage differences between the results of q and v as compared with those determined in Problem 7.14.

7.16   If the channel side slopes in Problem 7.14 were $3H:2V$ while other parameters remain unchanged, what would be the rate of flow q calculated on the basis of Eq. (7.9)?

7.17   Using Eq. (7.3), plot the curved perimeter of a water channel in which the top water-surface width B_w is 32.0 m and the maximum depth of water d_c at the center is 10.0 m. Determine also the rate of seepage q if the average hydraulic conductivity K is 1×10^{-2} cm/s. Plot also on the determined cross section of the channel the approximate best fit of a trapezoidal cross section. The pervious boundary is assumed at infinite depth. Use intervals of at least 5.0 m to calculate the ordinates of the cross section by means of Eq. (7.3).

7.18   Assuming in Problem 7.17 that an impervious rather than a pervious boundary is located at the infinite depth, calculate the rate of seepage q and plot the right-hand side free surface to a distance not greater than $x = 45.0$ m. (Calculate the ordinates at 5.0-m intervals.)

7.19   Assuming that the impervious boundary in Problem 7.18 has a definite depth of 30.0 m below the water level, find the rate of seepage q from a neatly drawn flow net. Compare the determined q with that obtained in Problem 7.18.

7.20   Resolve Example 7.1 when $t = 10$ days and $v_{am} = 0.2$ m per day while other parameters remain unchanged. If infiltration ceases after 10 days, determine the maximum height of the groundwater mound above the undisturbed water table after 17 days from the instant infiltration reached that water table.

7.21 Using all data given in Example 7.2, plot the groundwater-mound vertical cross section at $y = 0$ after 15 days from the start of infiltration and the cross section at $y = 20.0$ m after 21 days from the start of infiltration. Infiltration ceased after 15 days. Extend both plots to $x = 75$ m and calculate their ordinates at 10-m intervals.

REFERENCES

American Society of Civil Engineers: *Ground Water Management,* Manual of Engineering Practice 40, New York, 1972.

Back, W. R., et al.: "Carbon 14 Ages Related to Occurrence of Salt Water," *J. Hydraul. Div. Am. Soc. Civ. Eng.,* vol. 96, no. HY11, 1970, pp. 2325–2336.

Baffa, J. J.: "Injection Well Experience at Riverhead, N.Y.," *J. Am. Water Works Assoc.,* vol. 62, 1970, pp. 41–46.

Baier, D. C., and G. W. Wesner: "Reclaimed Waste for Groundwater Recharge," *Water Res. Bull.,* vol. 7, 1971, pp. 991–1001.

Banks, H. O.: "Utilization of Underground Storage Reservoirs," *Trans. Am. Soc. Civ. Eng.,* vol. 118, 1953, pp. 220–234.

Barksdale, H. C. and G. D. De Buchananne: "Artificial Recharge of Productive Ground-Water Aquifers in New Jersey," *Econ. Geol.,* vol. 41, no. 7, 1946, pp. 726–737.

Baumann, P.: "Technical Development in Ground Water Recharge," in V. T. Chow (ed.), *Advances in Hydroscience,* vol. 2, Academic, New York, 1965, pp. 209–279.

Beaver, J. A., and M. L. Frankel: "Significance of Ground-Water Management Strategy: A Systems Approach," *Ground Water,* vol. 7, no. 3, 1969, pp. 22–26.

Behnke, J. J.: "Clogging in Surface Spreading Operations for Artificial Ground-Water Recharge," *Water Resources Res.,* vol. 5, 1969, pp. 870–876.

Berend, J. E.: "An Analytical Approach to the Clogging Effect of Suspended Matter," *Bull. Int. Assoc. Sci. Hydrol.,* vol. 12, no. 2, 1967, pp. 42–55.

Bianchi, W. C., and D. C. Muckel: "Ground-Water Recharge Hydrology," Agricultural Research Service 41-161, U.S. Department of Agriculture, Washington, 1970.

Bittinger, M. W.: "The Problem of Integrating Ground-Water and Surface Water Use," *Ground Water,* vol. 2, no. 3, 1964, pp. 33–38.

Bittinger, M. W., and F. J. Trelease: "The Development and Dissipation of a Ground-Water Mound," *Trans. Am. Soc. Agric. Eng.,* vol. 8, 1965, pp. 103–104, 106.

Bouwer, H.: "Renovating Municipal Wastewater by High-Rate Infiltration of Groundwater Recharge," *J. Am. Water Works Assoc.,* vol. 66, 1974, pp. 159–162.

Bouwer, H.: *Groundwater Hydrology,* McGraw-Hill, New York, 1978.

Bouwer, H., et al.: "Land Treatment of Wastewater in Today's Society," *Civ. Eng.,* vol. 48, no. 1, 1978, pp. 78–81.

Brown, R. F., et al.: "Artificial Ground-Water Recharge as a Water-Management Technique on the Southern High Plains of Texas and New Mexico," Texas Department of Water Resources Report 220, 1978.

Bulten, B., C. Brandes, and J. van Puffelen: "Artificial Recharge with Wells: Report of a Trip to the United States of America" (in Dutch), Provincial Water Supply of North Holland, Netherlands, 1974.

California Department of Water Resources: "Planned Utilization of Ground Water Basins: Coastal Plain of Los Angeles County," Bulletin 104, Sacramento, 1968.

Conkling, H.: "Utilization of Groundwater Storage in Stream System Development," *Trans. Am. Soc. Div. Eng.,* vol. 111, 1946, pp. 275–305.

Conover, C. S.: "Ground-Water Resources: Development and Management," U.S. Geological Survey Circular 442, Washington, 1961.

Cooper, H. H., Jr., and M. I. Rorabaugh: "Ground-Water Movements and Bank Storage due to Flood Stages in Surface Streams," U.S. Geological Survey Water-Supply Paper 1536-J, Washington, 1963, pp. 343–366.

DeWiest, R. J. M.: *Geohydrology*, Wiley, New York, 1965.

Domenico, P. A.: *Concepts and Models in Groundwater Hydrology*, McGraw-Hill, New York, 1972.

Downing, R. A., et al.: "Regional Development of Groundwater Resources in Combination with Surface Water," *J. Hydrol.*, vol. 22, 1974, pp. 155–177.

Drost, W., et al.: "Point Dilution Methods of Investigating Ground Water Flow by Means of Radioisotopes," *Water Resources Res.*, vol. 4, 1968, pp. 125–146.

Dvoracek, M. J., and V. H. Scott: "Ground-Water Flow Characteristics Influenced by Recharge Pit Geometry," *Trans. Am. Soc. Agric. Eng.*, vol. 6, 1963, pp. 262–265, 267.

Franke, O. L., and N. E. McClymonds: "Summary of the Hydrologic Situation on Long Island, New York, as a Guide to Water-Management Alternatives," U.S. Geological Survey Professional Paper 627-F, Washington, 1972.

Gardner, W. R., and M. Fireman: "Laboratory Studies of Evaporation from Soil Columns in the Presence of a Water Table," *Soil Sci.*, vol. 85, 1958, pp. 244–249.

Gaspar, E., and M. Oncescu: *Radioactive Tracers in Hydrology*, Elsevier, Amsterdam, 1972.

Gatewood, J. S., et al.: "Use of Water by Bottom-Land Vegetation in Lower Safford Valley, Arizona," U.S. Geological Survey Water-Supply Paper 1103, Washington, 1950.

Glover, R. E.: "Ground Water Movement," U.S. Bureau of Reclamation Engineering Monograph 31, 1964.

Goss, D. W., et al.: "Fate of Suspended Sediment during Basin Recharge," *Water Resources Res.*, vol. 9, 1973, pp. 668–675.

Groot, C. R.: "Feasibility of Artificial Recharge at Newark, Delaware," *J. Am. Water Works Assoc.*, vol. 52, no. 6, 1960, pp. 749–755.

Hansen, H. J.: "Zoning Plan for Managing a Maryland Coastal Aquifer," *J. Am. Water Works Assoc.*, vol. 62, 1970, pp. 286–292.

Hantush, M. S.: "Preliminary Quantative Study of the Rosewell Groundwater Reservoir, New Mexico," New Mexico Institute of Mining and Technology, 1957.

Hantush, M. S.: "Growth and Decay of Groundwater-Mounds in Response to Uniform Percolation," *Water Resources Res.*, vol. 3, 1967, pp. 227–234.

Harpaz, Y.: "Artificial Ground-Water Recharge by Means of Wells in Israel," *J. Hydraul. Div. Am. Soc. Civ. Eng.*, vol. 97, no. HY12, 1971, pp. 1947–1964.

Hauser, V. L., and F. B. Lotspeich: "Artificial Groundwater Recharge through Wells," *J. Soil Water Conserv.*, vol. 22, 1967, pp. 11–15.

Hellwig, D. H. R.: "Evaporation of Water from Sand," *J. Hydrol.*, vol. 18, 1973, pp. 93–118.

International Association of Scientific Hydrology (IASH): "Artificial Recharge and Management of Aquifers," Publication 72, 1967.

Jacob, C. E.: "Correlation of Groundwater Levels and Precipitation on Long Island, New York," *Trans. Am. Geophys. Union*, vol. 24, 1943, pp. 564–573.

Jacob, C. E.: "Correlation of Groundwater Levels and Precipitation on Long Island, New York," *Trans. Am. Geophys. Union*, vol. 25, 1944, pp. 928–939.

Jansa, O. V.: "Artificial Ground-Water Supplies of Sweden," IASH Publication 33, 1951, pp. 227–237.

Jansa, O. V.: "Artificial Ground-Water Supplies of Sweden," IASH Report 2, Publication 37, 1954, pp. 269–275.

Johnson, A. I.: "Ground-Water Recharge on Long Island," *J. Am. Water Works Assoc.*, vol. 40, 1948, pp. 1159–1166.

Johnson, A. I., et al.: "Symposium on Planning and Design of Groundwater Data Programs," *Water Resources Res.*, vol. 8, 1972, pp. 177–241.

Johnston, R. H.: "Base Flow as an Indicator of Aquifer Characteristics in the Coastal Plain of Delaware," U.S. Geological Survey Professional Paper 750-D, Washington, 1971, pp. 212–215.

Kashef, A. I.: "On the Management of Ground Water in Coastal Aquifers," *Ground Water*, vol. 9, no. 2, 1971, pp. 12–20.

Kashef, A. I.: "Technical and Ecological Impacts of the High Aswan Dam," *J. Hydrol.*, vol. 53, 1981, pp. 73–84.

Kaufman, W. J., and D. K. Todd: "Application of Tritium Tracer to Canal Seepage Measurements, in *Tritium in the Physical and Biological Sciences*, International Atomic Energy Agency, Vienna, 1962, pp. 83–94.

Kazmann, R. G.: "River Infiltration as a Source of Ground-Water Supply," *Trans. Am. Soc. Civ. Eng.*, vol. 113, 1948, pp. 404–424.

Kazmann, R. G.: " 'Safe Yield' in Ground-Water Development, Reality or Illusion?" *Proc. Am. Soc. Civ. Eng.*, vol. 82, no. IR3, 1956, pp. 1103-1 to 1103-12.

Kazmann, R. G.: "Underground Hot Water Storage Could Cut National Fuel Needs 10%," *Civ. Eng.*, vol. 48, no. 5, 1978, pp. 57–60.

Keeley, J. W., and M. R. Scalf: "Aquifer Storage Determination by Radio-Tracer Techniques," *Ground Water*, vol. 7, 1969, pp. 17–22.

Kelly, T. E.: "Artificial Recharge at Valley City, North Dakota, 1932 to 1965," *Ground Water*, vol. 5, no. 2, 1967, pp. 20–25.

Klaer, F. H., Jr.: "Providing Large Industrial Water Supplies by Induced Infiltration," *Min. Eng.*, vol. 5, 1953, pp. 620–624.

Koch, E., et al.: "Design and Operation of the Artificial-Recharge Plant at Bay Park, New York," U.S. Geological Survey Professional Paper 751-B, Washington, 1973.

Kunkle, G. R.: "Computation of Ground-Water Discharge to Streams during Floods, or to Individual Reaches during Baseflow, by Use of Specific Conductance," U.S. Geological Survey Professional Paper 525-D, Washington, 1965, pp. 207–210.

Lee, C. H.: "The Determination of Safe Yield of Underground Reservoirs of the Closed Basin Type," *Trans. Am. Soc. Civ. Eng.*, vol. 78, 1915, pp. 148–151.

Malmberg, G. T.: "Available Water Supply of the Las Vegas Groundwater Basin, Nevada," U.S. Geological Survey Water Supply Paper 1780, Washington, 1965.

Malmberg, G. T.: "Reclamation by Tubewell Drainage in Rechna Doab and Adjacent Areas, Punjab Region, Pakistan," U.S. Geological Survey Water-Supply Paper 1608-0, Washington, 1975.

Mann, J. F., Jr.: "Safe Yield Changes in Groundwater Basins," in *Transactions of the International Geological Congress*, 11th session, pt. XX, 1960, pp. 17–23.

Mann, J. F., Jr.: "Factors Affecting the Safe Yield of Ground-Water Basins," *Trans. Am. Soc. Civ. Eng.*, vol. 128, pt. III, 1963, pp. 180–190.

Mann, J. F., Jr.: "Ground-Water Management in the Raymond Basin, California," *Eng. Geol. Case Histories*, Geological Society of America, vol. 7, 1969, pp. 61–74.

Marino, M. A.: "Artificial Groundwater Recharge: I. Circular Recharging Area," *J. Hydrol.*, vol. 25, 1975a, pp. 201–208.

Marino, M. A.: "Artificial Groundwater Recharge: II. Rectangular Recharging Area," *J. Hydrol.*, vol. 26, 1975b, pp. 29–37.

McCleskey, G. W.: "Problems and Benefits in Ground-Water Management," *Ground Water*, vol. 10, no. 2, 1972, pp. 2–5.

McWhorter, D. B., and J. A. Brookman: "Pit Recharge Influenced by Subsurface Spreading," *Ground Water*, vol. 10, no. 5, 1972, pp. 6–11.

Meinzer, O. E.: "Outline of Groundwater Hydrology, with Definitions," U.S. Geological Survey Water-Supply Paper 494, Washington, 1923.

Meyer, C. F.: "Status Report on Head Storage Wells," *Water Res. Bull.*, vol. 12, 1976, pp. 237–252.

Meyer, C. F., and D. K. Todd: "Conserving Energy with Heat Storage Wells," *Environ. Sci. Technol.*, vol. 7, 1973, pp. 512–516.

Moore, C. V., and J. H. Snyder: "Some Legal and Economic Implications of Sea Water Intrusion: A Case Study of Ground Water Management," *Nat. Res. J.*, vol. 9, 1969, pp. 401–419.

Morel-Seytoux, H. J.: "A Simple Case of Conjunctive Surface-Ground-Water Management," *Ground Water*, vol. 13, 1975, pp. 506–515.

Muckel, D. C.: "Replenishment of Ground Water Supplies by Artificial Means," Technical Bulletin 1195, Agricultural Research Service, U.S. Department of Agriculture, Washington, 1959.

Mundorff, M. J., et al.: "Hydrologic Evaluation of Salinity Control and Reclamation Projects in the Indus Plain, Pakistan: A Summary," U.S. Geological Survey Water-Supply Paper 1608-Q, Washington, 1976.

Muskat, M.: The Flow of Homogeneous Fluids through Porous Media, McGraw-Hill, New York, 1937; second printing, 1946, J. W. Edwards, Ann Arbor, Michigan.

Nace, R. L.: "Water Management, Agriculture, and Ground-Water Supplies," U.S. Geological Survey Circular 415, Washington, 1960.

Nightingale, H. I., and W. C. Bianchi: "Ground-Water Recharge for Urban Use: Leaky Acres Project," Ground Water, vol. 11, no. 6, 1973, pp. 36–43.

Orlob, G. T., and B. B. Dendy: "Systems Approach to Water Quality Management," J. Hydraul. Div. Am. Soc. Civ. Eng., vol. 99, no. HY4, 1973, pp. 573–587.

Parker, G. G., et al.: "Artificial Recharge and Its Role in Scientific Water Management, with Emphasis on Long Island, New York," Proceedings of the National Symposium on Ground-Water Hydrology, American Water Resources Association, 1967, pp. 193–213.

Peters, H. J.: "Groundwater Management," Water Res. Bull., vol. 8, 1972, pp. 188–197.

Philip, J. R.: "Theory of Infiltration," in V. T. Chow (ed.), Advances in Hydroscience, Academic Press, New York, 1969, pp. 216–296.

Pinder, G. F., and J. F. Jones: "Determination of the Ground-Water Component of Peak Discharge from the Chemistry of Total Runoff," Water Resources Res., vol. 5, 1969, pp. 438–445.

Polubarinova-Kochina, P. Ya.: The Theory of Ground-Water Movement, Princeton University Press, Princeton, N.J., 1962; translated from the original Russian version published in 1952 by Roger J. M. DeWiest.

Posey, S. W.: "Development and Management of Karst Groundwater Resources," in Guide to the Hydrology of Carbonate Rocks, UNESCO Studies and Reports in Hydrology No. 41, 1984.

Prill, R. C., and D. B. Aaronson: "Flow Characteristics of a Subsurface-Controlled Recharge Basin on Long Island, New York," U.S. Geol. Surv. J. Res., vol. 1, 1973, pp. 735–744.

Ragone, S. E.: "Geochemical Effects of Recharging the Magothy Aquifer, Bay Park, New York, with Tertiary-Treated Sewage," U.S. Geological Survey Professional Paper 751-D, Washington, 1977.

Rahman, M. A., et al.: "Effect of Sediment Concentration on Well Recharge in a Fine Sand Aquifer," Water Resources Res., vol. 5, 1969, pp. 641–646.

Ralston, D. R.: "Administration of Ground Water as Both a Renewable and Nonrenewable Resource," Water Res. Bull., vol. 9, 1973, pp. 908–917.

Rayner, F. A.: "Ground-Water Basin Management on the High Plains of Texas," Ground Water, vol. 10, no. 5, 1972, pp. 12–17.

Rebhun, M., and J. Schwartz: "Clogging and Contamination Processes in Recharge Wells," Water Resources Res., vol. 4, 1968, pp. 1207–1217.

Renshaw, E. F.: "The Management of Ground Water Reservoirs," J. Farm Econ., vol. 45, 1963, pp. 285–295.

Revelle, R., and V. Lakshminarayana: "The Ganges Water Machine," Science, vol. 188, 1975, pp. 611–616.

Richter, R. C., and R. Y. D. Chung: "Artificial Recharge of Ground Water Reservoirs in California," J. Irrig. Drain. Div. Am. Soc. Civ. Eng., vol. 85, no. IR4, 1959, pp. 1–27.

Ripley, D. P., and Z. A. Saleem: "Clogging in Simulated Glacial Aquifers due to Artificial Recharge," Water Resources Res., vol. 9, 1973, pp. 1047–1057.

Ripple, C. D., et al.: "Estimating Steady-State Evaporation Rates from Bare Soils under Conditions of High Water Table," U.S. Geological Survey Water-Supply Paper 2019-A, Washington, 1972.

Sasman, R. T., and R. J. Schicht: "To Mine or Not to Mine Groundwater?" J. Am. Water Works Assoc., vol. 70, 1978, pp. 156–161.

Schicht, R. J.: "Feasibility of Recharging Treated Sewage Effluent into a Deep Sandstone Aquifer," Ground Water, vol. 9, no. 6, 1971, pp. 29–35.

Schicht, R. J., and W. C. Walton: "Hydrologic Budgets for Three Small Watersheds in Illinois," Report Inv. 40, Illinois State Water Survey, Urbana, 1961.

Scott, V. H., and G. Aron: "Aquifer Recharge Efficiency of Wells and Trenches," *Ground Water*, vol. 5, no. 3, 1967, pp. 6–14.

Seaburn, G. E., and D. B. Aaronson: "Influence of Recharge Basins on the Hydrology of Nassau and Suffolk Counties, Long Island, New York," U.S. Geological Survey Water-Supply Paper 2031, Washington, 1974.

Selim, H. M., and D. Kirkham: "Unsteady Two-Dimensional Flow of Water in Unsaturated Soil above an Impervious Barrier," *Proc. Soil. Sci. Soc. Am.*, vol. 37, 1973, pp. 489–495.

Signor, D. C., et al.: "Annoted Bibliography on Artificial Recharge of Ground Water, 1955–67," U.S. Geological Survey Water-Supply Paper 1990, Washington, 1970.

Singh, K. P.: "Some Factors Affecting Baseflow," *Water Resources Res.*, vol. 4, 1968, pp. 985–999.

Sternau, R.: "Artificial Recharge of Water through Wells: Experience and Techniques," *Proceedings of the Symposium on Artificial Recharge and Aquifer Management*, UNESCO and IASH, 1967, pp. 91–100.

Suter, M.: "High-Rate Artificial Ground-Water Recharge at Peoria, Illinois," IASH Publication 37, 1954, pp. 219–224.

Suter, M., and R. H. Harmeson: "Artificial Ground-Water Recharge at Peoria, Illinois," Bulletin 48, Illinois State Water Survey, Urbana, 1960.

Task Group on Artificial Ground Water Recharge: "Experience with Injection Wells for Artificial Ground Water Recharge," *J. Am. Water Works Assoc.*, vol. 57, 1965, pp. 629–639.

Taylor, O. J., and R. R. Luckey: "Water Management Studies of Stream-Aquifer System, Arkansas River Valley, Colorado," *Ground Water*, vol. 12, 1974, pp. 22–38.

Terzaghi, K., and R. E. Peck: *Soil Mechanics in Engineering Practice*, Wiley, New York, 1967.

Thomas, H. E.: *The Conservation of Ground Water*, McGraw-Hill, New York, 1951.

Todd, D. K.: *Groundwater Hydrology*, Wiley, New York, 1959a.

Todd, D. K.: "Annotated Bibliography on Artificial Recharge of Ground Water through 1954," U.S. Geological Survey Water-Supply Paper 1477, Washington, 1959b.

Todd, D. K.: *Groundwater Hydrology*, 2d ed., Wiley, New York, 1980.

United Nations, Department of Economic and Social Affairs: "Ground-Water Storage and Artificial Recharge," Natural Resources Water Service 2., United Nations, New York, 1975.

U.S. Bureau of Reclamation: "Design of Small Dams," U.S. Department of Interior, Washington, 1965.

Valliant, J.: "Artificial Recharge of Surface Water to the Ogallala Formation in the High Plains of Texas," *Ground Water*, vol. 2, no. 2, 1964, pp. 42–45.

Vecchioli, J.: "Experimental Injection of Tertiary-Treated Sewage in a Deep Well at Bay Park, Long Island, N.Y.: A Summary of Early Results," *J. N. Engl. Water Works Assoc.*, vol. 86, 1972, pp. 87–103.

Veihmeyer, F. J., and F. A. Brooks: "Measurement of Cumulative Evaporation of Bare Soil," *Trans. Am. Geophys. Union*, vol. 35, 1954, pp. 601–607.

Visocky, A. P.: "Estimating the Ground-Water Contribution to Storm Runoff by the Electrical Conductance Method," *Ground Water*, vol. 8, no. 2, 1970, pp. 5–10.

Vogel, J. C.: "Carbon 14 Dating of Groundwater," in *Isotope Hydrology 1970*, International Atomic Energy Agency, Vienna, 1970, pp. 225–239.

Walton, W. C.: "Estimating the Infiltration Rate of Streambed by Aquifer-Test Analysis," International Association of Scientific Hydrology, General Assembly, Berkeley, Calif., 1963.

Walton, W. C.: *Groundwater Resource Evaluation*, McGraw-Hill, New York, 1970.

Water Research Association: *Proceedings, Artificial Groundwater Recharge Conference*, Medmenham, England, 1971.

Weschler, L. F.: "Water Resources Management: The Orange County Experience," California Government Series no. 14, University of California, Davis, 1968.

White, W. N.: "A Method of Estimating Ground-Water Supplies Based on Discharge by Plants and Evaporation from Soil," U.S. Geological Survey Water-Supply Paper 659, Washington, 1932, pp. 1–105.

Wiebenga, W. A., et al.: "Radioisotopes as Groundwater Tracers," *J. Geophys. Res.*, vol. 72, 1967, pp. 4081–4091.

Wigley, T. M. L.: "Carbon 14 Dating of Groundwater from Closed and Open Systems," *Water Resources Res.*, vol. 11, 1975, pp. 324–328.

Water and Power Resources Service: "Ground Water Manual," A Water Resources Technical Publication, U.S. Department of Interior, Washington, 1981.

Young, R. A.: "Safe Yield of Aquifers: An Economic Reformulation," *J. Irrig. Drain. Div. Am. Soc. Civ. Eng.*, vol. 96, no. IR4, 1970, pp. 377–385.

Zyvoloski, G., J. C. Bruch, Jr., and J. M. Sloss: "Solution of Equation for Two-Dimensional Infiltration Problems," *Soil Sci.*, vol. 122, 1976, pp. 65–70.

CHAPTER

8

WATER WELLS

8.1 Artesian Wells: Steady State

In the idealized confined aquifer shown in Fig. 8.1, an artesian well is pumped continuously at a constant rate Q. The horizontal undisturbed piezometric surface cc deflects around the well to a curved surface known as a *drawdown surface* (or *drawdown curve* or *cone of depression*). The entire surface moves gradually downward as the time of pumping increases, until it becomes *practically stationary* when steady-state flow is reached. It should be noted that a constant rate of pumping Q does not mean that the flow is steady. In a steady-state flow system, the magnitude and direction of water velocity at any point in the medium remains constant at all times; otherwise the flow would be transient.

At any time t since pumping started, the drawdown surface intersects the well surface (radius r_w) at the water level in the well at height $h_{w,t}$ above the impervious base aa. At a certain radius $r_{e,t}$ known as the *radius of influence at time t*, the drawdown is almost negligible. The radius $r_{e,t}$ increases with time, while $h_{w,t}$ decreases.

In idealized artesian aquifers, the direction of water flow is always horizontal. At any time t, the flow produced by pumping is confined within a cylinder of height D_q concentric with the well and with internal

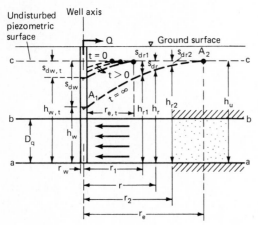

Figure 8.1 Changes in the drawdown curve due to pumping from an idealized artesian aquifer.

and external radii or r_w and $r_{e,t}$, respectively. The water beyond this cylinder at distance $r > r_{e,t}$ is theoretically unaffected by pumping. The circle of influence (radius $r_{e,t}$) is in effect a *groundwater divide*, which is the boundary between the zones that are affected and unaffected by pumping. At this radius of influence, the velocity of flow cannot be zero; it changes abruptly to some value. The velocity at $r_{e,t}$ *approaches* zero, which is similar to the velocity of approach in surface-water hydraulics. Theoretically speaking, the drawdown curve is asymptotic to the undisturbed piezometric plane cc. Through a period of pumping t, the water volume V_w pumped from the well is derived entirely from the zone between r_w and $r_{e,t}$. If there is no natural recharge, the pumped water is derived solely from storage (Sec. 6.3). After a very long period of time (theoretically $t = \infty$), the rate of compressibility of the soil material becomes negligible, leading to almost stationary values of $r_{e,t}$ and $h_{w,t}$. At this stage, steady-state flow is reached. Stationary values of r_e and h_w are produced mainly by the effect of natural recharge rather than by reaching 100 percent soil compressibility. The major portion of ultimate compressibility (for example, 90 percent) is reached after a definite time beyond which the change in drawdowns is negligible, $r_{e,t} = r_e$ and $h_{w,t} = h_w$ corresponding then to steady-state flow.

Applying Darcy's law at a radius r, we see that

$$Q = \underbrace{2\pi r D_q}_{\text{area}} \times \underbrace{Ki_g}_{\text{velocity}} \tag{8.1}$$

Because Q is constant, when r increases, $i_g = dh_r/dr$ decreases. At radius

r_e, the velocity of approach v_{ap} is given by

$$v_{ap} = \frac{Q}{2\pi r_e D_q} \tag{8.2}$$

The assumption of steady-state flow implies a recharge at radius r_e in order to satisfy Eq. (8.2) when the source of water from storage has been diminished. An ideal steady-state condition takes place when the water well is located in the center of an island of radius r_e surrounded by water at a level h_u above the impermeable plane aa (Fig. 8.1). A steady-state condition may be assumed to occur after a reasonably large time t beyond which no appreciable changes in drawdowns are measured.

At the present time, reasonably reliable procedures are available to analyze transient flow toward artesian wells. The use of these procedures requires an understanding of the cases analyzed earlier in this book under steady-state conditions. After long time periods, certain transient-flow systems are similar to a steady-state system (see Sec. 8.2). Furthermore, for some water-resources management problems, it may be necessary to study the most extreme conditions possible in order to err on the side of safety, such as, for example, in the evaluation of seepage through dams to find the maximum possible seepage losses. The analysis of steady-state cases is therefore informative and a prerequisite for comprehending analogous cases under transient states of flow.

Single Artesian Wells

Figure 8.1 shows an idealized artesian well pumping at a constant rate Q. After a long pumping time, it is assumed that a steady-state flow toward the well is established. Noting that i_g in Eq. (8.1) is equal to dh_r/dr, then we see that

$$\begin{aligned} Q &= 2\pi r D_q K \frac{dh_r}{dr} \\ &= 2\pi r T \frac{dh_r}{dr} \end{aligned} \tag{8.3}$$

Integrating this equation between various limits, the following formulas are obtained:

$$\frac{Q}{2\pi T} = \frac{h_u - h_w}{\ln (r_e/r_w)} = \frac{s_{dw}}{\ln (r_e/r_w)} \tag{8.4a}$$

$$\frac{Q}{2\pi T} = \frac{h_{r2} - h_{r1}}{\ln (r_2/r_1)} = \frac{s_{dr1} - s_{dr2}}{\ln (r_2/r_1)} = \frac{\Delta s_d}{2.3\Delta(\log r)} \tag{8.4b}$$

$$\frac{Q}{2\pi T} = \frac{h_u - h_r}{\ln (r_e/r)} = \frac{s_{dr}}{\ln (r_e/r)} \tag{8.4c}$$

and
$$\frac{Q}{2\pi T} = \frac{h_r - h_w}{\ln (r/r_w)} = \frac{s_{dw} - s_{dr}}{\ln (r/r_w)} \tag{8.4d}$$

where s_{dw}, s_{dr}, s_{dr1}, and s_{dr2} are the drawdowns at radii r_w, r, r_1, and r_2, respectively.

These equations are used mainly to find T. Selection of any of them depends on economics and the availability of field records. In addition to the needed records, the values of r_e and r_w are not known. Theoretically, r_e is at infinity, and r_w does not represent the nominal radius of the well r_c except in the absence of screens and gravel packs. The radius r_w is the effective radius of the well, which depends on the location and type of screens and the gradation and annular thickness of the gravel pack. If Q and r_e and r_w are known or evaluated, the use of Eq. (8.4a) needs no observation wells and only s_{dw} in the pumped well is recorded. Equation (8.4b) requires records at *two* observation wells; this, of course, is more expensive than using Eq. (8.4a), but Eq. (8.4b) does not include the unknowns r_e and r_w. Equation (8.4c) requires *one* observation well, but it includes the unknown r_e. Equation (8.4d) requires also one observation well, but it includes the unknown r_w. *Equation (8.4b) is recommended because it is similar to the unsteady-state equation at long time periods (see Sec. 8.2) and it does not include the unknowns r_e and r_w.*

The value of the effective radius is obtained by pumping tests, as explained later in this chapter. Once this is determined, the radius of influence r_e may be obtained from Eqs. (8.4a) and (8.4c) as follows:

$$r_e = \exp \frac{s_{dw} \ln r - s_{dr} \ln r_w}{s_{dw} - s_{dr}} \tag{8.5a}$$

(Records are taken at r and r_w.)

If T has been previously determined, r_e may be obtained from either Eq. (8.4a) or (8.4c) as follows:

$$r_e = \exp \left(\ln r_w + \frac{2\pi s_{dw} T}{Q} \right) \tag{8.5b}$$

(s_{dw} is recorded, and r_w is predetermined.) Alternatively,

$$r_e = \exp \left(\ln r + \frac{2\pi s_{dr} T}{Q} \right) \tag{8.5c}$$

(s_{dr} is recorded at an observation well at a distance r from the well.) The value r_e determined from any of these equations should be considered as some constant that satisfies Eq. (8.4a and c) rather than as the actual theoretical r_e. Using Eq. (8.5a, b, or c) is more realistic than assuming a value for r_e on the basis of experience, as is commonly done in practice. However, the errors in such assumptions do not have severe consequences, because r_e is always under the logarithmic signs (note that, for example, ln 500 = 6.214 and ln 1000 = 6.908).

Determination of r_e and r_w from field recordings The drawdown s_{dr} at radius r from the center of the well may be determined in terms of r_e and r_w from Eq. (8.4c and d) as follows:

$$s_{dr} = s_{dw}\left[1 - \frac{\ln{(r/r_w)}}{\ln{(r_e/r_w)}}\right] \tag{8.6}$$

This equation indicates that a semilog plot of s_{dr} versus log r should be a straight line (Fig. 8.2). If at least three observation wells are available, four records are obtained: s_{dw}, s_{dr1}, s_{dr2}, and s_{dr3} at radii r_w, r_1, r_2, and r_3, respectively. Plotting the three points corresponding to s_{dr1}, s_{dr2}, and s_{dr3}, a straight line ab can be fitted. The extension of this line to a zero drawdown (point b' in Fig. 8.2) gives the value of r_e. If we correct the field value s_{dw} by computing the screen losses (see Sec. 8.8), then the upward extension of line ba intersects the horizontal line passing by the

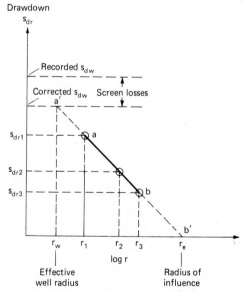

Figure 8.2 Graphic determination of the effective radius r_w of a well and its radius of influence r_e.

corrected s_{dw} at point a' corresponding to the effective radius r_w. The corrected s_{dw} is always less than the recorded one. If the well is uncased, as for example in a solid rock formation, no screens or gravel packs are needed, and the effective radius of the well r_w is equal to its nominal radius r_c.

Significance of T determined from a pumping test   Natural deposits are neither homogeneous nor isotropic. Therefore, assuming homogeneity and isotropy, the T value is determined from Eq. (8.4b) as follows:

$$T = \frac{Q(\ln r_2/r_1)}{2\pi(h_{r2} - h_{r1})} = \frac{Q(\ln r_2/r_1)}{2\pi(s_{dr1} - s_{dr2})} \tag{8.7}$$

The coefficient T thus determined should be looked on as a parameter that is compatible with the field conditions rather than with the absolute correct value $T = KD_q$. In most pumping tests, it is neither practical nor economical to drill the test well through the entire thickness D_q of the aquifer. Thus for partially penetrating wells, Eqs. (8.3), (8.4), and (8.5) cannot be considered exact. It should be recognized that although these equations are theoretical, they cannot be used without inserting the actual field records. These records reflect the actual conditions, such as heterogeneity, anisotropy, and the size of zones affected by pumping. The determined value $T = KD_q$ [Eq. (8.7)] represents a weighted average of both K and D_q within the zone affected by Q between r_w and r_e. If the value of Q from the same well is changed within the same formation, the flow zone also changes. For example, when Q increases, r_e increases and the flow domain also increases covering a larger zone of the aquifer that probably includes materials of different characteristics than the predominant materials. It is therefore preferable in a pumping test to use a well similar to planned productive wells and their anticipated Q.

These considerations should be taken into account in managing a water-resources project. In other fields of engineering, similar approximations are used. Planners may select parameters that represent extreme conditions in addition to factors of safety. At the present time, such factors have not been used in groundwater hydrology, although they are already in common use in the fields of seepage through and beneath dams (Secs. 5.4 and 5.5).

Flow nets in horizontal planes   The configuration of flow in a horizontal plane through the aquifer shown in Fig. 8.3 is the same as in any other horizontal plane. The flow is governed by the Laplace equation (Sec. 5.1), and therefore, a flow net can be traced. In such artesian wells, the curvilinear squares are formed by concentric circles with their centers

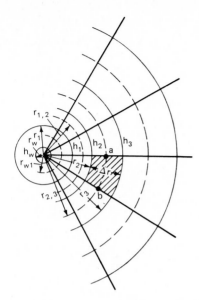

Figure 8.3 Part of a flow net in the horizontal plane in an artesian well.

coinciding with the well center, representing the equipotential lines and successive radii of the stream lines.

Instead of freehand sketching, the radii of these circles representing the equipotential lines can be computed if a certain number of radii representing the flow lines are assigned (Kashef, 1951). In order to have curvilinear squares, equal values of $\Delta Q = \Delta \psi$ are assumed between any two consecutive radii and a constant value of Δh is assumed between any consecutive concentric circles. In Fig. 8.3, four radii are shown: $r_w, r_1, r_2,$ and r_3, as well as the average radii $r_{w,1}$ (between r_w and r_1), $r_{1,2}$ (between r_1 and r_2), and $r_{2,3}$ (between r_2 and r_3). The intensities $h_w, h_1, h_2,$ and h_3 of the total heads are also indicated for the four equipotential surfaces. Then from Darcy's law,

$$\frac{Q}{2\pi K D_q} = \frac{Q}{2\pi T} = r_{w,1}\frac{h_1 - h_w}{r_1 - r_w} = r_{1,2}\frac{h_2 - h_1}{r_2 - r_1} = r_{2,3}\frac{h_3 - h_2}{r_3 - r_2}$$

$$= r_{n-1,n}\frac{h_n - h_{n-1}}{r_n - r_{n-1}} = \text{constant}$$

But $\qquad h_1 - h_w = h_2 - h_1 = h_3 - h_2 = h_n - h_{n-1} = \Delta h$

Then $\qquad \dfrac{r_1 - r_w}{r_{w,1}} = \dfrac{r_2 - r_1}{r_{1,2}} = \dfrac{r_3 - r_2}{r_{2,3}} = \dfrac{r_n - r_{n-1}}{r_{n-1,n}} = \text{a constant } C_w$

or $\qquad \dfrac{dr}{r} = d(\ln r) = C_w \qquad\qquad\qquad (8.8)$

The hatched field shown in Fig. 8.3 should be a curvilinear square when

$\Delta r = r_3 - r_2 = ab = 2\pi r_{2,3}/n_F$, where n_F is an assigned number of flow channels. Then

$$\frac{2\pi}{n_F} = \frac{r_3 - r_2}{r_{2,3}} = \frac{dr}{r}$$

and from Eq. (8.8),

$$\frac{2\pi}{n_F} = \frac{dr}{r} = d(\ln r) = \ln \frac{r_n}{r_{n-1}} = C_w \qquad (8.9)$$

This equation indicates that any r such as $r_{2,3}$ should be the logarithmic mean of r_2 and r_3 because $1/r_{2,3} = (\ln r_3 - \ln r_2)/(r_3 - r_2)$ and generally $1/r = (\ln r_n - \ln r_{n-1})/(r_n - r_{n-1})$. It can also be seen in the same equation that

$$\frac{r_n}{r_{n-1}} = \cdots = \frac{r_3}{r_2} = \frac{r_2}{r_1} = \frac{r_1}{r_w} = \exp C_w = \exp \frac{2\pi}{n_F} \qquad (8.10)$$

and

$$\frac{r_n}{r_w} = \frac{r_n}{r_{n-1}} \times \frac{r_{n-1}}{r_{n-2}} \times \frac{r_{n-2}}{r_{n-3}} \times \cdots = \exp \frac{2\pi}{n_F} = \exp nC_w \qquad (8.11)$$

Radii $r_1, r_2, \ldots, r_n$ would then have the values $r_w \exp C_w$, $r_w \exp 2C_w$, $r_w \exp 3C_w, \ldots, r_w \exp nC_w$, respectively.

The values of C_w and $\exp nC_w$ corresponding to n values between 1 and 8 are given in Table 8.1 for n_F values between 4 and 24. The minimum value of n_F given in Table 8.1 corresponds to four flow channels with $\Delta Q = \Delta \psi = Q/4$. In order to have a finer flow net, n_F should be greater than 4. If the selected value of n_F is greater than 24 or $n > 8$, Eqs. (8.9) through (8.11) should be used rather than Table 8.1. The last r_n value should be equal to or less than r_e. It is seldom that r_n exactly equals r_e. Therefore, the value of n_E would not necessarily be a whole number.

TABLE 8.1   Values of r_n/r_w*

		$\dfrac{r_n}{r_w} = \exp \pi C_w$ for n values:							
n_F	$C_w = \dfrac{2\pi}{n_F}$	1	2	3	4	5	6	7	8
4	1.570	4.81	23.14	111.32	535.50	2576.0	12,392	59,610	—
6	1.050	2.85	8.12	23.14	65.94	187.90	535.50	1526.00	4348
8	0.785	2.19	4.81	10.55	23.14	50.75	111.30	244.00	535.50
10	0.628	1.87	3.51	6.59	12.35	23.14	43.38	81.30	152.40
12	0.524	1.69	2.85	4.81	8.12	13.7	23.14	39.10	65.90
24	0.262	1.23	1.69	2.19	2.85	3.7	4.81	6.25	8.12

* See Eq. (8.11).

If $r_e = r_n$, n_E should be equal to n, but when $r_e > r_n$, n_E is determined by means of Eq. (8.11) as follows:

$$\frac{r_e}{r_w} = \exp n_E C_w = \exp \frac{2\pi n_E}{n_F}$$

or
$$n_E = \frac{n_F}{2\pi} \ln \frac{r_e}{r_w} \qquad (8.12)$$

From the principles of flow nets (Sec. 5.2),

$$Q = T(h_e - h_w)\frac{n_F}{n_E} = T s_{dw}\frac{n_F}{n_E}$$

Substituting the value of n_F/n_E [Eq. (8.12)] in the preceding equation, we get

$$Q = T s_{dw} 2\pi \frac{1}{\ln (r_e/r_w)}$$

or
$$\frac{Q}{2\pi T} = \frac{s_{dw}}{\ln (r_e/r_w)}$$

which is the same formula given by Eq. (8.4a).

EXAMPLE 8.1 Plot a flow net in the horizontal plane for an artesian well pumping at a rate of 3.6 m³/min. The values of r_w and r_e are, respectively, 0.15 and 600 m. The T value was found to be 1.35 m²/min from a field test.

Solution Assigning $n_F = 16$ (16 flow channels), then from Eq. (8.9),

$$C_w = \frac{2\pi}{n_F} = 0.3927$$

And from Eq. (8.12),

$$n_E = \frac{1}{0.3927} \ln \frac{600}{0.15} = \frac{8.294}{0.3927} = 21.12$$

Therefore, the highest value of n should be 21. From Eq. (8.11), r_1, r_2, r_3, . . . , r_{21} are calculated as shown in Table 8.2 for a $\Delta\psi = 3.3/16 = 0.21$ m³/min. Since the method is graphic, it may be sufficient to compute two radii only: $r_n = r_{21} = 572.252$ m and $r_{n-1} = r_{20} = 386.403$ m. The construction given in Fig. 8.4 can be followed to plot the rest of the circles representing the equipotential lines. Draw the circle passing by point a where $Oa = r_n$ and drop a vertical line from point c (where $Oc = r_{n-1}$) to meet this circle at point b. Join bO, and then

$$\frac{r_n}{r_{n-1}} = \exp \frac{2\pi}{n_F} = \operatorname{cosec} \alpha = \text{constant}$$

With O as center, draw an arc of a circle from c to intersect Ob at d and draw the vertical line de. Then Oe should be equal to r_{n-2} because $r_{n-1}/r_{n-2} =$

TABLE 8.2   Results of Solved Example 8.1

n	r_n, m°	n	r_n, m°
1	$r_1 = 0.222$	12	$r_{12} = 16.698$
2	$r_2 = 0.329$	13	$r_{13} = 24.729$
3	$r_3 = 0.486$	14	$r_{14} = 36.623$
4	$r_4 = 0.722$	15	$r_{15} = 54.238$
5	$r_5 = 1.069$	16	$r_{16} = 80.325$
6	$r_6 = 1.583$	17	$r_{17} = 118.959$
7	$r_7 = 2.344$	18	$r_{18} = 176.175$
8	$r_8 = 3.471$	19	$r_{19} = 260.911$
9	$r_9 = 5.141$	20	$r_{20} = 386.403$
10	$r_{10} = 7.613$	21	$r_{21} = 572.252$
11	$r_{11} = 11.275$		

° $r_n = r_w \exp nC_w = 0.15 \exp 0.3927n.$

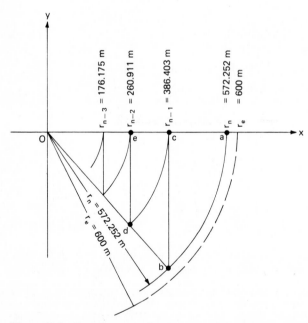

Figure 8.4   Construction of equipotential circles (Example 8.1).

cosec $\alpha = \exp(2\pi/n_F)$. Repeat the procedure to get the rest of the circles. The flow lines are radii making 16 flow channels. The angle between any two consecutive flow lines should be $2\pi/16 = \pi/8$ $(=22.5°)$.

If we use Eq. (8.4a) in this example, then

$$s_{dw} = \frac{Q}{2\pi T}\ln\frac{r_e}{r_w} = \frac{3.6}{2 \times 1.35\,\pi}\ln\frac{600}{0.15} = 3.52 \text{ m}$$

$$\Delta h = \frac{s_{dw}}{n_E} = \frac{3.52}{21.12} = \frac{1}{6}\text{ m}$$

and

$$\Delta\psi = \Delta q = \frac{3.6}{16\,D_q} = \frac{0.225}{D_q}\text{ m}^2/\text{min}$$

If h_u is numerically known, say, 100 m, then h_w $(=h_u - s_{dw}) = 100 - 3.52 = 96.48$ m and the intensity of all equipotential is known. The last circle r_n would have intensity $96.48 + 21 \times \frac{1}{6} = 99.98$ m (or $100 - (0.12/6) = 99.98$ m).

In curvilinear square flow nets, $\Delta\phi = \Delta\psi$. In this example,

$$\Delta\phi = K\Delta h = \frac{T}{D_q}\Delta h = \frac{1.35}{D_q}\times\frac{1}{6} = \frac{0.225}{D_q}\,(=\Delta\psi)$$

Combined Effects of Well Operation and Natural Uniform Flow

If the initial undisturbed piezometric surface is sloping with an angle α, then the natural darcian velocity $v_o = K\tan\alpha = K\,dh/dx$. Considering that axes x and y in the horizontal plane, with the origin at the center of the well, are as shown in Fig. 8.5 and that the direction of natural flow is in the negative x direction, then (Jacob, 1950):

For natural flow only,

$$\phi_o = K(h_u - h) + \text{constant} = K\frac{s_d}{x}x + \text{constant} = v_o x = \text{constant}$$

(8.13)

and $\psi_o = v_o y$

(8.14)

For a pumped well only,

$$\phi_{wd} = \frac{Q}{2\pi D_q}\ln\frac{r}{r_w} = \frac{Q}{4\pi D_q}\ln\frac{x^2 + y^2}{r_w^2} + \text{constant} \qquad (8.15)$$

and

$$\psi_{wd} = \frac{Q}{2\pi D_q}\theta = \frac{Q}{2\pi D_q}\tan^{-1}\frac{y}{x} \qquad (8.16)$$

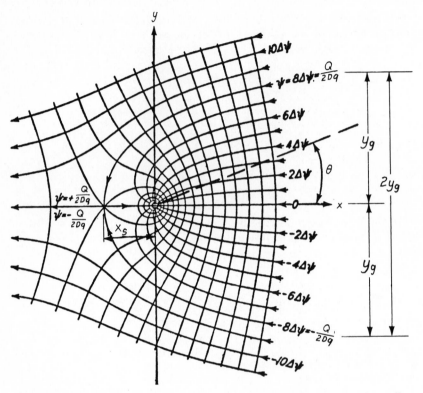

Figure 8.5 Distribution of stream lines and equipotential lines around a well in uniform flow. *(C. E. Jacob, "Flow of Ground Water," in H. Rouse (ed.), Engineering Hydraulics, Wiley, New York, 1950.)*

where θ is the angle between the radius passing by (x, y) and the x axis, and subscripts o and w refer, respectively, to the natural flow and the well. Equations (8.13) through (8.16) are written for the individual values of ϕ and ψ at a point $\equiv (x, y)$ before combining both effects. By superposition, the resultant flow would be described by

$$\phi = \phi_o + \phi_{wd} = v_o x + \frac{Q}{4\pi D_q} \ln \frac{x^2 + y^2}{r_w^2} + \text{constant} \qquad (8.17)$$

and

$$\psi = \psi_o + \psi_{wd} = v_o y + \frac{Q}{2\pi D_q} \tan^{-1} \frac{y}{x} \qquad (8.18)$$

The ϕ values can be written in terms of the total heads or drawdowns as $\phi = Kh = K(h_u - s_d)$. Plotting a square mesh on plane x, y (Fig. 8.6) and computing the values ϕ and ψ [Eqs. (8.17) and (8.18)] at each nodal point x, y, such as $o_1, a_1, b_1, \ldots$, then by interpolation, points of equal values of ϕ and ψ are found. Joining these points by smooth curves (Jacob, 1950), the flow configuration is obtained (Fig. 8.5). This figure is constructed by assigning values of $\Delta\phi = \Delta q = Q/16D_q$. In order to construct a curvilinear square flow net, $\Delta\phi = K\Delta h$ should be equal to $\Delta\psi$. In other words, $\Delta\phi = Q/16D_q$. The stream lines with intensities $\phi = \pm 8\Delta\psi = \pm Q/2D_q$ intersect at a point on the x axis with an x coordinate that is $-x_s$. The directions of flow are opposite to each other at point $\equiv (-x_s, o)$, indicating that the velocity at this point should be zero. Therefore, at $(-x_s, o)$, a *stagnation point* exists. Its location x_s is determined by equating y and v_x to zero, that is,

$$v_x = -\frac{\delta\phi}{\delta x} = -\frac{\delta}{\delta x}\left(v_o x + \text{constant} + \frac{Q}{4\pi D_q}\ln\frac{x^2}{r_w^2}\right)$$

$$= -v_o - \frac{Q}{2\pi D_q}\frac{x}{x^2 + 0} = 0$$

or $$x_s = -\frac{Q}{2\pi v_o D_q} \tag{8.19}$$

The stream lines $\psi = \pm 8\Delta\psi = \pm Q/2D_q$ passing through the stagnation point coincide with what is known as the *groundwater divide*. Flow lines surrounded by this divide terminate at the well; those beyond the divide join the natural flow. The rate of flow toward the well bounded by the groundwater divide should be equal to Q, which is the rate of pumping.

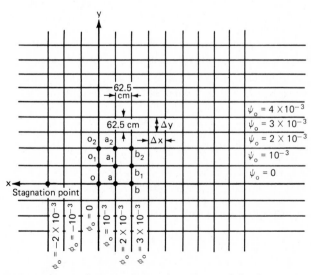

Figure 8.6 Grid system (Example 8.2). ψ and ϕ values are in square meters per minute.

EXAMPLE 8.2 Find the flow net resulting from superimposing the flow net of the individual well given in Example 8.1 and the flow net of a natural flow of velocity $v_o = 1.6 \times 10^{-3}$ m/min. Consider that $D_q = 225$ m (that is, $K = T/D_q = 1.35/225 = 6 \times 10^{-3}$ m/min).

Solution The selected drop in head for the well only is $\Delta h_w = s_{dw}/n_E = \frac{1}{6}$ m (Example 8.1) and $\Delta \phi_w = K \Delta h_w = \frac{1}{6} \times 6 \times 10^{-3} = 10^{-3}$ m²/min.

The selected stream function for the well is $\Delta \psi_{wd} = Q/16 D_q = 0.225/D_q = 0.225/225 = 1 \times 10^{-3}$ m²/min. If we select $\Delta \psi_o = \Delta \psi_{wd} = 1 \times 10^{-3}$ m²/min, then

$$\Delta y = \frac{\Delta \psi_o}{v_o} = \frac{1 \times 10^{-3}}{1.6 \times 10^{-3}} = 0.625 \text{ m} = 62.5 \text{ cm}$$

The flow net for natural flow only (unaffected by the well) would then consist of squares (Fig. 8.6) with a mesh size $\Delta y = \Delta x = 62.5$ cm. For the well only, the circles represent the s_d lines starting with zero at r_e ($\phi_w = K s_d = 0$) to its maximum value $s_{dw} = (\Delta h)_w \times n_E = 21.12/6 = 3.52$ m at the well surface; that is, $\phi_{wd} = K s_{dw} = 0.02$ m²/min. The equipotential lines of the natural flow are lines that are parallel to the y axis (Fig. 8.6). The line $\phi_o = 0$ coincides with the y axis, and the intervals $\Delta \phi_o = \Delta \psi_o = 1 \times 10^{-3}$ m²/min. If we superimpose the two flow nets, the resultant values at the points of intersection can be determined. If we join the points of equal intensities by smooth curves, the flow configuration is obtained. The intensities of the equipotential lines may be given in terms of either the total heads or the drawdowns or the velocity potential. The configuration of flow is similar to that shown in Fig. 8.5.

In extensive aquifers, the groundwater divide extends from $-x_s$ to the extreme boundaries of the aquifer with a theoretical $+x$ of infinity. If we substitute in Eq. (8.18) the coordinates $x = x_g = \infty$, $y = y_g$ of the point on the groundwater divide near the aquifer boundary at $+x = \infty$, and $\psi = Q/2D_q$, then

$$\frac{Q}{2D} = v_o y_g + \frac{Q}{2\pi D_q} \tan^{-1} \frac{y_g}{x_g}$$

or

$$\frac{2\pi D_q}{Q} \left(\frac{Q}{2D_q} - v_o y_g \right) = \tan^{-1} \frac{y_g}{x_g}$$

Then

$$\frac{y_g}{x_g} = \tan \left(\pi - \frac{2\pi v_o y_g D_q}{Q} \right) = -\tan \frac{2\pi v_o y_g D_q}{Q}$$

$$= 0 \quad \text{(because } x_g = \infty \text{)}$$

or

$$\frac{2\pi v_o y_g D_q}{Q} = \pi$$

Therefore,

$$2y_g = \frac{Q}{D_q v_o} \tag{8.20}$$

The groundwater divide is therefore asymptotic to the two horizontal lines $\pm y_g$ bordering the inflow zone. The natural flow directed to the well is equal to $2y_g v_o D_q = Q = 2\pi v_o x_s D_q$ [Eq. (8.19)], or

$$\text{Inflow zone} = 2y_g = 2\pi x_s \tag{8.21}$$

Using Eq. (8.21) in Example 8.2, we see that

$$2y_g = \frac{3.6}{225 \times 1.6 \times 10^{-3}} = 10 \text{ m} = 1000 \text{ cm}$$

Also from Eq. (8.21) we see that $x_s = 10/2\pi = 1.592$ m $= 159.2$ cm.

Combined Effects of Discharge and Recharge Wells

A recharge well has exactly the opposite effect of a discharge well. Recharge wells are used to pump water into an aquifer rather than pumping it out. They act as sources rather than as sinks. The equations are essentially the same except that $h_w > h_u$ and a cone of accretion occurs. The water level in a recharge well is above the undisturbed piezometric level. Recharge wells have many uses (see Chaps. 7 and 9).

From a practical point of view the analysis is interesting because we must introduce the concept of *images*, which is very useful in evaluating the flow configuration of wells pumped near barriers (impervious boundaries) or near streams cutting through aquifers (explained later). Figure 8.7 shows the flow pattern of a discharge well and a recharge well operating at the same time with the same rate of pumping $Q = qD_q$, where q is the rate of flow per unit thickness of the aquifer (Jacob, 1950). The equipotential lines are found to be a family of circles with radii $2x_w\sqrt{C_\phi}/(1 - C_\phi)$, where $+x_w$ and $-x_w$ are the horizontal coordinates of the discharge and recharge wells, respectively (Fig. 8.7), and C_ϕ is a constant for any specific equipotential line with intensity ϕ. Therefore,

$$C_\phi = \exp\frac{4\pi\phi}{q} = \frac{(x - x_w)^2 + y^2}{(x + x_w)^2 + y^2} \tag{8.22}$$

The centers of this family of circles lie on the x axis with the coordinates

$$x = x_w\frac{1 + C_\phi}{1 - C_\phi} \quad \text{and} \quad y = 0 \tag{8.23}$$

The equation for the equipotential lines is given by (Jacob, 1950)

$$y^2 + \left(x - x_w\frac{1 + C_\phi}{1 - C_\phi}\right)^2 = x_w^2\frac{4C_\phi}{(1 - C_\phi)^2} \tag{8.24}$$

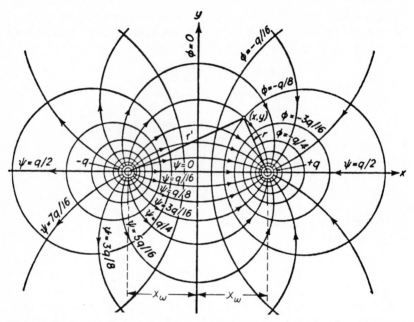

Figure 8.7 Stream lines and equipotential lines around discharge and recharge wells of equal pumping rates. *(C. E. Jacob, "Flow of Ground Water," in H. Rouse (ed.), Engineering Hydraulics, Wiley, New York, 1950.)*

The y axis is an equipotential line ($\phi = 0$, $C_\phi = 1$) that may be considered as one of the circles in the family with an infinite radius and a center at $(x = \infty, y = 0)$. The stream lines are given by the equation

$$x^2 + \left(y - \frac{x_w}{C_\psi}\right)^2 = x_w^2\left(1 + \frac{1}{C_\psi^2}\right) \tag{8.25}$$

where C_ψ is a constant for any specific stream line with an intensity ψ; that is,

$$C_\psi = \tan\frac{2\pi\psi}{q} = \frac{2x_w y}{x^2 - x_w^2 + y^2} \tag{8.26}$$

Equation (8.25) represents another family of circles; their centers lie on the y axis at points ($x = 0$, $y = x_w/C_\psi$), and their radii are equal to $x_w[1 + (1/C_\psi^2)]^{1/2}$. The segment of the x axis between the wells (from $x = -x_w$ to $x = +x_w$) is a stream line of intensity $\psi = 0$. The extensions to the right and to the left of this line constitute another stream line of a different intensity $\psi = \frac{1}{2}q$. It leaves the recharge well to infinity (theoretically) and returns from infinity to the discharge well. When $\psi = q/4$, C_ψ [Eq. (8.26)] equals infinity. The center of the circle corresponding to

$\psi = q/4$ lies on the origin and its radius is x_w. Theoretically speaking, there is complete circulation between the recharge and discharge wells; that is, all the water pumped into the recharge well returns to the discharge well and is withdrawn at a rate q. The shortest length of the flow path lies on the x axis between $x = -x_w$ and $x = +x_w$. The lengths of the other flow paths gradually increase to infinity.

In Fig. 8.7 it is clear that whenever the left portion is completely omitted, the portion of the flow net to the right of the y axis is a separate and complete flow net with the stream lines flowing normal to the y axis toward the well. This portion of the flow net may be regarded as a solution to another condition, that is, a discharge well pumping near a stream of constant ϕ along the y axis and withdrawing all its water from the stream, as long as the boundary conditions are satisfied. The recharge well is thus used as a tool to solve the problem of a discharge well pumping near a stream. It is called a *negative image* of the discharge well, with the stream boundary as a line of reflection.

Equations (8.24) and (8.25) define the water circulation between a recharge and a discharge well in an extensive aquifer when the undisturbed piezometric surface is horizontal and there is no initial natural groundwater flow. If there is a natural flow (Fig. 8.8) with a darcian velocity v_o, then there are three possible cases of circulation depending on the relative values of q, v_o, and x_w (which is equal to half the spacing between the discharge well and its recharge image). There is no circulation between the wells when

$$\frac{q}{\pi x_w v_o} < 1.0 \quad \text{(no circulation)} \qquad (8.27)$$

Figure 8.8a represents this case in which either well is not affected by the other. This means that the discharge well is deriving all its water from the aquifer and none from the stream. If q is increased and/or x_w is decreased while v_o remains unchanged, the following condition is reached:

$$\frac{q}{\pi x_w v_o} = 1.0 \quad \text{(limiting case)} \qquad (8.28)$$

This case is a limiting case (Fig. 8.8b), which means that there is no circulation between the wells. However, any slight change in the value of $q/\pi x_w v_o$ will result in no circulation between the wells when its value is less than 1.0 and in circulation between the wells when its value is greater than 1.0; that is,

$$\frac{q}{\pi x_w v_o} > 1.0 \quad \text{(circulation)} \qquad (8.29)$$

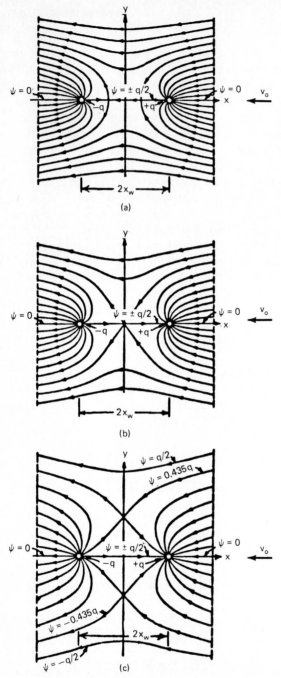

Figure 8.8 A discharge and a recharge well in uniform flow. *(a)* No circulation; *(b)* limiting case; *(c)* recirculation in diamond-shaped area between wells. *(C. E. Jacob, "Flow of Ground Water," in H. Rouse (ed.), Engineering Hydraulics, Wiley, New York, 1950.)*

This case is shown in Fig. 8.8c. The condition given in Eq. (8.29) may be produced from the limiting condition by increasing q or decreasing the spacing $2x_w$, or it takes place when the natural velocity v_o decreases.

In the case shown in Fig. 8.8c, the portion of water q that the well withdraws directly from a stream can be computed. If the rate of flow of water derived from the stream (circulating water) is q_{rs} per unit depth of the aquifer, then (Jacob, 1950)

$$\frac{q_{rs}}{q} = \frac{2}{\pi}\left(\tan^{-1} \sqrt{\frac{q}{\pi x_w v_o} - 1} - \frac{\pi x_w v_o}{q} \sqrt{\frac{q}{\pi x_w v_o} - 1} \right) \qquad (8.30)$$

The solution to this equation is given in graphic form in Fig. 8.9.

EXAMPLE 8.3 A discharge artesian well is pumping at a rate $q = 0.3$ m²/min per unit depth of an artesian aquifer, where the natural ground-water velocity $v_o = 1.99 \times 10^{-3}$ m/min. If the well is located at 12 m from a stream cutting through the entire depth of the aquifer, find the portion of the rate of flow q_{rs} withdrawn by the well from the stream.

Solution

$$\frac{q}{\pi x_w v_o} = \frac{0.3 \times 10^2}{11 \times 12 \times 1.99} = 4.0 \text{ (dimensionless)} > 1$$

Therefore, circulation takes place. From Fig. 8.9, $q_{rs}/q = 0.4 = 40$ percent. Rate of flow withdrawn from the stream is therefore $0.4 \times 0.3 = 0.12$ m²/min per unit depth of the aquifer.

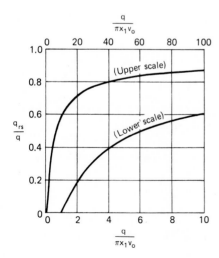

Figure 8.9 Graph showing relative recirculation between a discharge and a recharge well. (C. E. Jacob, "Flow of Ground Water," in H. Rouse (ed.), Engineering Hydraulics, Wiley, New York, 1950.)

EXAMPLE 8.4° The natural groundwater average velocity is 0.002 m/min, and its temperature is 12°C. If a recharged well injects water into the aquifer at a constant temperature of 30°C and a rate of 0.3 m²/min per unit thickness of the aquifer in order to cool it to at least 17°C and then the water is recovered by a discharge well pumping at the same rate and placed upgradient from the recharge well, determine the spacing between the two wells to satisfy the desired requirements.

Solution Neglecting the loss of heat by conduction, the heat-balance equation is

$$30°q_{rs} + 12°(q - q_{rs}) = 17°q$$

or
$$18q_{rs} = 5q \quad \text{or} \quad \frac{q_{rs}}{q} = 0.28$$

From the chart in Fig. 8.9, $q/\pi x_w v_o \approx 2.8$. Therefore,

$$x_w = \frac{0.3}{11 \times 0.002 \times 2.8} = 17.05 \text{ m}$$

The required spacing is $2x_w \approx 34$ m.

In Fig. 8.8a, two stagnation points lie on the x axis; their coordinates are given by (Jacob, 1950)

$$x_s = \pm x_w \sqrt{1 - \frac{q}{\pi x_w v_o}} \quad \text{(no circulation)} \quad (8.31)$$

Since the x axis has an intensity $\psi = 0$, the ψ_s corresponding to the groundwater divide passing through the stagnation point is $\psi_s = \pm q/2$.

When circulation occurs (Fig. 8.8c), two stagnation points lie on the y axis with coordinates

$$y_s = \pm x_w \sqrt{\frac{q}{\pi x_w v_o} - 1} \quad \text{(circulation)} \quad (8.32)$$

The stream line coinciding with the x axis has an intensity $\psi = 0$ when x is greater than $(x_w + r_w)$ or smaller than $-(x_w - r_w)$. Between $x = \pm(x_w - r_w)$, the ψ value is $\pm q/2$. The groundwater divide passing by the stagnation points has a ψ_s value given numerically by

$$\psi_s = \frac{q}{2} + v_o x_w \sqrt{\frac{q}{\pi x_w v_o} - 1} - \frac{q}{\pi} \tan^{-1} \sqrt{\frac{q}{\pi x_w v_o} - 1} \quad (8.33)$$

Equation (8.30) was derived from Eq. (8.33) by Jacob (1950).

° Example 8.4 is based on a similar example given by Jacob (1950), but the data have been converted to SI units.

Generally, the stream function ψ is given by

$$\psi = v_o y + \frac{q}{2\pi}\left(\tan^{-1}\frac{y}{x - x_w} - \tan^{-1}\frac{y}{x + x_w} \right) \qquad (8.34)$$

The horizontal component of the velocity at any point is given by

$$v_x = -\frac{\delta\psi}{\delta y} = -v_o - \frac{q}{2\pi}\left[\frac{x - x_w}{(x - x_w)^2 + y^2} - \frac{x + x_w}{(x + x_w)^2 + y^2} \right] \qquad (8.35)$$

From Eq. (8.35), the horizontal velocity $v_{x=0}$ at the origin ($x = 0$, $y = 0$) is given by

$$v_{x=0} = -v_o + \frac{q}{2\pi x_w} \qquad (8.36)$$

And the horizontal velocity at the surface of the discharge well ($x = x_w - r_w$, y) is given by

$$v_{x=(x_w-r_w)} = -v_o - \frac{qx_w}{\pi r_w(2x_w - r_w)} \qquad (8.37)$$

The groundwater divide of a discharge well pumping near a stream extends to a far distance where it is asymptotic to a natural stream line of intensity $\psi = \frac{1}{2}(q - q_{rs})$. Therefore,

$$y_g v_o = \tfrac{1}{2}(q - q_{rs})$$

or

$$2y_g = \frac{q - q_{rs}}{v_o} \qquad (8.38)$$

where the inflow zone $2y_g$ has the same physical significance as that in Eqs. (8.20) and (8.21).

Method of Images

It has been shown that a recharge well may be placed as a negative image of a discharge well pumping near a stream. The image proved to be valid when both wells have the same rate of pumping and whether the natural velocity v_o is equal to or greater than zero. The illustrated cases give the general concept of the method of images. However, before the solution is attempted, the validity of the proposed images can be tested. For example, in Fig. 8.10, a real discharge well ($+ W_{di}$) is close to a stream; in order to find the configuration of flow, an image recharge well ($- W_{re}$) is placed at an equal normal distance x_w in the opposite direction. Although it has been shown that this plan produces the required solution, it may be checked by the procedure explained in the following paragraphs in order

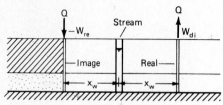

Figure 8.10 Vertical section for a stream boundary.

to apply the same general procedure in more complex situations with numerous boundaries and/or numerous wells.

If the discharge well in Fig. 8.11 is considered separately, the vector of velocity v_a at point a on the stream is along the radius $r_a = \overline{ao}$ and is equal to $v_a = Q/2\pi r_a D_q$. If a recharge well $(-W_{re})$ of the same intensity Q is placed opposite to the discharge well $(+W_{di})$ then the vector of velocity v'_a at the same point a (assuming the recharge well separately) will be along $r'_a = \overline{ao'}$. Owing to symmetry, $r'_a = r_a$; therefore, $v'_a = v_a$ and their vectors make the same angle β with $\overline{oo'}$. Now, assuming that both wells are in operation, by superimposing v_a and v'_a, their vertical components should be equal and opposite, canceling each other out. However, their horizontal components v_{ax} and v'_{ax} are equal and oriented to the

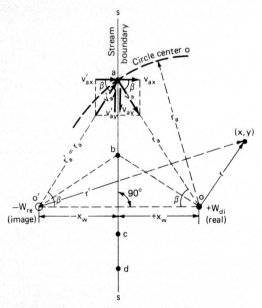

Figure 8.11 Use of image wells; plan for a discharge well near a stream.

right normal to *ss* (the stream boundary) at *a*. Therefore, the resultant of v_a and v_a' is equal to $2v_{ax} = (Q \cos \beta)/(\pi r_a D_q)$ normal to stream boundary *ss*. Since the direction of the resultant velocity at *a* is the same as the direction of the flow line at the same point, the boundary condition at *a* is satisfied. The same analysis holds true at any other point on the stream boundary *ss*, such as points *b*, *c*, and *d* (Fig. 8.11). Therefore, the well image ($-W_{re}$) located opposite to ($+W_{di}$) and of the same intensity gives the solution for a single discharge well pumping steadily near a stream.

If *ss* is a barrier boundary, such as an impervious boundary that resulted from a geologic fault (Fig. 8.12), a discharge image well ($+W_{di}$) is used. In this case, the real discharge well pumps near the barrier boundary and the positive image pumps at the same rate as the real well. The same procedure indicates that the velocity vector at *a* for the image is opposite to what is shown in Fig. 8.11. The horizontal velocities cancel out, and the resultant velocity is vertical along the direction of the barrier boundary, indicating that a flow line coincides with the barrier boundary, permitting no water to cross it. Hence the boundary condition is satisfied at point *a* or any other point on the boundary. The resultant effect of the real and image wells is determined by the principle of superposition.

If ϕ_{wd} and ϕ_{wr} represent the velocity potentials of a discharge and a recharge well, respectively, then from Eq. (8.15), the total velocity potential ϕ is given by

$$\phi = \phi_{wd} - \phi_{wr} = \frac{Q}{4\pi D_q}\left(\ln \frac{r^2}{r_w^2} - \ln \frac{r'^2}{r_w^2}\right) \qquad \text{(stream boundary)} \qquad (8.39)$$

where *r* and *r'* are the radial distances from point (x, y) to the discharge and recharge wells, respectively. Therefore,

$$\phi = \frac{Q}{4\pi D_q} \ln \frac{(x - x_w)^2 + y^2}{(x + x_w)^2 + y^2} \qquad \text{(stream boundary)} \qquad (8.40)$$

If there is a natural flow, its velocity potential ϕ_o [Eq. (8.13)] should be

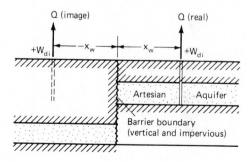

Figure 8.12 Vertical section for a barrier boundary.

added to ϕ given by Eq. (8.40). Since

$$\psi = \psi_{wd} - \psi_{wr} = \frac{Q}{2\pi D_q}\left(\tan^{-1}\frac{y}{x - x_w} - \tan^{-1}\frac{y}{x + x_w}\right)$$

$$\text{(stream boundary)} \quad (8.41)$$

then ψ_o of the natural flow should be added to ψ given by Eq. (8.41) if there is a natural flow.

In the case of an existing barrier near the discharge well,

$$\phi = \frac{Q}{4\pi D_q}\left(\ln\frac{rr'}{r_w^2}\right)$$

$$= \frac{Q}{4\pi D_q}\ln\frac{[(x - x_w)^2 + y^2][(x + x_w)^2 + y^2]}{r_w^2}$$

$$\text{(barrier boundary)} \quad (8.42)$$

and

$$\psi = \frac{Q}{2\pi D_q}\left(\tan^{-1}\frac{y}{x - x_w} + \tan^{-1}\frac{y}{x + x_w}\right)$$

$$\text{(barrier boundary)} \quad (8.43)$$

ϕ_o and ψ_o should be added to ϕ and ψ [Eqs. (8.42) and (8.43)] whenever a natural flow exists. The outcrop of the artesian aquifer is usually far away from the barrier boundary, and the natural flow is most likely directed toward that barrier boundary with no outlet. A backwater curve develops that is similar to that produced in the upstream of dams on rivers and streams. The barrier boundary, in effect, is an underground dam.

The method of images is very useful in analyzing a system of multiple wells operating near a boundary or a system of boundaries. Consider the three real discharge wells W_{di1}, W_{di2}, and W_{di3} located at distances x_{w1}, x_{w2}, and x_{w3}, respectively, from a stream ss (Fig. 8.13). Three images of recharge wells $-W_{re1}$, $-W_{re2}$, and $-W_{re3}$ are placed opposite these wells. The configuration of flow of these six wells gives the flow net to the right of ss for the three real discharge wells. The checking procedure is the same as that given for one discharge well, but any point a on ss is affected by six wells, and the resultant of the six individual velocities through a is normal to ss.

If one discharge well is operating near boundaries crossing at right angles, more than one image is needed. Figure 8.14a shows a discharging well pumped close to the intersection of a stream and a barrier boundary at a right angle. Image W_{re1} is a recharge well with the same normal distance to the stream as the real well, image W_{di1} is a discharge well with the same normal distance to the barrier boundary as the real well, and image W_{re2} is a recharge well that serves as the image of the discharge image W_{di2} with respect to the stream as well as the image of the recharge

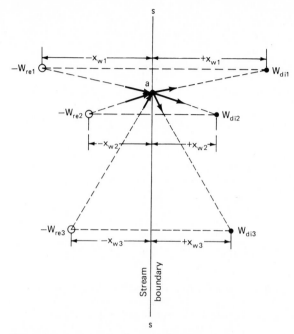

Figure 8.13 System of images for three discharge wells near a stream satisfying boundary conditions.

image W_{re1} with respect to the barrier boundary. There are three images for each real well. For example, if 3 real wells are operating, the flow net is the portion within the boundaries of the overall flow net determined for 12 wells. Although the procedure entails the superposition of simple cases, the solution is better programmed on a computer (see Chap. 9). Such a solution may include any number of real and image wells at any location from the boundaries.

The middle drawing in Fig. 8.14b shows two barrier boundaries intersecting at a right angle. In this case, all images are discharge wells. If the boundaries intersect at any angle α and the wedge boundaries are of like character, the angle α must be an aliquot part of $180°$. If they are not of like character, then α must be an aliquot part of $90°$ (Ferris et al., 1962). The exact number of image wells N_i is given by

$$N_i = \frac{360°}{\alpha} - 1 \qquad (8.44)$$

The image wells are located on a circle whose center is at the apex of the wedge and whose radius is equal to the distance from the real well to the apex of the wedge. Any image well should be located at a normal distance

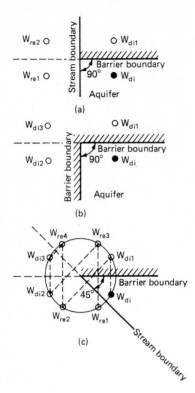

Figure 8.14 Plans of image-well system. *(a)* Stream normal to a barrier boundary; *(b)* two perpendicular barrier boundaries; *(c)* stream and barrier boundaries intersecting at 45° angle. Closed circles (●) represent real wells and open circles (○) image wells. W_{di} and W_{re} are discharge and recharge wells, respectively. *(J. G. Ferris et al., "Theory of Aquifer Tests," U.S. Geological Survey Water Supply Paper 1536-E, Washington, 1962.)*

from the boundary equal to that of the well for which the image is considered.

In the case shown in Fig. 8.14c, $\alpha = 45°$; then by use of Eq. (8.44), $N_i = 7$. If, for example, $\alpha = 90°$, $N_i = 3$ (such as the cases given in Fig. 8.14a and b). If the angle is 40°, Eq. (8.44) is not valid because 40° is not an aliquot part of 90°. However, 40° may for practical purposes be considered to be 45° to obtain a solution. If the wedge boundaries are not of like character, such as a stream intersecting a barrier boundary, α must be an aliquot part of 180°.

Figure 8.15 shows the images for two cases: parallel boundaries extending to a very large distance and a boundary intersecting normally these two parallel boundaries. The images are repeated to infinity. However, some of these far images do not affect the flow medium around the real wells and therefore should be eliminated.

Figure 8.16 shows the pattern of images for a well operating in a rectangular zone surrounded by crossing boundaries (Ferris et al., 1962). It is more convenient to find the drawdowns $s_{d(x,y)}$ at points (x, y) rather than ϕ. Since $\phi = -Ks_d$, the drawdowns can be determined using any of the previous equations for ϕ. The results of three cases follow:

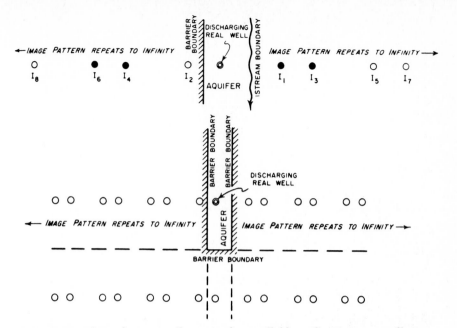

Figure 8.15 Plans of image-well systems for parallel boundaries. Image wells *I* are numbered in the sequence in which they were considered and located. Open circles (O) represent discharging wells and closed circles (●) recharging wells. *(J. G. Ferris et al., "Theory of Aquifer Tests," U.S. Geological Survey Water Supply Paper 1536-E, Washington, 1962.)*

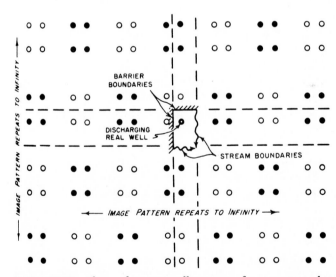

Figure 8.16 Plans of image-well systems for a rectangular aquifer. Open circles (O) represent discharging wells and closed circles (●) recharging wells. *(J. G. Ferris et al., "Theory of Aquifer Tests," U.S. Geological Survey Water Supply Paper 1536-E, Washington, 1962.)*

For a discharge well pumping near a stream,

$$s_{d(x,y)} = \frac{Q}{4\pi T} \ln \frac{(x + x_w)^2 + (y - y_w)^2}{(x - x_w)^2 + (y - y_w)^2} \tag{8.45}$$

where (x_w, y_w) are the coordinates of the pumped well.

For a discharge well pumping near the corner of two perpendicular streams,

$$s_{d(x,y)} = \frac{Q}{4\pi T} \ln \frac{[(x - x_w)^2 + (y + y_w)^2][(x + x_w)^2 + (y - y_w)^2]}{[(x - x_w)^2 + (y - y_w)^2][(x + x_w)^2 + (y + y_w)^2]} \tag{8.46}$$

For a discharge well pumping within a strip of width $2a$ bounded by two parallel streams (Todd, 1980),

$$s_{d(x,y)} = \frac{Q}{4\pi T} \ln \left[\cosh^2 \frac{\pi(y - y_w)}{2a} - \cos^2 \frac{\pi(x + x_w)}{2a} \right] \tag{8.47}$$

where the angles are expressed in radians. In the preceding three equations, $y_w = 0$ when the x axis is selected to pass through the pumped well.

8.2 Artesian Wells: Transient Conditions

It has been shown that steady-state flow takes place under one or both of the following conditions: (1) after time t approaches ∞ or a large time beyond which the changes in drawdown would be insignificant, and (2) if there is enough recharge to the aquifer or leakage such that the drawdown remains stationary after a certain definite time. After pumping starts, the pumping rate Q is derived entirely from storage owing to the compressibility of the aquifer. In Fig. 8.17, if the drawdown curve corresponding to time interval t_1 since the start of pumping is considered, then after time t_1, the zone affected by pumping lies within a cylindrical volume of internal radius r_w and external radius $r_{e,t1}$, which is the radius of influence after time t_1. Theoretically, $r_{e,t1} = \infty$ corresponds to a drawdown of zero. This can be thought of as the finite radius at which the drawdown is equal to a very small value, say, 2.0 to 5.0 cm. All the water pumped by the well is assumed to be instantly released from storage within the cylinder. The aquifer is assumed to be extensive, receiving no recharge; thus any annular cylinder of thickness Δr releases water from storage at a rate equal to

$$\Delta Q = 2\pi r \Delta r S \frac{\Delta h_{r,t}}{\Delta t}$$

where $\Delta h_{r,t}/\Delta t = \Delta s_d/\Delta t =$ the change in $h_{r,t}$ or s_d during a time interval

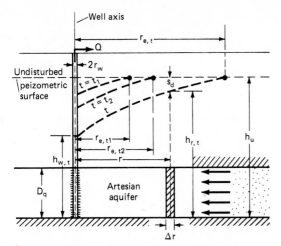

Figure 8.17 Transitional stages of piezometric surfaces resulting from pumping an artesian well.

Δt. Thus Q after time t_1 is

$$Q = -\int_{r=r_w}^{r_{e,t1}} 2\pi r S \frac{\delta s_d}{\delta t} dr \qquad (8.48)$$

Theis Formula

The equation governing flow toward an artesian well is given by (Chap. 6)

$$\frac{\delta^2 s_d}{\delta r^2} + \frac{1}{r}\frac{\delta s_d}{\delta r} = \frac{S}{T}\frac{\delta s_d}{\delta t} \qquad (6.11a)$$

The assumptions implied in deriving this equation have been given in Chap. 6. Theis (1935) solved this equation, adding the assumption that the artesian well is a vertical sink ($r_w = 0$) penetrating the entire depth of the aquifer and screened throughout that depth. Moreover, the rate of pumping Q is assumed constant (it really varies from 0 to Q over a very small time interval); then

$$h_{r,t} = h_u - \frac{Q}{4\pi T}\int_{r^2 S/4Tt}^{\infty} \frac{e^{-u}}{u} du \qquad (8.49a)$$

$$h_{r,t} = h_u - \frac{Q}{4\pi T} W(u) \qquad (8.49b)$$

where the integral is the exponential integral $-Ei(-u)$ known among groundwater workers as the *well function* $W(u)$, where u is given by

$$u = \frac{r^2 S}{4Tt} \quad \text{(dimensionless)} \tag{8.50}$$

Therefore,

$$s_d = \frac{Q}{4\pi T} W(u) \quad \text{(Theis' equation)} \tag{8.51}$$

where $W(u)$ = well function = $-Ei(-u)$
$\quad s_d$ = drawdown $(h_u - h_{r,t})$ at radius r and time t
$\quad h_{r,t}$ = the total head at radius r and time t
$\quad u$ = value given by Eq. (8.50) that depends on r and t
$\quad T$ = average transmissivity of the aquifer
$\quad t$ = time elapsed since the start of pumping
$\quad S$ = storage coefficient

$W(u)$, the well function, is dimensionless and depends on u as given by Table 8.3 or its condensed form Table 8.4. It can be expanded to the following convergent series:

$$s_d = \frac{Q}{4\pi T}\left(-0.577216 - \ln u + u - \frac{1}{2 \cdot 2!}u^2 + \frac{1}{3 \cdot 3!}u^3 - \cdots\right) \tag{8.52}$$

The graphic relationship between $1/u$ and $W(u)$ is given in Fig. 8.18 and that between u and $W(u)$ is shown in Figs. 8.19 and 8.20. All these curves are drawn on log-log plots. The curves shown in Fig. 8.20 are more detailed than the same relationship illustrated in Fig. 8.19. These curves are dimensionless and are applied to any rate of pumping under any physical or hydrologic conditions. In Fig. 8.19, for example, u is in fact $r^2/t \times \text{constant} = r^2/t \times S/4T$ [Eq. (8.50), and $W(u) = s_d \times \text{constant} = s_d \times 4\pi T/Q$. This means that the theoretical curve shown in Fig. 8.19 should be similar to a curve obtained from field records: log r^2/t versus log s_d. These theoretical curves and Table 8.3 or 8.4 may be used to find the drawdown s_d at any radius r and after any time t if the parameters S and T are predetermined or approximately known (Sec. 6.5). Figure 8.19 covers a range of u between 0.0001 and 3.2. For values of u less than 0.0001, curve III in Fig. 8.20 is used for the range between 10^{-4} and 10^{-8}. Curve I of the same figure is used when $u > 3.2$ for a range between 3.2 and 7.1.

The curves are used also to determine S and T by certain graphic methods explained later. Therefore, a large-scale chart based on Table 8.3 should be prepared by those working continuously in such evaluation if more precision is needed.

TABLE 8.3 Values of $W(u)$ for Values of u between 10^{-15} and 9.5

u / N	$N \times 10^{-15}$	$N \times 10^{-14}$	$N \times 10^{-13}$	$N \times 10^{-12}$	$N \times 10^{-11}$	$N \times 10^{-10}$	$N \times 10^{-9}$	$N \times 10^{-8}$	$N \times 10^{-7}$	$N \times 10^{-6}$	$N \times 10^{-5}$	$N \times 10^{-4}$	$N \times 10^{-3}$	$N \times 10^{-2}$	$N \times 10^{-1}$	N
1.0	33.9616	31.6590	29.3564	27.0538	24.7512	22.4486	20.1460	17.8435	15.5409	13.2383	10.9357	8.6332	6.3315	4.0379	1.8229	0.2194
1.5	33.5561	31.2535	28.9509	26.6483	24.3458	22.0432	19.7406	17.4380	15.1354	12.8328	10.5303	8.2278	5.9266	3.6374	1.4645	0.1000
2.0	33.2684	30.9658	28.6632	26.3607	24.0581	21.7555	19.4529	17.1503	14.8477	12.5451	10.2426	7.9402	5.6394	3.3547	1.2227	0.04890
2.5	33.0453	30.7427	28.4401	26.1375	23.8349	21.5323	19.2298	16.9272	14.6246	12.3220	10.0194	7.7172	5.4167	3.1365	1.0443	0.02491
3.0	32.8629	30.5604	28.2578	25.9552	23.6526	21.3500	19.0474	16.7449	14.4423	12.1397	9.8371	7.5348	5.2349	2.9591	0.9057	0.01305
3.5	32.7088	30.4062	28.1036	25.8010	23.4985	21.1959	18.8933	16.5907	14.2881	11.9855	9.6830	7.3807	5.0813	2.8099	0.7942	0.006970
4.0	32.5753	30.2727	27.9701	25.6675	23.3649	21.0623	18.7598	16.4572	14.1546	11.8520	9.5495	7.2472	4.9482	2.6813	0.7024	0.003779
4.5	32.4575	30.1549	27.8523	25.5497	23.2471	20.9446	18.6420	16.3394	14.0368	11.7342	9.4317	7.1295	4.8310	2.5684	0.6253	0.002073
5.0	32.3521	30.0495	27.7470	25.4444	23.1418	20.8392	18.5366	16.2340	13.9314	11.6280	9.3263	7.0242	4.7261	2.4679	0.5598	0.001148
5.5	32.2568	29.9542	27.6516	25.3491	23.0465	20.7439	18.4413	16.1387	13.8361	11.5330	9.2310	6.9289	4.6313	2.3775	0.5034	0.0006409
6.0	32.1698	29.8672	27.5646	25.2620	22.9595	20.6569	18.3543	16.0517	13.7491	11.4465	9.1440	6.8420	4.5448	2.2953	0.4544	0.0003601
6.5	32.0898	29.7872	27.4846	25.1820	22.8794	20.5768	18.2742	15.9717	13.6691	11.3665	9.0640	6.7620	4.4652	2.2201	0.4115	0.0002034
7.0	32.0156	29.7131	27.4105	25.1079	22.8053	20.5027	18.2001	15.8976	13.5950	11.2924	8.9899	6.6879	4.3916	2.1508	0.3738	0.0001155
7.5	31.9467	29.6441	27.3415	25.0389	22.7363	20.4337	18.1311	15.8280	13.5260	11.2234	8.9209	6.6190	4.3231	2.0867	0.3403	0.0000658
8.0	31.8821	29.5795	27.2769	24.9744	22.6718	20.3692	18.0666	15.7640	13.4614	11.1589	8.8563	6.5545	4.2591	2.0269	0.3106	0.0000376
8.5	31.8215	29.5189	27.2163	24.9137	22.6112	20.3086	18.0060	15.7034	13.4008	11.0982	8.7957	6.4939	4.1990	1.9711	0.2840	0.0000216
9.0	31.7643	29.4618	27.1592	24.8566	22.5540	20.2514	17.9488	15.6462	13.3437	11.0411	8.7386	6.4368	4.1423	1.9187	0.2602	0.0000124
9.5	31.7103	29.4077	27.1051	24.8025	22.4999	20.1973	17.8948	15.5922	13.2896	10.9870	8.6845	6.3828	4.0887	1.8695	0.2387	0.0000071

SOURCE: L. K. Wenzel, "Methods of Determining Permeability of Water-Bearing Materials, with Special Reference to Discharging-Well Methods," U.S. Geological Survey Water Supply Paper 887, Washington, 1942.

TABLE 8.4 Values of $W(u)$ for Values of u (Abbreviated Version of Table 8.3)

u	1.0	2.0	3.0	4.0	5.0	6.0	7.0	8.0	9.0
$\times 1$	0.219	0.049	0.013	0.0038	0.0011	0.00036	0.00012	0.000038	0.000012
$\times 10^{-1}$	1.82	1.22	0.91	0.70	0.56	0.45	0.37	0.31	0.26
$\times 10^{-2}$	4.04	3.35	2.96	2.68	2.47	2.30	2.15	2.03	1.92
$\times 10^{-3}$	6.33	5.64	5.23	4.95	4.73	4.54	4.39	4.26	4.14
$\times 10^{-4}$	8.63	7.94	7.53	7.25	7.02	6.84	6.69	6.55	6.44
$\times 10^{-5}$	10.94	10.24	9.84	9.55	9.33	9.14	8.99	8.86	8.74
$\times 10^{-6}$	13.24	12.55	12.14	11.85	11.63	11.45	11.29	11.16	11.04
$\times 10^{-7}$	15.54	14.85	14.44	14.15	13.93	13.75	13.60	13.46	13.34
$\times 10^{-8}$	17.84	17.15	16.74	16.46	16.23	16.05	15.90	15.76	15.65
$\times 10^{-9}$	20.15	19.45	19.05	18.76	18.54	18.35	18.20	18.07	17.95
$\times 10^{-10}$	22.45	21.76	21.35	21.06	20.84	20.66	20.50	20.37	20.25
$\times 10^{-11}$	24.75	24.06	23.65	23.36	23.14	22.96	22.81	22.67	22.55
$\times 10^{-12}$	27.05	26.36	25.96	25.67	25.44	25.26	25.11	24.97	24.86
$\times 10^{-13}$	29.36	28.66	28.26	27.97	27.75	27.56	27.41	27.28	27.16
$\times 10^{-14}$	31.66	30.97	30.56	30.27	30.05	29.87	29.71	29.58	29.46
$\times 10^{-15}$	33.96	33.27	32.86	32.58	32.35	32.17	32.02	31.88	31.76

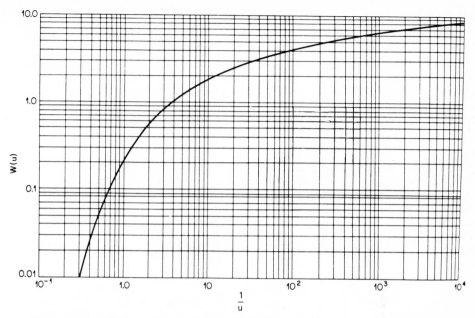

Figure 8.18 Relationship between log $(1/u)$ and log $W(u)$. *(W. C. Walton, Groundwater Resource Evaluation, McGraw-Hill, New York, 1970.)*

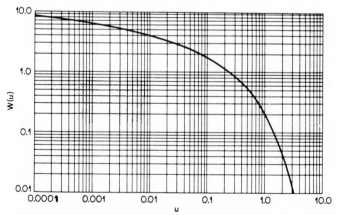

Figure 8.19 Relationship between log u and log $W(u)$. *(W. C. Walton, Groundwater Resource Evaluation, McGraw-Hill, New York, 1970.)*

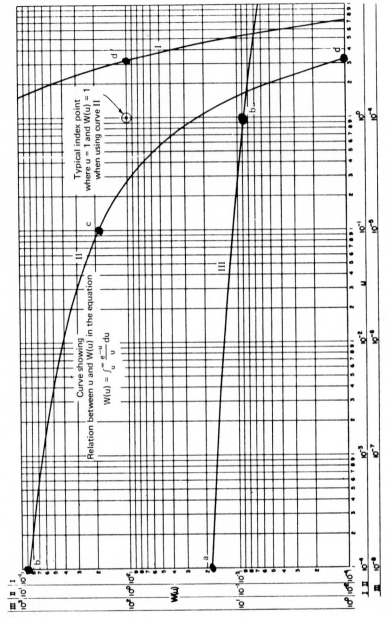

Figure 8.20 Type curve resulting from the plotting of u versus $W(u)$. Segments ab and bc lie within the range of the straight-line method of Jacob [see Eq. (8.56)]. (*U.S. Bureau of Reclamation, Ground Water Manual, U.S. Department of Interior, Washington, 1977.*)

Curve showing
Relation between u and W(u) in the equation

$$W(u) = \int_u^\infty \frac{e^{-u}}{u}\, du$$

Typical index point
where u = 1 and W(u) = 1
when using curve II

EXAMPLE 8.5 A fully penetrating artesian well is pumped at a rate $Q = 1600$ m³ per day (about 295 gal/min) from an aquifer whose S and T values are 4×10^{-4} and 0.145 m²/min, respectively. Find the drawdowns at a distance 3 m from the production well after 1 h of pumping and at a distance of 350 m after 1 day of pumping.

Solution

At $r = 3$ and $t = 1$ h:

$$u = \frac{r^2 S}{4Tt} \quad \text{[Eq. (8.50)]}$$

or

$$u = \frac{9 \times 4 \times 10^{-4}}{4 \times 0.145 \times 60} = 1.03 \times 10^{-4}$$

Then $W(u) = 8.62$, from Fig. 8.19 or Fig. 8.20 or Table 8.4 (by interpolation). From Eq. 8.51,

$$s_d = \frac{1600 \times 8.62}{24 \times 60 \times 4\pi \times 0.145} = 5.256 \text{ m} = 525.6 \text{ cm}$$

At $r = 350$ m and $t = 1$ day:

$$u = \frac{r^2 S}{4Tt} = \frac{350^2 \times 4 \times 10^{-4}}{4 \times 0.145 \times 24 \times 60} = 5.87 \times 10^{-2}$$

Then $W(u) = 2.316$, and

$$s_d = \frac{1600 \times 2.316}{4\pi \times 0.145 \times 60 \times 24} = 1.412 \text{ m} = 141.2 \text{ cm}$$

EXAMPLE 8.6 Find the approximate values of the radius of influence in the preceding example after 1 h and 1 day of continuous pumping.

Solution Assume that $s = 2.5$ cm $= 0.025$ m (practically zero) at $r_{e,t}$. Then by use of Eq. (8.51),

$$W(u) = \frac{4\pi T s_d}{Q} = \frac{4\pi \times 0.145 \times 0.025}{1600/(24 \times 60)} = 0.0410$$

From Fig. 8.19 (or Table 8.4), $u \approx 2.12$, and from Eq. (8.50),

$$r = r_{e,t} = \sqrt{\frac{4Ttu}{S}}$$

When $t = 1.0$ h $= 60$ min,

$$r_{e,t} = \sqrt{\frac{4 \times 0.145 \times 60 \times 2.12}{4 \times 10^{-4}}} = 429.46 \text{ m}$$

When $t = 1.0$ day $= 1440$ min,

$$r_{e,t} = \sqrt{\frac{4 \times 0.145 \times 1440 \times 2.12}{4 \times 10^{-4}}} = 2103.94 \text{ m}$$

Determination of Aquifer Parameters by Pumping Tests

In order to determine the parameters S and T, the drawdowns in a pumped well and in installed observation wells are recorded at different time intervals. In order to match the theoretical curves (Figs. 8.18 and 8.19), plots of these records are drawn on log-log paper to the same scale as the theoretical curves. The s_d values are plotted as ordinates, whereas the abscissas may be r^2 (at constant t), t, r^2/t, or t/r^2. If observation wells are not available, the records are taken in the pumped well and plotted as log s_{dw} versus log t, or log s_{dw} versus log r_w^2/t, or log s_{dw} versus log $1/t$. In order to use the theoretical curve, the recorded drawdowns in the pumped well should be adjusted before they are used. Each drawdown should be decreased in magnitude by the amount of screen losses (Sec. 8.8) and also should be adjusted to remove the tidal and barometric effects if any (Sec. 6.5).

The analysis and procedure of the evaluation of the parameters S and T are explained in the following. From Eqs. (8.50) and (8.51),

$$\log s_d = \left[\log \frac{Q}{4\pi T}\right] + \log W(u) \qquad (8.53)$$

and

$$\log \frac{r^2}{t} = \left[\log \frac{4T}{S}\right] + \log u \qquad (8.54)$$

The terms in brackets in the preceding two equations are constants. Thus the relationship between log s_d and log r^2/t is similar to the relationship between log $W(u)$ and log u given in Fig. 8.19. This means that if the field records are plotted as log s_d versus log r^2/t (using different observation wells) on a plot with the same size log intervals as that shown in Fig. 8.19 (or an enlarged curve), the two curves should be similar. If the field curve is plotted on transparent paper and overlayed on the theoretical curve, the two curves should be parallel if they have the same scale. They should therefore coincide if the field curve is moved in both the horizontal and vertical directions in order to eliminate the constants shown in the brackets in Eqs. (8.53) and (8.54). However, discrepancies are expected between theory and practice, and the field curve should match only a segment of the theoretical curve that usually covers the early times before the effects of later factors, such as leakage, start to be effective (see Sec. 8.3). If we select a point known as a *match point* on the matched segments, its four coordinates correspond to specific values of u, r^2/t, $W(u)$, and s_d. If we substitute the coordinates s_d and $W(u)$ in Eq. (8.51), T is determined. Then substituting the coordinates u and r^2/t and the computed T in Eq. (8.50), S is determined. Once a match point is located, the four coordinates of any other point away from the curves should give the same values of S and T.

By using the same analysis, it can be shown that a field plot of $\log s_d$ versus $\log t/r^2$ is similar to the theoretical curve given in Fig. 8.18. Since r is included in these field plots, the recordings should be taken at several observation wells in addition to the pumped well. If only *one* observation well is used, its distance r should be constant; the field plot of $\log s_d$ versus $\log 1/t$ should correspond to the curve $\log u$ versus $\log W(u)$ shown in Fig. 8.19, and the field plot of $\log s_d$ versus $\log t$ should correspond to Fig. 8.19. The pumped well recordings may be treated in the same manner as those of an observation well, substituting $r = r_w$; however, the recordings should be adjusted to remove the screen losses and other effects (Sec. 8.8).

EXAMPLE 8.7° An artesian well is pumped at a constant rate $Q = 1.888$ m³/min. Three observation wells are installed at the following distances from the pumping well: $r_1 = 91.44$ m, $r_2 = 121.92$ m, and $r_3 = 243.84$ m. The recorded drawdowns s_d at different times t are given in Table 8.5. Find S and T.

Solution The field recordings of the three observation wells 1, 2, and 3 are plotted on log-log paper (Fig. 8.21) with the same scale as that of the theoretical curve $1/u$ versus $W(u)$ (Fig. 8.18). The drawing is copied on transparent paper on a copying machine. Each one of the three curves is then matched on the theoretical curve and the match points are indicated by I, II, and III for wells 1, 2, and 3, respectively, in Fig. 8.21. The overlay of both curves is shown in Fig. 8.22 for observation well 2 only.

Observation Well 1: Coordinates of match point I are as follows:

$$\frac{1}{u} = 10 \qquad s_d = 0.338 \text{ m} \qquad t = 2.5 \text{ min} \qquad W(u) = 1.8229$$

Then $\quad T = \dfrac{QW(u)}{4\pi s_d} = \dfrac{1.888 \times 1.8229}{4\pi \times 0.338} = 0.8103 \text{ m}^2/\text{min}$

and $\quad S = \dfrac{4Ttu}{r^2} = \dfrac{4 \times 0.8103 \times 2.5 \times 0.1}{91.44^2} = 0.000097$

Observation Well 2: Coordinates of match point II are as follows (Fig. 8.22):

$$\frac{1}{u} = 10 \qquad s_d = 0.341 \text{ m} \qquad t = 10 \text{ min} \qquad W(u) = 1.8229$$

Then $\quad T = \dfrac{1.888 \times 1.8229}{4\pi \times 0.341} = 0.8032 \text{ m}^2/\text{min}$

and $\quad S = \dfrac{4 \times 0.8032 \times 10 \times 0.1}{121.92^2} = 0.00022\dagger$

° Example 8.7 is based on actual recordings given by Lohman (1972), but the data have been converted to SI units.

† Coincidentally in this case, the overlay of the curves (Fig. 8.22) gives the same abscissa values ($t = 4Tt/r^2 s_d = 1/u$ or $S = 4T/r^2$). Therefore, $S = (4 \times 0.8032)/121.92^2 = 0.00022$.

TABLE 8.5   Data for Examples 8.7 and 8.8

Time t since pumping started, min	Observation well 1: $r_1 = 91.44$ m		Observation well 2: $r_2 = 121.92$ m		Observation well 3: $r_3 = 243.84$ m	
	s_d, m	$\dfrac{r^2}{t}$, m²/min	s_d, m	$\dfrac{r^2}{t}$, m²/min	s_d, m	$\dfrac{r^2}{t}$, m²/min
1.0	0.201	3.716	0.049	14,838	0.001	59,546
1.5	0.265	2,477	0.082	9,871	0.006	39.676
2.0	0.302	1,858	0.116	7,419	0.012	29,676
2.5	0.338	1,484	0.140	5,935	0.021	23,870
3.0	0.369	1,239	0.162	4,935	0.027	19,999
4.0	0.415	929	0.204	3,710	0.049	14,838
5.0	0.454	742	0.235	2,968	0.067	11,935
6.0	0.485	619	0.265	2,464	0.082	9,935
8.0	0.533	465	0.302	1,852	0.113	7,419
10	0.567	372	0.341	1,484	0.140	5,955
12	0.600	310	0.369	1,239	0.162	4,968
14	0.634	265	0.384	1,129	0.180	4,258
18	0.671	206	0.436	826	0.219	3,290
24	0.719	155	0.482	619	0.265	2,477
30	0.759	124	0.518	494	0.290	2,000
40	0.808	93	0.573	371	0.341	1,484
50	0.847	74	0.610	297	0.375	1,194
60	0.878	62	0.643	246	0.402	994
80	0.927	46	0.683	185	0.454	742
100	0.963	37	0.725	148	0.494	595
120	1.000	31	0.759	124	0.518	497
150	1.042	25	0.799	99	0.558	397
180	1.070	21	0.829	83	0.591	329
210	1.100	18	0.856	71	0.619	284
240	1.119	16	0.878	62	0.643	248

Observation Well 3: Coordinates of match point **III** are as follows:

$$\frac{1}{u} = 3 \qquad s_d = 0.14 \text{ m} \qquad t = 10 \text{ min} \qquad W(u) = 0.83$$

Then
$$T = \frac{1.888 \times 0.83}{4\pi \times 0.14} = 0.8907 \text{ m}^2/\text{min}$$

and
$$S = \frac{4 \times 0.8907 \times 10 \times 1}{3(243.84)^2} = 0.00020$$

Discussion of Example 8.7   Once the field curve is overlayed on the theoretical curve, the match point does not necessarily have to be selected to lie on both curves. Any other point off the curves may be taken as a match point. For example, in Fig. 8.22, another match point (**IIa**) is selected for curve **II**; its coordinates are

$$s_d = 10 \text{ m} \qquad t = 10 \text{ min} \qquad \frac{1}{u} = 10 \qquad W(u) = 56$$

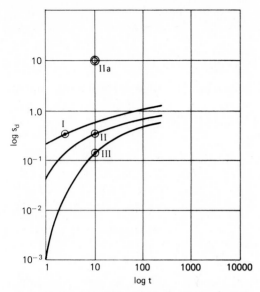

Figure 8.21 Graphic solution for Example 8.7.

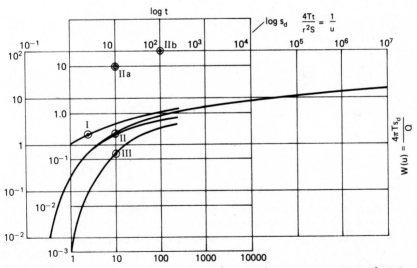

Figure 8.22 Overlay of field curves and theoretical type curve in Example 8.7.

Therefore, $T = \dfrac{1.888 \times 56}{4\pi \times 10} = 0.8414 \ \text{m}^2/\text{min}$

and $S = \dfrac{4 \times 0.8414 \times 10 \times 0.1}{121.92^2} = 0.000226$

Point IIb can also be a match point: $1/u = 100$, $W(u) = 100$, $t = 100$ min, and $s_d = 11.2$ m.

If the match point corresponds to one of the field records in Table 8.5, the coordinates s_d and t are better taken directly from the table rather than from the graph for more precision. The precision of reading the coordinates of $1/u$ or $W(u)$ depends on the scale of the theoretical chart. Similarly, if either $1/u$ or $W(u)$ coincides with a major axis (10^2, 10^{-4}, . . . , etc.), the other value would better be determined from Table 8.3 or 8.4 for more precision. This was used in determining the $W(u)$ of the match points in observation wells 1 and 2 corresponding to $1/u = 10$. However, if the match point is off the curve (as IIa or IIb in Fig. 8.22), the values of $1/u$ and $W(u)$, or s_d and t, *should be read graphically* because the match point is usually selected at the intersection of two major coordinates on either the theoretical curve or the field curve.

In Example 8.7, it is indicated that, excluding well 1, the average $T \simeq 0.85 \ \text{m}^2/\text{min}$ and the average $S \simeq 0.00022$. This means that more than one observation well is required to determine better weighted averages. Instead of analyzing one well at a time, it is preferable to plot a field curve of log s_d versus log t/r^2. Such a curve would group all recordings of the three wells (Table 8.5) in a single curve that should be matched by the theoretical curve (Fig. 8.18). If we plot log s_d versus log r^2/t rather than t/r^2, then S and T are determined using the theoretical curve log u versus log $W(u)$ (Fig. 8.19 or Fig. 8.20). If we solve Example 8.7 using this latter approach, a match point off the curve exists corresponding to (graph is not shown)

$$s_d = 0.17 \ \text{m} \qquad \frac{r^2}{t} = 1774 \ \text{m}^2/\text{min} \qquad u = 0.1 \qquad W(u) = 1.0$$

Then $T = \dfrac{QW(u)}{4\pi s_d} = \dfrac{1.888 \times 1.0}{4\pi \times 0.17} = 0.8838 \ \text{m}^2/\text{min}$

and $S = \dfrac{4Tu}{r^2/t} = \dfrac{4 \times 0.8838 \times 0.1}{1774} = 0.000199 = 0.0002$ (rounded)

It is to be noted that the recorded s_d value at any time t_1 at a distance r_1 are the same as s_d at time t_2 and distance r_2 when $r_1^2/t_1 = r_2^2/t_2$ for equal u values [see Eqs. (8.50) and (8.51)]. In Example 8.7, r_3 is twice the value of r_2; thus s_d in observation well 3 (radius r_3) at any time t is the same as that for observation well 2 at time $t/4$. When $t = 12$ min, for example, s_d in well 3 = 0.116 m, which is the same as for well 2 when $t = 12/4 = 3$ min (Table 8.5).

In observation well 3, the field recordings were terminated at time $t = 240$ min. The extension of the curve to later times may be approximately plotted on the basis of the recordings of observation wells 1 and 2. The drawdown $s_d = 0.643$ m in well 3 at time $t = 240$ min is the same as the drawdowns recorded for wells 1 and 2 at times $t = 14$ and 60 min, respectively. The drawdowns in well 2 from $t = 80$ to $t = 240$ min correspond to the same drawdowns in well 3, but from $t = 80(r_3/r_2)^2 = 80 \times 4 = 320$ min to $t = 240 \times 4 = 960$ min. The drawdowns in observation well 3 may thus be predicted beyond 240 min on the basis of recordings taken in well 1 until $t = 240(r_3/r_1)^2 = 240(243.84/91.44)^2 = 1706.7$ min.

Straight-Line Method

When $u = r^2 S/4Tt$ becomes very small, the terms in the series given by Eq. (8.52) become negligible after the second term (Jacob, 1950). Thus when t is relatively large and/or r is small, Eq. (8.52) reduces to

$$s_d = \frac{Q}{4\pi T}(-0.577216 - \ln u) \tag{8.55}$$

or

$$s_d \approx \frac{2.3Q}{4\pi T} \log \frac{2.25Tt}{r^2 S} \tag{8.56}$$

If one observation well is available, the relationship between s_d and $\log t$ is linear, because from Eq. (8.56)

$$\frac{\Delta s_d}{\Delta(\log t)} = \frac{2.3Q}{4\pi T} \tag{8.57}$$

or

$$(\Delta s_d)_o = \frac{2.3Q}{4\pi T} \quad \text{(for one log cycle of } t\text{)} \tag{8.57a}$$

where $(\Delta s_d)_o$ is the difference in the drawdowns over *one* log cycle of t.

The slope of the straight line plotted for s_d versus $\log t$ is given by $\Delta s_d/\Delta(\log t) = (\Delta s_d)_o$, which is the left-hand side of Eq. (8.57). This slope is determined graphically by fitting a straight line on a semilog plot. Equating this slope to $2.3Q/4\pi T$ [Eq. (8.57)], T is calculated. By substituting T and the coordinates s_d and t at any point on the fitted straight line in Eq. (8.56), S can be calculated.

The preceding method cannot be used unless sufficient points lie on a straight line. If the recordings do not fit a straight line, then the time of pumping is relatively short and the Theis graphic method explained earlier should be used.

It has been recommended (Jacob, 1950) that the straight-line method be used only when $u < 0.01$. It has been found, however, that the method may still be used when $u < 0.1$, as evidenced from the relationship between $W(u) = 4\pi T s_d/Q$ versus $\log u$ ($=\log r^2 S/4Tt$) drawn on a semilog plot (Fig. 8.23) using Table 8.3. The slope of the straight-line portion for all values of u less than 0.1 is equal to 2.30 (from the graph) or

$$2.30 = \frac{\Delta[W(u)]}{\Delta(\log u)} = \frac{4\pi T(\Delta s_d)/Q}{\Delta(\log u)} = \frac{4\pi T(\Delta s_d)_o}{Q} \tag{8.58}$$

which is the same as Eq. (8.57a).

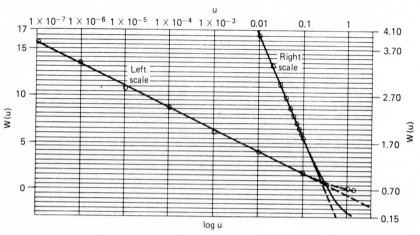

Figure 8.23 Semilog plot for $W(u)$ versus log u.

Once T is determined, S can *also* be determined on the basis of Table 8.3 by selecting any point (s_{d1}, t_1) on the fitted straight line of the field plot; then $W(u)_1 = 4\pi T s_{d1}/Q$ is calculated. The corresponding value u_1 is then determined from Table 8.3, and S is calculated from $S = 4Tt_1u_1/r^2$, where r is the distance of the observation well from the pumped well.

If no observation wells are available, the same method applies, but r_w is substituted for r; the recorded drawdowns s_{dw} in the pumped well are used after being adjusted to remove the screen and other losses (Secs. 6.5 and 8.8).

If recordings from several observation wells (at least two) are available, a composite straight line is fitted on a plot of s_d versus log r^2/t. The slope of this line is determined from Eq. (8.56) as

$$\frac{\Delta s_d}{\Delta(\log r^2/t)} = -\frac{2.3Q}{4\pi T} \tag{8.59}$$

or

$$(\Delta s_d)_o = -\frac{2.3Q}{4\pi T} \tag{8.59a}$$

where $(\Delta s_d)_o$ is the Δs_d value corresponding to one log cycle of r^2/t.

When several observation wells are available and one drawdown is recorded in each of them at the same time t since pumping has started, then following the same approach, the relationship between s_d and log r is a straight line (t is constant) with a slope

$$\frac{(\Delta s_d)}{\Delta(\log r)} = -\frac{2.3Q}{2\pi T} \tag{8.60}$$

or
$$(\Delta s_d)_o = -\frac{2.3Q}{2\pi T} \qquad (8.60a)$$

where $(\Delta s_d)_o$ is the Δs_d value corresponding to one log cycle of r. Once T is calculated from either Eq. (8.59a) or (8.60a), S is determined as previously explained.

It is to be noted that Eq. (8.60) is the same as the steady-state equation given in Sec. 8.1 as Eq. (8.4b). As previously stated, steady-state conditions can never be attained unless there is recharge or leakage into the aquifer or when the drawdowns do not change appreciably beyond a certain time t. Thus using the steady-state equation [Eq. (8.4b)], T can more or less be accurately calculated. The S value can then be determined by the procedure explained below Eq. (8.58).

Once T and S are determined, the radius of influence $r_{e,t}$ is determined by equating s_d to 0 in Eq. (8.56); then

$$0 = \frac{2.3Q}{4\pi T} \log \frac{2.25Tt}{r_{e,t}^2 S}$$

or
$$\frac{2.25Tt}{r_{e,t}^2 S} = 1$$

That is,
$$r_{e,t} = \sqrt{\frac{2.25Tt}{S}} \qquad (8.61)$$

If the value of s_d given by Eq. (8.56) is substituted into Eq. (8.5c), which gives r_e in the steady-state case, the same expression of $r_{e,t}$ given by Eq. (8.61) is obtained:

$$r_e = \exp \left[\ln r + \frac{2\pi T}{Q} \left(\frac{2.3Q}{4\pi T} \log \frac{2.25Tt}{r^2} \right) \right]$$

$$= \exp \left(\ln r + \tfrac{1}{2} \ln \frac{2.25Tt}{r^2 s} \right) = \left(\frac{2.25Tt}{S} \right)^{1/2}$$

EXAMPLE 8.8 Using the same recordings given in Table 8.5, find S and T using the straight-line method. Also find the radius of influence $r_{e,t}$ at $t = 30$ min and $t = 240$ min.

Solution Taking the late time recordings when $t \geq 10$ min, a semilog plot of s_d versus $\log t$ is drawn for the three wells in Fig. 8.24. Straight lines are fitted through the later time points and $(\Delta s_d)_o$ values are obtained.

Observation Wells 1 and 2:

$$(\Delta s_d)_o = 0.405 \qquad \text{(Fig. 8.24)}$$

Using Eq. (8.57a),

$$T = \frac{2.3Q}{4\pi(\Delta s_d)_o} = \frac{2.3 \times 1.888}{4\pi \times 0.0405} = 0.8532 \text{ m}^2/\text{min}$$

Observation Well 3:

$$(\Delta s_d)_o = 0.39 \quad \text{and} \quad T = 0.8860 \text{ m}^2/\text{min}$$

$$\text{Average } T = \frac{2 \times 0.8532 + 0.8860}{3} = 0.8641 \text{ m}^2/\text{min}$$

(Compare with Example 8.7.)

To find S, consider a point on any fitted straight line, say, point **II** (Fig. 8.24), such that $s_d = 0.725$ m and $t = 100$ min. Substituting in Eq. (8.56) the values of T, s_d, t, and $r_2 = 121.92$ m,

$$0.725 = \frac{1.888}{4\pi \times 0.8641} \ln \frac{2.25 \times 0.8641 \times 100}{121.92^2 S}$$

or

$$4.17 = \frac{0.0131}{S}$$

Therefore, $S = 0.00020$ (compare with Example 8.7).

The validity of the method can be checked by calculating the time t corresponding to $u = 0.1 = r^2S/4Tt$:

$$t = \frac{r^2S}{0.4T} = r^2 \frac{0.0002}{0.4 \times 0.8641} = 0.0005786r^2$$

The method is thus valid after time $t = 4.8$, 8.6, and 34.4 min, respectively, for observation wells 1, 2, and 3. (If the conservative criterion $u < 0.01$ is used, the times are, respectively, 48, 86, and 344 min.) The time used for the three wells to construct any straight line (Fig. 8.24) is more than

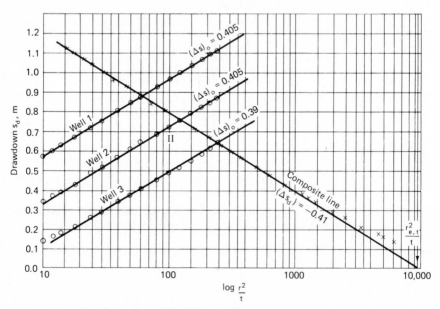

Figure 8.24 Graphic solution for Example 8.8.

10 min. This satisfies the condition in observation wells 1 and 2. In observation well 3, the time should be more than 34.4 min in order to justify the use of the straight-line method. This has been adopted already where the straight line was fitted at the late time points. (It should be noted that in Fig. 8.24 the first five points, until time $t = 24$ min, lie above the fitted straight line for observation well 3.)

At $t = 30$ min,

$$r_{e,t} = \sqrt{\frac{2.25Tt}{S}} \qquad \text{[Eq. (8.61)]}$$

or

$$r_{e,t} = \sqrt{\frac{2.25 \times 0.8641 \times 30}{0.0002}} = 540 \text{ m}$$

At $t = 240$ min,

$$r_{e,t} = 1527 \text{ m}$$

Discussion of Example 8.8 If we plot a composite graph for s_d versus $\log r^2/t$ for all recordings of the three wells in Example 8.8 and fit a straight line (Fig. 8.24), then

$$(\Delta s_d)_o = 0.41 \text{ (from figure)} = -\frac{2.3Q}{4\pi T} \qquad \text{[Eq. (8.59a)]}$$

Then

$$T = \frac{2.3 \times 1.888}{4 \times \pi \times 0.41} = 0.8428 \text{ m}^2/\text{min}$$

S is obtained in the same way as previously explained.

A plot can be drawn for s_d versus $\log r$ because from Eq. (8.59), $\Delta s_d/\Delta(\log r) = 2.3Q/2\pi Tt$, where t is a constant. In this example, there are only three points at any specific time. For example, when $t = 240$ min, the drawdowns s_d are 1.119, 0.878, and 0.643 m (Table 8.5) at radii $r = 91.44$, 121.92, and 243.84 m, respectively. These three points are plotted in Fig. 8.25. Since they do not lie on a straight line, any point may be excluded and there are three possible values of $(\Delta s_d)_o$. Since the answer is already known, it is obvious that the straight line joining points 2 and 3 gives a more or less accurate value of T. Such a plot should therefore be used only when *more than three observation wells* are available. This method is not, however, recommended because of the expense involved.

S may be determined in a different manner than that explained earlier, i.e., on the basis of Eq. (8.51). For example, point II in Fig. 8.24 has the coordinates $s_d = 0.725$ m and $t = 100$ min; then

$$W(u) = \frac{4\pi T}{Q} s_d = \frac{4 \times 11 \times 0.8641 \times 0.725}{1.888} = 4.1697$$

From Table 8.3, the corresponding u value is 8.75×10^{-3}. By using Eq. (8.50),

$$u = 8.75 \times 10^{-3} = \frac{r_2^2 S}{4Tt} = \frac{121.92^2 S}{4 \times 0.8641 \times 100}$$

or

$$S = 0.000204$$

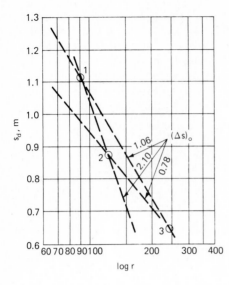

Figure 8.25 Graphic assessment of Example 8.8.

S also can be determined by extending the straight line (s_d versus $\log r^2/t$) to zero drawdown to find $r_{e,t}^2$ graphically. Then from Eq. (8.61),

$$S = \frac{2.25Tt}{r_{e,t}^2}$$

From Fig. 8.24, $r_{e,t}^2/t = 9600$, corresponding to a zero drawdown. From Eq. (8.61),

$$S = \frac{2.25 \times 0.8641}{9600} = 0.000203$$

If more than three observation wells are available and a plot s_d versus $\log r$ rather than $\log r^2/t$ is drawn at a certain time t for all values of s_d, then by extending the straight line to zero drawdown, $r_{e,t}$ can be determined graphically. S can be calculated from Eq. (8.61).

The straight-line method may be used if drawdown recordings are available after the pump is shut off. The method is known as a *recovery test*. The residual drawdowns, recorded after the pump is shut off, at any instant in an observation well distance r from the pumped well are the same as if the well had continued to discharge without shutting off the pump and an imaginary recharge well of equal Q had been introduced at the same instant the pump stopped. If t is the time since pumping started and t' is the time since pumping stopped, then applying Eq. (8.56), the residual drawdown s_d' is given by

$$s_d' = \frac{2.3Q}{4\pi T} \log \frac{2.25Tt}{r^2 S} - \frac{2.3Q}{4\pi T} \log \frac{2.25Tt'}{r^2 S}$$

or

$$s'_d = \frac{2.3Q}{4\pi T} \log \frac{t}{t'} \qquad (8.62)$$

and

$$\frac{\Delta s'_d}{\Delta (\log t/t')} = \frac{2.3Q}{4\pi T} \qquad (8.63)$$

If we plot s'_d versus $\log t/t'$, a straight line can be fitted through the points, and its slope $[=(\Delta s'_d)_o]$ over one log cycle of t/t' is determined. From Eq. (8.63),

$$T = \frac{2.3Q}{4\pi(\Delta s'_d)_o} \qquad (8.64)$$

The S value can be determined from recordings of the drawdowns before shutdown of the pump. If these recordings are not available and only the time t from the start of pumping until shutdown is known, an approximate method can be used to find S, as explained in the following example.

EXAMPLE 8.9 A well is pumped at a rate $Q = 4.613$ m³/min. The pump is shut off after 800 min of pumping. Drawdowns were recorded in an observation well at $r = 30.48$ m from the pumped well until the pump was shut off, and residual drawdowns were recorded thereafter (Table 8.6). Find S and T.

Solution Line 1 is plotted for s_d versus $\log t$ from $t = 10$ min to $t = 800$ min (Fig. 8.26). On the same plot, points s'_d versus $\log t/t'$ are fitted by line 2. $(\Delta s_d)_o$ and $(\Delta s'_d)_o$ over one log cycle are almost the same and equal to 0.29 m. Using Eq. (8.64),

$$T = \frac{2.3 \times 4.613}{4\pi \times 0.29} = 2.91 \text{ m}^2/\text{min}$$

The points fitted on line 2 correspond to lower values of t/t' or higher values of t (in order that u be less than 0.1). In this example, the well pump is shut down after 800 min. It is assumed in this method that pumping continues after the 800 min, but when t reaches 800 min, a recharge well of the same Q is assumed to be in operation. It is shown in Fig. 8.26 that the early points of residual drawdown s'_d are off the straight line at high values of t/t'.

The S value is determined as discussed earlier if line 1 is used. However, if recordings before the shutdown of the pump are not available, but the time $t = 800$ min, after which pumping ceased, is known as well as the residual drawdowns s'_d thereafter, S may be *approximately* determined using Theis' formula [Eq. (8.51)] for the last drawdown $s_d = 0.567$ m corresponding to $t = 800$ min:

$$W(u) = \frac{4\pi T}{Q} s_d = \frac{4\pi \times 2.91}{4.613} \times 0.567 = 4.502$$

From Table 8.3,

$$u = 6.25 \times 10^{-3} = \frac{r^2 S}{4Tt} = \frac{30.48^2 S}{4 \times 2.91 \times 800}$$

or $\qquad S = 0.063$

The u value should be less than 0.1 when using the straight-line solution. In order to check the corresponding time, substitute r, S, T, and $u = 0.1$ into $u = r^2 S/4Tt$; then $0.1 = 30.48^2 \times 0.063/(4 \times 2.91t)$, or $t = 50.2$ min. The points corresponding to $t < 50.2$ do not lie on line 1 in Fig. 8.26. If we consider that a real recharge well is in operation after pumping stops, then using the same formula, $t' = 50.2$ min. Therefore, $t = 50.2 + 800 = 850.2$ min. From Table 8.6, $t = 850.2$ corresponds to $t/t' = 17$. Values of $t/t' > 17$ will then be off straight line 2 in Fig. 8.26.

TABLE 8.6   Recovery Test Data for Example 8.9*

Pumping data		Recovery data			
t, min	s_d, m	t, min	t', min	$\dfrac{t}{t'}$	s_d', m
5	0.024	800	0	—	0.567
10	0.067	805	5	161.00	0.543
15	0.101	810	10	81.00	0.500
20	0.125	815	15	54.30	0.466
25	0.152	820	20	41.00	0.442
30	0.168	825	25	33.30	0.418
40	0.201	830	30	27.70	0.402
50	0.223	840	40	21.00	0.372
60	0.244	850	50	17.00	0.351
70	0.262	860	60	14.30	0.332
80	0.280	870	70	12.40	0.314
90	0.293	880	80	11.00	0.296
100	0.305	890	90	9.86	0.287
110	0.317	900	100	9.00	0.274
120	0.326	910	110	8.27	0.265
180	0.378	920	120	7.67	0.259
240	0.411	960	180	5.44	0.213
300	0.442	1040	240	4.33	0.186
360	0.463	1100	300	3.67	0.165
420	0.485	1160	360	3.22	0.149
480	0.503	1220	420	2.90	0.140
540	0.521	1280	480	2.67	0.122
600	0.527	1340	540	2.48	0.110
660	0.539	1400	600	2.33	0.110
720	0.552	1460	660	2.21	0.104
800[†]	0.567	1520	720	2.11	0.094
		1600	800	2.00	0.088

* Rate of pumping $Q = 4.613$ m³/min; observation well at $r = 30.48$ m.
† Pump was shut off after $t = 800$ min.

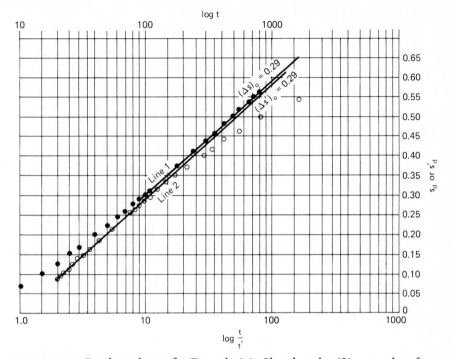

Figure 8.26 Graphic solution for Example 8.9. Closed circles (●) are a plot of s_d versus log t and circled points (○) a plot of s'_d versus log (t/t').

The recovery test may be used for recordings taken from the pumped well if no observation wells are available by substituting r_w for r. If r_w is not known, S cannot be determined.

Constant Drawdown Tests for Flowing Artesian Wells

In naturally flowing wells, the rate of flow changes over time when water is discharged freely without pumping. Tests can be made directly on flowing wells after maintaining constant drawdowns s_{dw}. This is achieved after capping the well long enough to stabilize the pressure inside the well. Then the well is opened and allowed to flow for a period of 2 to 4 h (Lohman, 1972), during which time measurements are made for the declining rates of flow Q at different times. Under these conditions, the value of u becomes less than 0.1 very rapidly because r_w is small:

$$u = \frac{r_w^2 S}{4Tt} < 0.1 \qquad (8.65)$$

The straight-line method is then applied (Lohman, 1972):

$$\frac{1}{Q} = \frac{2.3}{4\pi s_{dw} T} \log \frac{2.25Tt}{r_w^2 S} \tag{8.66}$$

(Q is variable, and s_{dw} is constant.)

Equation (8.66) is similar to Eq. (8.56), but in Eq. (8.66), Q is variable and s_{dw} is constant. Differentiating Eq. (8.66) with respect to log t, then

$$\frac{d(1/Q)}{d(\log t)} = \frac{2.3}{4\pi s_{dw} T} \tag{8.67}$$

If we measure Q at different times t, a straight line is obtained if $1/Q$ is plotted versus log t; then

$$T = \frac{2.3}{4\pi s_{dw}[\Delta(1/Q)_o]} \tag{8.68}$$

where $\Delta(1/Q)_o$ is measured over one log cycle of t. If we locate any point a with coordinates $(1/Q)_a$ and t_a on the straight line, then S is determined from Eq. (8.66) as

$$S = \frac{2.25Tt_a/r_w^2}{\log^{-1}(4\pi s_{dw} T/2.3Q_a)} \tag{8.69}$$

If there is doubt about the magnitude of r_w, S should not be determined from Eq. (8.69).

EXAMPLE 8.10° A flowing artesian well has a radius $r_w = 8.4$ cm. The well is shut off for several days, and then the discharge is resumed by opening the valve. The rates of flow are measured by collecting volumes of water over recorded time intervals (Table 8.7). Find S and T if constant $s_{dw} = 28.142$ m.

Solution A plot of $1/Q$ in minutes per cubic meter versus log t (Table 8.7) is made, and the points are fitted to a straight line using the late time points (Fig. 8.27). $\Delta(1/Q)_o = 8.5$ min/m³ over one log cycle of t. From Eq. (8.68),

$$T = \frac{2.3}{4\pi \times 28.142 \times 8.5} = 7.65 \times 10^{-4} \text{ m}^2/\text{min} = 1.102 \text{ m}^2 \text{ per day}$$

Point I on the straight line (Fig. 8.27) is selected such that $t_a = 30$ min and $(1/Q)_a = 49$ min/m³; then by use of Eq. (8.69),

$$S = \frac{(2.25 \times 7.65 \times 10^{-4} \times 30)/0.084^2}{\log^{-1}(4\pi \times 28.142 \times 7.65 \times 10^{-4} \times 49/2.3 \times 1)}$$

$$= 1.26 \times 10^{-5}$$

° Example 8.10 is based on recordings given by Lohman (1972), but the data have been converted to SI units.

TABLE 8.7 Data for Example 8.10

Time of observation[a]	Flow interval Δt, min	Volume of water ΔV_w measured during time intervals Δt, L	Rate of flow $Q = \dfrac{\Delta V_w}{\Delta t}$, L/min	Cumulative time since flow started, min	$\dfrac{1}{Q}$, min/m^3
10:30 A.M.	1	27.56	27.56	1	36.28
10:31 A.M.	1	26.94	26.94	2	37.12
10:32 A.M.	1	26.04	26.04	3	38.40
10:33 A.M.	1	23.77	23.77	4	42.07
10:34 A.M.	1	23.54	23.54	5	42.48
10:35 A.M.	1	23.54	23.54	6	42.48
10:37 A.M.	2	45.05	22.53	8	44.39
10:40 A.M.	3	66.43	22.14	11	45.17
10:45 A.M.	5	107.13	21.43	16	46.66
10:50 A.M.	5	104.10	20.82	21	48.03
10:55 A.M.	5	101.07	20.21	26	49.48
11:00 A.M.	5	101.07	20.21	31	49.48
11:10.30 A.M.	10.5	207.48	19.76	41.5	50.61
11:20 A.M.	9.5	184.84	19.46	51	51.39
11:30 A.M.	10	193.43	19.34	61	51.71
11:45 A.M.	15	286.74	19.12	76	52.30
12:00 P.M.	15	283.91	18.93	91	52.83
12:12 P.M.	12	223.49	18.62	103	53.71
12:22 P.M.	11	203.20	18.47	113	54.14
TOTAL	114	TOTAL 2260	AVERAGE Q 19.82		

[a] Test started at 10:29 A.M.

The limiting time t beyond which the straight-line method is valid is checked.

$$u = 0.1 = \frac{r_w^2 S}{4Tt}$$

or $$t = \frac{r_w^2 S}{0.4T} = \frac{0.084^2 \times 1.26 \times 10^{-5}}{0.4 \times 7.65 \times 10^{-4}} = 0.00029 \text{ min}$$

(In less than a minute, u becomes less than 0.1.)

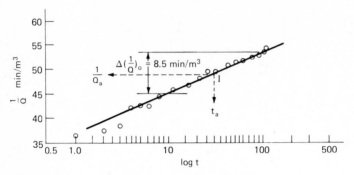

Figure 8.27   Graphic solution for Example 8.10.

8.3   Leaky Artesian Aquifers

The process of water leakage into or out of an artesian aquifer was explained in Sec. 6.4 with reference to Fig. 6.8. If it is assumed that the water table in the unconfined aquifer overlying the aquitard (Fig. 6.8) is initially at the same level as the piezometric surface of the aquitard, then pumping increases the drawdown at section ab from s_{d1} to s_{d2}. This gradually accelerates the downward leakage from the aquitard into the main artesian aquifer. The pumped water is derived from leakage as well as storage. If the rate of leakage increases until it is equal to the rate of pumping, the drawdown curve and the area of influence stabilize. In other words, a steady-state condition is feasible under leaky conditions.

Upward leakage can take place if line AB (Fig. 6.8) is initially lower than CD. After pumping, the zones of downward and upward leakage can be determined on the basis of the relative positions of the drawdown curve and the piezometric surface of the aquitard. This latter surface is affected by pumping after a relatively long time. In most theories of leakage, it is assumed that line AB (Fig. 6.8) coincides initially with the undisturbed piezometric surface of the aquitard and that the latter is not affected by pumping.

Leakage can be transmitted through numerous overlying aquitards and/or an aquitard system existing below the main aquifer. Aquitards are assumed to be more or less incompressible if they are thought of only as transmitting media. However, they usually consist of fine-grained materials that are more compressible (Sec. 6.2) than the materials of the main aquifer. Therefore, the coefficient of storage S_a of an aquitard is greater than zero. In the following subsections, various conditions are discussed, such as steady and transient leakage either considering or neglecting the storativity of the aquitard.

Steady-State Flow

The general equation for steady-state flow is given by Eq. (6.35). The solution is given by Hantush and Jacob (1955) as follows:

$$s_{df} = \frac{Q}{2\pi T} K_o\left(\frac{r}{B_k}\right) \tag{8.70}$$

where s_{df} = equilibrium drawdown at distance r from the well, cm
          B_k = leakage factor, cm

$$K_o\left(\frac{r}{B_k}\right) = \text{modified Bessel function of the second kind and zero order}$$

$$B_k = \sqrt{R_a D_q K} = \sqrt{R_a T} = \sqrt{\frac{D_a T}{K_a}} \tag{6.33}$$

Other symbols have been defined earlier. The relationship between $K_o(r/B_k)$ and r/B_k is given graphically in Fig. 8.28 on the basis of the tabulated values in Table 8.8.

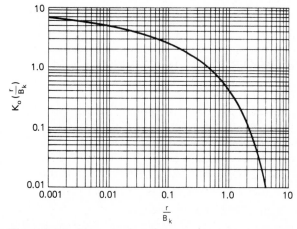

Figure 8.28 Distance-drawdown type curve considering leakage but neglecting storage in aquitards. (W. C. Walton, "Selected Analytical Methods for Well and Aquifer Evaluation," Illinois State Water Survey Bulletin 49, 1962.)

TABLE 8.8 Values of $K_o(r/B_k)$

N	$N \times 10^{-3}$	$N \times 10^{-2}$	$N \times 10^{-1}$	N
		$\dfrac{r}{B_k}$		
1.0	7.0237	4.7212	2.4271	0.4210
1.5	6.6182	4.3159	2.0300	0.2138
2.0	6.3305	4.0285	1.7527	0.1139
2.5	6.1074	3.8056	1.5415	0.0623
3.0	5.9251	3.6235	1.3725	0.0347
3.5	5.7709	3.4697	1.2327	0.0196
4.0	5.6374	3.3365	1.1145	0.0112
4.5	5.5196	3.2192	1.0129	0.0064
5.0	5.4143	3.1142	0.9244	0.0037
5.5	5.3190	3.0195	0.8466	
6.0	5.2320	2.9329	0.7775	0.0012
6.5	5.1520	2.8534	0.7159	
7.0	5.0779	2.7798	0.6605	0.0004
7.5	5.0089	2.7114	0.6106	
8.0	4.9443	2.6475	0.5653	
8.5	4.8837	2.5875	0.5242	
9.0	4.8266	2.5310	0.4867	
9.5	4.7725	2.4776	0.4524	

SOURCE: M. S. Hantush, "Analysis of Data from Pumping Tests in Leaky Aquifers," *Trans,. Am. Geophys. Union,* vol. 37, no. 6, 1956, pp. 702–714.

The method of matching curves is used to determine T and K_a (Walton, 1970). From Eqs. (6.33) and (8.70), the following two equations are determined:

$$\log s_{df} = \log \frac{Q}{4\pi T} + \log K_o\left(\frac{r}{B_k}\right)$$

$$\log r = \tfrac{1}{2} \log \frac{TD_a}{K_a} + \log \frac{r}{B_k}$$

or

$$\log s_{df} = C_1 + \log K_o\left(\frac{r}{B_k}\right) \tag{8.71}$$

and

$$\log r = C_2 + \log \frac{r}{B_k} \tag{8.72}$$

where C_1 and C_2 are constants.

If we plot the field recordings of s_{df} and r on log-log paper (s_{df} as ordinate), the field curve should more or less match the curve given in Fig. 8.28. A match point is selected with the four coordinates s_{df}, r,

$K_o(r/B_k)$, and r/B_k. T is then determined from Eq. (8.70), and K_a is determined from Eq. (6.33), that is, $K_a = D_a T/B_k^2$. The rate of leakage at a is $v_a = K_a(s_{df}/D_a)$ [Eq. (6.30)].

In some cases, when r/B_k values are less than 0.05, the solution does not require a matching procedure (Hantush, 1956, 1964b). If we plot s_{df} versus log r on semilog paper, a straight line is obtained; when r/B_k values are less than 0.05, Eq. (8.70) reduces to

$$s_{df} \approx \frac{2.3Q}{2\pi T} \log 1.12 \frac{B_k}{r} \qquad (8.73)$$

The slope of the straight line $\Delta s_{df}/\Delta(\log r)$ is determined from the field plot and equated to the same slope $\delta s_{df}/\delta(\log r)$ determined by differentiating Eq. (8.73):

$$\frac{\delta s_{df}}{\delta(\log r)} = -\frac{2.3Q}{2\pi T} \qquad (8.74)$$

The value of T is then determined from the preceding equation.

Extending the straight line of the field plot to $s_{df} = 0$ and reading the corresponding r_e value, from Eq. (8.73) we see that

$$0 = \frac{2.3Q}{2\pi T} \log 1.12 \frac{B_k}{r_e}$$

Therefore, log $1.12(B_k/r_e) = 0$, and thus $1.0 = 1.12(B_k/r_e)$, or

$$B_k = \frac{r_e}{1.12} \qquad (8.75)$$

B_k and T are found, then by the use of Eq. (6.33), K_a is calculated.

It should be noted that the two methods used require numerous observation wells (at least three) at different distances r. Since these methods are based on the steady-state conditions, the entire rate of flow Q is derived from leakage sources only and S of the aquifer cannot be obtained (see Example 8.11).

Transient Conditions

If there is leakage under transient conditions, the compressibility of the main artesian aquifer is taken into account. An aquitard exhibits more compressibility than the main aquifer, but its compressibility is generally very slow. Thus it takes a long time for an appreciable amount of water to be released from storage within an aquitard. Two types of solutions are available:

1. Assuming that the aquitards are incompressible. This means that

the pumping time is not long enough to allow the effects of compressibility to be detected.

2. Taking the compressibility of aquitards into account.

Incompressible aquitards Equation (6.34) governs this case. The solution has been given by Hantush and Jacob (1955):

$$s_d = \frac{Q}{4\pi T} W\left(u, \frac{r}{B_k}\right) \tag{8.76}$$

where $u = r^2S/4Tt$ and the values of the well function $W(u, r/B_k)$ are listed in Table 8.9 (Hantush, 1956) for the practical ranges of u and r/B_k and plotted graphically (Walton, 1970) in Fig. 8.29.

By means of a matching procedure, T, S, and K_a are determined. From Eq. (8.76) and since $u = r^2S/4Tt$,

$$\log s_d = \log \frac{Q}{4\pi T} + \log W\left(u, \frac{r}{B_k}\right)$$

and

$$\log t = \log \frac{r^2S}{4T} + \log \frac{1}{u}$$

or

$$\log s_d = C_1 + \log W\left(u, \frac{r}{B_k}\right) \tag{8.77}$$

and

$$\log t = C_2 + \log \frac{1}{u} \tag{8.78}$$

where C_1 and C_2 are constants.

If we plot a field curve of $\log s_d$ versus $\log t$ for an observation well, it should match one of the set of theoretical curves given in Fig. 8.29. These curves give the relationship between $\log W(u, r/B_k)$ and $\log 1/u$ for various values of r/B_k. A match point corresponds to five coordinates: $W(u, r/B_k)$, $1/u$, r/B_k, s_d, and t. From Eq. (8.76), T is calculated, and then S is calculated from $S = 4Tt/(1/u)r^2$. Since r/B_k is known at the match point, then B_k is determined, and from Eq. (6.33), K_a is computed. If the match point does not lie on one of the shown r/B_k curves (Fig. 8.29), r/B_k is determined by interpolation between the two theoretical curves above and below the match point.

EXAMPLE 8.11°   An artesian well is pumping at a rate of $Q = 0.0946$ m³/min $= 136.3$ m³ per day (25 gal/min) from an aquifer of average thickness $D_q = 2.438$ m (8 ft) overlain by an aquitard of average thickness $D_a = 4.267$ m (14 ft). Drawdowns at various times were recorded in an ob-

---

° Example 8.11 is based on data from Walton (1970, pp. 233–237), but the data have been converted to SI units.

TABLE 8.9 Values of $W(u, r/B_k)$

u \ r/B_k	0.01	0.015	0.03	0.05	0.075	0.10	0.15	0.2	0.3	0.4	0.5	0.6	0.7	0.8	0.9	1.0	1.5	2.0	2.5
0.000001	9.4413																		
0.000005																			
0.00001	9.4176	8.6313																	
0.00005	8.8827	8.4533	7.2450																
0.0001	8.3983	8.1414	7.2122	6.2282	5.4228														
0.0005	6.9750	6.9152	6.6219	6.0821	5.4062	4.8530													
0.001	6.3069	6.2765	6.1202	5.7965	5.3078	4.8292	4.0595	3.5054											
0.005	4.7212	4.7152	4.6829	4.6084	4.4713	4.2960	3.8812	3.4567	2.7428	2.2290									
0.01	4.0356	4.0326	4.0167	3.9795	3.9091	3.8150	3.5725	3.2875	2.7104	2.2253	1.8486	1.5550	1.3210	1.1307					
0.05	2.4675	2.4670	2.4642	2.4576	2.4448	2.4271	2.3776	2.3110	1.9283	1.7075	1.4927	1.3115	1.2955	1.1210	0.9700	0.8409			
0.1	1.8227	1.8225	1.8213	1.8184	1.8128	1.8050	1.7829	1.7527	1.6704	1.5644	1.4422	1.2955	1.1791	1.0505	0.9297	0.8190	0.4271	0.2278	
0.5	0.5598	0.5597	0.5596	0.5594	0.5588	0.5581	0.5561	0.5532	0.5453	0.5344	0.5206	0.5044	0.4860	0.4658	0.4440	0.4210	0.3007	0.1944	0.1174
1.0	0.2194	0.2194	0.2193	0.2193	0.2191	0.2190	0.2186	0.2179	0.2161	0.2135	0.2103	0.2065	0.2020	0.1970	0.1914	0.1855	0.1509	0.1139	0.0803
5.0	0.0011	0.0011	0.0011	0.0011	0.0011	0.0011	0.0011	0.0011	0.0011	0.0011	0.0011	0.0011	0.0011	0.0011	0.0011	0.0011	0.0010	0.0010	0.0009

SOURCE: M. S. Hantush, "Analysis of Data from Pumping Tests in Leaky Aquifers," *Trans. Am. Geophys. Union*, vol. 37, no. 6, 1956, pp. 702–714.

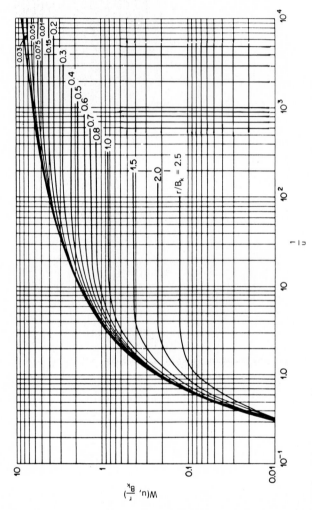

Figure 8.29 Time-drawdown type curves considering leakage but neglecting storage in aquitards. (*W. C. Walton, "Selected Analytical Methods for Well and Aquifer Evaluation," Illinois State Water Survey Bulletin 49, 1962.*)

TABLE 8.10   Data for Example 8.11°

Time since the start of pumping t, min	Drawdown s_d, m
5	0.232
28	1.006
41	1.094
60	1.244
75	1.338
244	1.667
493	1.817
669	1.862
958	1.911
1,129	1.951
1,185	1.957

° Drawdowns at an observation well distance $r = 29.261$ m at the indicated times.
SOURCE: Adapted from W. C. Walton, *Groundwater Resource Evaluation*, McGraw-Hill, New York, 1970. (Units have been changed to SI units.)

servation well at a distance $r = 29.261$ m (96 ft) from the pumped well. These recordings are given in Table 8.10. After a relatively long period of time, when $t = 1185$ min, recordings of drawdowns were taken in three observation wells at distances $r = 71.323, 29.221,$ and 28.043 m, respectively (Table 8.11). Find the average values of T, S, and K_a. Also find the rate of leakage at a radius $r = 30$ m.

Solution If we plot log s_d versus log t from the data given in Table 8.10 and match it on the theoretical curve (Fig. 8.29), the match point (Fig. 8.30)

TABLE 8.11   Data for Example 8.11°

Distance from pumped well r, m	Drawdown s_d, m
71.323	0.991
29.261	1.957
28.042	2.042

° Drawdowns after 1185 min in three observation wells at the distances shown.
SOURCE: Adapted from W. C. Walton, *Groundwater Resource Evaluation*, McGraw-Hill, New York, 1970. (Units have been changed to SI units.)

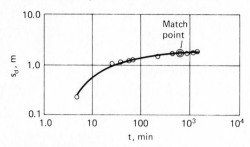

Figure 8.30 Graphic solution by matching curves (Example 8.11).

corresponds to the following (the overlay of the two curves is not shown):

$$W\left(u, \frac{r}{B_k}\right) = 3.3 \qquad \frac{1}{u} = 220 \qquad t = 669 \text{ min}$$

$$s_d = 1.862 \text{ m} \qquad \frac{r}{B_k} = 0.22$$

(by interpolation within r/B_k curves, Fig. 8.29). Using Eq. (8.76),

$$1.862 = \frac{0.0946}{4\pi T} \times 3.3$$

or

$$T = 0.0133 \text{ m}^2/\text{min}$$

and

$$K = \frac{T}{D_q} = \frac{0.0133}{2.438} = 0.0055 \text{ m/min}$$

Since $u = r^2 S/4Tt$, then

$$S = \frac{4Tt}{r^2(1/u)} = \frac{4 \times 0.0133 \times 669}{29.261^2 \times 220} = 1.9 \times 10^{-4}$$

Since $r/B_k = 0.22$, then

$$B_k = \frac{2}{0.22} = \frac{29.261}{0.22} = 133 \text{ m}$$

From Eq. (6.33),

$$K_a = \frac{TD_a}{B_k^2} = \frac{0.0133 \times 4.267}{133^2} = 3.2 \times 10^{-6} \text{ m/min}$$

Steady-State Solution The late three drawdown recordings given in Table 8.10 may not be sufficient to plot a drawdown-distance curve to match the one in Fig. 8.28. The straight-line method is therefore used (Fig. 8.31). The slope of the line is -2.51 (determined from the graph, as Δs_{df} over one log cycle of r equals $0.59 - 3.1 = -2.51$). From Eq. (8.74),

$$-2.51 = -\frac{2.3 \times 0.0946}{2\pi T} \qquad \text{or} \qquad T = 0.0138 \text{ m}^2/\text{min}$$

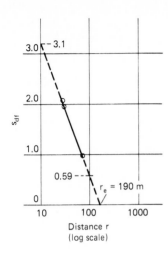

Figure 8.31 Graphic solution using straight-line method (Example 8.11).

From Fig. 8.31, $r_e = 190$ m (corresponding to $s_{df} = 0$); then, from Eq. (8.75),

$$B_k = \frac{190}{1.12} = 169.64 \text{ m}$$

and from Eq. (6.33),

$$K_a = \frac{TD_a}{B_k^2} = \frac{0.0138 \times 4.267}{169.64^2} = 2 \times 10^{-6} \text{ m/min}$$

Note: Although the preceding values of T and K_a (steady state) are satisfactorily comparable with the transient-state solution, the straight-line method should not be used unless $r/B_k < 0.05$. This condition is not satisfied for the three values of r given in Table 8.11. The steady-state solution given above should then be regarded as only an exercise in applying the procedure. More than three recordings are needed at later times.

Compressible aquitards   Equation (6.36) is the governing equation for cases of leakage taking into account water released from storage from within an aquitard. The assumption made in the preceding section that aquitards are incompressible is a practical one. The compressibility of the aquitards exists at all times, but it is not detected at early stages of pumping because of the time lag involved in leakage. The solution of Eq. (6.36) is given in the following form (Hantush, 1964a):

$$s_d = \frac{Q}{4\pi T} W(u, \psi) \tag{8.79}$$

where

$$u = \frac{r^2 S}{4Tt} \quad \text{or} \quad S = \frac{4Tt}{r^2(1/u)} \tag{8.80}$$

TABLE 8.12 Values of $W[u, \psi_k]^\circ$

u \ ψ_k	(−3)			(−2)			(−1)		
	1	2	5	1	2	5	1	2	5
1(−6)	11.9842	11.4237	10.5908	9.9259	9.2469	8.3395	7.6497	6.9590	6.0463
5(−6)	10.8958	10.4566	9.7174	9.0866	8.4251	7.5284	6.8427	6.1548	5.2459
1(−5)	10.3739	9.9987	9.3203	8.7142	8.0657	7.1771	6.4944	5.8085	4.9024
5(−5)	9.0422	8.8128	8.3171	7.8031	7.2072	6.3523	5.6821	5.0045	4.1090
1(−4)	8.4258	8.2487	7.8386	7.3803	6.8208	5.9906	5.3297	4.6581	3.7700
5(−4)	6.9273	6.8375	6.6024	6.2934	5.8561	5.1223	4.4996	3.8527	2.9933
1(−3)	6.2624	6.1969	6.0193	5.7727	5.4001	4.7290	4.1337	3.5045	2.6650
5(−3)	4.6951	4.6649	4.5786	4.4474	4.2231	3.7415	3.2483	2.6891	1.9250
1(−2)	4.0163	3.9950	3.9334	3.8374	3.6669	3.2752	2.8443	2.3325	1.6193
5(−2)	2.4590	2.4502	2.4243	2.3826	2.3040	2.1007	1.8401	1.4872	0.9540
1(−1)	1.8172	1.8116	1.7949	1.7677	1.7157	1.5768	1.3893	1.1207	0.6947
5(−1)	0.5584	0.5570	0.5530	0.5463	0.5333	0.4969	0.4436	0.3591	0.2083
1(0)	0.2189	0.2184	0.2169	0.2144	0.2097	0.1961	0.1758	0.1427	812(−4)
5(0)	115(−5)	114(−5)	114(−5)	112(−5)	110(−5)	104(−5)	934(−6)	763(−6)	423(−6)
10(0)	415(−8)	414(−8)	411(−8)	407(−8)	399(−8)	375(−8)	339(−8)	277(−8)	153(−8)

u \\ ψ_k	(0)			(1)			(2)		
	1	2	5	1	2	5	1	2	5
1(−6)	5.3575	4.6721	3.7756	3.1110	2.4671	1.6710	1.1361	0.6879	0.2698
5(−6)	4.5617	3.8836	3.0055	2.3661	1.7633	1.0574	0.6256	0.3091	787(−4)
1(−5)	4.2212	3.5481	2.6822	2.0590	1.4816	0.8285	0.4519	0.1978	388(−4)
5(−5)	3.4394	2.7848	1.9622	1.3943	0.8994	0.4024	0.1685	494(−4)	405(−5)
1(−4)	3.1082	2.4658	1.6704	1.1359	0.6878	0.2698	963(−4)	222(−4)	107(−5)
5(−4)	2.3601	1.7604	1.0564	0.6252	0.3089	787(−4)	166(−4)	169(−5)	129(−7)
1(−3)	2.0506	1.4776	0.8271	0.4513	0.1976	388(−4)	590(−5)	361(−6)	
5(−3)	1.3767	0.8915	0.4001	0.1677	493(−4)	403(−5)	205(−6)	228(−8)	
1(−2)	1.1122	0.6775	0.2670	955(−4)	221(−4)	106(−5)	274(−7)		
5(−2)	0.5812	0.2923	755(−4)	160(−4)	164(−5)	126(−7)			
1(−1)	0.3970	0.1789	359(−4)	552(−5)	340(−6)				
5(−1)	0.1006	325(−4)	288(−5)	151(−6)	171(−8)				
1(0)	365(−4)	993(−5)	547(−6)	151(−7)					
5(0)	167(−6)	309(−7)							
10(0)									

° The numbers in parenthesis are powers of 10 by which the other numbers are multiplied, e.g., $488(−4) = 0.0488$.
SOURCE: M. S. Hantush, "Hydraulics of Wells," in V. T. Chow (ed.), *Advances in Hydroscience*, vol. 1, Academic Press, New York, 1964, pp. 281–432.

and
$$\psi_k = \frac{1}{4} \sqrt{\frac{S_a}{S}} \frac{r}{B_k} = \frac{r}{4} \sqrt{\frac{K_a S_a}{TSD_a}} \qquad (8.81)$$

S and S_a are, respectively, the coefficient of storage of the main aquifer and the aquitard.

The values of the well function $W(u, \psi_k)$ are given in Table 8.12 (Hantush, 1964) and presented in graphic form in Fig. 8.32 (Lohman, 1972). Time drawdown curves of field recordings are drawn on a log-log plot of the same scale as the theoretical chart given in Fig. 8.32. Matching both curves, the coordinates of the match point have specific values of s_d, t, $1/u$, $W(u, \psi_k)$, and ψ_k. If we substitute $W(u, \psi_k)$ and s_d into Eq. (8.79), then T is calculated. If we substitute t, $1/u$, and the computed T

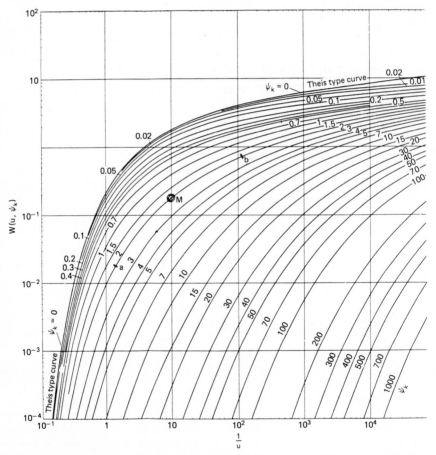

Figure 8.32 Family of type curves for $1/u$ versus $W(u, \psi_k)$. (*S. W. Lohman, "Groundwater Hydraulics," U.S. Geological Survey Professional Paper 708, Washington, 1972.*)

into Eq. (8.80), S is found. Applying Eq. (8.81), the match point coordinate ψ_k is substituted in the equation together with the determined values T and S (r and D_a are known); then the product $K_a S_a$ is determined.

If consolidation tests are performed on undisturbed samples taken from the aquitards, then c_v and m_v (Sec. 6.2) can be determined. From Eqs. (3.19) and (6.19),

$$K_a S_a = D_a c_v m_v^2 \gamma_w^2 \qquad \text{(neglecting water compressibility } \beta_w) \qquad (8.82)$$

The experimental value $K_a S_a$ can be checked against that determined from the field. However, if the value of S_a is required, it may be convenient to find K_a from a consolidation test [Eq. (3.19)] and then compute S_a from the already determined product $K_a S_a$. The value of K_a also can be determined from the field test using the early recordings and matching them on Fig. 8.29 (the aquitard is considered incompressible during the early times).

When the main aquifer under investigation is overlain and underlain by other aquifers from which it is separated by confining aquicludes, a pumping test is used to find S and T of the main aquifer. Then from the coordinate ψ_k of the match point, the value of $\sqrt{K_a S_a/D_a}$ is obtained from Eq. (8.81):

$$\sqrt{\frac{K_a S_a}{D_a}} = \frac{4\psi_k}{r}\sqrt{TS} \qquad (8.83)$$

This value $\sqrt{K_a S_a/D_a}$ represents a weighted average of the properties of the aquitards. Once a test is performed and the parameters and ψ_k are determined, the drawdown s_d anywhere in the aquifer can be evaluated by using Eq. (8.79).

EXAMPLE 8.12° A well is pumping from an artesian aquifer at a rate of 2839 m³/min. The artesian aquifer is confined by a clay layer of a thickness $D_a = 1.829$ m. An unconfined aquifer 60.96 m thick overlies the clay layer. Time-drawdown measurements in an observation well at a distance $r = 426.72$ m from the pumping well are shown in Table 8.13. Find S and T of the main aquifer and the value of $K_a S_a$ for the confining clay layer. Also find the anticipated drawdowns at $r = 100$ and 300 m after 5 h of pumping.

Solution The time-drawdown measurements (Fig. 8.33) are plotted on log-log paper of the same scale as the type used for the curves in Fig. 8.32. The curves are matched, and a match point M is selected as shown in Fig. 8.34 (which is the overlay of Figs. 8.33 and 8.32). The coordinates of the match

---

° Example 8.12 is based on recordings given by Lohman (1972), but the data have been converted to SI units.

TABLE 8.13   Time-Drawdown Measurements°

Time t since pumping began, min	Drawdown, m	Time t since pumping began, min	Drawdown, m
6.37	0.003	90	0.229
8.58	0.006	100	0.250
10.23	0.009	137	0.317
11.90	0.012	150	0.341
12.95	0.015	160	0.357
14.42	0.018	173	0.378
15.10	0.021	184	0.387
16.88	0.024	200	0.411
17.92	0.030	210	0.427
21.35	0.037	278	0.512
21.70	0.040	300	0.536
22.70	0.043	315	0.558
23.58	0.046	335	0.570
24.65	0.052	365	0.607
29	0.064	390	0.640
30	0.067	410	0.649
32	0.073	430	0.671
34	0.079	450	0.680
36	0.085	470	0.698
38	0.091	490	0.707
41	0.101	510	0.728
44	0.110	560	0.756
47	0.116	740	0.890
50	0.128	810	0.930
54	0.140	890	0.972
60	0.158	1,255	1.116
65	0.171	1,400	1.161
70	0.183	1,440	1.177
80	0.198	1,485	1,189

° See Example 8.12
SOURCE: Adapted from S. W. Lohman, "Groundwater Hydraulics," U.S. Geological Survey Professional Paper 708, Washington, 1972. (Units have been changed to SI units.)

point M are

$$s_d = 0.28 \text{ m} \qquad t = 115 \text{ min} \qquad \frac{1}{u} = 10 \qquad W(u, \psi_k) = 0.19 \qquad \psi_k = 2$$

(The segment ab in Fig. 8.33 matches exactly a segment of the curve $\psi_k = 2$ in Fig. 8.32.) From Eq. (8.79),

$$0.28 = \frac{2.839}{4\pi T} \times 0.19$$

or

$$T = 0.1533 \text{ m}^2/\text{min}$$

From Eq. (8.80),

$$S = \frac{4 \times 0.1533 \times 115}{426.72^2 \times 10} = 3.87 \times 10^{-5}$$

And from Eq. (8.81),

$$K_a S_a = \left(\frac{4\psi_k}{r}\right)^2 TSD_a$$

$$= \left(\frac{4 \times 2}{426.72}\right)^2 \times 0.1533 \times 3.87 \times 10^{-5} \times 1.829$$

or $\qquad K_a S_a = 3.8 \times 10^{-9}$ m/min

The anticipated drawdowns at $r = 100$ and 300 m after 5 h (300 min) are determined as follows:

$$u = \frac{r^2 S}{4Tt} = \frac{100^2 \times 3.87 \times 10^{-5}}{4 \times 0.1533 \times 300} \qquad \text{or} \qquad \frac{1}{u} = 475.35$$

Applying Eq. (8.81),

$$\psi_k = \frac{100}{4}\sqrt{\frac{3.8 \times 10^{-9}}{0.1533 \times 3.87 \times 10^{-5} \times 1.829}} = 0.47$$

The charts are entered in Fig. 8.34 with the determined values of $1/u$ and ψ_k; $W(u, \psi_k)$ is found to be approximately 2.2. Then from Eq. (8.79),

$$s_d = \frac{2.839}{4\pi \times 0.1533} \times 2.2 = 3.242 \text{ m}$$

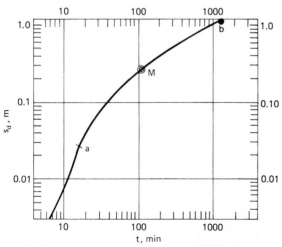

Figure 8.33 Graphic solution for Example 8.12.

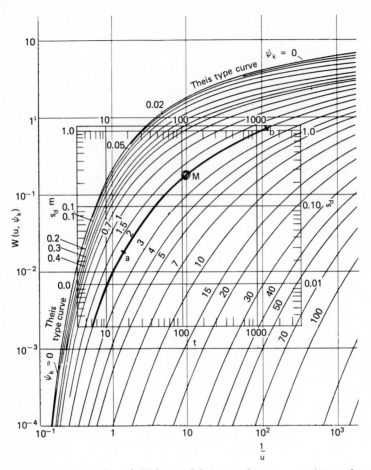

Figure 8.34 Matching field data and theoretical type curves (Example 8.12).

By a similar procedure at $r = 300$ m and $t = 300$ min, $1/u = 52.8$, $\psi_k = 1.41$, and $W(u, \psi_k) \simeq 0.7$, and from Eq. (8.79),

$$s_d = \frac{2.839}{4\pi \times 0.1533} \times 0.7 = 1.032 \text{ m}$$

8.4   Comments on Graphic Matching

It has been shown that the matching procedures are used in connection with several curves and charts. If a Theis type curve is used, it is assumed that pumping time is not so long that leakage will start to develop; otherwise the field curve will not match the type curve. If the time is long but leakage is not effective, the Theis curve and Jacob's straight-line method are valid. It is sometimes difficult to decide whether the time is long enough to justify use of the straight-line method. Thus it is always advisable to plot the recordings on both log-log and semilog plots. The straight-line portion on a semilog plot is preferable to the graphic matching procedure.

In practice, it is not advisable to go back to the field at a later time to repeat the field tests or complete them. The field recordings should be well planned in advance. Once these recordings are available, they should be used as soon as possible. For example, if the semilog plot does not reveal a linear portion, the match procedure should be used.

One should not be restricted to use of the Theis curve without checking for the possibility of leakage. Therefore, it may be advisable to use Figs. 8.29 and 8.34 rather than Figs. 8.18, 8.19, and 8.20. If it is found that the field curve approaches the extreme upper curve (a Theis type curve), then it should be matched on the Theis type curve. Matching of the field curve should be done carefully, and the best match takes place when a large segment of the field curve matches a type theoretical curve. It is also recommended to plot s_d versus log t (or log t/r^2 or log r^2/t or log r) and to use any straight-line portion as a check.

In all cases, one should remember that the matching procedures are approximate. Strictly speaking, a perfect match is not expected because of the differences between the assumptions made in the theories and the actual conditions in the field, where perfect homogeneity and isotropy of soil materials are impossible. In addition, the depths of formations are not uniform, as assumed in theory, and it is not economical nor warranted to determine the exact extent of these depths.

On many occasions, the lower portions of both the field and the Theis type curves match nicely, but the upper portion of the field curve deviates from the theoretical curve. If the upper portion of the field curve lies above the Theis type curve, it signifies the existence of an unnoticed impermeable barrier boundary near the pumped well (see Sec. 8.5). However, if the field curve lies below the Theis type curve, three conditions are possible:

1. Leakage has developed.

2. A recharge stream boundary lies within the area of influence of the pumped well (Sec. 8.5).

3. The pumped well is a gravity well (Sec. 8.10) rather than an artesian well.

The problem is rather complex if more than one of these conditions exists (for example, leakage and a stream boundary). In such cases, the evaluation of S and T require proper judgment and experience as well as thorough investigation of the geology and hydrology of the site where the wells are tested and used.

Large-scale charts should be used in order to employ the matching procedures properly. One should not rely on the small-size figures usually produced in textbooks. It is preferable to plot these charts from their proper tables unless large-size charts are already available (see Lohman, 1972).

8.5 Use of Image Wells: Transient Flow

The use of image wells is necessary to find the configuration of flow around wells pumping near natural boundaries. The layout of image wells is exactly the same as for the steady-state conditions (Sec. 8.1); however, the applied equations are those governing the transient flow system.

Locating Single Boundaries

The theoretical transient flow equations [Eqs. (8.51) and (8.56)] imply that the aquifers extend over an infinite area. If a barrier boundary or boundaries exist or when recharging stream boundaries cut through the aquifer at relatively close distances from the pumping wells, image wells should be introduced. Under actual field conditions, a stream is seldom fully penetrating and the hydraulic continuity between the stream and the aquifer may be restricted by partial clogging of the streambed by fine materials (U.S. Bureau of Reclamation, 1977). Stream and barrier boundaries do not follow straight lines. Theoretical solutions are therefore approximate, and in cases where the boundaries are unnoticed, location of the boundary that satisfies the solution may be shifted from the actual location. In many instances, sufficient geologic and hydrologic information delineate these boundaries. When such information is not available, the following procedures assist only in locating single boundaries. The stream boundaries are usually obvious, and therefore, the following procedures are directed mainly toward determination of concealed or unnoticed barrier boundaries.

Use of the Theis formula If a boundary is close to a discharge well, the effects on drawdown are noticed in a relatively short time. However, if the boundary is at a great distance, the effects are felt after a longer time or never felt at all when the duration of pumping is not sufficient to cause the influenced zone to come closer to the boundary. Using subscripts r

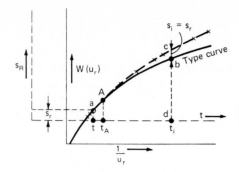

Figure 8.35 Graphic explanation for locating a barrier boundary using the Theis formula.

and i for the real and imaginary wells, respectively, then

$$s_r = \frac{Q}{4\pi T} W(u_r) \quad \text{and} \quad u_r = \frac{r_r^2 S}{4Tt_r}$$

and

$$s_i = \frac{Q}{4\pi T} W(u_i) \quad \text{and} \quad u_i = \frac{r_i^2 S}{4Tt_i} \tag{8.84}$$

where r_r and r_i are, respectively, the distances of an observation well from the pumping well and its image. Other symbols were defined previously. Generally, the resultant drawdown $s_R = s_r \pm s_i$, where the plus and minus signs correspond, respectively, to the barrier and stream boundaries:

$$s_R = s_r \pm s_i = \frac{Q}{4\pi T} [W(u_r) \pm W(u_i)] = \frac{Q}{4\pi T} \sum W(u) \tag{8.85}$$

A field plot of log s_R versus log t is then matched on the theoretical type curve (Fig. 8.35). Any value s_r on the matched curve *below* point A (where the two curves deviate) and at time $t_r < t_A$ is selected. Then the exact value s_r is scaled vertically between the type curve and the field curve, thus locating time $t_i > t_A$. The time t_i corresponds to $s_r = s_i$ for the imaginary well.[*] It follows from Eq. (8.84) that $W(u_r) = W(u_i)$ and $u_r = u_i$ for points a and b (Fig. 8.35). Thus,

$$\frac{r_r^2}{t_r} = \frac{r_i^2}{t_i}$$

or

$$r_i = r_r \sqrt{\frac{t_i}{t_r}} \tag{8.86}$$

---

[*] In Fig. 8.35 at time t_i, $S_R = bd + cb = s_r$ (at time t_i) + s_i (at time t_i). Thus by this construction s_i at time $t_i = s_r$ at time t_r. If each well is operating separately, this means that $W(u_r)$ at time $t_r = W(u_i)$ at time t_i.

The calculated r_i from Eq. (8.86) is the distance between the image well and the observation well. However, this information is not sufficient to find the distance between the pumped well and the image well. If three observation wells 1, 2, and 3 are available, then following the same procedure, three values of r_i [Eq. (8.86)] are obtained: r_{i1}, r_{i2} and r_{i3}. If the location of observation well 1 (Fig. 8.36) is used as the center, an arc

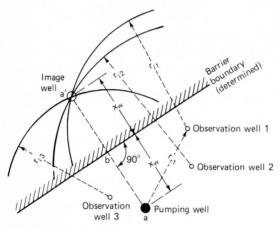

Figure 8.36 Location of a barrier boundary using Eq. (8.86).

of a circle with radius r_{i1} can be drawn. In order to locate the image well, another arc with radius r_{i2} (center at observation well 2) is drawn. The point of intersection of these two arcs should give the location of the image well. However, a third observation well is required to accommodate the approximations used in the graphic matching and field records and to accommodate the heterogeneities of the aquifer material. If a line is drawn between the located image well and the pumped well, the line that perpendicularly bisects this line gives the location of the barrier boundary sought (Fig. 8.36).

Use of Jacob's formula For long time periods and/or small radii of observation wells from the pumped well, the drawdown-time curve plots as a straight line [Eq. (8.56)], and

$$s_r = \frac{2.3Q}{4\pi T} \log \frac{2.25Tt_r}{r_r^2 S}$$

$$s_i = \frac{2.3Q}{4\pi T} \log \frac{2.25Tt_i}{r_i^2 S}$$

(8.87)

Again, subscripts r and i refer to the real well and its image, respectively.

The final chart is similar to that given in Fig. 8.37. The effect of the concealed barrier boundary develops beyond point A (Fig. 8.37). The full effect takes place at time $t_{A'}$ corresponding to point A' beyond which a straight line is obtained. If we select a drawdown s_r in the observation

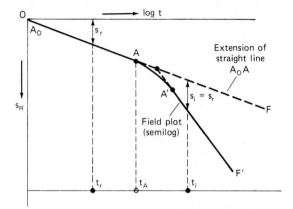

Figure 8.37 Graphic explanation for locating a barrier boundary using Jacob's formula.

well radius r from the pumping well at time $t_r < t_A$ on the early portion of the plot and then scale the same value as $s_i = s_r$ between the second straight-line portion and the extension of the first (Fig. 8.37), then time t_i corresponding to $s_r = s_i$ due to the image well is determined. From Eq. (8.87), when $s_r = s_i$, then

$$\frac{2.25 T t_r}{r_r^2 S} = \frac{2.25 T t_i}{r_i^2 S} \qquad \text{or} \qquad r_i = r_r \sqrt{\frac{t_i}{t_r}}$$

which is the same result given by Eq. (8.86). If the procedure is repeated on two other observation wells, the location of the image can be determined as explained in the previous procedure (Fig. 8.36).

Use of Stallman's matching curves   The plot shown in Fig. 8.38 was prepared by Stallman (1963) in order to find a concealed barrier boundary as well as the location of a recharging stream. The Theis type curve is shown in a log-log plot of $\Sigma W(u)$ versus $1/u_r$, where $\Sigma W(u)$ is given by Eq. (8.85) and u_r is defined in Eq. (8.84). The set of curves above the type curve are drawn for various ratios of r_i/r_r, and they are used to find the barrier boundary by means of a matching procedure. The set of curves below the type curve are used to locate a recharging stream if so desired. A field plot is prepared for an observation well in which the recorded resultant drawdowns s_R [Eq. (8.85)] are drawn versus t on log-log paper

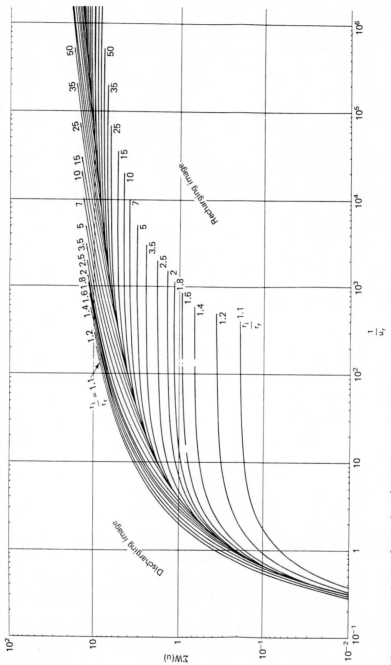

Figure 8.38 Logarithmic plot of $\Sigma W(u)$ versus $1/u_r$ for various values of r_i/r_r. (R. W. Stallman, "Type Curves for the Solution of Single Boundary Problems," in Short-cuts and Special Problems in Aquifer Tests, U.S. Geological Survey Water Supply Paper 1545-C, Washington, 1963)

that has the same scale as the theoretical chart (Fig. 8.38). If both curves are matched in such a way that their early portions almost coincide, a match point is located on one of the r_i/r_r curves (or between two consecutive curves). The coordinates of the match point are s_R, t, $\Sigma W(u)$, $1/u_r$, and r_i/r_r (read directly from the curves or by interpolation). Substituting the appropriate values in Eq. (8.85), T is determined. Substituting T, t, r_r, and u_r in the u_r expression given in Eq. (8.84), S is calculated. From the coordinate r_i/r_r of the match point, r_i is determined. The procedure is repeated on at least two additional observation wells, and the barrier boundary is located by following the same procedure indicated on Fig. 8.36.

The method can also be used to locate the stream boundary if the matching curves are below the Theis type curve (Fig. 8.38). The previous two methods also may be applied (Theis and Jacob) by following the same procedure. If the Theis formula is used, the field plot shown in Fig. 8.35 will be below the type curve beyond point A. If Jacob's equation is used to locate the stream boundary, s_i should be scaled above AF (Fig. 8.39). Note that in this figure, line $A'F'$ is almost horizontal, meaning that a steady-state condition is reached. This can also be proved by subtracting s_i from s_r [Eq. (8.87)], keeping t the same.

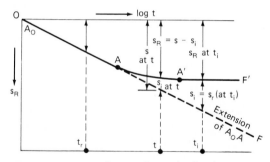

Figure 8.39 Graphic explanation for locating a stream boundary using Jacob's formula.

$$s_R = s_r - s_i = \frac{2.3Q}{4\pi T}\left(\log \frac{2.25Tt}{r_r^2 S} - \log \frac{2.25Tt}{r_i^2 S}\right)$$

or $$s_R = \frac{2.3Q}{4\pi T}\log \frac{r_i^2}{r_r^2} \tag{8.88}$$

where the time t is eliminated. The preceding equation is essentially the same as the steady-state equation [Eq. (8.39)].

Streams Completely Penetrating
Artesian Aquifers

A stream has no effect on a pumping well if at any time t the radius of influence $r_{e,t}$ is less than the distance x_w between the well and the stream. However, if the well is close enough to the stream or t is large enough, the well would derive a certain portion Q_{rs} from the stream. Generally, when x_w^2/t decreases, the ratio Q_{rs}/Q increases.

In a nonleaky aquifer, the resultant drawdown s_R for a well pumping near a stream is determined by applying Eq. (8.85) with a minus sign and a common time t, or

$$s_R = s_r - s_i = \frac{Q}{4\pi T}[W(u_r) - W(u_i)] \tag{8.85a}$$

Using Jacob's equation [Eq. (8.87)] for a common time t,

$$s_R = s_r - s_i = \frac{2.3Q}{4\pi T}\left(\log \frac{2.25Tt_t}{r_r^2 S} - \log \frac{2.25Tt_i}{r_i^2 S}\right)$$

or
$$s_R = \frac{2.3Q}{2\pi T} \log \frac{r_i}{r_r} = s_{df} \tag{8.88a}$$

It is clear that Eq. (8.88a) gives the values of $s_R = s_{df}$ at equilibrium as t is eliminated. Equations (8.85a) and (8.88a) correspond to the condition of complete circulation when $Q_{rs}/Q = 1.0$ (Sec. 8.1) if two real discharge and recharge wells are operating. When an image well is used as a tool to find the flow pattern of a discharge well pumping near a stream, then $Q_{rs}/Q = 1.0$ indicates that all the pumped water is derived from the stream.

The ratio Q_{rs}/Q becomes less than 1.0 when there is natural flow (Sec. 8.6) in the aquifer. On the basis of Eq. (8.85a), Theis (1941) developed the relationship between Q_{rs}/Q and $u_{xw} = x_w^2 S/4Tt$ under transient flow systems. The results are given in a chart form (Fig. 8.40).

8.6   Effects of Natural Flow

The configuration of the flow of wells can be determined if natural flow exists as analyzed in Sec. 8.1 for steady-state conditions. When a well is pumping under transient conditions of flow and at a constant rate Q, the natural flow affects the resultant values of the drawdowns. If pumping time is relatively short, the natural flow can be considered as steady-state flow. The changes in natural flow within 1 year may be less than the effect of pumping during the early hours (Kashef, 1975). Practically, the draw-

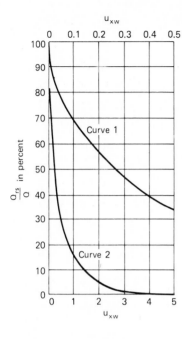

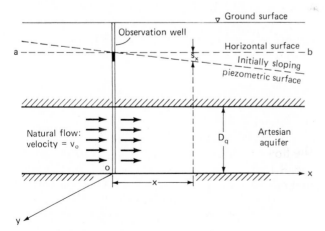

Figure 8.40 Graph for determining percentage of pumped water (transient condition) being diverted from a source of recharge (stream). The x axis for curve 1 is the upper (*top*) scale, and the x axis for curve 2 is the lower (*bottom*) scale. (*C. V. Theis, "The Effect of a Well on the Flow of a Nearby Stream," Trans. Am. Geophys. Union, vol. 22, 1941, p. 737.*)

downs of a discharge well under transient conditions of flow can be superimposed on those produced by the *steady* natural flow to find the resultant drawdown at any time t since pumping began. If the undisturbed piezometric surface is sloping (Fig. 8.41), then the drawdowns s_x (no pumping or any disturbances) are

Figure 8.41 Drawdown due to steady-state natural groundwater flow in an artesian aquifer.

$$s_x = \pm \frac{v_o}{K} x = \pm \frac{v_o x}{T} D_q \tag{8.89}$$

where v_o is the natural darcian velocity in the artesian aquifer and $T = KD_q$; x is positive and to the right of the normal line to the cross section passing by the origin o (Fig. 8.41). If x is to the left of the origin, s_x will be negative, i.e., above the horizontal plane ab. The drawdown s_r below ab of a pumping well installed at the origin (Fig. 8.41) is determined by the Theis formula [Eq. (8.51)], or

$$s_r = \frac{Q}{4\pi T} W(u) \tag{8.90}$$

The resultant drawdown s_R is determined by combining Eqs. (8.89) and (8.90):

$$s_R = s_r + s_x = \pm \frac{v_o}{T} x D_q + \frac{Q}{4\pi T} W(u) \tag{8.91}$$

Note that $r \neq x$ except in the vertical plane passing through the axis of the well and oriented along the direction of v_o. If we draw a rectangular or square grid system in the horizontal plane (x, y), then s_R can be determined at any node of the grid by means of Eq. (8.91), noting that $r = \sqrt{x^2 + y^2}$. Contour lines can then be traced for lines of equal s_R.

The magnitude and direction of the darcian velocity $v_{x,y}$ at any point x, y can be determined as follows:

$$s_r = \frac{Q}{4\pi T} \int_{r^2 S/4Tt}^{\infty} \frac{e^{-u}}{u} du \qquad \text{[see Eq. (8.49a)]}$$

Then

$$\frac{\delta s}{\delta r} = \frac{\delta s}{\delta u} \frac{\delta u}{\delta r} = \frac{e^{-u}}{u} \frac{Q}{4\pi T} \frac{\delta u}{\delta r} \tag{8.92}$$

but

$$\frac{\delta u}{\delta r} = \frac{\delta}{\delta r}\left(\frac{r^2 S}{4Tt}\right) = \frac{rS}{2Tt}$$

By substitution in Eq. (8.92),

$$\frac{\delta s}{\delta r} = \frac{Qe^{-u}}{2\pi rT} \tag{8.93}$$

At any cylindrical surface of radius r and height D_q, the rate of flow $Q_{r,t}$ at time t is less than Q:

$$Q_{r,t} = 2\pi r D_q K \frac{\delta s}{\delta r} = 2\pi r T \frac{\delta s}{\delta r} \tag{8.94}$$

Combining Eqs. (8.93) and (8.94), then

$$Q_{r,t} = Qe^{-u} \tag{8.95}$$

The radial velocity $v_{r,t}$ due to the well effect only is

$$v_{r,t} = \frac{Qe^{-u}}{2\pi r D_q}$$

The components of $v_{r,t}$ in the x and y directions are $v_{x,t}$ and $v_{y,t}$, and

$$v_{x,t} = \frac{xQe^{-u}}{2\pi D_q \sqrt{x^2 + y^2}} \qquad (8.96a)$$

and

$$v_{y,t} = \frac{yQe^{-u}}{2\pi D_q \sqrt{x^2 + y^2}} \qquad (8.96b)$$

Combining the natural velocity v_o with $v_{r,t}$, the resultant darcian velocity $v_{x,y}$ is

$$v_{x,y} = \sqrt{(v_{x,t} + v_o)^2 + v_{y,t}^2}$$

or

$$v_{x,y} = \sqrt{v_o^2 + \frac{v_o x Q e^{-u}}{\pi D_q \sqrt{x^2 + y^2}} + \left(\frac{Qe^{-u}}{2\pi D_q}\right)^2} \qquad (8.97)$$

The direction α with the x axis of the velocity vector $v_{x,y}$ is

$$\alpha = \tan^{-1} \frac{v_{y,t}}{v_{x,t} + v_o} = \tan^{-1} \frac{yQe^{-u}}{xQe^{-u} + 2\pi D_q v_o \sqrt{x^2 + y^2}} \qquad (8.98)$$

The directions of the velocities can thus be traced at all nodes of the selected grid.

8.7 Multiple Wells and Boundaries

An equation can be written for any number of discharge or recharge wells and their images. If the same general procedures indicated in Eqs. (8.85), (8.88), and (8.91) are followed, any additional expressions for any number of discharge and/or recharge wells or their images may be added. It would be better in these cases to write a computer program that could carry out the calculations at the nodes of a selected grid. The grid could be made finer, and any additional nodes could be added later in the program. The drawdown contours can be traced and the magnitudes of the velocities and their directions at the nodes can be found.

The principle of superposition is permissible in artesian aquifers (because the mathematical equations written for the drawdowns are linear and the flow medium remains the same), and thus the resultant drawdown at any point should be equal to the algebraic sum of the individual effects of natural flow and of wells and their images. If the wells are relatively close, there will be interference among their zones of influence (Fig. 8.42). The wells can have different rates of flow, and their operations do not have to start at the same time. Equations for determining the flow pattern have been developed for different well arrangements (Muskat, 1937; Rao et al., 1971; see also Chap. 9).

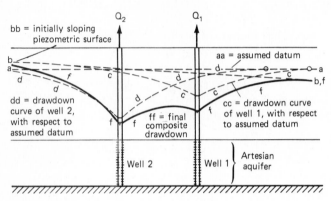

Figure 8.42 Composite drawdown curves due to natural flow and well pumping.

For water-supply purposes, it is recommended that interference be avoided by spacing the wells far apart. On some occasions it may be more economical to allow some interference (Walton, 1970) to offset other costs, such as those of land and connecting pipelines and electric equipment. For drainage purposes, the situation is different, and well interference is required. In some cases, upward leakage to aquitards above artesian aquifers produces waterlogging of the aquitards, which are used as agricultural land. In order to minimize leakage, drainage by artesian wells can be used to lower the pressures in main artesian aquifers and thus interference of wells is desirable. The same is true for some sites of engineering structures constructed above the aquitards. Interfering wells should be planned to lower the artesian pressures beneath the planned structures in order to preserve the stability of the temporary excavations made for the foundations. If deep borings are not made and the artesian aquifers and their pressures are not recognized, the bottoms of these excavations in the confining layers can be subjected to an upward hydraulic flow (boiling) that disturbs the natural soil and ultimately may cause the excavations to collapse.

Various control methods have been developed for certain special cases of interference (Hantush, 1964a and Rao et al., 1971). For example, Theis (1941) gave a formula to determine the permissible distance between production and disposal wells in an extensive isotropic aquifer. Theis (1957) also derived a formula to determine the spacing between two production wells in order to get optimum production. His formula included the costs of raising water, equipment, maintenance, depreciation, etc., as well as Q of each production well and T of the aquifer.

From experience, Walton (1970) suggested that for small values of Q and T, two production wells should at least be spaced at $2D$ (twice the depth of the aquifer). When a multiple-well system is used, the spacings between the wells should be at least about 80 m (Walton, 1970). A better water yield is obtained from production wells if they are spaced parallel to but far away from barrier boundaries or as near to the center of a buried valley as possible. In the case of a recharge stream boundary, the wells should be parallel to the boundary and as close as possible. Legal restrictions should be observed at all times. Well spacings are to be observed in order to minimize the effects of interference on such production wells as those for water supply and irrigation. However, as mentioned earlier, close wells should be planned if it is necessary to increase the drawdown, such as in preventing waterlogging in agricultural land produced by upward leakage from shallow artesian aquifers and in relieving artesian pressures beneath foundation excavations. In the latter case, close wells are needed to surround the planned excavations, and they are operated continuously until the foundations are constructed.

8.8   Losses, Specific Capacity, and Well Efficiency

The *specific capacity* of a well is defined as the ratio of Q to the drawdown s_{dw} measured in the pumping well; therefore,

$$\text{Specific capacity} = \frac{Q}{s_{dw}} \tag{8.99}$$

Drawdown s_{dw} is the drawdown s_d at $r = r_w$. If the well is not cased (for example, if it is drilled in sandstone, where the well bore does not need any casings, screens, or gravel packs), then s_{dw} may be determined from the equations given in Secs. 8.1 through 8.6 whether in the steady or transient state (Fig. 8.43a). If the well is screened throughout the depth of the aquifer, screen losses are produced by friction during water entry to the well (Fig. 8.43b). When a gravel pack is needed, the piezometric surface near the well is subjected to additional lowering due to the additional head losses resulting from the turbulent flow within the gravel pack surrounding the well (Fig. 8.43c). The additional drawdown in Fig. 8.43a is called *formation losses*, whereas the additional losses in Fig. 8.43b and c are known as *entry losses*. Formation losses are expressed as

$B_f Q$, whereas entry losses are expressed as $C_f Q^n$ (Jacob, 1950). Thus,

$$s_{dw} = B_f Q + C_f Q^n \qquad (8.100)$$

where C_f = a constant under steady-state and transient conditions of flow that depends on the construction and conditions of the well and its effective radius (see Table 8.14)

n = a power to Q

In Eq. (8.100),

$$B_f = \frac{\ln (r_e/r_w)}{2\pi T} \qquad \text{(under steady-state conditions)} \qquad (8.101a)$$

or $$B_f = \frac{2.30}{4\pi T} \log \frac{2.25 Tt}{r_w^2 S} \qquad \text{(under transient conditions)} \qquad (8.101b)$$

Under the steady-state condition, n was approximately given as 2.0 (Jacob, 1947), but it may deviate considerably from 2.0 (Rorabaugh, 1953). Pumping tests should be made to evaluate n as well as the coefficients B_f and C_f. Moreover, in the case of a well with a gravel pack (Fig. 8.43c), the radius of the well cannot be the nominal well radius. After the wells are constructed, usually they are *developed* (water is pumped in to

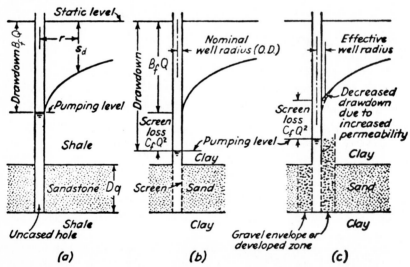

Figure 8.43 Distribution of drawdown around typical wells in a confined uniform sand. (*a*) Ideal well without screen or casing; (*b*) screened or cased wells, sand undisturbed by drilling; (*c*) developed or gravel-walled well, sand improved or replaced, effective well radius increased. (*C. E. Jacob, "Drawdown Test to Determine the Effective Radius of an Artesian Well," Trans. Am. Soc. Civ. Eng., vol. 112, 1947.*)

TABLE 8.14   Relationship between C_f [Eq. (8.100)] and the Well Condition

C_f, min²/m⁵	Well condition
<0.5	Properly designed and developed
0.5–1.0	Mild deterioration or clogging
1.0–4.0	Severe deterioration or clogging
>4.0	Difficult to restore well to original capacity

SOURCE: W. C. Walton, "Selected Analytical Methods for Well and Aquifer Evaluation," Illinois State Water Survey Bulletin no. 49, 1962.

stabilize the fine materials); this disturbs the aquifer materials in the vicinity of the well and affects the nominal radius of the well. The effective radius r_w of the well should therefore be determined. A carefully planned pumping test should be adequate to determine B_f, C_f, n, and the effective radius r_w in addition to the aquifer parameters S and T.

The specific capacity of a well Q/s_{dw} is constant when the flow is steady because B_f, C_f, and n are constants [Eqs. (8.100) and (8.101a)]. Under transient states of flow, Q/s_{dw} decreases over time because B_f depends on the time of pumping [Eq. (8.101b)]. If Q increases, the specific capacity decreases (Fig. 8.44).

Losses increase in old wells over time because of clogging and deterioration of well screens. If the power n [Eq. (8.100)] is considered as 2.0,

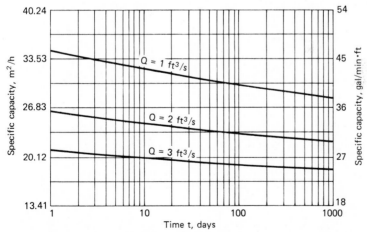

Figure 8.44 Variation of the specific capacity of a typical pumping well with discharge and time: $T = 1.24$ m²/min, $S = 2.6 \times 10^{-5}$ (Example 8.13). (*C. E. Jacob, "Drawdown Test to Determine the Effective Radius of an Artesian Well," Trans. Am. Soc. Civ. Eng., vol. 112, 1947.*)

the dimension of C_f will be min^2/m^5 because the unit of $C_f Q^n$ is meters. Walton (1962) gave the relationships between C_f and the well conditions listed in Table 8.14. Various methods have been developed to find B_f, C_f, and n for steady and transient conditions (Jacob, 1947; Rorabaugh, 1953; Sheahan, 1971; Lennox, 1966; Labadie and Helweg, 1975). These methods are, in general, based on step-drawdown pumping tests in which s_{dw} is measured for successively increasing values of Q.

In Jacob's method, it is suggested that several drawdown tests (three or more) be conducted changing the pumping rate Q. After each test, a period of rest should be sufficient to ensure almost complete recovery. Recordings are plotted on a semilog plot as shown in Fig. 8.44. Generally, straight lines are obtained; the gentle slopes of these lines indicate that the specific capacity of the well changes only slightly over time.

From Eq. (8.100), the specific capacity Q/s_{dw} is given by the following equation:

$$\frac{Q}{s_{dw}} = \frac{1}{B_f + C_f Q^{n-1}} \tag{8.102}$$

Introducing B_f into Eq. (8.101b), we see that

$$\frac{Q}{s_{dw}} = \frac{1}{(2.30/4\pi T) \log 2.25 Tt/r_r^2 S + C_f Q^{n-1}} \tag{8.103}$$

The unknowns in this equation are S, T, r_w, C_f and n. Therefore, five drawdown tests are needed to write five simultaneous equations. However, it is preferable to evaluate S and T from the recordings of a nearby observation well unaffected by the well losses; then only three drawdown tests would be needed to find r_w, C_f, and n. Also, if n is assumed to be equal to 2.0, only two tests would be required.

The specific capacity of a well may be used to evaluate T (Fig. 8.45),

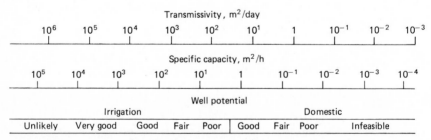

Figure 8.45 Comparison of transmissivity, specific capacity, and well potential. Transmissivity $T = K\overline{D}_q$, where K = permeability (hydraulic conductivity) and $\overline{D}_q$ = saturated thickness of the aquifer. Specific capacity values are based on a pumping period of approximately 8 h but are otherwise generalized. *(U.S. Bureau of Reclamation, Ground Water Manual, U.S. Department of Interior, Washington, 1977.)*

and of course, if T has been predetermined, the specific capacity can be evaluated from Fig. 8.45.

EXAMPLE 8.13 Figure 8.44 shows the results of three drawdown tests (Jacob, 1947). If $T = 1.24$ m²/min and $S = 0.000026$ (determined separately), find r_w (effective well radius) and the coefficients C_f and n [Eq. (8.100)].

Solution Three points are selected at time $t = 10$ days on the three lines in Fig. 8.44 corresponding to $Q = 1.7$, 3.4, and 5.1 m³/min. The three Q/s_{dw} ordinates are, respectively, 0.5128, 0.407, and 0.3317 m²/min, as determined from the figure.

If we substitute these values into Eq. (8.102) and then arrange three simultaneous equations in B_f, C_f, and n, we get

$$1.95 = B_f + C_f(1.7)^{n-1} \tag{i}$$
$$2.458 = B_f + C_f(3.4)^{n-1} \tag{ii}$$
$$3.105 = B_f + C_f(5.1)^{n-1} \tag{iii}$$

The B_f values [Eq. (8.101b)] should be the same for the three selected points in the preceding equations because t is the same (10 days). Subtracting Eq. (ii) from Eq. (i) and Eq. (iii) from Eq. (ii) and dividing the resulting equations gives

$$0.912 = \frac{1 - (0.5)^{n-1}}{(1.5)^{n-1} - 1}$$

This equation is solved by trial and error, and n is found to be equal to 2.18. Substituting $n = 2.18$ in Eqs. (i), (ii), and (iii), we have

$$1.95 = B_f + 1.87C_f \tag{ia}$$
$$2.458 = B_f + 4.238C_f \tag{iia}$$
$$3.015 = B_f + 7.838C_f \tag{iiia}$$

By subtracting Eq. (iia) from (iiia), C_f is determined:

$$C_f = \frac{2.458 - 1.95}{4.238 - 1.87} = 0.214 \text{ min}^{2.18}/\text{m}^{5.54}$$

(Note that if $n = 2.0$, the unit of C_f is min²/m⁵ and generally the unit of C_f is min^n/m^{3n-1}.)

If we add Eqs. (ia), (iia), and (iiia), then $7.423 = 3B_f + 12.946C_f$, and if we substitute $C_f = 0.214$, B_f is

$$B_f = \tfrac{1}{3} (7.423 - 12.946 \times 0.214) = 1.55 \text{ min/m}^2$$

If we substitute in Eq. (8.101b) the values $t = 10$ days $= 144 \times 10^4$ min, $T = 1.24$ m²/min, and $S = 0.000026$, then

$$1.55 = \frac{2.30}{4\pi \times 1.24} \log \frac{2.25 \times 1.24 \times 1.44 \times 10^4}{0.00026 r_w^2}$$

or $\quad r_w$ (effective well radius) $= 0.221$ m $= 22$ cm

It should be noted that the values of Q/s_{dw} are taken from a small-scale figure and so approximate answers should be expected. In addition, the simultaneous equations are generally very sensitive to the approximate values. Any small changes in the T value would affect appreciably the result of the small value of r_w in Eq. (8.101b). It is therefore recommended that other times be used (depending on the accuracy of field recordings) near the recordings coinciding on the fitted straight lines. Other methods previously cited should also be considered.

The efficiency of a well is a measure of the extent of well losses at a specified period of pumping. When T and S are known, the theoretical specific capacity is given by

$$\frac{Q}{s_{dw}} = \frac{1}{B_f} = \frac{1}{\dfrac{2.3}{4\pi T} \log \dfrac{2.25Tt}{r_w^2 S}}$$

using Jacob's formula [Eq. (8.56)]. And

$$\frac{Q}{s_{dw}} = \frac{4\pi T}{W(u_w)}$$

where $u_w = r_w^2 S / 4Tt$ using the Theis formula [Eq. (8.51)]. If s_{dw} is recorded in the field at the specified time t, it should be greater than the theoretical s_{dw} because of entry losses; thus the field specific capacity should be equal to $Q/(B_f Q + C_f Q^n)$ at time t. The efficiency E_{wf} of the well is given as the percentage of the ratio of the theoretical and field specific capacities, or

$$
\begin{aligned}
E_{wf} &= \frac{100 B_f}{B_f + C_f Q^{n-1}} \\
&= 100 \times \frac{\text{computed drawdown}}{\text{measured drawdown}}
\end{aligned}
$$

(8.104)

In practice, an inefficient well is recognized from the initial recovery rate when pumping is stopped. When the well losses are relatively high, the rate of recovery is rapid. A rough rule of thumb is given by Todd (1980, page 159): "If a pump is shut off after 1 hour of pumping and 90% or more of the drawdown is recovered after 5 minutes, it can be concluded that the well is unacceptably inefficient."

8.9   Partially Penetrating Artesian Wells

In the previous sections, the artesian wells were assumed to fully penetrate the entire depth of the aquifer. This means that water enters the well through a surface area of $2\pi r_w D_q$. In shallow aquifers, this can be achieved by perforated casing, screens, and/or gravel packs throughout the aquifer depth D_q. This becomes rather expensive with deep aquifers. The screens are usually designed for certain segments (Todd, 1980; Walton, 1970; UOP Johnson Division, 1966; U.S. Bureau of Reclamation, 1977). The well may be provided by one or more segments depending on the locations of the most productive layers of the aquifer. The design of screens and their construction are beyond the scope of this book, and the reader is referred to the cited references.

When an idealized artesian well is screened throughout the depth of the aquifer, the stream lines are all horizontal (Fig. 8.46a). The configuration of flow is deformed whenever the vertical length of the water entry is smaller than D_q (Fig. 8.46b and c) or when the well penetrates the aquifer for just few centimeters (Fig. 8.46d). The deflected flow lines in the case of partially penetrating wells are longer than those for wells of complete penetration. Therefore, if it is assumed that the boundary conditions and the rate of flow Q are the same for both cases, more frictional

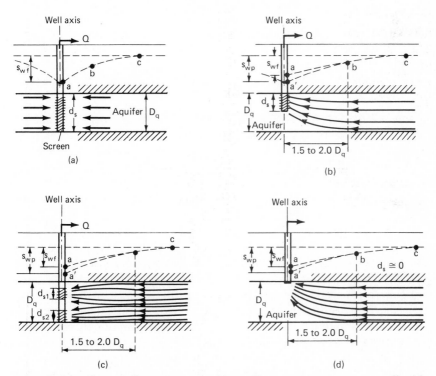

Figure 8.46 Flow configuration toward partially penetrating artesian wells. (*a*) Screen length $d_s = D_q$ (full penetration); (*b*) screen length $d_s < D_q$; (*c*) two segments of screens; (*d*) well penetrating the aquifer for a few centimeters.

losses are anticipated, leading to drawdowns $s_{wp} > s_{wf}$ at the well surface (s_{wf} corresponds to fully penetrating wells). Thus the specific capacity Q/s_{wp} for partially penetrating wells is less than that for fully penetrating wells. In order to maintain $s_{wp} = s_{wf}$, the rate of pumping of the partially penetrating well should be increased.

Partially penetrating wells have been theoretically investigated for steady and transient states by various investigators (see Hantush, 1961a, 1961b, 1964a; Walton, 1970; Nisle, 1958; Brons and Marting, 1961; Muskat, 1937; Huisman, 1972). A computer solution is given by Kirkham (1959), and studies by electric analogs are given by Li et al. (1954) and Franke (1967). A simple approach is given by Sternberg (1973) that has been checked with several complex approaches for transient states.

This approach and its limitations are explained as follows: Using Jacob's formula [Eq. (8.56)],

$$s_{dw} = s_{wf} = \frac{Q}{4\pi T} \ln \frac{2.25Tt}{r_w^2 S}$$

Then
$$s_{wp} = s_{wf} + \Delta s_{dw} \tag{8.105}$$

and
$$s_{wp} = \frac{Q}{4\pi T}\left(\ln \frac{2.25Tt}{r_w^2 S} + 2s_p \right) \tag{8.106}$$

where
$$s_p = \frac{2\pi T \Delta s_{dw}}{Q}$$

The relationship between s_p and $\log D_q/r_w$ for various d_s/D_q ratios is shown in Fig. 8.47, where d_s is the length of the screen. It is clear that the relationship between s_p and $\log D_q/r_w$ is linear except when r_w is relatively large and the screen is relatively short (note the lower portion of curves $d_s/D_q = 0.1$ and 0.2 in Fig. 8.47). The method is developed when the screen location is as shown in Fig. 8.48.

Lowering of the drawdown of well due to partial penetration results in a drawdown curve that is lower than that for a fully penetrating well only for a distance $a'b \simeq 1.5D_q$ to $2.0D_q$ (Fig. 8.46b, c, and d), beyond which the curve is the same (Todd, 1980). It is therefore preferable to take field recordings at observation wells located not closer than $1.5D_q$ from the well in order to avoid the effect of partial penetration when S and T are evaluated. As stated by Todd (1980, p. 152), "Common field situations often reduce the practical importance of partial penetration."

The cited analytical methods are based on formulas in which the aquifer parameters are included. These parameters are commonly evaluated by means of pumping tests. If the pumped well is a partially penetrating well, the tests determine the screen losses as well as the aquifer losses. The latter includes the effects of partial penetration, and there is

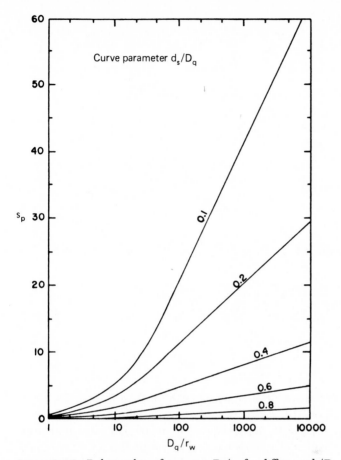

Figure 8.47   Relationship of s_p versus D_q/r_w for different d_s/D_q ratios. *(Y. M. Sternberg, "Efficiency of Partially Penetrating Wells," Ground Water, vol. 11, no. 3, 1973.)*

no need to use any analytical formula. The values of Δs_{dw} [Eq. (8.105)] are automatically included in the B_f factor of Eq. (8.101b). As stated earlier, the parameters T and S should be based on recordings at observation wells not closer than $1.5D_q$ to $2.0D_q$ from the pumped well.

In anisotropic alluvial deposits, the length of the screen is errone-ously considered as equivalent to the saturated thickness of the aquifer D_q when it is necessary to evaluate K from the value of T determined from pumping tests. This practice is not valid because (as in the case shown in Fig. 8.46d) it gives an erroneous value of $K = \infty$.

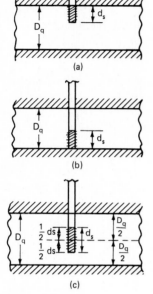

Figure 8.48 Locations of screens for which Fig. 8.47 is valid (*a*) Screen at top of aquifer; (*b*) screen at bottom of aquifer; (*c*) screen centered with center of aquifer.

8.10 Gravity Wells

A water well withdrawing water from an unconfined aquifer is known as a *gravity well* (or *water-table well* or *ordinary well*). The pattern of water movement toward such wells differs from that in artesian wells. The flow lines are not horizontal, except the stream line along the lower impervious boundary (Fig. 8.49). At any point (r, y) on a flow line, the velocity has a vertical component v_y. In idealized artesian wells, the flow lines are all horizontal, except in the case of partially penetrating wells in the zones near the well (Sec. 8.9).

The analyses for gravity wells are more complex than those for artesian wells because the upper boundary of the flow zone (free surface *ab* in Fig. 8.49) is unknown before the solution is complete. This is the same situation as in seepage through earth dams (Sec. 5.5). Many approaches have been used for finding the true free surface for steady and transient states of flow. Most of the rigorous solutions are not suitable for practical use. For this reason, various assumptions have been introduced. Even in the more rigorous solutions, the free surface is located, but the distribution of heads within the flow medium, including those along the impervious base, are generally not determined. Solutions employing rigorous numerical methods are comprehensive, but they require certain skills.

In this section, practical approaches based on extensive research are explained. They can be used to locate the free surface, the base pressures along the impervious beds, and the pressure or total heads at any point within the flow medium. These approaches can also be used to analyze the interference between two or more wells under steady or transient states of flow.

Steady-State Conditions

Dupuit's equation Applying Dupuit's assumptions (see Sec. 5.5), the following equations are obtained:

$$\frac{Q}{\pi K} = \frac{h_u^2 - h_w^2}{\ln r_e/r_w} \quad \text{(exact)} \tag{8.107}$$

and $\quad \dfrac{Q}{\pi K} = \dfrac{h_{r1}^2 - h_{r2}^2}{\ln r_1/r_2} = \dfrac{h_r^2 - h_w^2}{\ln r/r_w} = \dfrac{h_u^2 - h_r^2}{\ln r_e/r}$ (approximate) (8.108)

The various h and r values are shown in Fig. 8.49. It is obvious from Eq. (8.107) that when Q increases, h_w decreases until the maximum value Q_{max} is reached when $h_w = 0$ (Fig. 8.49), or

$$\frac{Q_{max}}{\pi K} = \frac{h_u^2}{\ln r_e/r_w} \quad \text{(exact)} \tag{8.109}$$

Any of the expressions given in Eq. (8.108) can be used to plot the curve h_r versus r, known as *Dupuit's curve* (Figs. 8.49 and 8.50). This

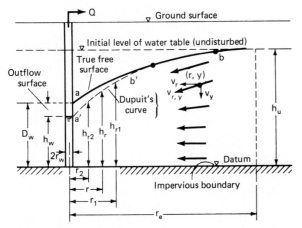

Figure 8.49 True free surface in a gravity well and its location with respect to Dupuit's curve.

curve was erroneously thought of as the true free surface. Dupuit's curve intersects the surface of the well at point a' (Fig. 8.49) at a height h_w above the datum. This means that when $h_w = 0$, Dupuit's curve intersects the surface of the well at point a (Fig. 8.50), allowing no water to flow to

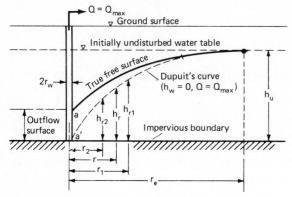

Figure 8.50 True free surface and Dupuit's curve corresponding to maximum possible rate of flow of a gravity well.

the well (the outflow surface disappears). Thus Eq. (8.108) should not be used to find the true free surface, which lies above Dupuit's curve, allowing an outflow surface aa' (Fig. 8.49) above the water level in the well (assuming no losses). The point of intersection a of the true surface with the surface of the well is at a height $D_w > h_w$ (Muskat, 1937; Polubarinova-Kochina, 1962; Kashef, 1965a).

Although Eqs. (8.107) and (8.109) are based on Dupuit's assumptions, in which the vertical flow is neglected, it has been shown experimentally by Muskat (1937) that they give the exact values of Q. The same conclusion was later proved mathematically by Charney in 1951 (according to Polubarinova-Kochina, 1962). Therefore, if h_w, r_w, Q, and h_u are known, the K value is *correctly* determined by

$$K = \frac{Q \ln r_e/r_w}{\pi(h_u^2 - h_w^2)} \tag{8.110}$$

Equations (8.107) and (8.109) as shown do not allow determination of the true free surface. Various methods have been developed to locate the free surface, and two of the more practical methods are explained in the following section.

Experimental method By means of experiments based on analogies with electric systems, Babbitt and Caldwell (1948) developed the following equation for a single water-table well:

$$D_r = h_u - \frac{C_r}{h_u} \frac{h_u^2 - h_w^2}{\ln r_e/r_w} \ln \frac{r_e}{0.1 h_u} \qquad (8.111)$$

or

$$\overline{D}_r = 1 - C_r \frac{1 - m_w^2}{\ln r_e/r_w} \ln \frac{\bar{r}_e}{0.1} \qquad (8.111a)$$

where D_r = height of free surface above the impervious boundary
$\quad\quad C_r$ = a coefficient determined from Fig. 8.51

In Eqs. (8.111) and (8.111a),

$$\overline{D}_r = D_r/h_u \qquad \bar{r}_e = r_e/h_u \qquad m_w = h_w/h_u$$

The curve shown in Fig. 8.51 should have been refined in such a way that the abscissa reads $(r - r_w)/(r_e - r_w)$ rather than r/r_e. However, the curve is reliable, and the method supersedes other approximation methods currently used. The elevation of the free surface D_w at r_w is obtained by substituting $C_r = 0.6$ (Fig. 8.51) corresponding to $r/r_e = 0$ in Eq. (8.111):

$$D_w \simeq h_u - \frac{0.6}{h_u} \frac{h_u^2 - h_w^2}{\ln r_e/r_w} \ln \frac{r_e}{0.1 h_u} \qquad (8.112a)$$

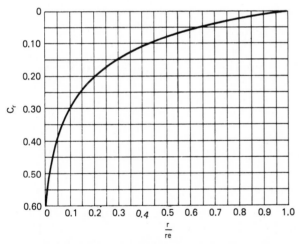

Figure 8.51 Relationship between r/r_e and C_r (steady-state flow toward a gravity well). (H. E. Babbitt and D. H. Caldwell, "The Free Surface Around, and Interference between Gravity Wells," Univ. Illinois Bull., vol. 45, no. 30, 1948.)

and by the use of the exact Eq. (8.107):

$$\overline{D}_w = 1 - \frac{0.6\,Q}{\pi K h_u^2} \ln \frac{\overline{r}_e}{0.1} \qquad (8.112b)$$

where

$$\overline{D}_w = \frac{D_w}{h_u}$$

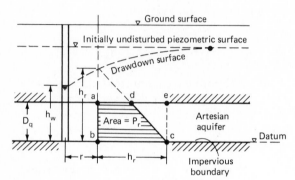

Figure 8.52 Pressure diagram in an idealized artesian aquifer due to pumping (steady-state flow).

Unified well formulas If the areas of the pressure-head diagrams P_r across vertical sections are determined, the following well formulas (Kashef, 1965a, 1970, 1971) are valid for artesian wells (Fig. 8.52), gravity wells (Fig. 8.53), or overpumped artesian wells (Fig. 8.54) where the upper portion of the main artesian aquifer is drained in a region around the well because of excessive pumping that changes the flow in that region to gravitational. Therefore,

$$\frac{Q}{2\pi K} = \frac{P_e - P_r}{\ln r_e/r} \qquad (8.113a)$$

$$\frac{Q}{2\pi K} = \frac{P_r - P_w}{\ln r/r_w} \qquad (8.113b)$$

$$\frac{Q}{2\pi K} = \frac{P_e - P_w}{\ln r_e/r_w} \qquad (8.113c)$$

$$P_r = P_e - \frac{Q \ln r_e/r}{2\pi K} \qquad (8.113d)$$

where P_e, P_r, and P_w are, respectively, the areas of the pressure-head diagrams at radii r_e, r, and r_w. Therefore,

$$P_e = \tfrac{1}{2}h_u^2 \qquad \text{and} \qquad P_w = \tfrac{1}{2}h_w^2 \qquad (8.114)$$

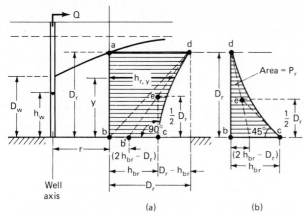

Figure 8.53 Total-head and pressure diagram in a gravity well (steady-state flow). (*a*) Total-head diagram; (*b*) pressure-head diagram.

P_r for any type of well can be evaluated by referring to Figs. 8.52, 8.53, and 8.54. In the case of artesian wells (Fig. 8.52), the area of the pressure-head diagram at a radius r is

$$P_r = \tfrac{1}{2}D_q[(h_r - D_q) + h_r]$$

or
$$P_r = D_q h_r - \tfrac{1}{2}D_q^2 \qquad \text{(artesian wells)} \qquad (8.115)$$

where D_q = depth of the aquifer
h_r = total head at radius r with respect to the shown datum
$h_r = h_{br}$, which is the base pressure at point b

The pressure-head diagram *abcd* is linear in this case.

In the case of water-table wells (Fig. 8.53*a*), the distribution of the vertical velocity component v_y at the vertical section *ab* is linear (Polubarinova-Kochina, 1962), producing a parabolic distribution of the

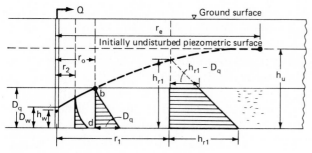

Figure 8.54 Overpumped artesian well.

total-head diagram $abcd$ (Kashef, 1965a, 1965b, 1970a, 1971). The vertical velocity decreases gradually from $(v_y)_{max}$ at the free surface (at point a, section ab) downward to $v_y = 0$ at point b. The tangent at c in the total-head diagram is vertical, and the tangent at d intersects the base at a distance $bb' = 2h_{br} - D_r$, where h_{br} is the base pressure head at b (Kashef, 1968). Since the parabola is a second-degree type and the area of the pressure-head diagram bcd is equal to the area of the total-head diagram $abcd$ minus the area of the position-head diagram abd, then

$$P_r = [D_r^2 - \tfrac{2}{3}(D_r - h_{br})D_r] - \tfrac{1}{2}D_r^2$$

or
$$P_r = \tfrac{2}{3}h_{br}D_r - \tfrac{1}{6}D_r^2 \quad \text{(gravity wells)} \tag{8.116}$$

If the pressure-head diagram is drawn with a vertical datum (bd in Fig. 8.53b), then the tangent at c makes a 45° angle with the horizontal, because the pressure-head diagram in Fig. 8.52a has been rotated 45°.

If pumping is extensive in relatively shallow artesian aquifers, then overpumping leads to the case shown in Fig. 8.54. At a vertical section distance r_1 from the center line of the well, the pressure-head diagram is linear and is similar to that shown in Fig. 8.52. At radius r_2, the P_r diagram is parabolic and resembles that shown in Fig. 8.53b. At radius r_o, the boundary between the artesian and gravity flow conditions, the P_r diagram is approximately linear and $P_o = \tfrac{1}{2}D_q^2$. The vertical boundary bd should in fact be slightly curved to accommodate the transition of the horizontal flow lines into curved ones (Kashef, 1971).

It should be noted that in developing the formulas given by Eq. (8.113), the vertical velocity components are considered and Dupuit's assumptions are eliminated. However, there is a relationship between the height h_r of the Dupuit curve at radius r and the magnitude of P_r at the same section. Although h_r is not the height of the true free surface, it has the following physical significance (Kashef, 1968):

$$h_r = \sqrt{2P_r} \tag{8.117}$$

General equations have been developed for the free surface, the base pressure heads, and the hydraulic head at any point in the flow medium. Ninety cases covering a wide range of practical parameters have been programmed and solved numerically by computer (Kashef, 1971) to locate the free surface and to determine the values of the base pressure heads for each case. The results are fitted in Eq. (8.111) to find refined values for C_r. These are tabulated in Table 8.15 and plotted in Fig. 8.55. When the values of $\bar{r}_e = r_e/h_u$ are 1.0 or less, the values of C_r depend on the value of $m_w = h_w/h_u$. In Table 8.15, the C_r values are given for these cases when $m_w = 0$ and $m_w = 0.8$. For any intermediate value of m_w, C_r can be determined by interpolation. Owing to the fact that cases where $\bar{r}_e < 4.0$ are not common in practice, they are not shown in Fig. 8.55.

TABLE 8.15 Tabulated Values of C_r

| | r_e | | | | | | | | | | |
$\dfrac{r - r_w}{r_e - r_w}$	0.25°	0.25†	0.5°	0.5†	1.0°	1.0†	3†	4†	6†	12†	40†
1.00	0.0000	0.0000	0.0000	0.0000	0.0000	0.0000	0.0000	0.0000	0.0000	0.0000	0.0000
0.60	0.0641	0.1322	0.0914	0.1274	0.0984	0.1058	0.0746	0.0701	0.0633	0.0543	0.0434
0.40	0.1239	0.2434	0.1666	0.2243	0.1758	0.1886	0.1360	0.1275	0.1155	0.0991	0.0790
0.30	0.1564	0.3050	0.2087	0.2839	0.2252	0.2442	0.1799	0.1689	0.1537	0.1320	0.1050
0.20	0.1875	0.3665	0.2524	0.3510	0.2842	0.3161	0.2404	0.2277	0.2087	0.1797	0.1426
0.14	0.2045	0.4018	0.2782	0.3942	0.3242	0.3693	0.2903	0.2787	0.2574	0.2229	0.1769
0.08	0.2191	0.4345	0.3018	0.4367	0.3658	0.4322	0.3561	0.3525	0.3312	0.2921	0.2324
0.04	0.2266	0.4516	0.3147	0.4627	0.3915	0.4763	0.4108	0.4235	0.4080	0.3756	0.3036
											(0.2761)†§
0.02	0.2294	0.4581	0.3198	0.4732	0.4019	0.4969	0.4400	0.4658	0.4448	0.4462	0.3733
											(0.3321)†§
0.00	0.2314	0.4620	0.3230	0.4814	0.4087	0.5153	0.4982	0.5020	0.5080	0.5522	0.5474
											(0.4657)†§

Note: The values in this table are plotted in Fig. 8.55

° $m_w = 0$.

† $m_w = 0.8$.

‡ For all values of m_w.

§ Shown dashed in Fig. 8.55.

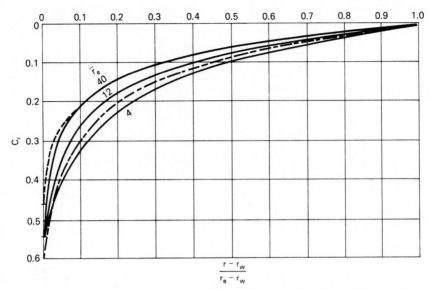

Figure 8.55 Relation between $(r - r_w)/(r_e - r_w)$ versus C_r. The solid-line curves (—) represent $\bar{r}_e$ values 4, 12, and 40 for all values of m_w, the dashed-line curve (----) represents $\bar{r}_e = 40$ when $m_w = 0.8$ and the line-dashed curve (—·—·—·—) represents the Babbit and Caldwell curve.

When $\bar{r}_e$ exceeds 1.0, it has been found that C_r values are almost independent of the m_w values. Three curves are shown in Fig. 8.55 corresponding to $\bar{r}_e = 4$, 12, and 40. For any other $\bar{r}_e$ value between these limits, the C_r values can be interpolated. When $\bar{r}_e$ reaches a high value, such as 40 and generally more than 25, the C_r curve depends on m_w for a range of small r values. The solid portion of the curve $\bar{r}_e = 40$ corresponding to values of $(r - r_w)/(r_e - r_w)$ between 0 and 0.1 (Fig. 8.55) corresponds to $m_w = 0$, and the dashed portion for the same range corresponds to $m_w = 0.8$. Apparently, these small differences can be ignored, and the solid curve can still be used for all values of r and m_w.

Equation (8.111) can thus be applied, but the C_r values should be obtained from Table 8.15 or Fig. 8.55 rather than from Fig. 8.51. Then,

$$\bar{D}_r = 1 - C_r \frac{1 - m_w^2}{\ln r_e/r_w} \ln \frac{\bar{r}_e}{0.1} \tag{8.118}$$

or

$$\bar{D}_r = 1 - \frac{C_r Q}{\pi K h_u^2} \ln \frac{\bar{r}_e}{0.1} \tag{8.118a}$$

Babbitt and Caldwell's curve is also plotted for comparison in Fig. 8.55, assuming that $r/r_e \approx (r - r_w)/(r_e - r_w)$. The curve lies between the curves corresponding to $\bar{r}_e = 4$ and 12, and its C_r value at $r = 0$ is 0.6, which is slightly larger than a range from 0.5 to 0.55, as shown in Table 8.15 and Fig. 8.55. When $\bar{r}_e \geqslant 4$,

$$\bar{D}_w \simeq 1 - \frac{(0.5 \text{ to } 0.55)Q}{\pi K h_u^2} \ln \frac{\bar{r}_e}{0.1} \qquad (8.119)$$

The base pressure heads can be determined after the free surface is located. Once D_r is determined from Eq. (8.118), h_{br} is determined from Eq. (8.116) as follows:

$$h_{br} = \frac{P_r + \frac{1}{6}D_r^2}{\frac{2}{3}D_r} = \frac{3P_r}{2D_r} + \frac{1}{4}D_r \qquad (8.120)$$

If in Eq. (8.113d), $\frac{1}{2}h_u^2$ is substituted for P_e, then P_r in Eq. (8.120) can be expressed as

$$P_r = \frac{h_u^2}{2} - \frac{Q \ln r_e/r}{2\pi K}$$

The total head $h_{r,y}$ at any point r, y (Fig. 8.53) is determined as follows (Kashef, 1968):

$$h_{r,y} = h_{br} + \frac{y^2(D_r - h_{br})}{D_r^2} \qquad (8.121)$$

In order to determine P_r, h_{br}, and $h_{r,y}$ at any section, Eqs. (8.113d), (8.120), and (8.121) are applied. The magnitude of K necessary to apply these equations can be determined from Eq. 8.107 in terms of the known values of Q, r_e and r_w and the recorded values of h_u and h_w.

Another well-known unified well equation used for both artesian and gravity wells is derived from Dupuit's formula [Eq. (8.108)] as follows:

$$\frac{Q}{\pi K} = \frac{h_{r1}^2 - h_{r2}^2}{\ln r_1/r_2} = \frac{(h_{r1} - h_{r2})(h_{r1} + h_{r2})}{\ln r_1/r_2}$$

or
$$\frac{Q}{2\pi K} = \frac{\frac{1}{2}(h_{r1} + h_{r2})(h_{r1} - h_{r2})}{\ln r_1/r_2} \qquad (8.122)$$

If h_{r1} and h_{r2} are recorded from the field and $\frac{1}{2}(h_{r1} + h_{r1})$ is equated to $\bar{D}_q$, which is an equivalent thickness of an artesian aquifer, then Eq. (8.122) reduces to the well-known Thiem equation:

$$\frac{q}{2\pi K \bar{D}_q} = \frac{h_{r1} - h_{r2}}{\ln r_1/r_2} \qquad (8.123)$$

where $\bar{D}_q = (h_{r1} + h_{r2})/2$ in dealing with gravity wells and $\bar{D}_q$ is the artesian aquifer thickness D_q in dealing with artesian wells. Equation (8.123) has the same form as Eq. (8.4b) used for artesian wells. However, it should be noted that the recordings of h_{r1} and h_{r2} from shallow observation wells installed in an unconfined aquifer are in fact those of D_{r1} and D_{r2}. If the observation wells are relatively deep and have their open bottom ends at (r_1, y_1) and (r_2, y_2), the field recordings would be equivalent to $h_{(r1,y1)}$ and $h_{(r2,y2)}$, representing the total heads at the bottoms of these observation wells rather than the heights of the free surface. If the observation wells are screened throughout their depths, the recordings cannot be used because they represent the average total heads throughout their depths.

Using Eq. (8.108), Theis (1963) made the following analysis:

$$T = Kh_u = \frac{Q \ln r_1/r_2}{2\pi(h_{r1}^2 - h_{r1}^2)/2h_u} = \frac{Q \ln r_1/r_2}{2\pi[(h_{r1}^2/2h_u + h_u/2) - (h_{r2}^2/2h_u + h_u/2)]}$$

(8.124)

Substituting the drawdowns s $(=h_u - h_r)$, then

$$s_{dr1}^2 = h_u^2 - 2h_{r1}h_u + h_{r1}^2$$

or

$$\frac{s_{dr1}^2}{2h_u} = \tfrac{1}{2}h_u - h_{r1} + \frac{h_{r1}^2}{2h_u}$$

and

$$s_{dr2}^2 = h_u^2 - 2h_{r2}h_u + h_{r2}^2$$

or

$$\frac{s_{dr2}^2}{2h_u} = \tfrac{1}{2}h_u - h_{r2} + \frac{h_{r2}^2}{2h_u}$$

From the preceding two equations, Eq. (8.124) may be written as

$$T = \frac{Q \ln r_1/r_2}{2\pi[(s_{dr1}^2/2h_u + h_{r1}) - (s_{dr2}^2/2h_u + h_{r2})]}$$

Since $h_{r1} - h_{r2} = s_{dr2} - s_{dr1}$, then

$$T = \frac{Q \ln r_1/r_2}{2\pi[(s_{dr2} - s_{dr2}^2/2h_u) - (s_{dr1} - s_{dr1}^2/2h_u)]}$$

(8.125)

In other words, applying the artesian well formulas to gravity wells, the recorded drawdowns s_d should be adjusted by subtracting $s_d^2/2h$. This correction is used only when $s_d < 25$ percent of $\overline{D}_q$. It is also used in gravity wells under transient conditions, in order to use the Theis and Jacob equations developed for artesian wells. This gives a clear idea about the origin of this widely used adjustment. It is recommended, however, that the method summarized in Table 8.15 and Fig. 8.55 be used. This method can also be used for transient states under certain restrictions, as explained later.

Overpumped artesian wells It has been shown that the unified well formulas [Eq. (8.113)] are used for artesian, gravity, and overpumped artesian wells by substituting the appropriate value of P_r [Eqs. (8.115) and (8.116)]. Overpumped artesian wells (Fig. 8.54) combine both artesian and gravity flow systems (Kashef, 1971). At a distance $r_o < r \leqslant r_e$, the P_r value is given by Eq. (8.115). The boundary between the two systems of flow is at radius r_o (Fig. 8.54) and $P_o = \frac{1}{2}D_q{}^2$, where D_q is the depth of the artesian aquifer. As stated earlier, the boundary at r_o is slightly curved (Kashef, 1971), but in practice it is assumed vertical. The flow zone between r_w and r_o may be looked on as identical to a gravity well drilled in the center of an island of radius r_o and surrounded by a water body of depth D_q above the lower impervious boundary.

Applying the unified well equation [Eq. (8.113)] for the gravity flow system between r_o and r_w,

$$\frac{Q}{2\pi K} = \frac{P_o - P_r}{\ln r_o/r} \tag{8.126a}$$

Here $r_w \leqslant r \leqslant r_o$, and P_r is given by Eq. (8.116). In addition,

$$\frac{Q}{2\pi K} = \frac{P_e - P_r}{\ln r_e/r} \tag{8.126b}$$

Here $r_o \leqslant r \leqslant r_e$, and P_r is given by Eq. (8.115). Thus,

$$\frac{Q}{2\pi K} = \frac{P_e - P_w}{\ln r_e/r_w} \tag{8.126c}$$

and

$$\frac{Q}{2\pi K} = \frac{P_o - P_w}{\ln r_o/r_w} \tag{8.126d}$$

The value of K for the aquifer can be determined from Eq. (8.126c), noting that $P_e = h_u D_q - \frac{1}{2}D_q^2$, $P_w = \frac{1}{2}h_w^2$, and $P_o = \frac{1}{2}D_q^2$. Using these values of P and Eq. (8.126d), the radius r_o of the boundary between the artesian and gravity flow systems is given by

$$r_o = r_w \exp\left[\frac{\pi K}{Q}(D_q^2 - h_w^2)\right] \tag{8.127}$$

Multiple gravity wells The individual drawdowns resulting from the effects of several artesian wells are added algebraically because the governing equation (Laplace equation) is linear and the *flow medium remains unchanged* (because it is bounded by confining layers). In gravity flow systems, the flow domain changes in each well as a result of the water drainage above the free surface. Although the flow is governed by the same linear equation, the drawdowns due to individual wells cannot be added to find the resultant drawdown (Kashef, 1970a, 1970b). This conclusion can be visualized in the analysis of a gravity flow system in two-dimensional cases. The earth section shown in Fig. 8.56 is subjected once to a water body on the left side (Fig. 8.56*a*) and then to a water body on the right side (Fig. 8.56*b*). Although the sections are the same geometric shape, the flow domain is different in each of the two cases. If both cases are superimposed (Fig. 8.56*c*) by adding the drawdowns, the resultant free surface is lower than plane $d_1 d_2$. It is obvious that if both water bodies exist at the same time, a hydrostatic condition develops and the free surface coincides with surface $d_1 d_2$ (Fig. 8.56*c*). Therefore, a restricted form of superposition should be applied in the case of multiple gravity wells. This is performed by adding the decrements $\Delta h_{br} = h_u - h_{br}$ of the base pressure heads due to the individual effects of each well. Across any vertical section, the total pressure is equal to $\gamma_w P_r$, where γ_w is the unit weight of water and P_r is the area of the pressure-head diagram across the vertical section. The total pressures can then be superimposed as force vectors. Following this reasoning, the resultant effect of multiple gravity wells can be obtained, as explained in the following paragraphs.

A proper grid system is drawn to represent a horizontal plane covering all wells and extending over all their individual influence zones; that is, the boundary of such a grid should include all the circles of influence of the individual wells (Kashef, 1970*a*). If we select axes *x* and *y* in the

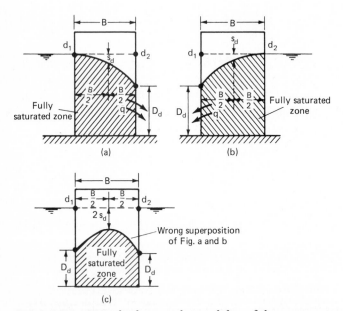

Figure 8.56 Example showing the invalidity of the superposition of drawdowns in gravity flow systems. (*a*) Seepage through a vertical dam (upstream to the left); (*b*) seepage through same vertical dam (upstream to the right); (*c*) water levels the same on both sides of the dam.

grid, at a point (x, y), the change in the base pressure head h_{br} from the original h_u will be (for *one* well)

$$\Delta h_{br} = h_u - h_{br} \qquad (8.128)$$

h_{br} is given by Eq. (8.120), in which P_r is calculated from Eq. (8.113d) and D_r from Eq. (8.118). The distance between any point (x, y) in the grid to the well under consideration gives the values of r in Eq. (8.113d). The change $\Delta P_r = P_e - P_r$ at the vertical section through point (x, y) can be calculated from Eq. (8.113a):

$$\Delta P_r = P_e - P_r = \frac{Q}{2\pi K} \ln \frac{r_e}{r} \qquad (8.129)$$

At point (x, y), the resultant value of the base pressure head $h_{b(x,y)}$ due to all wells is equal to

$$h_{b(x,y)} = h_u - \sum_{i=1}^{n_w} (\Delta h_{br})_i \qquad (8.130)$$

where n_w is equal to the number of wells. Also, the resultant value of the areas of the pressure-head diagrams at a vertical section through point (x, y) is given by

$$P_{x,y} = P_e - \sum_{i=1}^{n_w} (\Delta P_r)_i \qquad (8.131)$$

The r and Q values in Eqs. (8.129) and (8.131) are different depending on the distance r of each well from a certain point (x, y) and the pumping rate Q of each individual well.

If we introduce the assumption of parabolic distribution of the pressure-head diagram, then the height of the final free surface $D_{x,y}$ can be computed from Eq. (8.120) as follows:

$$h_{b(x,y)} = \frac{3P_{x,y}}{2D_{x,y}} + \frac{D_{x,y}}{4}$$

Then, solving for $D_{x,y}$,

$$D_{x,y} = 2h_{b(x,y)} - \sqrt{4h_{b(x,y)}^2 - 6P_{x,y}} \qquad (8.132)$$

The procedure is repeated at all nodes of the grid. If more refinement is needed, the mesh size of the grid should be reduced. The computations are best performed by writing an appropriate computer program for rapid and efficient computations. Although the method is approximate, it is very helpful in groundwater management to find the optimum conditions of a group of gravity wells. The computer analysis is essential if alternative solutions are sought by changing the number, location, pattern, and capacities of the group of wells.

EXAMPLE 8.14 Well I (Fig. 8.57) is pumping at a rate $Q = 1.25$ m³/min (1800 m³ per day) from an unconfined aquifer in which the height above the lower impervious boundary of the initial water table is $h_u = 30$ m. The flow reaches steady state when the drawdown s_{dw} in the well is recorded as 3.542 m. If $r_w = 0.25$ m and the T value from a previous pumping test is 0.41 m²/min, determine (a) the radius of influence r_e, (b) the drawdowns at vertical sections through a, b, and c at distances 25, 75, and 125 m, respectively, from the center line of the well, and (c) the depth of the outflow surface. If two other wells II and III are placed at 50 m from the well in such a way that the three wells are along a straight line, find the resultant drawdown at the vertical sections through points a and b (Fig. 8.57).

Solution

WELL I

PART a: Using the correct form of Dupuit's equation [Eq. (8.107)], then

$$r_e = r_w \exp\left[\frac{\pi K}{Q}(h_u^2 - h_w^2)\right]$$

$$= 0.25 \exp\left[\frac{\pi \times 0.41}{1.25 \times 30}(\overline{30}^2 - \overline{26.458}^2)\right] = 240.432 \text{ m}$$

(Note that $K = T/h_u = 1.367 \times 10^{-2}$ m/min and $h_w = 30 - 3.542 = 26.458$ m.)

PART b: Since $\bar{r}_e = r_e/h_u = 240.432/30 \approx 8$, C_r is obtained by interpolation from Fig. 8.55 between the values corresponding to $\bar{r}_e = 4$ and 12 (or from Table 8.15 between $\bar{r}_e = 6$ and 12):

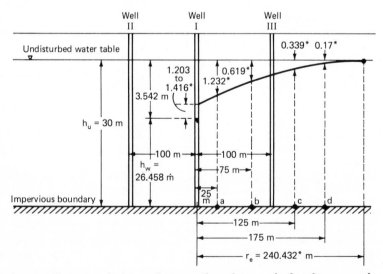

Figure 8.57 Data for Example 8.14. The values marked with an asterisk in this figure were determined from the solution.

r, m	$\dfrac{r - r_w}{r_e - r_w}$	$\bar{r}_e = 12$	$\bar{r}_e = 4$	$\bar{r}_e = 8$ (interpolation)
			C_r	
25	0.103	0.262	0.320	0.290
75	0.311	0.130	0.162	0.146
125	0.52	0.070	0.090	0.080
175	0.728	0.035	0.044	0.040

Applying Eq. (8.118) and introducing the appropriate C_r values from above $[m_w = h_w/h_u = 26.458/30 = 0.882$ and $(1 - m_w^2) = 0.222]$, then at point a distanced 25 m from the well,

$$\bar{D}_r = 1 - \frac{0.29 \times 0.222}{\ln 961.728} \ln \frac{8}{0.1} = 0.959 \qquad \text{and} \qquad D_r = 28.77 \text{ m}$$

Then $s_{25} = 30 - 30 \times \bar{D}_r = 1.232$ m. Similarly,

At point b: $\quad s_{75} \;= 30 - 30 \times 0.979 = 0.619$ m $\qquad$ and $\qquad D_r = 29.37$ m
At point c: $\quad s_{125} = 30 - 30 \times 0.989 = 0.339$ m $\qquad$ and $\qquad D_r = 29.67$ m
At point d: $\quad s_{175} = 30 - 30 \times 0.994 = 0.17$ m $\qquad$ and $\qquad D_r = 29.83$ m

PART C: Applying Eq. (8.119),

$$\bar{D}_w \simeq 1.0 - \frac{(0.5 \text{ to } 0.55) \times 1.25}{\pi \times 0.41 \times 30} \ln \frac{8}{0.1} = 0.929 \text{ to } 0.922$$

The depth of the outflow surface is $h_u \times \bar{D}_w - h_w$, and therefore, it ranges from 1.203 to 1.416 m.

WELLS I, II, AND III

From Fig. 8.57, point a is at $r = 25$, 125, and 75 m from wells I, II, and III, respectively, and point b is at $r = 75$, 175, and 25 m from wells I, II, and III, respectively.

Since the wells are operating at the same rate in the same aquifer, the values of P_r, ΔP_r, h_{br}, and Δh_{br} are tabulated below for the distances 25, 75, 125, and 175 m:

r, m	D_r, m°	ΔP_r, m²†	P_r, m²‡	h_{br}, m§	Δh_{br}, m¶
25	28.77	32.94	417.06	28.937	1.063
75	29.37	16.95	433.05	29.459	0.541
125	29.67	9.52	440.48	29.686	0.314
175	29.83	4.62	445.38	29.853	0.147

° From part b.
† From Eq. (8.129).
‡ $P_r = P_e - P_r = \frac{1}{2}h_u^2 - \Delta P_r$.
§ From Eq. (8.120).
¶ $\Delta h_{br} = h_u - h_{br}$.

Applying Eq. (8.130),

At point a: $h_{b(x,y)} = 30 - (1.063 + 0.314 + 0.541) = 28.082$ m
At point b: $h_{b(x,y)} = 30 - (0.541 + 0.147 + 1.063) = 28.249$ m

Applying Eq. (8.131),

At point a: $P_{x,y} = \dfrac{900}{2} - (32.94 + 9.52 + 16.95) = 390.59$ m²

At point b: $P_{x,y} = \dfrac{900}{2} - (16.95 + 4.62 + 32.94) = 395.49$ m²

Applying Eq. (8.132),

At point a: $D_{x,y} = 27.688$ m

and the resultant drawdown equals $30 - 27.688 = 2.312$ m.

At point b: $D_{x,y} = 27.878$ m

and the resultant drawdown equals $30 - 27.878 = 2.122$ m.

Note: If the superposition is used to add up the individual drawdowns, the drawdown at a would be $1.232 + 0.339 + 0.619 = 2.19$ m and that at b would be $0.619 + 0.17 + 1.232 = 2.021$ m. This means that the differences in the results are 12.2 cm (about 5 in) at a and 10.1 cm (about 4 in) at b. However, the order of magnitudes of the differences may be greater at other points or under other conditions.

The method can be extended to cases of overpumped wells after delineating the boundaries between the artesian and gravity flow systems. The problem is initially solved as if all the wells are artesian (Sec. 8.7); the contours with P_r intensities $\leqslant \frac{1}{2}D_q^2$ (D_q is the thickness of the artesian aquifer) delineate the regions within which gravity flow takes place. Wells within these zones are treated as gravity wells.

Transient Conditions

The drawdown curve that is the free surface in a gravity well changes over time under transient conditions. At time $t = 0$, the water table is undisturbed, and at times $t_1, t_2, \ldots$, etc., the free surface is lowered to the successive locations shown in Fig. 8.58. If we assume a vertical prism *abcd* in the figure within the saturated zone with volume $V_I = \pi(r_2^2 - r_1^2)h_u$, then at time $t = t_1$ when the saturation height is decreased, the saturated volume of the prism decreases to $V_I - \Delta V_I = \pi(r_2^2 - r_1^2)(h_u - s_{d1})$. At time $t = t_2$, it decreases to $V_I - \Delta V_I - \Delta V_{II} = \pi(r_2^2 - r_1^2)(h_u - s_{d2})$, and so on. Between times $t = 0$ and $t = t_1$, volume ΔV_I is drained by gravity, yielding an additional amount of water over that derived from storage within the prism due to the decline in head.

In the prism shown in Fig. 8.58, the water released due to a head

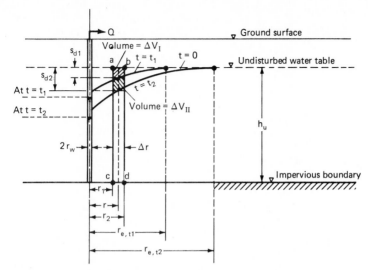

Figure 8.58 Graphic explanation of the specific yield resulting from lowering of the free surface in gravity wells.

decline s_{d1} consists of two portions: one portion is related to S_y as a result of drainage by gravity (Sec. 6.3) and the other is released from storage and related to the coefficient of storage S as if the aquifer were artesian. The total water yield is equal to $\Delta V_1(S_y + S) = \pi(r_2^2 - r_1^2)s_{d1}(S_y + S)$. As stated earlier (Sec. 6.5), S_y values range from about 0.1 to 0.30, while S values range from about 10^{-5} to 10^{-3}. In most cases, the S value is negligible as compared with the value of S_y.

Assuming that the aquifer is extensive and that there is no source of recharge or leakage, steady-state conditions can never be reached. The free surface continues to drop at a decreasing rate by continuous pumping, thus expanding the zone of influence over time. As stated in Sec. 6.3, gravity drainage does not take place immediately, especially in stratified sediments, and the time lag due to the delayed response should be reflected in the recording of water levels during the early stages of pumping. The coefficient of storage during this early period is approximately equal to S, equivalent to an artesian aquifer of the same material as that of the unconfined aquifer. As time passes, the rate of drop of the drawdown curve decreases when the drained water starts to reach the free surface. After a relatively long time, the rate of drawdown increases when gravity drainage keeps pace with the declining water levels (Walton, 1970).

Field recordings of log s_d versus log t (Fig. 8.59) produce a sigmoid drawdown curve on a log-log plot. It indicates three distinct segments. The early segment of the curve within a period of a few minutes may fit

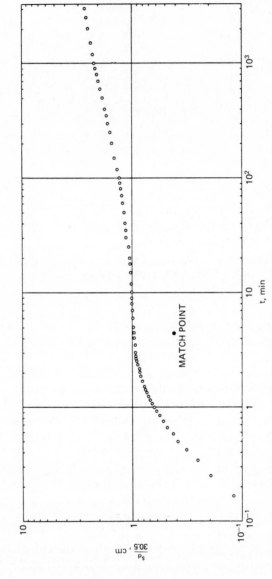

Figure 8.59 Logarithmic plot of drawdown versus time for observation well 139, near Fairborn, Ohio. At the match point $(4\pi Ts)/Q = 1.0$, $(4\pi Tt)/(r^2 S_y) = 1.0$, $t = 4.4$ min, $s_d = 12.9$ cm, and $r/D_t = 0.4$. (W. C. Walton, "Application and Limitation of Methods Used to Analyze Pumping Test Data," Water Well J., pts. 1 and 2, vol. 14, 1960.)

the Theis type curve for artesian aquifers (Fig. 8.18). In the case shown in Fig. 8.59, this period did not exceed about 6 min when the drawdown reached about 30 cm at a rate of about 5 cm/min. The intermediate segment displays a relatively flat portion of the curve, indicating that gravity drainage is still in the process of moving downward to reach the free surface. During this intermediate period, the drawdowns become more or less stationary, an almost steady-state condition. Between $t = 6$ min and $t = 350$ min (Fig. 8.5), the drawdown dropped only an additional 22 cm at a rate of 0.064 cm/min (compare with the earlier rate of 5 cm/min). During the time period of the third segment, gravity drainage increases and the storage coefficient S reaches the specific yield S_y. At this stage the delayed gravity drainage ceases to influence the drawdown of the gravity well. The minimum time t_{min} required to reach this latter stage is determined by various methods explained later. Beyond time t_{min}, the true water-table conditions exist.

Delayed yield and the flow toward gravity wells under transient conditions have been investigated by Boulton (1954a, 1954b, 1963, 1964), Stallman (1961a, 1961b), Prickett (1965), Kashef (1970), Neuman (1972), Ehlig and Halepaska (1976), and several others. The effects of anisotropy and partial penetration of pumped and observation wells have been investigated by Stallman (1965), Neuman (1975), and others.

Delayed yield also may be affected by the rate at which air can move into the zone immediately above the water table to occupy the drained volume. Sometimes air movement is restricted, such as in the case of initially wet soil layers above the water table. In such cases, negative air pressures develop, producing a delay in the drainage procedure (Bouwer, 1978).

Boulton's solution   The material in the references cited in the preceding paragraph is compiled and simplified in this section and presented as a single method. Although the method is usually accredited to Boulton, it includes simplifications, graphs, and findings originated by several others. The drawdown s_d is given by (Boulton, 1954b) as either

$$s_d = \frac{Q}{4\pi T} W\left(u_y, \frac{r}{D_t}\right) \quad \text{(late time)} \qquad (8.133a)$$

or
$$s_d = \frac{Q}{4\pi T} W\left(u, \frac{r}{D_t}\right) \quad \text{(early time)} \qquad (8.133b)$$

In Eqs. (8.133a and b), $T = Kh_u$, where h_u is the height of undisturbed water table above the impervious boundary. In addition,

$$u = \frac{r^2 S}{4Tt} \qquad (8.134)$$

where S is an early time coefficient of storage, which is equivalent to an artesian aquifer. Also,

$$u_y = \frac{r^2 S_y}{4Tt} \tag{8.135}$$

where S_y is the true specific yield (corresponding to the true gravity flow conditions). And

$$D_t = \sqrt{\frac{t_d T}{S_y}} \tag{8.136}$$

where $t_d =$ is known as the delay index. Its value depends on the soil type (Prickett, 1965) and may be estimated from Fig. 8.60a or from the results of a pumping test.

Equations (8.133a and b) are represented in graphic form as families of delayed-yield curves in Fig. 8.61. The curves are drawn on log-log paper for various r/D_t values between 0.1 and 3.0 with ordinates $4\pi T s_d/Q$ and two different abscissas: $1/u = 4Tt/r^2S$ (upper scale) for type A curves to the left of the break (where the numbers are written) and $1/u_y = 4Tt/r^2S_y$ (lower scale) for type Y curves to the right of the break. Both types A and Y curves are asymptotic to two Theis type curves: the left curve (Fig. 8.61) corresponds to the artesian condition correspond-

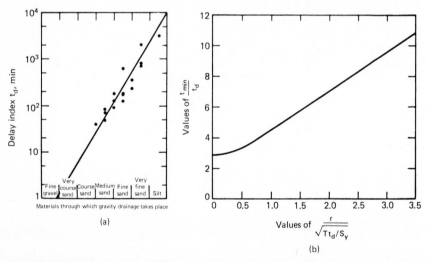

(a)

(b)

Figure 8.60 Empirical method for estimating the minimum length of a pumping test in an unconfined aquifer. (a) Delay index versus soil materials; (b) curve for estimating t_{min} at which effects of delayed gravity cease to influence drawdown of a pumping gravity well. (*T. A. Prickett, "Type-Curve Solution to Aquifer Tests under Water-Table Conditions," Ground Water, vol. 3, no. 3, 1965.*)

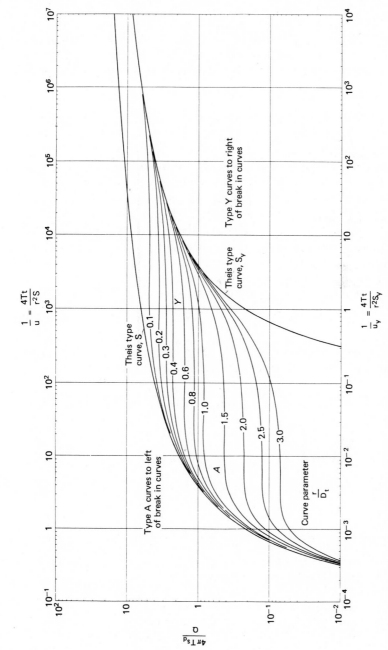

Figure 8.61 Delayed-yield type curves. (*N. S. Boulton, "Analysis of Data from Non-equilibrium Pumping Tests Allowing for Delayed Yield from Storage," Proc. Inst. Civ. Eng. (Lond.), vol. 26, 1963.*)

Theis type curve, S

Type A curves to left of break in curves

$\frac{1}{u} = \frac{4Tt}{r^2S}$

0.1
0.2
0.3
0.4
0.6
0.8
1.0
1.5
2.0
2.5
3.0
A
Y

Curve parameter $\frac{r}{D_t}$

$\frac{Q}{4\pi T s_d}$

Theis type curve, S_y

Type Y curves to right of break in curves

$\frac{1}{u_y} = \frac{4Tt}{r^2S_y}$

ing to the coefficient of storage S, whereas the right curve corresponds to S_y.

In order to evaluate S, T, and S_y, the matching method is used. The field plot of log s_d versus log t (Fig. 8.59) is drawn on log-log paper of the same scale as that of the theoretical curves (Fig. 8.61). Keeping the axes parallel, the early time segment is matched on type A curves and a match point is located. The coordinates of this match point are s_{d1}, t_1, $W(u, r/D_t)$, r/D_t, and $1/u$. Using Eq. (8.133b), T is determined. If we substitute this value of T and the coordinates t_1 and u of the match point into Eq. (8.134), S can be calculated. This value of S corresponds to the early time when the gravity drainage has not had enough time to develop. Moving the field curve to the right and along the curve corresponding to the match value r/D_t (or an interpolated curve), another match point is obtained with a new set of coordinates: s_{d2}, t_2, $W(u_y, r/D_t)$, and $1/u_y$. Substituting s_{d2} and $W(u_y, r/D_t)$ into Eq. (8.133a), T is again calculated, and it should be almost equal to the value of T corresponding to the first match point. Substituting T, u_y, and t_2 in Eq. (8.135), S_y is obtained. The specific yield S_y corresponds to the third segment of the curve (type Y) and represents the true value of the storage coefficient of the unconfined aquifer. Pumping tests are then conducted over a long period of time in order to plot the third segment of the curve.

EXAMPLE 8.15   The field plot shown in Fig. 8.59 was given by Lohman (1972) for an observation well at radius $r = 22.25$ m from a well pumping at a constant rate $Q = 4.09$ m³/min (1080 gal/min). Find S_y and t_{min}.

Solution   The early-time match point (not shown on Fig. 8.59) gives an S value of 3×10^{-3}. The late-time match point (shown in Fig. 8.59) has the coordinates $s_d = 12.9$ cm, $W(u_y, r/D_t) = 1.0$, $1/u_y = 1.0$, $t = 4.4$ min, and $r/D_t = 0.4$. Then, from Eq. (8.133a),

$$0.129 = \frac{4.09}{4\pi T} \times 1.0 \quad \text{or} \quad T = 2.523 \text{ m}^2/\text{min}$$

which is equal to that determined from the first match point.

The specific yield S_y is calculated from Eq. (8.135) as

$$S_y = \frac{4Tt}{r^2(1/u_y)} = \frac{4 \times 2.523 \times 4.4}{22.25^2 \times 1.0} = 0.09$$

The delay index t_d and the time t_{min} necessary to reach the true gravity-well conditions are determined for this example as follows:

The second match point has the coordinate $r/D_t = 0.4$, and since $r = 22.25$ m, then $D_t = 55.625$ m. From Eq. (8.136),

$$\text{Delay index } t_d = \frac{D_t^2 S_y}{T} = \frac{55.625^2 \times 0.09}{2.523} = 110.4 \text{ min}^\circ$$

$^\circ$ Figure 8.60a shows that the aquifer material is medium sand. In this example, the real aquifer material is glacial sand and gravel (Lohman, 1972).

Figure 8.60b gives the relationship between t_{min}/t_d and $r/D_t = r/\sqrt{Tt_d/S_y}$ as developed by Prickett (1965) on the basis of data on unconfined aquifers in four states. The graph gives a reasonable value for t_{min} that can be used in planning a pumping test period. Since $r/D_t = 0.4$ (from the coordinates of the match point), the point on the field curve that merges into the Y-type curves (Fig. 8.60) corresponds to t_{min} from Fig. 8.60b:

$$\frac{t_{min}}{t_d} = 3.15 \quad \left(\text{corresponding to } \frac{r}{D_t} = 0.4 \right)$$

Therefore, $\quad t_{min} = 3.15 t_d = 3.15 \times 110.4 = 348$ min(5.8 hr)

After 5.8 h, the effects of delayed yield are negligible and the true gravity-well conditions prevail. For each observation well, t_{min} should be different because r changes.

Once T, S_y, and t_{min} are determined from a pumping test, as has been explained, a Theis type curve can be used directly (Walton, 1970) to trace the free surface after time $t > t_{min}$ by using the determined values of T and S_y and one of the appropriate type curve (Figs. 8.18, 8.19, and 8.20).

Additional details on the analysis of gravity wells under transient conditions In constructing the families of curves given in Fig. 8.61, the tangents at the break points are horizontal. Theoretically, the equations from which Eq. (8.133) is derived are strictly valid when N_s approaches infinity, where

$$N_s = \frac{S + S_y}{S} \tag{8.137}$$

In practice, when $N_s > 100$, the tangents at the break points (Fig. 8.60) are still almost horizontal with very gentle slopes. Prickett (1965) indicated that the curves may still be valid for N_s values as small as 6.5.

It should be noted that in matching the field curve of log s_d versus log t on the early type A curves, the first time period is very short (just a few minutes). The water-level measurements during this early period require special techniques (Walton, 1970).

The U.S. Bureau of Reclamation (1977) used the thickness of aquifer h_u rather than the parameter D_t in Boulton's solution. This approach changes the magnitude of the delay index to $t_d = h_u S_y / K$ and the parameter r/D_t to r/h_u. This approximation introduces no obvious benefit to the procedure.

It should also be noted that type A curves (Fig. 8.61) are almost the same as those for leaky confined aquifers (Fig. 8.29) if the leakage factor B_k is replaced by the parameter D_t.

The determined values of S_y and T are in fact average values of the characteristics of the aquifer material in the drained regions above the drawdown produced by the pumping tests. The accuracy of the aquifer test results are affected only slightly by the idealized assumptions, such as uniform thickness of the aquifer, horizontal boundaries, and homogeneity of aquifer materials. Although most aquifers are anisotropic, the effect on drawdowns is minor (U.S. Bureau of Reclamation, 1977) when the test well and its observation wells are fully penetrating. However, if the aquifer is strongly anisotropic or the wells are partially penetrating, other solutions are available, and some of these are summarized in graphic form. The reader should examine the references cited at the beginning of this section in addition to the references cited in Walton (1970) for more details concerning these methods. If the observation wells are selected with distances $r > 1.5$ to $2.0h_u$, the effects of partial penetration and screen losses will be negligible.

The assumption of an infinitely extensive aquifer is practically never realized because of the existence of boundaries (recharge streams and vertical barrier boundaries). In these cases, the method of images explained in Sec. 8.5 for artesian wells is used for gravity wells with certain restrictions: (1) the drawdown in the well should be no more than 50 percent of the original saturated thickness h_u of the aquifer (Hantush, 1965; Walton, 1970) and (2) the recorded drawdowns should be adjusted before use by means of the following equations:

$$s_{dj} = s_d - \frac{s_d^2}{2h_u} \qquad (8.138)$$

where s_{dj} = adjusted drawdown at radius r

s_d = measured drawdown in an observation well distance r from the pumped well

h_u = undisturbed height of the water table above the impervious boundary

The origin of Eq. (8.138) has been illustrated by Eq. (8.125) on the basis of steady states of flow. However, it is also used for transient conditions when s_d/h_u varies between 0.1 and 0.25. When $s_r/h_e < 0.1$, the artesian formulas can be used without any adjustments of the drawdown. When s_d/h_u exceeds 0.25, Boulton's method is recommended (U.S. Bureau of Reclamation, 1977). Even when Boulton's solution is applied, the records are still commonly adjusted using Eq. (8.138) before being used (Lohman, 1972; and Walton, 1970) in order to compensate for variations in T due to decreases in the saturated depth.

In most analyses of gravity wells, it is assumed that all water pumped from a well comes from storage within the aquifer and/or water yield due to gravity drainage of the zones above the continuously lowering free surface. Aquifers are seldom isolated, and usually they are recharged by various means, such as precipitation, irrigation water, and leakage and seepage from water bodies such as streams and lakes. The effects of these recharges as well as discharges are usually added by superposition.

Pumping tests are usually conducted to determine the following (U.S. Bureau of Reclamation, 1977; Walton, 1970):

1. Aquifer parameters and location of its boundaries
2. The general efficiency of the well before its completion
3. The correct pump capacity and setting of the well immediately after completion
4. The deterioration of the well following a period of extensive use

Multiple gravity wells under transient conditions After the aquifer parameters are determined, a group of gravity wells can be analyzed approximately to find the resultant base pressures and drawdowns at any vertical section. The wells in a group may be discharge or recharge wells or their images. The method presented in the following is based on the restricted method of superposition presented (Sec. 8.7) for wells flowing at a steady state.

A grid system is used, and computerized calculations are preferred. In the analysis, the time of pumping of each gravity well should exceed $t = t_{min}$ (Fig. 8.60b).

It has been shown (Kashef, 1970b) that for a single gravity well at time $t > t_{min}$,

$$\frac{Q}{2\pi K} = \frac{P_e - P_{r,t}}{W(u_y)} \tag{8.139}$$

where $W(u_y)$ = the well function corresponding to $u_y = r^2 S_y / 4Tt$ [Eq. (8.135)]

P_e = the area of the pressure-head diagram at time t and radius $r_{e,t}$ °

$P_{r,t}$ = the area of the pressure-head diagram at radius r and time t

$$P_{r,t} = \tfrac{2}{3} h_{b(r,t)} D_{r,t} - \tfrac{1}{6} D_{r,t}^2 \tag{8.140}$$

where $h_{b(r,t)}$ = base pressure at r and t

$D_{r,t}$ = height of free surface above the impervious base at r and t

After a relatively long period of time (Kashef, 1970b), Eq. (8.139) changes to

$$\frac{Q}{2\pi K} = \frac{P_e - P_{r,t}}{\ln \left[(2.25Tt/S)/r\right]^{1/2}} \tag{8.141a}$$

or

$$\frac{Q}{2\pi K} = \frac{P_e - P_{r,t}}{\ln r_{e,t}/r} \tag{8.141b}$$

° $P_e = \tfrac{1}{2} h_u^2$ [see Eq. (8.114)]

where $r_{e,t}$ is the radius of influence at time t:

$$r_{e,t} = \left(\frac{2.25Tt}{S}\right)^{1/2} \tag{8.142}$$

The height of the water level in the well at time t and radius r_w is given by

$$h_{w,t} = \sqrt{h_u^2 - \frac{Q}{2\pi K} W(u_w)} \tag{8.143a}$$

where

$$u_w = \frac{r_w^2 S}{4Tt}$$

After a relatively long period of time, $h_{w,t}$ is given by

$$h_{w,t} = \sqrt{h_u^2 - \frac{Q}{\pi K} \ln \frac{r_{e,t}}{r_w}} \tag{8.143b}$$

Using Eq. (8.141b), $P_{r,t}$ can be obtained at any r at time t. The height $D_{r,t}$ can be obtained from Eq. (8.118) by replacing D_r with $D_{r,t}$ and r_e with $r_{e,t}$. The value of $m_w = h_{w,t}/h_u$ is given by Eq. (8.143b). At the vertical section r, $h_{b(r,t)}$ is obtained from Eq. (8.140) after $P_{r,t}$ and $D_{r,t}$ are calculated. In other words, the problem reduces to a steady-state problem at a certain time t by considering r_e to equal $r_{e,t}$ [Eq. (8.142)]. The same general analysis given in Sec. 8.7 can thus be followed.

If more precision is required, Eq. (8.139) is used to find $P_{r,t}$. The type Y Theis curve (Fig. 8.61) is used to find $D_{r,t}$ ($=h_u - s_d$). The remainder of the procedure is essentially the same as explained in the previous paragraph, except that $h_{w,t}$ is determined from Eq. (8.143a). If recharge wells are among the group of wells or their images, the analysis is also essentially the same, except for changing the signs.

It should be emphasized again that S_y and T should be predetermined and that the time of pumping should exceed t_{min} in order to use the type Y Theis curve (Fig. 8.61).

The effects of natural groundwater flow on a group of wells can also be approximated. Usually, the natural flow is considered as steady-state flow during the relatively short period of well operation (Sec. 8.7). If we select the appropriate axes in the grid system, P_x and h_{bx} at x are obtained in a similar manner to that explained in Sec. 5.5. Since the distribution of the pressure diagram is parabolic, D_x can be computed in terms of P_x and h_{bx} by using an equation similar to Eq. (8.140).

8.11 Land Subsidence Due to Well Pumping

Pumping of wells leads to soil compaction, which eventually lowers the level of the ground surface; this is known as *land subsidence*. If the ground surface is loaded by a building or a levee (for example, a bank of a canal or an earth fill for a highway), it settles because of these loads, which are new additions to the original natural weight of the soil layers. The settlement results from the compressibility of the natural soil layers below the level of the ground surface as a result of any additional effective stresses (Secs. 3.9 and 6.2).

In artesian wells, pumping imposes added effective stresses that lead to compressibility of the main aquifer as well as the aquitards above and below the aquifer. As a result of pumping, the water pressure also decreases in the aquitards, thus increasing the effective stresses. Since most of the materials of aquitards are fine soils, the process of compression is very slow, but the ultimate compressibility is much greater than that of the main aquifer, which is usually formed of coarser materials. The overall land subsidence at any vertical line is the algebraic sum of the compressibilities of all soil layers including the aquifer and the aquitards.

Theoretically speaking, there should be a rebound after pumping stops, but such negative settlements (upward upheaval) are not fully developed (Sec. 6.2). If the water table in an unconfined aquifer or the piezometric surface in an artesian aquifer fluctuates periodically, the settlement may become important. According to test results (Terzaghi and Peck, 1967), the magnitude of the increment of settlement decreases with increases in the number of loading cycles and approaches zero at later stages. However, the final total settlement is many times greater than the settlement produced by the first cycle.

Unconfined aquifers suffer less compressibility than artesian aquifers. Pumping from these aquifers leads to land subsidence only when they are underlain by aquitards rather than aquicludes.

Land subsidence produces detrimental effects within its zone of occurrence, such as flood hazards, settlement of buildings (and subsequent cracking or failure), and damage to such other engineering structures as bridges, railroads, tunnels, pipelines, canal linings, and impounding dams. The damage is compounded by the uneven land subsidence due to well pumping. The settlement is roughly proportional to the drawdowns, and consequently, uneven settlements with patterns similar to those of the drawdown surfaces occur.

Progressive subsidence can be reduced by controlling pumping and/or recharging the aquifer during seasons of abundant water. These operations are important phases of groundwater-resources management. For example, changing the number, capacities, spacings, and patterns of wells may satisfy the objectives of water demands in water yield, and if these changes are well planned, they will reduce the subsidence to an acceptable level.

Meinzer and Hard (1925) were probably the first to recognize the compressibility of artesian aquifers resulting from a decline in water

TABLE 8.16 Areas of Major Land Subsidence due to Groundwater Overdraft

Location	Depositional environment and age	Depth range of compacting beds, m	Maximum subsidence, m	Area of subsidence, km²	Time of principal occurrence
Japan:					
Osaka	Alluvial and shallow marine; Quaternary	10–400	3	190	1928–1968
Tokyo	As above	10–400	4	190	1920–1970+
Mexico:					
Mexico City	Alluvial and lacustrine; late Cenozoic	10–50	9	130	1938–1970+
Taiwan:					
Taipei basin	Alluvial and lacustrine; Quaternary	10–240	1.3	130	1961–1969+
United States:					
Arizona, central	Alluvial and lacustrine; late Cenozoic	100–550	2.3	650	1948–1967
California:					
Santa Clara Valley	Alluvial and shallow marine; late Cenozoic	55–300	4	650	1920–1970
San Joaquin Valley (three subareas)	Alluvial and lacustrine; late Cenozoic	60–1000	2.9–9	11,000 (>0.3 m)	1935–1970+
Lancaster area	Alluvial and lacustrine; late Cenozoic	60–300(?)	1	400	1955–1967+
Nevada:					
Las Vegas	Alluvial; late Cenozoic	60–300	1	500	1935–1963
Texas:					
Houston-Galveston area	Fluvial and shallow marine; late Cenozoic	60–600(?)	1–1.5	6,860 (>0.15 m)	1943–1964+
Louisiana:					
Baton Rouge	Fluvial and shallow marine; Miocene to Holocene	50–600(?)	0.3	650	1934–1965+

SOURCE: J. F. Poland, "Subsidence and Its Control in Underground Waste Management and Environmental Implications," American Association of Petroleum Geologists Memoir 18, 1972, pp. 50–71.

pressures. The problem of land subsidence has now achieved wideworld attention; areas of major land subsidence due to groundwater overdraft are listed in Table 8.16 (Poland, 1972). There are several other areas of major land subsidence in addition to those on this list. The well-publicized subsidence in Venice, for example, reached 15 cm over a period of 43 years from 1930 to 1973 (Gambolati and Freeze, 1973, 1974). At San Jose, California, the trend of subsidence relative to the decline in the piezometric surface over 50 years indicated that (Poland and Davis, 1969) water levels dropped about 45 m in 45 years during which the land surface subsided 3.0 m at an average rate of 1.0 m of settlement for every 15-m drop in water level, that is, about 7 cm per year. Selecting a different period (between 1920 and 1935), the yearly rate increased to 9.3 cm per year. In order to understand the aggravated effects on long engineering structures within the zone of influence of pumped wells, these rates should be compared with the maximum allowable differential settlements required by the major building codes.

An approach explained in Sec. 6.2 can be used to evaluate the magnitude and rate of land subsidence due to well pumping. The effective stresses producing settlements are numerically equal to $\gamma_w \Delta h$, where Δh is the drop in head and γ_w is the unit weight of water assuming that the total stress σ_t remains unchanged. Figure 8.62 shows a diagrammatic sketch of an aquitard overlying an artesian aquifer. Before pumping, the water level in the aquitard is at the ground surface (Fig. 8.62a). After pumping (Fig. 8.62b), the water level drops Δh, and the effective stress

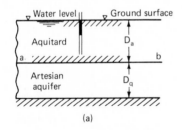

(a)

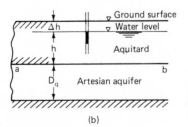

(b)

Figure 8.62 Compressibility of an aquitard. (a) Original water level of aquitard at ground surface; (b) water level dropped by Δh.

at *ab* is calculated as follows:

Before pumping: $\bar{\sigma}_1 = D_a \gamma'$
After pumping: $\bar{\sigma}_2 = \Delta h \gamma + h \gamma'$

where γ = unit weight of the aquitard material (saturated or moist)
      γ' = submerged unit weight of the same material, which is equal
        to $\gamma - \gamma_w$
      h = the value shown in Fig. 8.62*b*

Then
$$\Delta\bar{\sigma} = \bar{\sigma}_2 - \bar{\sigma}_1 = \Delta h \gamma + h \gamma' - D_a \gamma'$$
$$= \Delta h \gamma - \gamma'(D_a - h) = \Delta h \gamma - \gamma' \Delta h$$
or
$$\Delta\bar{\sigma} = \Delta h(\gamma - \gamma') = \Delta h \gamma_w \qquad (8.144)$$

The magnitude of $\Delta\bar{\sigma}$ at any other horizontal plane is also given by Eq. (8.144).

The compressibility of a fine-grained layer is given by

$$S_u = \Delta n D_a = \frac{\Delta e}{1 + e_o} D_a = \frac{C_c}{1 + e_o} D_a \log \frac{\bar{\sigma}_o + \Delta\bar{\sigma}_o}{\bar{\sigma}_o} \qquad (6.18)$$

The symbols used in this equation have already been defined in Sec. 6.2. In the preceding equation, the compression index C_c is determined by means of a consolidation test. The effective stress $\bar{\sigma}_o$ is determined at the center line of the aquiclude, and $\Delta\bar{\sigma}_o$ is given by Eq. (8.144). As explained in Sec. 6.2, Eq. (6.18) gives the ultimate settlement S_u at a theoretically infinite time. The magnitude of S_u and its rate are therefore determined as previously explained in Sec. 6.2. However, well pumping produces variable magnitudes for $\Delta\bar{\sigma}$ [Eq. (8.144)] depending on the time and the location of the point under investigation. In this type of analysis, the coefficient of consolidation c_v is determined from the field rather than by laboratory testing. Once the parameters T and S are determined, $c_v \simeq T/S$ [Eq. (6.27)].

A simpler approach is sometimes used rather than employing the logarithmic relationship given by Eq. (6.18). Since $m_v = -\Delta n / \Delta\bar{\sigma}$ [Eq. (6.15)] and $S_u = \Delta n D_a$ [Eq. (6.18)], then

$$S_u = D_a m_v \Delta\bar{\sigma} \qquad (8.145)$$

m_v is determined from the slope of the curve e versus $\bar{\sigma}$ drawn on a natural scale within the region of the $\Delta\bar{\sigma}$ [Eq. (8.144)] for which the settlement is sought. Since S_u/D_a is a strain and $\Delta\bar{\sigma}$ is a stress, then

$$\frac{\text{Stress}}{\text{Strain}} = \frac{\Delta\bar{\sigma}}{S_u/D_a} = \text{Young's modulus } E_s$$
$$\text{(or bulk modulus of compression)}$$

or
$$S_u = \frac{D_a}{E_s} \Delta\bar{\sigma} = D_a m_v \Delta\bar{\sigma} \tag{8.145a}$$

E_s is in fact equal to $1/m_v$. *Approximate* ranges of m_v and E_s are given by Jumikis (1962); however, it is recommended that E_s or m_v be determined by soil testing (Terzaghi and Peck, 1967).

From Eq. (6.19), $m_v D_a = S/\gamma_w - D_a n\beta_w$; then, by substitution in Eq. (8.145),

$$S_u = \Delta\bar{\sigma}\left(\frac{S}{\gamma_w} - D_a n\beta_w\right) \tag{8.146}$$

This equation was also given by Lohman (1972).

If water compressibility β_w is neglected,

$$S_u \simeq \Delta\bar{\sigma}\frac{S}{\gamma_w} \tag{8.147}°$$

EXAMPLE 8.15 An artesian aquifer 28 m thick is overlain by an aquitard 12 m thick (Fig. 8.63). The original water table in the aquitard was at the ground surface. After 200 days of continuous pumping, the water level dropped 2 m in the aquitard at section *acb*. During this period, the drawdown in the artesian aquifer dropped 7.5 m at the same section. Compute the approximate ultimate settlement and the settlement after 200 days. The following data are available.

Aquitard: $C_c = 0.3$ $c_v = 2.45 \times 10^{-4}$ cm²/s
               $m_v = 6.7 \times 10^{-5}$ cm²/g $K = 1.64 \times 10^{-8}$ cm/s
               $e_o = 0.84$ $\gamma = 1.76$ g/cm³ $\gamma_w = 1$ g/cm³

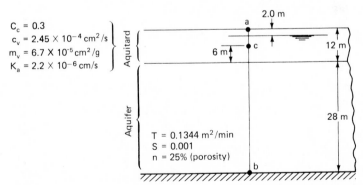

Figure 8.63 Compressibility of aquitard and aquifer (Example 8.15).

° Comparing Eqs. (8.147) and (8.145a), then

$$\frac{S}{\gamma_w} = \frac{D_a}{E_s} = D_a m_v \quad \text{or} \quad S_s \simeq m_v \gamma_w$$

Artesian aquifer: $T = 0.1344 \text{ m}^2/\text{min}$ Porosity $n = 25$ percent
$S = 0.001$ $\beta_w = 4.69 \times 10^{-8} \text{ cm}^2/\text{g}$

Solution

SETTLEMENT OF THE AQUIFER

This settlement usually takes place immediately; thus the total settlement is equal to the settlement after 200 days. Using Eq. (8.146),

$$S_u = (7.5 \times 100 \times 1)\left(\frac{0.001}{1} - 28 \times 100 \times 0.25 \times 4.69 \times 10^{-8}\right)$$

$$= 0.748 \text{ cm}$$

[Using Eq. (8.147) (i.e., neglecting β_w), $S_u = 0.750$ cm.]

SETTLEMENT OF THE AQUITARD

$$\Delta\bar{\sigma}_o = 2 \times 100 \times \gamma_w = 200 \text{ g/cm}^2 \quad \text{[Eq. (8.144)]}$$

$\bar{\sigma}_o$ is calculated at center line of the aquitard (point c in Fig. 8.63) and equals $600 \times 1.76 = 1056 \text{ g/cm}^2$. Applying Eq. (6.18), then

$$\text{Total } S_u = \frac{0.3}{1 + 0.84} \times 1200 \log \frac{1056 + 200}{1056} = 14.74 \text{ cm}$$

Using Eq. (8.145a), $S_u = D_q m_v \Delta\bar{\sigma} = 1200 \times 6.7 \times 10^{-5} \times 200 = 16.08$ cm. [Note that S_u calculated from Eqs. (6.18) and (8.145a) may not be that close in other cases.]

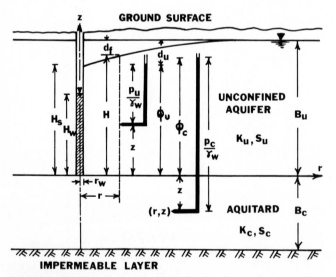

Figure 8.64 Diagrammatic sketch for the determination of land subsidence. *(A. I. Kashef and K. R. Chang, "Determination of Land Subsidence due to Well Pumping by Numerical Analyses," Proceedings of the 2d International Symposium on Land Subsidence, Anaheim, Calif., Dec., 1976.)*

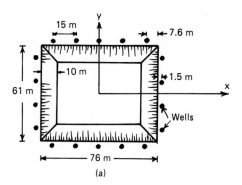

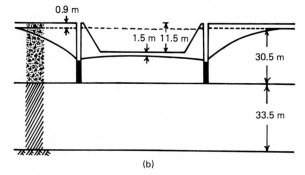

Figure 8.65 Layout of a solved example to determine land subsidence. (*a*) Layout of wells in plan; (*b*) vertical section. (*A. I. Kashef and K. R. Chang, "Determination of Land Subsidence due to Well Pumping by Numerical Analyses," Proceedings of the 2d International Symposium on Land Subsidence, Anaheim, Calif. Dec., 1976.*)

The settlement of the aquitard (14.74 cm) takes place at an infinite time (Sec. 6.2 and Fig. 6.2). In order to calculate the portion of S_u that takes place after 200 days, T_v is calculated from Eq. (6.14); it should be noted that $H_p = \frac{1}{2} \times 12$ m because the drainage of the aquitard is permitted upward as well as downward (see Sec. 6.2). Therefore,

$$T_v = \frac{c_v t}{(\frac{1}{2} \times 12 \times 100)^2} = \frac{2.45 \times 10^{-4} \times 200 \times 24 \times 60 \times 60}{360,000} = 0.0118$$

From Fig. 6.2, $U_c = 12$ percent when $T_v = 0.0118$. Therefore, the settlement after 200 days is $S_u \times 0.12 = 0.12 \times 14.74 = 1.79$ cm. And the total settlement of the aquifer and aquitard combined after 200 days is $1.79 + 0.748 = 2.538$ cm.

Discussion of Example 8.15 Analysis of this example is simplified in order to illustrate the use of the given equations. In geotechnical engineering, the aquitard is usually subdivided into a certain number of layers, rather than calculating $\bar{\sigma}_o$ at the center line. Also, in the preceding solution it is assumed that $\Delta\bar{\sigma}_o$ in the aquitard took place immediately rather than gradually. Several procedures are available to deal with the gradual increase in effective pressures. In order to be on the safe side, it is recommended that the total settlement of the aquitards (at $t = \infty$) be considered. In this example, total settlement at the vertical line ab (Fig. 8.63) is $0.748 + 14.74 = 15.488$ cm, which is the worst condition possible.

The analytical determination of land subsidence becomes more complex for a group of wells, especially when they are not continuously pumped. It has been shown that determinaton of the flow configuration of a group of wells is not a routine calculation. The conditions become more complicated when the effects within the aquitards (above and below the artesian aquifer) are considered. Aquitards existing below unconfined aquifers add to the complexity of the problem. Kashef and Chang (1976) presented a computerized numerical solution to illustrate such problems. The method is not a routine calculation having a closed-form solution. The investigated system is shown in Fig. 8.64. An example is solved for a group of 18 completely penetrating wells (in the uncon-

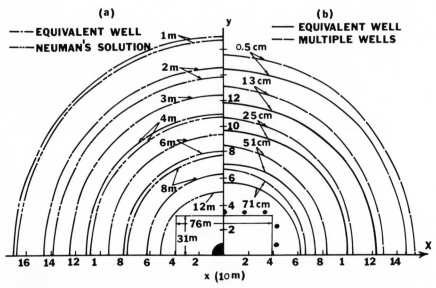

Figure 8.66 Land subsidence and drawdowns of the layout shown in Fig. 8.65. (*a*) Contours of drawdowns in meters for an equivalent well of the same capacity as the well group, determined by the authors' proposed method and compared with Neuman's approach (1975); (*b*) contours of land subsidence in centimeters, determined by the authors' proposed method, considering both the group effect of the wells and the equivalent well. (*A. I. Kashef and K. R. Chang, "Determination of Land Subsidence due to Well Pumping by Numerical Analyses," Proceedings of the 2d International Symposium on Land Subsidence, Anheim, Calif., Dec., 1976.*)

fined aquifer only) that are arranged around the site of a planned excavation (Figs. 8.65 and 8.66). The coefficients of storage of the aquifer and aquitard are, respectively, assumed to be 0.2 and 2×10^{-3}, and the hydraulic conductivities of the aquifer and aquitard are, respectively, assumed to be 5×10^{-5} and 5×10^{-7} m/s. In order to reach a drawdown of 1.5 m below the bottom of the excavation (see section in Fig. 8.65), each well is continuously pumped at a capacity of 4 m³/min for 15 days. The drawdowns are also computed by considering an equivalent well with a capacity of $18 \times 4 = 72$ m³/min and comparing with Neuman's solution (Neuman, 1975) as shown on the left-hand side of Fig. 8.66. On the right-hand side of Fig. 8.66, land-subsidence contours as shown for the multiple wells and for one equivalent well pumping at a rate equal to all capacities of the 18 wells.

PROBLEMS AND DISCUSSION QUESTIONS

8.1 Explain with neat sketches the stages of the drawdown curves from time $t = 0$ to $t \approx \infty$ resulting from a discharge well pumping continuously at a constant rate from the following
(a) Confined aquifer
(b) Unconfined aquifer

8.2 What is an overpumped artesian well? Explain by a sketch and show how to determine the r_e and r_o values under steady-state conditions of flow.

8.3 Define the radius of influence under both steady-state and transient conditions. Show how to determine the radii of influence for the following cases:
(a) Steady-state flow (artesian wells)
(b) Transient flow (artesian wells)
(c) Gravity wells under steady-state flow
(d) Gravity wells under transient flow
Discuss the difference between the theoretical and actual velocities at these radii.

8.4 Some hydrologists believe that steady-state flow around discharge wells can never be reached; others believe that it can be reached under certain conditions. What are these conditions?

8.5 After reading Chap. 8, disuss all the benefits gained from the steady-state analysis of water wells in relation to analysis under transient conditions.

8.6 An artesian well is pumping at a rate of 3.6 m³/min. The depth of the aquifer and its average hydraulic conductivity are, respectively, 32 m and 3×10^{-2} cm/s. If the flow is assumed steady and the effective radius of the well and its radius of influence are, respectively, 0.15 and 600 m, find the drawdowns at radii 0.15, 10, 100, and 400 m. [Use Eq. (8.4) and assume $h_u = 42$ m.]

8.7   Assume that the values for r_w and r_e were not given in Problem 8.6, and determine these values analytically and graphically on the basis of the calculated drawdowns at radii 10, 100, and 400 m.

8.8   Discuss the significance of the coefficient of transmissivity T. Explain why its determination in the field is preferable to the separate determination of hydraulic conductivity. Explain also the effects of time and the rate of pumping of a test well on the field results of the T value.

8.9   Plot a flow net in the horizontal plane for an artesian well pumping at a constant rate of 1.0 m³/min. The values of r_w and r_e are, respectively, 0.15 and 360 m. The coefficient of transmissivity T was found from a field test to be 1.05 m²/min. Use 12 flow channels.

8.10   If the h_u value in Problem 8.9 is 78 m, plot the drawdown curve on the basis of the drawn flow net.

8.11   If the average depth of the aquifer in Problem 8.9 is 68 m, determine the hydraulic conductivity. Also plot a flow net across a vertical section passing through the well axis, assuming five flow lines (i.e., four flow channels).

8.12   Explain why the flow net in Problem 8.11 does not consist of squares. On the basis of Darcy's law applied at the center of one of the fields, determine the necessary conditions that control the sizes of these fields.

8.13   If the groundwater in the aquifer in Problem 8.9 is initially moving at a rate of 2.5×10^{-4} m/min, plot a flow net (in the horizontal plane) for the combined effects of pumping and natural flow. Use the data and results obtained in Problems 8.9 and 8.10. Indicate the groundwater divide and the location of its asymptotes.

8.14   Groundwater is flowing at a darcian velocity of 0.4 m per day at a steady rate normal to the direction of a stream that is at the end boundary of an aquifer. The thickness of the aquifer is 30 m, and its average porosity is 26 percent. A well is planned to be pumped at a rate of 30 m³/h. Three locations have been suggested for the well: 10, 15, and 35 m normal to the stream. Determine the following for each suggested location assuming steady-state flow:
(a) Rate of flow withdrawn from the aquifer
(b) Location of stagnation point s
(c) Magnitude and direction of the seepage velocity at a point in the aquifer halfway between the well and the stream
(d) Same as part (c) but the point is on a line making a 45° angle to the normal but at the same distance from the well

8.15   An artesian aquifer 12 m thick is fed by a stream at a rate of 1.22 m per day. A small factory near the stream is disposing hot water into the aquifer through a recharge well at a constant rate of 270 m³ per day. What would be the minimum distance x from the well to the stream that would prevent the hot water from thermally polluting the stream? If the well is located at $\frac{1}{2}x$, determine the rate of flow of the hot water polluting the stream (assume steady-state flow).

8.16 A discharge artesian well is located in the corner zone between a stream and a barrier boundary. If the angle between the stream and the barrier boundary is 45°, draw a plan showing the necessary images required to determine the flow configuration around the well. The following data are available:
  (a) The distance of well from the point of intersection of the two boundaries is 77.5 m.
  (b) The normal distance from the well to the barrier boundary is 20 m.
  (c) The rate of flow of the pump Q is 1.0 m³/min.
  (d) The depth of the artesian aquifer is 65 m.
  (e) The average coefficient of transmissivity is 1.05 m²/min.
  (f) The undisturbed height of the piezometric surface above the upper boundary of the aquifer is 10 m.
  (g) The flow is assumed steady.
  Plot also a contour map for the piezometric heads in the zone between the boundaries and extending 200 m from their point of intersection.

8.17 An artesian aquifer 16 m thick consists of sand of an average porosity of 25 percent. Evaluate the coefficient of storage of the aquifer if the vertical compressibility of its material is equal to 1.022 cm²/g.

8.18 An artesian well is pumped at a constant rate Q of 1000 m³ per day from an extensive aquifer of an average thickness of 35 m. If the specific storage of the aquifer is 3×10^{-6} m⁻¹ and the hydraulic conductivity is 9.5 m per day, find the following:
  (a) The drawdown at a point 4 m from the well after 12 h of continuous pumping using both the Theis and Jacob methods
  (b) The time during which Jacob's method cannot be used at the same point in part (a)
  (c) The approximate radius of influence after 12 h of pumping (using all possible methods)
  (d) The residual drawdown after 20 h at the same point in part (a) if pumping is discontinued after 12 h

8.19 Make copies of the large plates given in Lohman (1972). These will give you more accurate results than the small-scale curves given in most textbooks.

8.20 An artesian well is pumping continuously at a rate Q of 250 m³/h. The coefficients of storage and transmissivity of the artesian aquifer are, respectively, 3×10^{-4} and 1.10 m²/min. After 15 h, the drawdown at an observation well installed at $r_1 = 30$ m from the pumped well is 0.3 m. Find the following:
  (a) The drawdown at a distance $r_2 = 1.5$ m from the pumped well after 15 h of continuous pumping
  (b) The slope of the straight line drawn on a semilog plot as drawdown versus log r^2/t using Jacob's method
  (c) The change in Q that would be compatible with the preceding recordings if the Theis equation were used

8.21 An artesian well is pumped at a constant rate Q of 56.65 m³/h. Recordings in an observation well located at a distance 174.42 m from the pumped well were as follows:

Time, min	Drawdown, m	Time, min	Drawdown, m	Time, min	Drawdown, m
3	0.041	20	0.171	120	0.322
4	0.058	24	0.185	160	0.342
5	0.070	28	0.192	200	0.363
6	0.081	36	0.218	240	0.380
8	0.102	48	0.241	300	0.400
10	0.118	60	0.259	360	0.415
12	0.133	80	0.286	420	0.428
16	0.151	100	0.305	480	0.439

Use the method of matching curves and find the coefficients of storage and transmissivity. Determine also the approximate radii of influence after 1.0, 2.0, and 8.0 h.

8.22   From the data given in Example 8.7 in the text and those given in Table 8.5, find the expected drawdowns in observation well 3 and the time of their occurrence beyond the 240 min of pumping. The recordings at observation wells 1 and 2 corresponding to all r^2/t values less than 248 (r^2/t value at observation well 3 corresponding to time $t = 240$ min) should be used. Tabulate your results.

8.23   Assume that t is a constant and differentiate Eq. (8.56) with respect to radius r and determine the hydraulic gradient at the effective well radius r_w. Prove that the same result can be obtained using Darcy's law.

8.24   On the basis of Jacob's method, find the ratio Q/T if the drawdowns in a pumped artesian well are 1.0 and 2.0 m after, respectively, 1.5 and 15 h. Also find the distance from the well at which a drawdown of 1.0 m occurs after 25 days of continuous pumping at a constant rate.

8.25   Using the recordings given in Problem 8.21, determine the coefficients of storage and transmissivity using the straight-line method. Also find the radii of influence after 1, 2, and 8 h.

8.26   An artesian well with a radius $r_w = 0.15$ m is pumped at a constant rate $Q = 2000$ m^3 per day from an extensive artesian aquifer of an average thickness of 50 m. If the coefficient of storage $S = 1.50 \times 10^{-4}$ and the hydraulic conductivity $K = 40$ cm/h, draw the following on semilog plots:
(a)   A diagram for the drawdowns versus log r at time $t = 15$ h using the Theis equation
(b)   The same as part (a) but using Jacob's equation
Both diagrams should cover the range between $r = r_w = 0.15$ m and $r = r_{e,t}$.

8.27   In the previous problem, find the distances from the pumped well of three observation wells at which the drawdowns are equal to $s_{dw}/4$, $s_{dw}/2$, and $3s_{dw}/4$, where s_{dw} is the drawdown at $r = r_w$. Apply both the Theis and Jacob equations.

8.28 Rework Example 8.9 by redrawing Fig. 8.26 using twice the scale of s_d. Study carefully the comments given in that example. Also plot the data in Table 8.6 in a graph of s_d versus t (natural scale).

8.29 An artesian aquifer of thickness $D_q = 30$ m is overlain by an aquitard 4 m thick. The coefficients of storage and transmissivity of the aquifer are, respectively, 0.00032 and 0.09 m²/min. A well pumped at a rate of 45 m³/h produced a drawdown of 50 cm in an observation well at a distance 108 m after time $t = 30$ min. Neglect the storage in the aquitard and determine the hydraulic resistance, leakage factor, hydraulic conductivity of the aquitard, and leakage rate.

8.30 The overlay of the field data with the theoretical curve (Fig. 8.29) was not shown in Example 8.11. Rework this example by selecting a match point off the curves.

8.31 Rework Example 8.12 selecting a match point off the curves.

8.32 Usually a field curve and the lower segment of a Theis type curve match well. However, the upper portion of the field curve deviates from the theoretical curve. Give all probable reasons when the upper portion of the field curve lies above or below the Theis type curve.

8.33 A discharging test well ($r_w = 0.2$ m) is pumping at a constant rate of 4.6 m³/min. An observation well is installed at 30 m north of the test well. A stream is located at about 50 m from the well. Verify the location of the boundary using the Theis and Jacob methods. Also check the result by using Stallman's curves (Fig. 8.38). The field data are tabulated as follows:

Test Well: Field Records

Time, min	Drawdown, cm	Time, min	Drawdown, cm	Time, min	Drawdown, cm	Time, min	Drawdown, cm
2	113	20	145	70	159	240	165
4	123	25	148	80	160	300	166
6	128	30	150	90	160	360	167
8	132	40	153	100	162	420	168
10	136	50	156	120	162	480	168
15	141	60	157	180	164	540	168

Observation Well: Field Records

Time, min	Drawdown, cm	Time, min	Drawdown, cm	Time, min	Drawdown, cm	Time, min	Drawdown, cm
5	2.4	40	18.6	100	25.0	360	29.3
10	6.7	50	20.4	110	25.3	420	29.6
15	9.8	60	21.6	120	25.9	480	29.9
20	12.5	70	22.9	180	27.4	540	29.9
25	14.3	80	23.8	240	28.3	600	30.2
30	16.2	90	24.4	300	29.0	660	30.2

8.34 A discharging test well ($r_w = 0.2$ m) is pumping at a constant rate of 4.6 m³/min. An observation well is installed at 30 m north of the test well.

Locate the barrier boundary using the Theis and Jacob methods. Also check the results using Stallman's curves (Fig. 8.38). The field data are tabulated as follows:

Test Well: Field Records

Time, min	Drawdown, cm	Time, min	Drawdown, cm	Time, min	Drawdown, cm	Time, min	Drawdown, cm
2	101	25	146	90	168	420	201
4	122	30	149	100	171	480	204
6	125	40	152	120	174	540	207
8	131	50	155	180	183	600	210
10	134	60	159	240	190	660	213
15	137	70	162	300	195	720	216
20	143	80	165	360	198	780	220

Observation Well: Field Records

Time, min	Drawdown, cm	Time, min	Drawdown, cm	Time, min	Drawdown, cm	Time, min	Drawdown, cm
0	0	40	20.4	110	36.6	480	67.1
5	2.4	50	23.5	120	38.1	540	69.5
10	6.7	60	25.9	180	46.0	600	71.9
15	9.8	70	29.0	240	51.8	660	75.0
20	12.5	80	30.8	300	57.0	720	76.2
25	14.9	90	32.9	360	60.7	840	80.2
30	17.1	100	34.7	420	64.0	960	84.4

8.35　In Problems 8.33 and 8.34, the test well was used to replace an observation well. Give your comments as to the accuracy of the method and what additional actions should have been carried out. Also tabulate the results of the three methods used and compute the percentage differences with respect to the results obtained from the Theis method.

8.36　Determine the percentage of pumped water being diverted from the stream by means of the discharge well in Problem 8.33. Determine the required values of S and T from the field recordings using either the Theis or Jacob approach.

8.37　Determine the rate of flow and the darcian velocity in the aquifer described in Problem 8.26 at radii $\frac{1}{3}(r_{e,t} - r_w)$ and $\frac{2}{3}(r_{e,t} - r_w)$ after 15 h of pumping.

8.38　If the undisturbed height h_u of the piezometric surface in Problem 8.26 was 60 m (above the lower impervious boundary) and the groundwater was initially flowing at a constant rate of 2.5×10^{-4} m/min, determine the drawdowns due to the combined effects of pumping and natural flow along the circumference of a circle centered at the pumping well that has a radius of 50 m. Select 12 equidistant points on the circumference of this circle. Find also the magnitude and direction of the darcian velocity at a point on this circle making a 60° angle with the direction of natural flow (assume natural flow under steady-state conditions).

8.39 A discharging well was installed at a distance 50 m from the well given in Problem 8.26. The additional well is similar to the original, but it started pumping 3 h later at the same capacity. Find the drawdown after 6 h of pumping the additional well at a middle section between the wells and at 25 m outside both wells. If the natural flow given in Problem 8.38 is added, determine the drawdowns at the same section and after the same time.

8.40 Rework Example 8.13 selecting time $t = 80$ days. Follow the same procedure given in that example to determine the coefficients B_f, C_f, n, and the effective well radius r_w. From the computed value of C_f, determine the well condition according to Table 8.14.

8.41 Assume the conditions given in Example 8.13 (and Problem 8.40) and determine the drawdown s_{dw} in the well if the rate of pumping is 5.2 m^3/min corresponding to the following cases:
   (a) Assume that the flow is steady and that there are no screen losses.
   (b) Assume that the flow is transient and neglect the screen losses ($t = 10$ days).
   (c) Assume the same conditions as in part (b) but consider the screen losses.
   (d) Assume that the screen length is 30 percent of the depth of the aquifer and that the flow is transient but neglect the screen losses ($t = 10$ days).
   (e) Assume the same conditions as in part (d) but consider the screen losses.

8.42 How does one avoid the effects of screen losses and partial penetration of wells in conducting pumping tests to determine the coefficients of transmissivity and storage?

8.43 A gravity well is pumping at a rate of 1.02 m^3/min from an unconfined aquifer in which the water table was initially at a height $h_u = 25$ m above the lower impervious boundary. The flow was assumed steady when the drawdown in the well was recorded as 2.97 m. If the radius of influence of the well was $r_w = 0.25$ m and the coefficient of transmissivity was assumed to be 0.4 m^2/min, determine
   (a) The radius of influence r_e
   (b) The drawdowns at distances 20, 50, and 100 m from the well
   (c) The depth of the outflow surface

8.44 Plot in Problem 8.43 the total-head diagrams across the vertical sections given in part (b) of the same problem. Plot also (using a vertical datum) the pressure-head diagrams across the same sections. Indicate in all diagrams the ordinates at $y = 0$, $y = \frac{1}{3}D_r$, $y = \frac{2}{3}D_r$, and $y = D_r$.

8.45 If two wells of the same characteristics as those given in Problem 8.43 are installed 40 m apart, plot a section showing the free surface between the two wells and extending 250 m beyond either well. Indicate the heights of the free surface at equal intervals of 20 m.

8.46 Field pumping tests in an unconfined aquifer indicate that the early coefficient of storage is equal to 0.001 and that the late specific yield is equal to 0.15. The coefficient of transmissivity T was found to be 0.09 m^2/min.

The test well was pumped continuously at a rate of 45 m³/h, and an observation well was installed at a distance of 25 m from the pumped well. After 16 min, the drawdown in the observation well was 52 cm. Find the following:
(a) The delay index
(b) The drawdown in the observation well after 4 h
(c) The coefficient of storage S using the Theis equation after making the necessary corrections in the drawdown determined in part (b)
(d) The time required for the well to start pumping water derived from gravity drainage

8.47 Rework Example 8.15 selecting your own match points. Find the values of the coefficient of storage, the specific yield, the delay index, and the time t_{min} necessary to reach the true gravity-well conditions.

8.48 Three gravity wells are installed 50 m apart forming an equilateral triangle. The undisturbed height of the water table is 30 m above the lower impervious boundary. The wells start pumping at the same time at equal rates Q of 4 m³/min. Previous pumping tests indicate that the average coefficient of transmissivity T is 2.5 m²/min and that the average specific yield S_y is 0.09. The effective radius r_w of each well is 0.25 m. Plot a contour map for the free surface over an area 200 × 200 m centered with the center of the well group after 15 days of continuous pumping using the following approaches:
(a) Use the curves shown in Figs. 8.18, 8.19, or 8.20 or Tables 8.3 and 8.4, assuming that $S = S_y$ and that the time (15 days) has already exceeded the time t_{min} necessary to reach the true gravity-well conditions. The drawdowns of each well are *approximately* superimposed.
(b) Use the method expressed by Eqs. (8.135) and (8.139).
(c) Use the method expressed by Eqs. (8.141a) and (8.141b), assuming that the period of 15 days is a relatively long period of time.

8.49 Determine the land subsidence at the midpoints of the sides of the equilateral triangle forming the well group in Problem 8.48, neglecting water compressibility.

8.50 If the wells in Problem 8.48 were installed to drain the site in order to excavate for construction of the foundation of an engineering building in the dry, determine the extent of the deepest possible excavation and the elevation of its bottom that can be made after 15 days. Using the three approaches listed in Problem 8.48, select the most conservative excavation.

REFERENCES

Babbitt, H. E., and D. H. Caldwell: "The Free Surface Around, and Interference between Gravity Wells," *Univ. Illinois Bull.*, vol. 45, no. 30, 1948.
Boulton, N. S.: "Unsteady Radial Flow to a Pumped Well Allowing for Delayed Yield from Storage," International Association of Scientific Hydrology Publication 37, 1954a, pp. 472–477.

Boulton, N. S.: "The Drawdown of the Water Table under Nonsteady Conditions Near a Pumped Well in an Unconfined Formation," *Proc. Inst. Civ. Eng. (Lond.)*, pt. 3, 1954*b*, pp. 564–579.

Boulton, N. S.: "Analysis of Data from Nonequilibrium Pumping Tests Allowing for De-layed Yield from Storage," *Proc. Inst. Civ. Eng. (Lond.)*, vol. 26, 1963, pp. 469–482.

Boulton, N. C.: Closure Discussion of "Analysis of Data from Nonequilibrium Pumping Tests Allowing for Delayed Yield from Storage" *Proc. Inst. Civ. Eng. (Lond.)*, vol. 28, 1964, pp. 603–610 (discussions by R. W. Stallman, W. C. Walton, and J. Ineson and reply by author).

Bouwer, H.: *Groundwater Hydrology*, McGraw-Hill, New York, 1978.

Brons, F., and V. E. Marting: "The Effect of Restricted Fluid Entry on Well Productivity," *Pet. Technol.*, vol. 13, no. 2, 1961, pp. 172–174.

Davis, S. N., and R. J. M. DeWiest: *Hydrogeology*, Wiley, New York, 1966.

Ehlig, C., and J. C. Halepaska: "A Numerical Study of Confined-Unconfined Aquifers Including Effects of Delayed Yield and Leakage," *Water Resources Res.*, vol. 12, 1976, pp. 1175–1183.

Ferris, J. G., D. B. Knowles, R. H. Brown, and R. W. Stallman: "Theory of Aquifer Test," U.S. Geological Survey Water Supply Paper 1536-E, Washington, 1962, pp. 69–174.

Franke, O. L.: "Steady-State Discharge to a Partially Penetrating Artesian Well: An Elec-trolyte-Tank Model Study," *Ground Water*, vol. 5, no. 1, 1967, pp. 29–34.

Gambolati, G., and R. A. Freeze: "Mathematical Simulation of the Subsidence of Venice," *Water Resources Res.*, vol. 9, 1973, pp. 721–733, and vol. 10, 1974, pp. 563–577.

Hantush, M. S.: "Analysis of Data from Pumping Tests in Leaky Aquifers," *Trans. Am. Geophys. Union*, vol. 37, 1956, pp. 702–714.

Hantush, M. S.: "Drawdown Around a Partially Penetrating Well," *Proc. Am. Soc. Civ. Eng.*, vol. 87, no. HY4, 1961*a*, pp. 83–98.

Hantush, M. S.: "Aquifer Tests on Partially Penetrating Wells," *Proc. Am. Soc. Civ. Eng.*, vol. 87, no. HY5, 1961*b*, pp. 171–195.

Hantush, M. S.: "Aquifer Tests on Partially Penetrating Wells," *Trans. Am. Soc. Civ. Eng.*, vol. 127, pt. 1, 1962, pp. 284–308.

Hantush, M. S.: "Hydraulics of Wells," in V. T. Chow (ed.), *Advances in Hydroscience*, vol. 1, Academic Press, New York, 1964*a*, pp. 281–432.

Hantush, M. S.: "Drawdown Around Wells of Variable Discharge," *J. Geophys. Res.*, vol. 69, 1964*b*, pp. 4221–4235.

Hantush, M. S.: "Wells Near Streams with Semipervious Beds," *J. Geophys. Res.*, vol. 70, 1965, pp. 2829–2838.

Hantush, M. S., and C. E. Jacob: "Nonsteady Radial Flow in an Infinite Leaky Aquifer," *Trans. Am. Geophys. Union*, vol. 36, no. 1, 1955, pp. 93–100.

Huisman, L.: *Groundwater Recovery*, Winchester Press, New York, 1972.

Jacob, C. E.: "Radial Flow in a Leaky Artesian Aquifer," *Trans. Am. Geophys. Union*, vol. 27, no. 2, 1946, pp. 198–205.

Jacob, C. E.: "Drawdown Test to Determine the Effective Radius of an Artesian Well," *Trans. Am. Soc. Civ. Eng.*, vol. 112, 1947, pp. 1047–1070.

Jacob, C. E.: "Flow of Ground Water," in H. Rouse (ed.), *Engineering Hydraulics*, Wiley, New York, 1950, pp. 321–386.

Jacob, C. E.: "Determining the Permeability of Water-Table Aquifers," in *Methods of Determining Permeability, Transmissibility and Drawdown*, U.S. Geological Survey Water Supply Paper 1536-I, Washington, 1963, pp. 245–271.

Jumikis, A. R.: *Soil Mechanics*, Van Nostrand, Princeton N.J., 1962.

Kashef, A. I.: "Numerical Solutions of Steady State and Transient Flow Problems," Ph.D. thesis, Purdue University, West Lafayette, Ind., 1951.

Kashef, A. I.: "Exact Free Surface of Gravity Wells," *Proc. Am. Soc. Civ. Eng.*, vol. 91, no. HY4, 1965*a*, pp. 167–184.

Kashef, A. I.: "Seepage through Earth Dams," *J. Geophys. Res.*, vol. 70, no. 24, 1965*b*, pp. 6121–6128.

Kashef, A. I.: "Dispersion and Diffusion in Porous media: Salt-Water Mounds in Coastal Aquifers," Water Resources Research Institute, University of North Carolina, Raleigh, report no. 11, 1968.

Kashef, A. I.: "Interference between Gravity Wells: Steady-State Flow," *Water Res. Bull.*, vol. 6, no. 4, 1970*a*, pp. 617–630.

Kashef, A. I.: "Multiple Gravity Wells under Transient States of Flow," *Water Resources Res.*, vol. 6, no. 6, 1970*b*, pp. 1729–1736.

Kashef, A. I.: "Overpumped Artesian Wells among a Well Group," *Water Res. Bull.*, vol. 7, no. 5, 1971, pp. 981–990.

Kashef, A. I.: "Management of Retardation of Salt Water Intrusion in Coastal Aquifers," Office of Water Research and Technology, U.S. Dept. of Interior, Washington, 1975.

Kashef, A. I., and K. R. Chang: "Determination of Land Subsidence due to Well Pumping by Numerical Analysis," *Proceedings of the 2d International Symposium on Land Subsidence*, Anaheim, Calif., Dec., 1976, pp. 167–178.

Kirkham, D.: "Exact Theory of Flow into a Partially Penetrating Well," *J. Geophys. Res.*, vol. 64, 1959, pp. 1317–1327.

Labadie, J. W., and O. J. Helweg: "Step-Drawdown Test Analysis by Computer," *Ground Water*, vol. 13, 1975, pp. 438–444.

Lennox, D. H.: "Analysis and Application of Step-Drawdown Test," *Proc. Am. Soc. Civ. Eng.*, vol. 92, no. HY6, 1966, pp. 25–48.

Li, W. H., P. Block, and G. Benton: "A New Formula for Flow into Partially Penetrating Wells in Aquifers," *Trans. Am. Geophys. Union*, vol. 35, 1954, pp. 806–811.

Lohman, S. W.: "Groundwater Hydraulics," U.S. Geological Survey Professional Paper 708, Washington, 1972.

Meinzer, O. E., and H. A. Hard: "The Artesian Water Supply of the Dakota Sandstone," U.S. Geological Survey Water Supply Paper 520-E, Washington, 1925, pp. 73–95.

Muskat, M.: The Flow of Homogeneous Fluids through Porous Media, McGraw-Hill, New York, 1937; second printing, 1946, by J. W. Edwards, Ann Arbor, Mich.

Neuman, S. P.: "Theory of Flow in Unconfined Aquifers Considering Delayed Response of the Water Table," *Water Resources Res.*, vol. 8, 1972, pp. 1031–1045.

Neuman, S. P.: "Analysis of Pumping Test Data from Anisotropic Unconfined Aquifers Considering Delayed Gravity Response," *Water Resources Res.*, vol. 11, 1975, pp. 329–342.

Nisle, R. G.: "The Effect of Partial Penetration in Pressure Buildup in Oil Wells," *Trans. Am. Inst. Min. Metal. Pet. Eng.*, vol. 213, 1958, pp. 85–90.

Poland, J. F.: "Subsidence and Its Control in Underground Waste Management and Environmental Implications," American Association of Petroleum Geologists Memoir 18, Tulsa, Okla., 1972, pp. 50–71.

Poland, J. F., and G. H. Davis: "Land Subsidence due to Withdrawal of Fluids," in D. J. Varnes and G. Kiersch (eds.), *Reviews in Engineering Geology*, vol. 2, Geological Society of America, Boulder, Colo., 1969, pp. 187–269.

Polubarinova-Kochina, P. Y.: *Theory of Groundwater Movement*, Princeton Univ. Press, Princeton, N.J., 1962; translated by R. J. M. DeWiest from the Russian edition, 1952.

Prickett, T. A.: "Type-Curve Solution to Aquifer Tests under Water-Table Conditions," *Ground Water*, vol. 3, no. 3, 1965, pp. 5–14.

Rao, D. B., et al.: "Drawdown in a Well Group along a Straight Line," *Ground Water*, vol. 9, no. 4, 1971, pp. 12–18.

Rasmussen, W. C., and G. E. Andreasen: "Hydrologic Budget of the Beaverdam Creek Basin, Maryland," U.S. Geological Survey Water Supply Paper 1472, Washington, 1959.

Rorabaugh, M. I.: "Graphical and Theoretical Analysis of Step-Drawdown Test of Artesian Well," *Proc. Am. Soc. Civ. Eng.*, vol. 79, 1953.

Sheahan, N. T.: "Type-Curve Solution of Step-Drawdown Test," *Ground Water*, vol. 9, no. 1, 1971, pp. 25–29.

Stallman, R. W.: "Boulton's Integral for Pumping-Test Analysis," in *Short Papers in the Geologic and Hydrologic Sciences*, U.S. Geological Survey Professional Paper 424-C, Washington, 1961*a*, pp. C24–C29.

Stallman, R. W.: "The Significance of Vertical Flow Components in the Vicinity of Pumping Wells in Unconfined Aquifers," in *Short Papers in the Geologic and Hydrologic Sciences*, U.S. Geological Survey Professional Paper 424-B, Washington, 1961*b*, pp. B41–B43.

Stallman, R. W.: "Type Curves for the Solution of Single Boundary Problems," in *Shortcuts and Special Problems in Aquifer Tests*, U.S. Geological Survey Water Supply Paper 1545-C, Washington, 1963, pp. C45–C47.

Stallman, R. W.: "Effects of Water-Table Conditions on Water-Level Changes near Pumping Wells," *Water Resources Res.*, vol. 1, no. 2, 1965, pp. 295–312.

Sternberg, Y. M.: "Efficiency of Partially Penetrating Wells," *Ground Water*, vol. 11, no. 3, 1973, pp. 5–8.

Terzaghi, K., and R. B. Peck: *Soil Mechanics in Engineering Practice*, 2d ed., Wiley, New York, 1967.

Theis, C. V.: "The Relation between the Lowering of the Piezometric Surface and the Rate and Duration of Discharge of a Well Using Groundwater Storage," *Trans. Am. Geophys. Union*, vol. 16, 1935, pp. 519–524.

Theis, C. V.: "The Effect of a Well on the Flow of a Nearby Stream," *Trans. Am. Geophys. Union*, vol. 22, 1941, pp. 734–738.

Theis, C. V.: "The Spacing of Pumped Wells," U.S. Geological Survey, Ground Water Branch, Ground Water Notes, Hydraulics, no. 31, open-file report, Washington, 1957.

Theis, C. V.: "Estimating the Transmissibility of a Water-Table Aquifer from the Specific Capacity of a Well," in *Methods of Determining Permeability, Transmissibility, and Drawdown*, U.S. Geological Survey Water Supply Paper 1536-I, Washington, 1963, pp. 332–336.

Todd, D. K.: *Ground Water Hydrology*, 2d ed., Wiley, New York, 1980.

UOP Johnson Division: *Ground Water and Wells*. St. Paul, Minn., 1966.

U.S. Bureau of Reclamation: *Ground Water Manual*, U.S. Department of Interior, Washington, 1977.

Walton, W. C.: "Application and Limitation of Methods Used to Analyze Pumping Test Data," *Water Well J.*, pts, 1 and 2, vol. 14, 1960.

Walton, W. C.: "Selected Analytical Methods for Well and Aquifer Evaluation," Illinois State Water Survey Bulletin no. 49, 1962.

Walton, W. C.: *Groundwater Resource Evaluation*, McGraw-Hill, New York, 1970.

Wenzel, L. K.: "Methods of Determining Permeability of Water-Bearing Materials, with Special Reference to Discharging-Well Methods," U.S. Geological Survey Water Supply Paper 887, Washington, 1942.

SALTWATER INTRUSION

This chapter evaluates seawater (or ocean water) intrusion into coastal aquifers. There are many coastal areas in the United States (Newport, 1977) and around the world where the problem is severe. The geologic occurrence of salt water inland or in coastal zones as well as the appearance of salt water from other sources than seawater intrusion, such as agricultural activities, are beyond the scope of this chapter.

9.1 Ghyben-Herzberg Interface

If the seaward boundary of an unconfined or a confined aquifer is in contact with a saltwater body, such as a sea or an ocean, saltwater encroachment may take place. The freshwater body will overlie the saltwater body because the unit weight of fresh water (about 1 g/cm³) is less than that of salt water (about 1.022 to 1.031 g/cm³). The boundary surface between both types of water is known as the *saltwater-freshwater interface* or the *interface*. Actually, this boundary is a transitional zone of varying salinity because of molecular diffusion and convectional dispersion between salt and fresh waters. However, in most analyses, a sharp interface is considered as a distinct curve in the aquifer cross section,

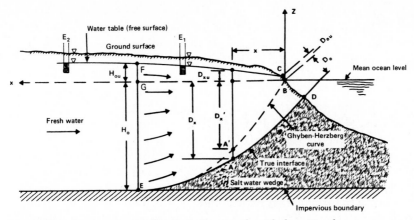

Figure 9.1 Diagrammatic sketch indicating the Ghyben-Herzberg curve as compared with the true interface in an unconfined aquifer.

assuming that the two fluids are immiscible (Fig. 9.1). If the coastal aquifer consists of two or more distinct aquifers, each aquifer will have an independent interface (Fig. 9.2). The extent of intrusion depends on several factors, such as climatic conditions, the aquifer boundaries and its hydrologic and structural properties, changes in seaward natural flow, human activities such as discharge and recharge wells, barometric changes, and tidal effects.

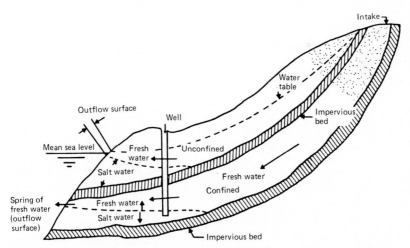

Figure 9.2 Saltwater intrusion in two-aquifer system according to d'Andrimont. *(J. S. Brown, "Study of Coastal Ground Water with Special Reference to Connecticut," U.S. Geological Survey Water Supply Paper 537, Washington, 1925.)*

Although saltwater intrusion was recognized early in history (Brown, 1925), the first publication on the subject is that of Braithwaite (1855), who described salinity problems caused by well pumping in London and Liverpool. He suggested that the infiltration of seawater was caused by lowering the groundwater level below that of the sea. It is now well established that saltwater intrusion takes place even if the water-table level is slightly above the seawater level (Sec. 9.2).

Ghyben (1888–1889) developed the location of the sharp interface analytically, and Herzberg (1901) arrived independently at the same conclusion. Thus the curve is now identified in the literature as the *Ghyben-Herzberg interface.* At the time, the outflow surfaces (seepage surfaces above and below sea level) (Fig. 9.1) were not recognized. It has been found that the actual natural interface is lower than the Ghyben-Herzberg interface. However, the exact locations of the interface and the transitional zone are difficult to determine either by theoretical methods or by field testing if all possible factors are included (such as convectional dispersion, molecular diffusion, tidal effects, barometric changes, evaporation, leakage from and into confining beds, earthquakes and other seismic waves, transient flow, and human activities). The interface should therefore be considered to be a first approximation in dealing with an actual groundwater-management problem in coastal aquifers.

As shown in Fig. 9.1, the equilibrium at point A' at a distance x from the shoreline on the interface (neglecting the velocity of the natural flow) results in

$$(D'_x + D_{xu})\gamma_w = D'_x\gamma_{st}$$

where γ_w and γ_{st} are, respectively, the unit weights of fresh and salt water. Then

$$D'_x = \frac{D_{xu}}{\alpha_s} \tag{9.1}$$

where

$$\alpha_s = \frac{\gamma_{st} - \gamma_w}{\gamma_w} \tag{9.2}$$

(Since $\gamma_w = 1.0$, $\alpha_s = \gamma_{st} - \gamma_w = \gamma_{st}$, and $\gamma_{st} = \alpha_s + 1$.) Thus the total thickness of the freshwater lens is equal to

$$(D'_x + D_{xu}) = (1 + \alpha_s)\frac{D_{xu}}{\alpha_s} \tag{9.3}$$

If $\gamma_{st} \cong 1.025$ g/cm^3 and $\gamma_w \cong 1.0$ g/cm^3 are substituted in Eq. (9.2), then $1/\alpha_s = 40$. In other words, the depth of the fresh water below sea level is 40 times as much as the height of the fresh water above the same level.

The value of γ_{st} at 15°C varies from as low as 1.022 g/cm^3 to as high as 1.031 g/cm^3 (salinity of about 41.5 parts per thousand). The average salinity of the water of the oceans is about 35 parts per thousand; thus its average density at 15°C is about 1.026, and $1/\alpha_s$ would then vary between 37 and 45. The lower value was given by Herzberg (1901), although he noticed that Eq. (9.1) does not hold in all cases and is strongly influenced by the fineness or coarseness of the sand dunes that were the media of his field tests. Other investigators checked the Ghyben-Herzberg findings (d'Andrimont, 1902, 1905a, 1905b). Herzberg (1901) recognized that water levels rise and fall as a result of tidal effects, that water levels lag 3 to 4 h behind the tide, and that salinity increases during dry seasons and periods of heavy pumping. These findings are consistent with our present knowledge.

It should be realized that Eq. (9.1) is based on the assumption that freshwater and saltwater bodies are stagnant. In reality, the movement of fresh water is faster than that of salt water; the latter may be regarded as stagnant. However, the effect of freshwater movement should be considered (Sec. 9.3). In fact, Eq. (9.1) is based on the field measurement of D_{xu}, which is influenced by the movement of the groundwater toward the sea. The resulting interface should then be a curve, since the free surface (height D_{xu}) is a curve. If hydrostatic conditions prevail, the interface should be horizontal, and this may be rare, such as when an aquifer is isolated between the sea and some vertical impervious boundary inland.

9.2 Limiting Conditions of Saltwater Intrusion

An artesian (confined) aquifer is considered in this analysis. Conclusions may be extended to unconfined aquifers, as explained later. Figure 9.3 represents a finite zone of a confined coastal aquifer near the sea or ocean. Salt water intrudes into this zone to maintain the equilibrium conditions of water pressure. The source of the salt water is the saltwater body that is in direct contact with the terminal boundary of the aquifer. Salt water also may be derived at a lower rate from the rise of salt water occupying the relatively impervious lower beds when they are connected to the main saltwater body. The common assumption that these beds are practically impervious is used mainly to simplify analytical solutions. If these beds consist of rock formations, they generally contain fissures, cracks, and/or joints through which water percolates at very low velocities. However, if they consist of very fine soil sediments, water also percolates through the voids, again at very low velocities. Thus if the hydraulic conditions necessitate such movements, salt water may start to rise from these beds, and eventually salt water will be in balance with the fresh water.

As shown in Fig. 9.3, one of three states of water occurrence may take place

1. The aquifer may contain only fresh water.

2. The aquifer may be entirely invaded by salt water.

3. Fresh water and salt water may concurrently exist. In this case the freshwater zone is above the saltwater zone because of the fresh water's smaller specific gravity (neglecting the transitional zone).

If the saltwater velocity is assumed to be negligible and the effects of tides and barometric pressure are disregarded, the saltwater pressure p_{st} at the bottom of the main aquifer is given by

$$p_{st} = (D_q + \eta)\gamma_{st} \qquad (9.4)$$

D_q and η are as indicated in Fig. 9.3. For the aquifer to be void of any salt water (state 1 in the preceding list) at the location of an observation well, the following condition must be satisfied:

$$(z + D_q)\gamma_w \geq (D_q + \eta)\gamma_{st} \qquad (9.5)$$

Thus if there is no salt water at the location of the observation well, the height z from Eq. (9.5) is

or
$$\begin{aligned} z &\geq \eta\gamma_{st} + D_q(\gamma_{st} - \gamma_w) \\ z &\geq \alpha D_q + \eta(1 + \alpha_s) \end{aligned} \qquad (9.6)$$

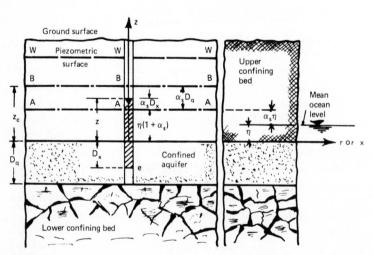

Figure 9.3 Saltwater and freshwater occurrence in confined aquifers.

If plane AA (Fig. 9.3) is drawn at a distance $\eta(1 + \alpha_s)$ above the x axis and plane BB at a distance $z = z_c = \alpha_s D_q + \eta(1 + \alpha_s)$, a limiting condition exists when the water level in the observation well coincides with plane BB in such a way that a further drop in the water level will be accompanied by a rise of salt water from the bottom of the well. Thus if $z \geq z_c$, the aquifer will have only fresh water at the location of the observation well, and

$$z \geq z_c = \alpha_s D_q + \eta(1 + \alpha_s) \qquad \text{(fresh water only)} \qquad (9.7)$$

If $z < z_c$, then by examining Eqs. (9.6) and (9.7) as well as Fig. 9.3, D_q can assume another value, $D_x < D_q$, and

$$z = \alpha_s D_x + \eta(1 + \alpha_s) \qquad \text{(fresh and salt water)} \qquad (9.8)$$

The total freshwater height will be $z + D_x = (D_x + \eta)(1 + \alpha_s)$, and the depth of saltwater intrusion will be equal to $D_q - D_x$.

When z decreases, D_x also decreases [Eq. (9.8)], and finally, when $D_x = 0$ (theoretically), $z = \eta(1 + \alpha_s)$. The aquifer will contain only salt water when the water level is at or lower than plane AA (Fig. 9.3).

These principles are valid whether the piezometric surface WW (Fig. 9.3) is horizontal or curved (deflected) as a result of, say, well pumping. The level of water in an observation well can disclose the extent of saltwater intrusion at the well location in relation to the stationary planes AA and BB (Fig. 9.3). If the recorded water elevation is at or above BB, there will be no salt water in the aquifer at the recording station. If the water level is at any position between planes AA and BB, saltwater intrusion will develop. If the water level is at AA or lower, the aquifer will be void of any fresh water at the location of the observation well, and the pores will be completely saturated with salt water. If several observation wells are available, the interface will pass by points e (Fig. 9.3) whenever $0 < D_x \leq D_q$.

When a pumped well is operating in a coastal aquifer, it is usually recommended that the drawdown at any point not reach an elevation lower than the mean sea level in order to prevent saltwater pumping. According to this recommendation, the entire drawdown curve should be above sea level. Strictly speaking, this is one of the misconceptions in groundwater hydrology. According to the previous analysis, the pumped water will be entirely fresh when the water level in the pumped well (lowest point on the drawdown curve) is higher than plane AA (Fig. 9.3). The difference in elevations between the mean sea level and plane AA depends on the location of the top boundary of the confined aquifer with respect to sea level and the unit weights of salt and fresh waters. If, for example, $\gamma_{st} = 1.025$ g/cm^3, $\gamma_w = 1.0$ g/cm^3, and $\alpha_s = 0.025$, plane AA will be higher than sea level by a distance 0.025η. If η is as small as 10 m, plane *AA* will be above sea level by 25 cm (9.84 in). However, if $\eta =$ 100 m (328 ft), the difference in elevations will be 2.5 m (8.2 ft). Once the lowest drawdown in a pumped well reaches plane *AA* (and not sea level), the quality of the pumped water rapidly deteriorates, even though a large portion of the drawdown curve remains above *AA* (Kashef, 1967, 1972, 1977).

9.3 Hydrodynamic Effects on the Interface

Mathematical techniques such as conformal mapping (including hodograph method) were introduced to locate the interface (Bear, 1979; Bear and Dagan, 1964*a*). Mathematical approximations were also attempted, as well as laboratory and field experiments, in order to locate the interface, taking into consideration the moving natural flow. Hubbert (1940, 1969), Glover (1964), Cooper et al. (1964), and Henry (1959) recognized the analogy between the free-surface conditions in earth dams (under steady-state conditions) and the interface (Fig. 9.4). For simplicity, the outflow surface is designated as either vertical or horizontal rather than as the true sloping boundary. Because of the limited dimensions of the widths of earth dams, the analogy may be stated more realistically as existing between gravity flow in two-dimensional earth sections (Kashef, 1977) and saltwater intrusion. Accordingly, the solutions used for gravity-flow problems (Sec. 5.7) can be used (with certain modifications) to find the interface under steady-state freshwater flow. In most of the solutions for finding the interface, by analogy to the free surface in

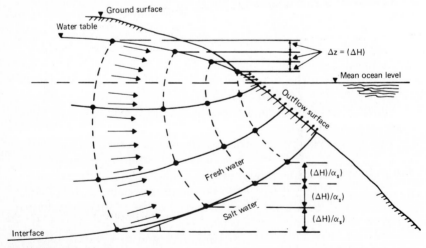

Figure 9.4 Hydraulic conditions at the interface.

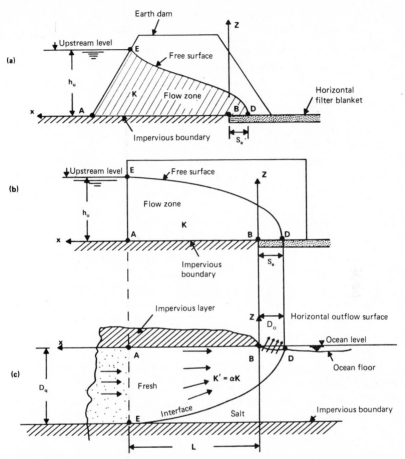

Figure 9.5 Analogy between free surface and interface in artesian aquifers assuming a horizontal outflow surface. (*a*) Section in an earth dam with a horizontal filter blanket; (*b*) fictitious wide rectangular earth dam with a horizontal filter blanket; (*c*) interface *ED* analogous to free surface in part (*b*). Rates of flow in parts (*b*) and (*c*) are not equal. (*A. I. Kashef, "Management and Control of Salt-Water Intrusions in Coastal Aquifers," CRC Crit. Rev. Environ. Control, vol. 7, no. 3, 1977.*)

earth dams, the hydraulic conductivity K is replaced by

$$K' = \alpha_s K \tag{9.9}$$

This means that the rate of flow is smaller than its equivalent in an earth dam with the same geometric boundaries (Figs. 9.5, 9.6, and 9.7).

In unconfined aquifers, the consensus among various investigators (Charmonman, 1965; Cooper et al., 1964; Kashef, 1968*b*; and Verruijt,

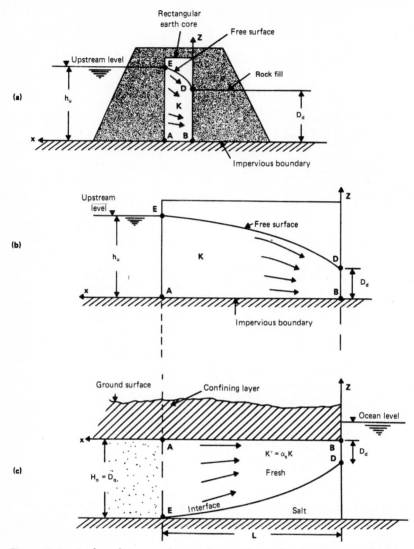

Figure 9.6 Analogy between free surface and interface in artesian aquifers assuming a vertical outflow surface. (*a*) Rectangular earth core in a rock-fill dam; (*b*) fictitious wide rectangular earth dam; (*c*) interface *ED* analogous to free surface in part (*b*). Rates of flow in parts (*b*) and (*c*) are not equal. (*A. I. Kashef, "Management and Control of Salt-Water Intrusions in Coastal Aquifers," CRC Crit. Rev. Environ. Control, vol. 7, no. 3, 1977.*)

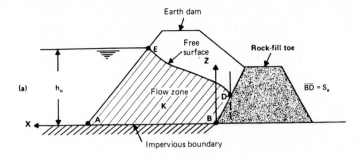

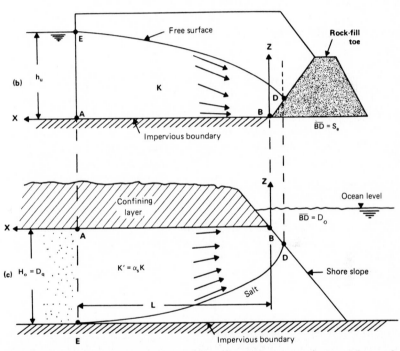

Figure 9.7 Analogy between free surface and interface in artesian aquifers with a sloping outflow surface. (*a*) Earth dam with a rock-fill toe; (*b*) fictitious wide earth dam with a vertical upstream face and a rock-fill toe; (*c*) interface *ED* analogous to free surface in part (*b*). Rates of flow in parts (*b*) and (*c*) are not equal. (*A. I. Kashef, "Management and Control of Salt-Water Intrusions in Coastal Aquifers," CRC Crit. Rev. Environ. Control, vol. 7, no. 3, 1977.*)

1968) is that the zone beneath a horizontal plane (*AB* in Fig. 9.8) coinciding with sea level can be treated as a confined aquifer and the relatively small zone above that plane has a free surface with a height D_{xu} approximately equal to the heads along *AB*. In hydraulic terms, this assumption is equivalent to assuming that in the real flow system there is a horizontal streamline with a height equal to H_o (along *AB* in Fig. 9.8) above the lower impervious boundary. Generally, the portion of total rate of flow above sea level is negligible, and this assumption is justified. Thus if the total rate of flow in an unconfined aquifer is q, the rate of flow above plane *AB* (coinciding with the sea level) will be about $q/(1 + \alpha_s)$ (Kashef, 1968*b*).

The main solutions for finding the interface have been grouped (Kashef, 1976*b*) in the following unified formula, which can be used for unconfined or confined aquifers with horizontal or vertical outflow surfaces:

$$D_x^2 = A_s \frac{q_a}{K'} x + B_s \left(\frac{q_a}{K'}\right)^2 \tag{9.10}$$

where D_x = thickness of freshwater zone at a distance x from the sea front in a confined aquifer or the portion of the freshwater zone below sea level in an unconfined aquifer

x = distance from the sea front in an aquifer

A_s, B_s = constants given in Table 9.1 according to the indicated conditions

q_a = steady rate of flow in a confined aquifer or the portion of the total rate of flow q below the plane coinciding with sea level in unconfined aquifers

The term q_a can be expressed as

$$q_a = \frac{q}{1 + \alpha_s} \tag{9.11}$$

x is positive left of the origin, which is located at the sea front in an unconfined aquifer. In confined aquifers, the x axis coincides with the upper impervious boundary (and at sea level in unconfined aquifers).

In a confined aquifer, Eq. (9.10) and Table 9.1 are used, taking into account that q_a is the total rate of natural flow in the aquifer. If the outflow surface is assumed vertical, D_o is the same as D_x at $x = 0$. If the outflow surface is assumed horizontal, D_x at $x = 0$ [Eq. (9.10)] gives the depth of the interface at the origin.

Since unconfined aquifers are treated as confined aquifers below the x axis (Fig. 9.8), assuming that the x axis is a horizontal streamline, then q_a in Eq. (9.10) is the portion of the total rate of flow q and can be

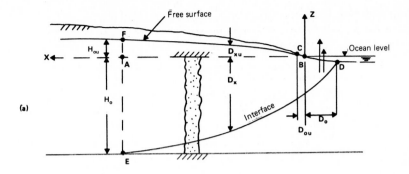

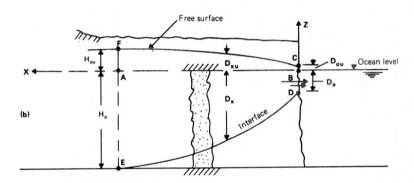

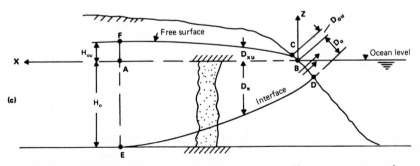

Figure 9.8 Interfaces in unconfined aquifers determined by assuming equivalent artesian aquifers (*AB* is assumed impervious). (*a*) Horizontal outflow surface; (*b*) vertical outflow surface; (*c*) sloping outflow surface. (*A. I. Kashef, "Management and Control of Salt-Water Intrusions in Coastal Aquifers," CRC Crit. Rev. Environ. Control, vol. 7, no. 3, 1977.*)

TABLE 9.1 **Main Solutions for Locating the Interface with Reference to Eq. (9.10)**

Method of solution and investigator	Assumed outflow surface	Value of A_s in Eq. (9.10)	Value of B_s in Eq. (9.10)	Length of outflow surface D_o
Conformal mapping: Charmonman (1965) and Verruijt (1968)	Horizontal	2.0	$(1 - \alpha_s^2)$	$-\frac{1}{2}(1 - \alpha_s)\frac{q}{K'} \approx -\frac{1}{2}(1 - \alpha_s^2)\frac{q}{K'}$
Conformal mapping: Glover (1964) and Henry (1959) using Kozeny's solution	Horizontal	$2(1 + \alpha_s)$	$(1 + \alpha_s^2)$	$-0.5\frac{q}{K'} \approx -0.5(1 + \alpha_s)\frac{q}{K'}$
Hydraulic forces: Kashef (1967)	Vertical	2.0	0.5	$0.704\frac{q_a}{K'} = \frac{0.704q}{(1 + \alpha_s)K'}$
Mathematical approximation using Dupuit's assumption but introducing the outflow depth determined by Polubarinova-Kochina (using conformal mapping)	Vertical	2.0	0.5491	$0.741\frac{q_a}{K'} = \frac{0.741q}{(1 + \alpha_s)K'}$

SOURCE: A. I. Kashef, "Theoretical Developments and Practical Needs in the Field of Salt Water Intrusion," in *Advances of Groundwater Hydrology*, American Water Resources Association, 1976b, pp. 228–242.

determined from Eq. (9.11). The portion of the rate of flow above the x axis (Fig. 9.8) amounts to about 2.5 percent of q [Eq. (9.11); and Kashef, 1977]. In practice, a free surface having a height above the x axis equal to 2.5 percent of D_x [Eq. (9.10)] may be plotted. This is justified from the following conclusions given by Charmonman (1965), who compared the results of solutions for unconfined and confined aquifers. Horizontal outflow surfaces were assumed and the value of γ_{st} was 1.025 g/cm³.

1. The level of the piezometric surface at any point along the assumed confining aquifer (x axis in Fig. 9.8) is lower than the true free surface by an amount less than 1.3 percent.

2. The interface of the assumed artesian aquifer is deeper than the interface of the true unconfined aquifer. The error increases steadily from less than 1.3 percent at infinite distance from the shoreline to less than 2.6 percent at the shoreline.

The length L of the intruded saltwater wedge in all aquifers is given by (Kashef, 1968b)

$$L = \frac{K'H_o^2}{2q_a} \tag{9.12a}$$

or

$$L = \frac{K'H_o^2(1 + \alpha_s)}{2q} \tag{9.12b}$$

Equation (9.12) has a different basis than Eq. (9.10). However, if the latter is used to find L by substituting $D_x = H_o$, $x = L$, $\alpha_s = 0.025$, $A_s = 2$, and $B_s = 0.5491$ (from the last row of Table 9.1); L would therefore have the value given by Eq. (9.12) less a negligible amount for all practical purposes (that is, $L = K'H_o^2/2q_a - 0.275q_a/K'$).

In most rigorous analyses, the segment of the outflow surface above sea level D_{ou} (Fig. 9.8) is not accounted for. Although such a segment exists, it is very small and is approximately equal to $0.025D_{x=0}$, where $D_{x=0}$ is D_x at $x = 0$. If the outflow surface is considered vertical, then $D_o = D_{x=0} = 0.741q_a/K'$ and $D_{ou} = 0.025D_o$. When the outflow surface is assumed horizontal, then from Eq. (9.10), $D_{x=0} = \sqrt{B_s}(q_a/K')$ and $D_{ou} = 0.025\sqrt{B_s}q_a/K'$ or $D_{ou} \cong 0.025q_a/K'$. (The value of B_s is considered to be equal to $1 - \alpha_s^2$, as given in the first row of Table 9.1.)

9.4   Modified Ghyben-Herzberg Curve

The Ghyben-Herzberg curve has been modified to be compatible with the rigorous solutions (Kashef, 1983). The attractiveness of the Ghyben-Herzberg principle to practitioners was the main reason for this attempt to provide simplicity without introducing appreciable errors. The symbols D_{xv} and D_{xh} to express D_x (Fig. 9.1) corresponding respectively to assumed vertical and horizontal outflow surfaces were introduced, and the following dimensionless equations were derived; they give essentially the same results as the rigorous solutions.

$$\frac{D_x'}{H_o} \cong \frac{D_{xu}}{\alpha_s H_o} \cong \sqrt{x} \qquad \text{(based on Ghyben-Herzberg principle)} \tag{9.13}$$

$$\frac{D_{xv}}{H_o} \cong \left[\left(\frac{D_{xu}}{\alpha_s h_o}\right)^2 + \frac{0.1375(1 - \bar{x})}{\bar{L}^2}\right]^{1/2} \tag{9.14}$$

$$\frac{D_{xh}}{H_o} \cong \left[\left(\frac{D_{xu}}{\alpha_s H_o}\right)^2 + \frac{1}{4\bar{L}^2}\right]^{1/2} \tag{9.15}$$

where $\qquad \bar{x} = \dfrac{x}{L} \qquad$ and $\qquad \bar{L} = \dfrac{L}{H_o} = \dfrac{K'}{2v_o}$

and v_o is the darcian natural velocity within the aquifer.

In applying Eqs. (9.14) and (9.15), D_{xu} values are determined from the field rather than analytically by the use of Eq. (9.14). Instead of applying Eqs. (9.13) to (9.15), Table 9.2 has been prepared for a range of $\bar{L}$ between 1.5 and 10 and for values of $\bar{x}$ between 0.1 and 0.9 and compared with the Ghyben-Herzberg solution. Whenever $\bar{L}$ values increase, the differences decrease until the three values determined from Eqs. (9.13) to (9.15) become very close to the Ghyben-Herzberg values for high values of $\bar{x}$ and for all values of $\bar{L} \geqslant 4.0$.

The values of D_{xv} and D_{xh} are always larger than D'_x. In addition, the values of D_{xh} are always greater than D_{xv}. The values of D_{xh} and D_{xv} [Eqs. (9.14) and (9.15)] are based on the rigorous solutions, and the outflow surfaces are also idealized (either vertical or horizontal), as in the rigor-

TABLE 9.2   Determination of Interface: Modified Ghyben-Herzberg Approach

$\bar{L}$		$\bar{x}$ 0	0.1	0.2	0.3	0.4	0.5	0.6	0.7	0.8	0.9	1.0
1.5	(a)	0.247	0.394	0.499	0.585	0.661	0.728	0.790	0.848	0.901	0.952	1.000
	(b)	0.333	0.459	0.558	0.641	0.715	0.782	0.843	0.901	0.954	1.005	1.054
2.0	(a)	0.185	0.362	0.467	0.569	0.649	0.719	0.783	0.843	0.898	0.950	1.000
	(b)	0.249	0.402	0.512	0.602	0.680	0.750	0.814	0.873	0.928	0.981	1.031
3.0	(a)	0.124	0.337	0.461	0.557	0.640	0.712	0.779	0.839	0.896	0.949	1.000
	(b)	0.167	0.357	0.477	0.573	0.654	0.726	0.792	0.853	0.910	0.963	1.014
4.0	(a)	0.093	0.328	0.455	0.553	0.637	0.710	0.777	0.838	0.895	0.949	1.000
	(b)	0.126	0.341	0.465	0.562	0.645	0.718	0.785	0.846	0.903	0.957	1.008
6.0	(a)	0.062	0.322	0.451	0.550	0.634	0.708	0.776	0.837	0.895	0.949	1.000
	(b)	0.084	0.327	0.455	0.554	0.638	0.712	0.779	0.841	0.898	0.952	1.003
8.0	(a)	0.046	0.319	0.449	0.549	0.633	0.708	0.775	0.837	0.895	0.949	1.000
	(b)	0.063	0.322	0.452	0.551	0.636	0.710	0.777	0.839	0.897	0.951	1.002
10.0	(a)	0.037	0.318	0.448	0.549	0.633	0.708	0.775	0.837	0.895	0.949	1.000
	(b)	0.050	0.320	0.450	0.550	0.634	0.709	0.776	0.838	0.896	0.950	1.001
All values of $\bar{L}$	(c)	0	0.316	0.447	0.547	0.632	0.707	0.774	0.837	0.894	0.949	1.000

Note: (a), values of D_{xv}/H_o [Eq. (9.14)]; (b), values of D_{xh}/H_o [Eq. (9.15)]; (c), values of D'_x/H_o [Eq. (9.13), independent of $\bar{L}$].

SOURCE: A. I. Kashef, "Harmonizing Ghyben-Herzberg Interface with Rigorous Solutions," *Ground Water*, vol. 21, no. 2, 1983, pp. 153–159.

ous solutions. The design value should be intermediate between D_{xh} and D_{xv} for a sloping outflow face. Owing to a lack of sufficient solutions for the latter case, the selection of an appropriate value is based on interpolation (see the analogous case in seepage through earth dams, Sec. 5.7).

Practically, the differences in results between the rigorous solutions and the Ghyben-Herzberg approach may in fact be less than the effect of small errors in recording water levels in the field, as illustrated in the following example.

EXAMPLE 9.1° The lower impervious boundary of a coastal unconfined aquifer is at elevation -107 m below sea level (Fig. 9.1). Two observation wells are installed at E_1 and E_2. The ground-surface elevations are 2.433 m at E_1 and 3.81 m at E_2, and the water levels are 1.803 m at E_1 and 2.743 m at E_2. The observation well at E_1 is closer to the shore, and the distances of both wells are to be determined from the solution. In addition, $\alpha_s = 0.025$, $K = 2 \times 10^{-2}$ cm/s (57.6 ft per day), and $v_o = 5 \times 10^{-5}$ cm/s (0.144 ft per day). Find L and the thicknesses of the freshwater lens at E_1 and E_2.

Solution

$$\text{Natural gradient} \quad i_g = \frac{\alpha_s}{2\bar{L}} = \frac{v_o}{K} = \frac{\alpha_s H_o}{2L} = \frac{5 \times 10^{-5}}{2 \times 10^{-2}} = 0.0025$$

Therefore, $\quad L = 535$ m $\quad \bar{L} = \dfrac{L}{H_o} = \dfrac{535}{107} = 5.0 \quad \alpha_s H_o = 2675$ m

Water level at $E_2 > \alpha_s H_o$; therefore, E_2 is located beyond the intruded length L by a distance $(2.743 - 2.675)/i_g = 27.2$ m or a distance $(535 + 72.2) = 562.2$ m from the shoreline.

Water level at $E_1 < \alpha_s H_o$; therefore, E_1 is located in the region above the saltwater-intruded wedge:

$$D'_x = \frac{1}{\alpha_s} D_{xu} \text{ [Eq. (9.13)]} = \frac{1.803}{0.025} = 72.12 \text{ m} = H_o \sqrt{\bar{x}}$$

$$\bar{x} = 0.454$$

Therefore, E_1 is at a distance $0.454 \times 535 = 243$ m from the origin and the distance between E_1 and E_2 is 319.2 m. Applying Eqs. (9.14) and (9.15) at E_1,

$$D_{xv} = 72.36 \text{ m} \quad \text{and} \quad D_{xh} = 72.91 \text{ m}$$

It should be noted in this example that the entire interface can be plotted from Eqs. (9.13), (9.14), and (9.15) or by the use of Table 9.2. The free-surface curve also can be plotted on the basis of just *one* recording at observation well E_1 using Eq. (9.13).

The differences between D'_x and either D_{xv} or D_{xh} in this example are negligible. These differences would be greater for lesser values of $\bar{x}$ and larger values of v_o. The differences at observation well E_1 are as follows:

$$D_{xv} - D'_x = 0.24 \text{ m} \quad D_{xh} - D'_x = 0.79 \text{ m} \quad D_{xh} - D_{xv} = 0.55 \text{ m}$$

° Example 9.1 is based on a similar example given by Kashef (1983).

From a practical point of view, such small differences justify the use of the Ghyben-Herzberg principle. The differences produced from the sensitivity of the water-level records may be much greater. For example, if it is expected that there will be an error of ± 5.0 cm only in the recording at E_1, this would lead to differences in the values of D'_x, D_{xv}, and D_{xh} on the order of ± 1.95 m, which is much greater than the differences between Ghyben-Herzberg and the rigorous solutions. This conclusion is very useful for simplifying management problems in coastal aquifers (see Secs. 9.7 and 9.8).

9.5 Model Studies

Model studies were and are still used to verify certain theories in the general field of groundwater flow or to determine the values of certain parameters. In complex cases, model studies are extremely useful, especially when an analytic solution does not exist. Each type of model has several drawbacks. In the saltwater-intrusion field, viscous flow models have been used extensively. Models dealing with gravity flow systems and problems in oil fields (multiple-fluid systems) have features that are analogous to some phases of seawater intrusion.

The fundamental physical laws of nature are strikingly similar in many fields. Darcy's law for groundwater flow systems is analogous to Fourier's law for heat transfer, Ohm's law for electric conduction, and other fundamental laws in such other fields as electrostatics and magnetism. These analogies can thus be applied in specially designed models to simulate groundwater flow patterns. Physical models as well as models consisting of electrolytic tanks, stretched membranes, electric conduction sheets, viscous flow between closely spaced plates, magnetic flux systems, heat transfer experiments, and others have been used. Glass beads, spherical shots, or sand have been used in physical models. Mathematical models based on numerical analysis have also been powerful tools for solving complex cases that cannot be solved using classical mathematical procedures (Bear, 1979).

Viscous models date back to 1897 (Hele-Shaw, 1898, 1899) and have since been used by several investigators. The analogy that forms the basis for these models is that a viscous fluid flowing through a narrow channel follows the basic hydrodynamic laws of a fluid flowing through porous media. The narrow channel is usually formed by two closely spaced, transparent, vertical or horizontal plates that simulate a segment of an aquifer (Kashef, 1970). The geometric shapes at both ends of the model should be the same as the prototype. The model can be used to analyze various cases of steady and transient groundwater flow systems. It can be constructed to study radial as well as two-dimensional problems. Variation in soil permeabilities can also be accounted for in such models by observing the model scales (Bear, 1960).

Santing (1957) extended the horizontal Hele-Shaw model to study steady and transient states of flow. Storage in an artesian aquifer was simulated by connecting vertical storage vessels to the model. Santing used his viscous model to study water losses by withdrawal or evaporation and water recharge by rainfall or recharge wells. His model contains many simplifications that must be verified. The simulation of water wells in such a model and the disturbing effects of the storage vessels on the very small flow rate within the model are questionable.

Dvoracek and Scott (1963) used Hele-Shaw models to study the effects of recharge pits on the groundwater system. They used 72×18 in rectangular plates with a trapezoidal segment to simulate the recharge pit. The small spacing (0.0365 in = 0.093 cm) allowed them to use distilled water. Marino (1967) studied the growth and decay of groundwater ridges using rectangular plates $243 \times 45 \times 1.27$ cm with closed ends and spaced at 0.13 cm. Shell Tellus oil 72 was used as the fluid and was introduced by capillary tubes.

Columbus (1965) used a Hele-Shaw model to study the steady-state case of seawater intrusion in coastal aquifers with vertical outflow surfaces. His model consisted of two 237×30 cm transparent Lucite plates. The inlet and outlet reservoirs consisted of cylinders joined by closely spaced brass tubes at each end of the model to avoid disturbing the fresh water entering the model. Brass tubes extending across the inlet reservoir of salt water were used to collect the fresh water from the model. The plates were spaced at 0.1 cm; this allowed the use of salt water and distilled water as the fluids as opposed to simulating fluids.

The viscous model used by Kashef (1970) was made of two parallel plates of Lucite acrylic plastic 1.25 cm thick, 240 cm long, and 64 cm high. The spacing between the plates was kept uniform with the aid of 0.6-cm brass washers arranged in a square grid pattern (mesh size of 27×27 cm) and held in place by means of brass bolts extending through the washers and plates. A square grid was drawn on the front of the channel, and the shape of the intruded saltwater wedge (or the free surface using one fluid to calibrate the model) was recorded by a movie camera. Fresh water was simulated by a viscous fluid with a specific gravity of 1.0 at 26.6°C (80°F), which is almost equal to that of distilled water at the same temperature (specific gravity of 0.9967), by mixing fresh water with 0.4 percent by weight of carboxymethylcellulose (CMC), 7 high premium (7HP). The CMC was found to have the advantages of efficient thickening to any desired viscosity in a water solution; easy solubility in hot or cold water; resistance to greases, oils, and organic matters; and physiological inertness. Salt water was simulated by a solution of the same percentage of CMC mixed with 1.5 percent by weight of regular sugar and 2.25 percent by weight of table salt to produce a fluid with a specific gravity of 1.023. Diphenyl fast blue was added to the simulated fresh water to obtain a distinct interface in the recording films. The spacing of the plates was relatively large as compared with other models because it was found that the tolerance in the thicknesses of commercial plate exceeds many of the previously reported spacings. In the study of transient-state conditions, the freshwater flow was increased or decreased (Kashef, 1970). When the rate of freshwater flow was increased, the transitional zone was practically eliminated (Kashef, 1970). However, when the rate of freshwater flow was decreased, the saltwater wedge advanced inward, forming a distinct transitional zone.

Viscous flow models are generally preferred to physical sand models, in which entrapped air and capillarity create experimental errors. Furthermore, the flow of the liquids, the free surface, and/or the interface are visible through the transparent plates of the model and can be photographically recorded. However, the following experimental disadvantages are experienced using these models:

1. The effect of changes in temperature during the test and the subsequent gradual changes in the viscosities can be a problem, especially in transient-state tests.

2. The tolerance in the plates is, in many cases, of the same order of magnitude or more than the planned spacing between the plates. A relatively narrow spacing is always desirable in order to allow the use of small quantities of fluids with relatively low viscosities. Models with extremely narrow spacings are questionable.

3. The measurement of the changing rates of flow poses some problems. It requires continuous fast recording or more than one person performing the test. Automatic sensing devices are therefore recommended.

4. It is sometimes difficult to keep the two fluids separated at the outflow surfaces.

5. If the model is used to investigate various factors at the same time (discharge, recharge, storage, evaporation, well effects, anisotropy, etc.), theoretical oversimplifications are usually made, such as using a two-dimensional model to simulate a three-dimensional prototype.

6. The selection of the fluid or fluids sometimes creates problems. It is one of the major decisions in constructing a Hele-Shaw model, yet the choice of fluid is often based solely on convenience. The fluids should have sufficient viscosity to accommodate the selected spacing and should be of such type as to minimize experimental errors (Kashef, 1970).

9.6 Deviation from Idealization

The governing mathematical equations for general flow and dispersion include assumptions that are introduced in order to develop or solve these equations under idealized hydraulic and geometric boundary conditions. Deviation from idealization leads to highly complex mathematical equations that can only be solved numerically using computers or expensive testing (Bear, 1979; Gelhar and Collins, 1971; Oakes and Edworthy, 1976; Ogata, 1970; Scheidegger, 1961).

In the previous analysis, the flow medium was assumed to be homogeneous and isotropic. Limited cases of anisotropy can be implied in these analytical solutions when the extreme values of the hydraulic conductivities are normal to each other and have the same value everywhere in the medium. Many investigators have attempted to solve other complex problems for heterogeneous layered media, introducing irregular boundaries closer to actual conditions. Although such solutions may be beneficial in the long run, the numerous physical parameters assumed in the solutions are rather difficult to obtain in practice. For example, evaporation losses and water recharge from rainfall can easily be included in most of the hydrodynamic equations. These rates, which are erratic in reality, are usually assumed constant or subject to a prescribed pattern that may never be achieved in practice. Field investigations are commonly performed to verify analytic methods (Lusczynski and Swarzenski, 1966; U.S. Geological Survey, 1966; Kohout, 1960; Cooper et al., 1964; Kashef, 1967). However, the vagaries of nature are not compatible with theoretical idealization. For this reason, analytic solutions are of value only if they include the most probable ranges of natural conditions. If different solutions corresponding to various extreme conditions are available, individual judgment and experience are necessary to select the most appropriate results from a wide range of analytic solutions. In the following, a few factors are presented to illustrate that the Ghyben-Herzberg or other interfaces are indeed idealized.

Earth Tides, Ocean Tides, and Seismic Waves

In the groundwater field, observation and water wells are usually used to find the physical parameters of the aquifer and to detect any effects resulting from barometric changes, ocean tides, earth tides, earthquakes, and other types of seismic waves (see Sec. 6.5). These factors disturb the natural freshwater-saltwater interface in coastal aquifers. For example, it was reported in 1913 that earth tides caused a relatively large fluctuation (in excess of 6 cm) in a well near Cradock, South Africa (Bredehoeft, 1967). It is also well established that earthquakes, detonation of explosives, and even passing trains or trucks affect the water levels in observation wells, at least temporarily. The intensity of seismic-wave effects depends on the distance from the source, the depth and characteristics of the observation well, and the characteristics of soil layers within the domain of these waves. Some reported recordings are rather unusual. For example, it was reported (Cooper et al., 1965) that the water level in an open artesian well near Perry, Florida fluctuated over a double amplitude of as much as 4.6 m (about 15 ft) in response to the Alaskan earthquake of March 27, 1967. This change is equivalent to a change in effective pressure of about 0.47 kg/cm² (940 lb/ft²) at the bottom level of an observation well. Considering the distance between Alaska and Florida, this change in effective pressure is rather extraordinary. It was also reported by the same authors that the response in other wells penetrating the same aquifer was negligible.

Barometric changes and the effects of ocean tides have been investigated by Jacob (1950) and others. A decrease in barometric pressure will cause a slight rise in the ocean level and a consequent rise in the interface. The effects of tides and barometric changes were discussed in Chap. 6 (Sec. 6.5).

Dispersion and Diffusion of Salt Water

In the previous idealization, fresh and salt waters were assumed immiscible, although they are in fact miscible (Muskat, 1937; Polubarinova-Kochina, 1962; DeWiest, 1969; Bear, 1979). Although numerous investigations dealing with dispersion and diffusion have been conducted, the results are still inconclusive and cannot be applied by groundwater managers (Scheidegger, 1961; Gelhar and Collins, 1971; Oakes and Edworthy, 1976; Bear, 1979).

Under steady-state conditions, Verruijt (1971) proved that the transitional zone between the salt water and the fresh water (known also as the *dispersion* or *diffusion zone*) is not of great importance. However, field evidence in two rather exceptional cases, in Pearl Harbor, Hawaii, and the Cutler area near Miami, Florida (Fig. 9.9), indicated that the transitional zones are relatively large (Kohout, 1960; Cooper et al., 1964). Rumer and Harleman (1963) attempted to study dispersion at the interface both analytically and experimentally. They concluded that the stationary freshwater-saltwater interface is subject to dispersion in the absence of tidal motion as a result of lateral flow of fresh water parallel to the interface. They found that the width of the transitional zone in this case grows from zero at the wedge toe (in an artesian aquifer), where the velocity is zero, to a maximum at the aquifer boundary at the ocean. They also investigated tidal effects and concluded that the transitional zone is a sigmoid shape. Rumer and Harleman (1963) believed that the transi-

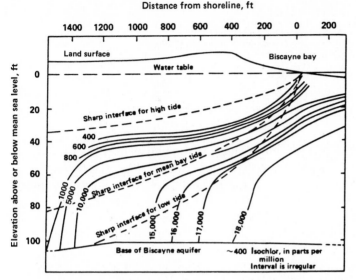

Figure 9.9 Section through the Cutler Area near Miami, Florida showing isochlors on September 18, 1958 and interfaces computed as if conditions were steady at high, mean, and low tides. *(H. H. Cooper et al., "Sea Water in Coastal Aquifers," U.S. Geological Survey Water Supply Paper 1613, Washington, 1964.)*

tional zone is greater in anisotropic media than in isotropic media and that the rate of dispersion produced by the tide decreases landward. They concluded that in nature, tides do not appear to produce large zones of dispersion in homogeneous aquifers.

Generally, the water density ρ near the interface changes over time because of the increase or decrease in salinity and its average is given by (mass of solution)/(volume of solution). The average concentration c of a substance (salt in this case) is defined by $c =$ (mass of substance)/(mass of solution). If a substance is introduced into a flow system over an area or at a point, it will move away from its source and be distributed throughout the flow medium by combined mechanisms leading to a concentration change over location and time. The combined mechanisms in a two-dimensional flow system consist of molecular diffusion and sorption as well as convectional dispersion (due to the fluid velocity), which consists of longitudinal dispersion (along the flow direction) and lateral dispersion (normal to the flow direction).

Longitudinal dispersion ϵ_s and lateral dispersion ϵ_n are expressed as follows (Rumer and Harleman, 1963):

$$\epsilon_s = \epsilon_m + \lambda_s (v_s^n)_s \qquad (9.16)$$

$$\epsilon_n = \epsilon_m + \lambda_n (v_s^n)_n \qquad (9.17)$$

where v_s = the seepage velocity
         n = a certain power of v_s
         ϵ_m = the molecular diffusion, cm²/s
    λ_s, λ_n = constants

In actual problems, the power n is considered as unity, and therefore,

$$\epsilon_s \simeq \lambda_s(v_s)_s + \epsilon_m \tag{9.16a}$$

$$\epsilon_n \simeq \lambda_n(v_s)_n + \epsilon_m \tag{9.17a}$$

where λ_s and λ_n (length dimensions) are known as *intrinsic dispersivity coefficients* and are functions of the solid structure of the flow medium only. For practical purposes, the rate of gain of substance within a certain volume due to leaching from porous material or chemical reactions is disregarded. The salt-balance equation for an element should thus be written in such a way that the time rate of change of salt within the voids of the element plus the net rate of efflux of salt from this element by convection (due to average seepage velocity v_s) is equal to the net rate of efflux of salt from the pore volume by dispersion and molecular diffusion. The hydrodynamic dispersion equation in a steady saturated flow of a homogeneous fluid through a homogeneous isotropic porous medium is thus

$$\frac{\delta C_{st}}{\delta t} = \frac{\delta}{\delta s}\left(\epsilon_s \frac{\delta C_{st}}{\delta s} - v_s C_{st}\right) + \left(\epsilon_n \frac{\delta C_{st}}{\delta n} - v_n C_{st}\right) \tag{9.18}$$

where s and n denote directions along and normal to the flow lines, respectively, $C_{st} = c/c_o$ is the relative concentration of salt ($0 \leqslant C_{st} \leqslant 1$), and c_o is the initial concentration of salt. The elastic properties of the fluids and the flow medium are ignored in Eq. (9.18).

Various forms of salt-balance equations have been developed, and various problems under specified conditions have been solved mathematically, numerically, or experimentally [see, for example, de Josselin de Jong (1968) and Pinder and Cooper (1970)]. In most of these analyses, theoretical and experimental simplifications of various types have been introduced.

Henry, in his studies of dispersion (Cooper et al., 1964), introduced an oversimplification by assuming the dispersion coefficient ϵ_c to be a scalar quantity of the same magnitude in all directions. He solved a case in which $\bar{L} = L/H_o = 2.0$, $q/KH_o = 0.263$, and $\epsilon_c/q = 0.1$. His results can be summarized as follows:

1 The sharp interface approximately separates the freshwater flow from the circulating salt water.

2. The isochlors approach the base vertically (Fig. 9.9), as recorded in field tests by Kohout (Cooper et al., 1964).

3. The extent of saltwater intrusion L is less than that determined analytically (assuming that the fluids are immiscible).

4. The lowest streamline of the idealized analytic solution passes through the center of the transitional zone.

Cooper et al. (1964) assumed that ϵ_c has the same value in all directions and gave its value due to ocean tides as follows (disregarding the molecular diffusion ϵ_m):

$$\epsilon_c = \frac{4\lambda_c A_t}{t_{ot}} \qquad (9.19)$$

where A_t is the amplitude of the tide, t_{ot} is the period of the displacement of water in the aquifer caused by ocean tides, and $\lambda_c \approx \lambda_s \approx \lambda_n$ [Eqs. (9.16) and (9.17)]. As reported by Cooper et al. (1964), λ_c increases with increases in the uniformity coefficient; for example, $\lambda_c = 0.063$ cm for Ottawa sand and 0.13 cm for Monterey sand and becomes as high as 2.79 cm for a sand with a uniformity coefficient of 3.88.

Field records are limited; many are unpublished. Two field cases were extensively investigated by Kohout (Cooper et al., 1964), who indicated that there is a moving cycle from the ocean floor into the transitional zone and back to the ocean. The geohydrology of these two cases (one in Hawaii and the other in Florida) is exceptional; it differs from most of the coastal aquifers in the United States and elsewhere. Terrestrial springs as well as other exceptional geologic features around Pearl Harbor affect the uniform flow pattern. In the Florida site (Fig. 9.9), the aquifer is essentially cavernous limestone with exceptionally high gradients and hydraulic conductivities. The value of the latter ranged between 2.3 and 3.3 cm/s, and thus Darcy's law, which is implied in all governing formulas, is not valid here. Other observations in the Florida site confirmed these unusual conditions. For example, after rainfall in the Silver Bluff area of Biscayne Bay, salt water was rapidly expelled from the aquifer. Such rapid changes cannot occur under normal circumstances because of the very slow process of retardation of salt water by freshwater recharge (Sec. 9.8).

The reader can consult the work of Bear (DeWiest, 1969; Bear, 1979) for a thorough discussion of hydrodynamic dispersion. It should be realized that saltwater intrusion problems, including the actual field conditions of continuously changing interfaces and time-dependent dispersion, are very complex to analyze.

Layered Systems

Deep aquifers usually consist of interbedded impervious layers (clay layers, for example) or pockets. The most reliable field methods are not adequate to detect such layers or pockets and cannot give a perfect picture of the aquifer system. Even when deep borings are done, many of these pockets are missed. Continuous and very thin clay layers or beds also may be missed by boring. In deep aquifers, the location of the lower impervious boundary is almost a matter of evaluation rather than of exact knowledge. The boundary may cover a zone of various formations or even fissured rock that has a high hydraulic conductivity. This practical difficulty was overcome in conducting permeability pumping tests by introducing the coefficient of transmissivity T ($=KD_q$) rather than by trying to find its components K and D_q (Chap. 6).

Bear (1979) analyzed the condition of a two-layered system (Fig. 9.10). The impervious bed gg' separates the main unconfined aquifer into two separate aquifers; the upper is unconfined and the lower is confined. Two saltwater wedges are developed, one in each aquifer. The total rate of flow q in the main aquifer before its separation is divided into q_1 above gg' and q_2 below gg' (Fig. 9.10). Assuming the hydraulic conductivities K_1 for the upper unconfined aquifer and K_2 for the lower confined aquifer, then (Bear, 1979)

$$q_1 = \frac{K_1 H_{o1}}{L_1}(D_{gu} - \tfrac{1}{2}\alpha_s H_{o1}) = \frac{T_1}{L_1}(D_{gu} - \tfrac{1}{2}\alpha_s H_{o1}) \tag{9.20}$$

$$q_2 = \frac{K_2 H_{o2}}{L_2}(D_{gu} - \tfrac{1}{2}\alpha_s H_{o2} - \alpha_s H_{o1}) = \frac{T_2}{L_2}(D_{gu} - \tfrac{1}{2}\alpha_s H_{o2} - \alpha_s H_{o1}) \tag{9.21}$$

$$q = q_1 + q_2 \tag{9.22}$$

$$\frac{q_1}{q_2} = \frac{K_1 H_{o1}}{K_2 H_{o2}}\frac{L_2}{L_1}\frac{q + \alpha_s H_o K_2 H_{o2}/2L_2}{q - \alpha_s H_o K_1 H_{o1}/2L_1}$$

$$\text{or } \frac{q_1}{q_2} = \frac{T_1 L_2}{T_2 L_1}\frac{q + \alpha_s H_o T_2/2L_2}{q + \alpha_s H_o T_1/2L_1} \tag{9.23}$$

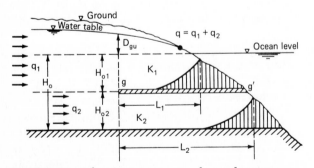

Figure 9.10 Saltwater intrusion in a layered system.

Symbols that have not as yet been defined in these equations are shown in Fig. 9.10. Once q_1 and q_2 have been determined, each aquifer is treated as a separate aquifer to find its saltwater wedge using the methods previously discussed.

The main difficulties in applying Eqs. (9.20), (9.21), and (9.23) are practical ones; the determination of L_1, L_2, T_1, T_2, q, and the exact location of gg' in order to record D_{gu} are not simple. In practice, most likely one average value for T is used, and owing to the difficulty (or expense involved) in finding the shape of the slope below sea level, L_1 and L_2 are considered equal to one value L''. Accordingly, Eqs. (9.20) through (9.23) are approximated as follows:

$$\frac{q_1}{q_2} \simeq \frac{2L''q + \alpha_s H_o T}{2L''q - \alpha_s H_o T} \simeq \frac{D_{gu} - \frac{1}{2}\alpha_s H_{o1}}{D_{gu} - \frac{1}{2}\alpha_s H_{o2} - \alpha_s H_{o1}} \qquad (9.24)$$

In order to use Eq. (9.24), q, H_o, and T or D_{gu}, H_{o1}, and H_{o2} should be predetermined.

Analyzing the preceding equations, Bear (1979) concluded that

1. As H_o increases, the values of L increase, with a larger value of intrusion in the lower aquifer than in the upper one.

2. As the length of the separating layer gg' increases, seawater intrusion increases in the upper aquifer and decreases in the lower one.

9.7 Disturbance of the Natural Interface

The sharp natural interface developed under the influence of the steady-state flow of fresh groundwater has already been discussed. The natural flow cannot remain unchanged, and it is subject to natural and/or artificial changes. Water pumping of wells, drainage by horizontal drains, water recharge through channels, pits, wells, or galleries, and water flooding above the ground surface are some of the major factors that disturb the shape of the natural interface. One or more of these factors can produce a three-dimensional rugged interface or the so-called saltwater mound. The natural flow is also subject to seasonal fluctuations that may advance or retreat the natural interface. However, under the most common field conditions, seasonal fluctuations are negligible. For example, in eastern coastal areas in the United States, the piezometric surfaces and water tables fluctuate between less than 30 cm (1.0 ft) and 2.0 m (6.0 ft) throughout the year. If the rise in the piezometric surface is the same everywhere, the gradients will be the same and no increase in the rates of flow will take place. However, in practice, a small increase in the rate of flow occurs gradually after the rainy season. This indicates that the assumption of a steady natural flow is practical and valid. The transient locations of the natural interfaces may be reasonably treated as successive steady-state cases corresponding to various rates of flow. However, different analytical and experimental approaches have been developed by various investigators to study movement of the interface (see, for example, Rumer and Harleman, 1963; Bear and Dagan, 1964*b*; and Kashef, 1970).

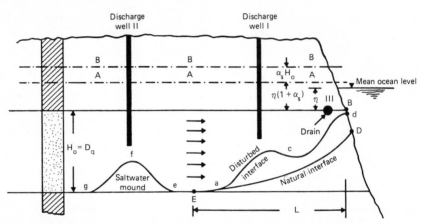

Figure 9.11 Disturbance of natural interface as a result of discharge wells and horizontal drains (sinks). *(A. I. Kashef, "Management and Control of Salt-Water Intrusions in Coastal Aquifers," CRC Crit. Rev. Environ. Control, vol. 7, no. 3, 1977.)*

Water wells (discharge or recharge) and horizontal drains installed to control saltwater intrusion produce a maximum disturbance of the natural interface at or below their locations. The interface below discharge wells or drains moves upward, forming saltwater mounds. The hydraulic conditions of these mounds are known as *upconing* (or *upwelling*), as shown in Fig. 9.11. In this figure, an artesian aquifer is diagrammatically shown in which planes AA and BB have the same significance as those in Fig. 9.3. It is assumed in this figure that the original interface is ED and that the interfaces $Eacd$ and gfe resulted from the operation of two vertical wells (I and II) and a sink or horizontal drain (III). Well I and drain III are constructed above the natural saltwater wedge of intruded length L. Both these activities lead to the new position $Eacd$ of the interface. However, well II is constructed beyond length L, causing a local upconing beneath the well due to heavy pumping. If the lower boundary of the aquifer is highly impervious, the water level in the well may drop below plane AA (Fig. 9.11) without developing saltwater upconing. Because soils and rocks are porous and their voids are continuous in the system, these lower beds are connected to the ocean. Local upconing *gfe* (Fig. 9.11) develops, but *only* after a long period of excessive pumping (see Sec. 9.2). Local upconing beyond the original distance *L* from the shoreline may or may not develop, depending on the distance of the well from the shoreline, the pumping rate, the duration of pumping, and the characteristics of the lower confining beds.

Many attempts have been made to develop solutions for the effects of wells, drains, and other causes of natural-interface disturbance. Among these is determination of the interface beneath a horizontal drain (Strack, 1972; Bear, 1979). Hantush (1968) gave analytic expressions for the growth and decay of freshwater lenses under the effects of drains, rectangular recharge areas, infinite strip recharge areas, and injection wells. Three types of unconfined coastal aquifers were considered: infinite, semi-infinite and close semi-infinite, and finite. Numerous mathematical approximations were introduced in this work.

The groundwater field as a whole is closely related to the control and management of saltwater intrusion. It has been found, for example, that the flow pattern toward an artesian well in coastal aquifers is analogous to that of a freshwater gravity well (Kashef, 1967). As previously stated, seepage through earth dams is analogous to certain conditions of saltwater intrusion (Figs. 9.5, 9.6, and 9.7).

Upconing Due to Drains

Bear and Dagan (1964) investigated various cases involving horizontal drains using the hodograph method. They studied two cases in which the drains were located at the upper boundary of an artesian aquifer (Fig. 9.12). In one case, the drain was intercepting half its flow q_{dr} from each

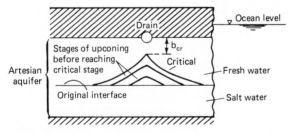

Figure 9.12 Upconing below a drain assuming an initial horizontal interface.

side, while in the other case, the drain was closer to the shore and intercepting the total amount q_{dr} from the inland side. In the first case, Bear and Dagan found that when the saltwater body was stagnant, the critical depth b_{cr} between the drain and the highest point in the upconing curve was given by

$$\frac{b_{cr}K}{\alpha_s q_{dr}} = 0.222 \tag{9.25}$$

At a certain value of b_{cr}, the rate of flow q_{dr} becomes critical such that when it is slightly increased, salt water enters the drain. The interface (upper curve in Fig. 9.12) was plotted by Bear and Dagan (1964) in a curve that cannot be defined except when a point on the interface is known. No definite conclusions were reached for the second case, where the drain intercepted the entire q_{dr} value from one side (similar to drain III in Fig. 9.11).

In the preceding analyses, the interface was assumed to be a horizontal plane. Muskat (1937) made the same assumption in solving a similar problem and gave the following equation:

$$\frac{q_{dr}}{2\alpha_s K D_x} \sin \pi \left(1 - \frac{b_{cr}}{D_x}\right) = \cosh\left[\pi\left(1 - \frac{b_{cr}}{D_x}\right)\right] + 1 \tag{9.26}$$

where b_{cr} and q_{dr} correspond to the critical values, and D_x is the original depth below the drain (located at the upper impervious boundary) and the interface. Bear (1979) also gave solutions using Dupuit's assumption when the drain is located below the upper impervious boundary in confined aquifers as well as when the drain is installed in an unconfined aquifer.

Other approximate solutions have been given by several investigators (Bear, 1979). In all these solutions, the rise in the interface should be small; otherwise the methods cannot be used. It has been recommended that the maximum disturbance in the original interface not exceed a range between one-fourth to one-third the initial distance from the interface to the drain. Generally, in unconfined coastal aquifers, once the free surface due to a drain or a well is obtained by available methods, the Ghyben-Herzberg approach can be used to find the upconing curves. Upconing only takes place where the water table is below line *BB* (Fig. 9.19). The conditions become critical whenever the water table approaches this line.

Upconing Due to Discharge Artesian Wells

Schmorak and Mercado (1969) used field investigations to study the upconing mechanism of the interface due to well pumping. They defined a certain elevation above the initial interface (which they assumed to be horizontal) and called it the *critical rise.* They found that as pumping increases, the interface forms an expanding mound with a maximum

height below the vertical axis of a partially penetrating well. Once the
maximum height reaches the critical rise, a sudden rise of salt water
occurs in the well. On the basis of these studies, Schmorak and Mercado
(1969) prepared two design monographs from which the maximum permissible pumping and rise of the interface can be obtained. Dagan and
Bear (Bear, 1979) also gave mathematical solutions for local interface
upconing. A simple approach is explained in the following paragraph.

The transitional stages in pumping fresh water from an artesian well
until salt water reaches the well at the critical condition (Fig. 9.13),
assuming an initially horizontal piezometric surface, are best understood
with reference to planes AA and BB (Figs. 9.3, 9.11, and 9.13). If the
entire drawdown curve $a_1 c_1$ lies between planes WW (the initially undisturbed piezometric surface) and BB after pumping, no saltwater upconing takes place (see Sec. 9.2) and D_w (Fig. 9.13) is equal to D_q. The salt
water starts to intrude upward once point a_1 drops below plane BB. The
drawdown curve $a_2 b_2 c_2$ (Fig. 9.13), which represents a later-time drawdown curve than that corresponding to $a_1 c_1$, intersects plane BB at b_2.
The segment $a_2 b_2$ on this curve corresponds to the upconing curve $a_2' b_2'$
and $0 < D_w < D_q$. Further pumping will develop the drawdown curve
$a_3 b_3 c_3$, where a_3 lies on plane AA and b_3 lies on plane BB. In this case, the
height D_w in the well will be reduced to zero and the upconing curve $a_3' b_3'$
will develop. The radius $R_c = \overline{o b_3'}$ is the critical radius of upconing. The

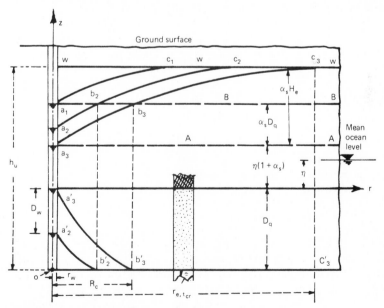

Figure 9.13 Upconing due to a discharge artesian well.

radius of influence at this critical stage corresponding to the critical t_{cr} is $r_{e,t,cr}$. When time t exceeds t_{cr}, salt water rather than fresh water is pumped out. The curves $a_2'b_2'$ and $a_3'b_3'$ are identical to the drawdown curves in gravity wells (Kashef, 1968a). Meanwhile, the transient flow toward a gravity well can be treated as a steady-state flow problem at a specific time if the proper time-dependent radius of influence $r_{e,t}$ is written as [Eq. (8.61)] follows:

$$r_{e,t} = (2.25vt)^{1/2} = \left(2.25\frac{Tt}{S}\right)^{1/2} \tag{9.27}$$

where v is the soil diffusivity, which equals $T/S = K/S'$, and $S' = S/D_q$, which is specific storage. Accordingly (Kashef, 1970), when $D_w = 0$ (Fig. 9.13),

$$\frac{Q}{\alpha_s K} = \frac{Q}{K'} = \frac{2\pi D_q(H_e - D_q)}{\ln r_{e,t,cr}/R_c} = \frac{2\pi D_q(H_e - D_q)}{\ln (2.25vt_{cr})^{1/2}/R_c} = \frac{D_q^2}{\ln R_c/r_w} \tag{9.28}$$

where $H_e = [h_u - D_q - \eta(1 + \alpha_s)]/\alpha_s$, r_w is the effective radius of the well, and Q is the rate of pumping (constant). The critical radius of upconing R_c is then given by

$$R_c = r_w \exp \frac{\alpha_s \pi KD_q^2}{Q} \tag{9.29}$$

and the critical time t_{cr} is given by

$$t_{cr} = \frac{R_c^2}{2.25v} \exp \frac{4\pi\alpha_s KD_q(H_e - D_q)}{Q} \tag{9.30a}$$

or $\qquad t_{cr} = \frac{r_w^2}{2.25v} \exp \frac{2\alpha_s \pi KD_q^2(2H_e/D_q - 1)}{Q} \tag{9.30b}$

The shape of the upconing curve is determined as a drawdown for a gravity well (Sec. 8.10) or approximated using the Ghyben-Herzberg equation with reference to curve a_3b_3 (Fig. 9.13).

Growth of Saltwater Zone Due to Well Pumping (Artesian)

The critical state of saltwater upconing beneath discharge wells has already been explained assuming that the initial piezometric surface is horizontal. In this section the initially deformed piezometric surface corresponding to the natural saltwater-freshwater interface is considered. The growth of the saltwater zone due to various rates and durations of pumping has been studied with a computer program. The boundaries between the totally freshwater zones and the intruded zones are determined for certain selected cases. A discharge well located at $x > L$ may or may not develop local upconing, as previously discussed (Fig. 9.11).

The effects of a pumped well can be superimposed on the effects of natural flow in the aquifer. However, the principle of superposition cannot be valid where two fluids exist because of variations of the flow medium zones (Kashef, 1975) (see Sec. 8.10). In order to overcome this problem, expansion of the saltwater zone is determined by tracing the boundaries between the *totally* freshwater and the intruded zones; thus these boundaries can be determined on the basis of the principle of superposition because they lie at the end of a totally freshwater zone. The exact determination of the saltwater mounds within the intruded zone is highly complex because of the effects of both pumping and natural flow.

It can be shown easily that the hydraulic heads along the upper boundary of an artesian aquifer are slightly less than those along the natural interface. Thus the combination of natural and imposed effects such as water pumping should preferably be based on the heads along the upper boundary in order to be on the safe side. Accordingly, the natural conditions are defined by the following equations:

$$(H_b)_x \simeq \alpha_s D_q \sqrt{\bar{x}} \qquad \left(\frac{x}{L} = \bar{x} \leqslant 1.0\right) \tag{9.31}$$

and
$$(H_b)_x = \tfrac{1}{2}\alpha_s D_q (1 + \bar{x}) \qquad \left(\frac{x}{L} = \bar{x} \geqslant 1.0\right) \tag{9.32}$$

where $(H_b)_x$ is the total head along the upper boundary of an artesian aquifer.

Using the Theis (1935) equation for wells in confined aquifers,

$$s_d = \frac{Q}{4\pi T} W(u) \tag{9.33}$$

where s_d is the drawdown below WW (Fig. 9.13) due to the effect of pumping and $W(u)$ is the well function (Chap. 8). And,

$$u = \frac{r^2 S}{4Tt} = \bar{\delta}\frac{r^2}{L^2} = \bar{\delta}\bar{r}^2 = \bar{\delta}[(\bar{x}_w - \bar{x})^2 + (\bar{y}_w - \bar{y})^2]$$

where $\bar{\delta} = $ a dimensionless time factor
$r = $ distance from any point to the well center
$t = $ time elapsed since the start of pumping

In this equation,

$$\bar{\delta} = \frac{uL^2}{r^2} = \frac{L^2 S}{4Tt}$$

$$\bar{r} = \frac{r}{L}$$

$$\bar{x} = \frac{x}{L} \qquad \text{and} \qquad \bar{y} = \frac{y}{L}$$

$$\bar{x}_w = \frac{x_w}{L} \qquad \text{and} \qquad \bar{y}_w = \frac{y_w}{L}$$

$$r = [(x_w - x)^2 + (y_w - y)^2]^{1/2}$$
$$= L[(\bar{x}_w - \bar{x})^2 + (\bar{y}_w - \bar{y})^2]^{1/2}$$

where x and y are coordinates of any point in the horizontal plane and x_w and y_w are coordinates of the center of the well in the horizontal plane. Thus the resultant total head $(H_b)_{x,y}$ with reference to AA (Fig. 9.13) as datum can be determined from Eqs. (9.31), (9.32), and (9.33) as follows:

$$(H_b)_{x,y} = \alpha_s D_q \sqrt{\bar{x}} - \frac{Q}{4\pi T} W(u) \qquad (\bar{x} \leqslant 1.0) \qquad (9.34)$$

and $\qquad (H_b)_{x,y} = \tfrac{1}{2}\alpha_s D_q (1 + \bar{x}) - \frac{Q}{4\pi T} W(u) \qquad (\bar{x} \geqslant 1.0) \qquad (9.35)$

It should be noted that the superposition applied in Eqs. (9.34) and (9.35) combines two different states of flow. The natural groundwater flow system is in the steady state, while the discharge well is in the transient state. As explained earlier (Chap. 8), fluctuations in the natural piezometric heads usually occur over a relatively long period of time as compared with well drawdowns.

A grid is used to compute $(H_b)_{x,y}$ at each node in the horizontal plane (Fig. 9.14). The $\bar{x}$ values (positive to the left) are normal to the ocean front, and the $\bar{y}$ values are parallel to that front. The grid consists of a square mesh with a size $\Delta x/L = \Delta y/L = 0.1$. If we introduce the following dimensionless values of a pumping factor:

$$\bar{\beta} = \frac{Q}{4\alpha_s \pi K D_q^2} \qquad \text{and} \qquad H° = \frac{(H_b)_{x,y}}{\alpha_s D_q}$$

the following simple equations are obtained:

$$H° = \sqrt{\bar{x}} - \bar{\beta} W\{\bar{\delta}[(\bar{x}_w - \bar{x})^2 + (\bar{y}_w - \bar{y})^2]\} \qquad (\bar{x} \leqslant 1.0) \qquad (9.36a)$$

and $\quad H° = \tfrac{1}{2}(1 + \bar{x}) - \bar{\beta} W\{\bar{\delta}[(\bar{x}_w - \bar{x})^2 + (\bar{y}_w - \bar{y})^2]\} \qquad (\bar{x} \geqslant 1.0)$
$$\qquad\qquad (9.37a)$$

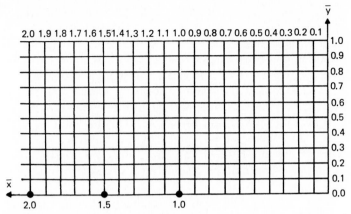

Figure 9.14 Half the grid used in computing the growth and decay of saltwater intrusion due to discharge and recharge wells. The closed circles (•) represent Well locations.

where $W\{\cdot\cdot\cdot\}$ is the well function. By selecting the various locations of the discharge wells along the $\bar{x}$ axis, Eqs. (9.36a) and (9.37a) reduce to

$$H^\circ = \sqrt{\bar{x}} - \bar{\beta}W\{\bar{\delta}[(\bar{x}_w - \bar{x})^2 + \bar{y}^2]\} \qquad (\bar{x} \leq 1.0) \qquad (9.36b)$$

and
$$H^\circ = \tfrac{1}{2}(1 + \bar{x}) - \bar{\beta}W\{\bar{\delta}[(\bar{x}_w - \bar{x})^2 + \bar{y}^2]\} \qquad (\bar{x} \geq 1.0) \quad (9.37b)$$

Kashef and Smith (1975) solved the cases shown in Table 9.3 using a computer program to apply Eqs. (9.36b) and (9.37b) at the nodes of the

TABLE 9.3 **Growth of Saltwater Zones Due to a Discharge Well: Solved Cases of Various $\bar{\beta}$, $\bar{\delta}$, and $\bar{x}_w$**

$\bar{\beta}$	$\bar{\delta}$	\multicolumn{6}{c}{$\bar{x}_w$}					
		1.0	1.5	2.0	3.0	5.0	9.0
10.0	9	°	°	°	°	°	°
	14.0625	°	°	°	°	°	°
	25	°	°	°	°	°	°
1.0	9	°	°	°	°		
	14.0625	°	°	°	°		
	25	°	°	°	°		
0.1	9	°					
	14.0625	°					
	25	°					

° Indicates solved cases.
SOURCE: A. I. Kashef and J. C. Smith, "Expansion of Salt-Water Zone due to Well Discharge," *Water Resour. Bull.*, vol. 11, no. 6, 1975, pp. 1107–1120.

grid shown in Fig. 9.14. The overall size of the grid has been taken between $\bar{x} = (\bar{x}_w - 1.0)$ and $\bar{x} = (\bar{x}_w + 1.0)$ and between $\bar{y} = +1.0$ and $\bar{y} = -1.0$. Since the upper half of this grid is symmetrical to the lower half, the computations have been carried out between $\bar{x} = (\bar{x}_w - 1.0)$ and $\bar{x} = (\bar{x}_w + 1.0)$ and $\bar{y} = 0$ and $\bar{y} = +1.0$. In order to have a practical range of $\bar{\delta}$ values, the practical ranges of the values of L, S, K, and D_q have been examined and the solved cases corresponding to the selected $\bar{\delta}$ and $\bar{\beta}$ values are shown in Table 9.3. From Eq. (9.27) and the value of $u = r^2 S/4Tt$, it is clear that the selected values of the time factor $\bar{\delta}$ (9, 14.0625, and 25) correspond, respectively, to $r_{e,t} = 0.25L$, $0.20L$, and $0.15L$. Other values of $\bar{\beta}$ and $\bar{\delta}$ may be selected if necessary using the same procedure.

Figures 9.15 to 9.17 present the results graphically when $\bar{\beta} = 10$ and the well is located at $(\bar{x}_w, \bar{y}_w) = (1.0, 0)$, $(1.5, 0)$, and $(2.0, 0)$, respectively. The same general pattern is obtained when $\bar{\beta} = 1.0$ (not shown). Figure 9.18 shows the difference between two extreme cases. The upper diagram (Fig. 9.18a) includes three cases: the upper right-hand hatched zone represents the growth in the intruded saltwater zone when $\bar{\beta} = 10$ and the well is located at $\bar{x}_w = 1.0$. The lower right-hand hatched zone represents the case of the same well when $\bar{x}_w = 1.5$. The isolated hatched zone on the left represents all points having $H^\circ \leqslant 1.0$ when $\bar{x}_w = 3.0$ and

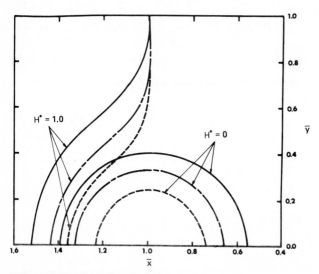

Figure 9.15 Growth of saltwater intrusion due to the indicated conditions: $(\bar{x}_w, \bar{y}_w) \equiv (1.0, 0.0)$. —, $\bar{\delta} = 9.0$; ——, $\bar{\delta} = 14.0625$; -----, $\bar{\delta} = 25$. $\bar{\beta} = 10$. (A. I. Kashef and J. C. Smith, "Expansion of Salt-Water Zone due to Well Discharge," Water Res. Bull., vol. 11, no. 6, 1975.)

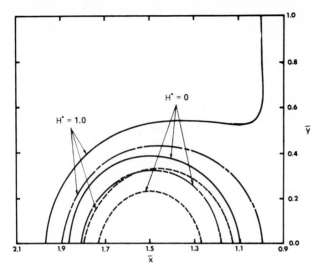

Figure 9.16 Growth of saltwater intrusion due to the indicated conditions: $(\overline{x}_w, \overline{y}_w) \equiv (1.5, 0.0)$. —, $\overline{\delta} = 9.0$; —·—, $\overline{\delta} = 14.0625$; -----, $\overline{\delta} = 25$. $\beta = 10$. *(A. I. Kashef and J. C. Smith, "Expansion of Salt-Water Zone due to Well Discharge," Water Res. Bull., vol. 11, no. 6, 1975.)*

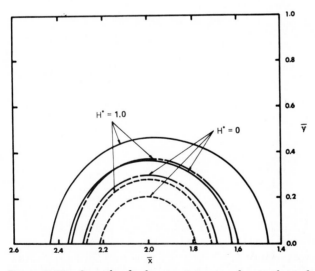

Figure 9.17 Growth of saltwater intrusion due to the indicated conditions: $(\overline{x}_w, \overline{y}_w) \equiv (2.0, 0.0)$. —, $\overline{\delta} = 9.0$; —·—, $\overline{\delta} = 14.0625$; -----, $\overline{\delta} = 25$. $\beta = 10$. *(A. I. Kashef and J. C. Smith, "Expansion of Salt-Water Zone due to Well Discharge," Water Res. Bull., vol. 11, no. 6, 1975.)*

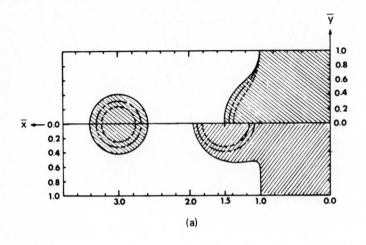

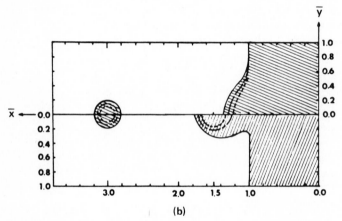

Figure 9.18 Growth of saltwater intrusion due to discharge wells at locations $\bar{x}_w = 1.0$, 1.5, and 3.0. —, $\bar{\delta} = 9.0$; ———, $\bar{\delta} = 14.0625$; -----, $\bar{\delta} = 25$. (*a*) Curves $H° = 1.0$ due to $\bar{\beta} = 10$ and other indicated conditions; (*b*) curves $H° = 1.0$ due to $\bar{\beta} = 1.0$ and other indicated conditions. (*A. I. Kashef and J. C. Smith, "Expansion of Salt-Water Zone due to Well Discharge," Water Res. Bull., vol. 11, no. 6, 1975.*)

$\bar{\beta} = 10$. Because of the small scale of the diagram, the boundary of this latter zone appears to be a circle; in fact, it is not quite a circle and it is not symmetrical to the line $\bar{x} = 3.0$. Figure 9.18*b* represents the same conditions included in Fig. 9.18*a* but for $\bar{\beta} = 1.0$. Figures 9.15 to 9.18 indicate that the original zones of saltwater intrusion grow when $\bar{x}_w$ ranges from 1.0 to 1.5. However, when $\bar{x}_w \geq 2.0$, local saltwater-intruded zones develop and are isolated from the original natural zone (which remains unchanged). These isolated zones have no effect on the quality of pumped water unless the permeability of the lower confining bed is relatively high and the pumping duration is relatively long, causing upward saltwater movement and saltwater upconing. One should remember that the equations are based on the Theis equation, in which time lag is neglected.

A recharge well has an effect on the interface that is opposite that of a discharge well; it results in decay rather than growth of the original saltwater wedge. The effects of a single recharge well can be obtained in a manner similar to that for a discharge well by changing the negative signs of the second terms on the right-hand side of Eqs. (9.36) and (9.37) to positive signs. These effects are explained in Sec. 9.8.

Upconing Due to Discharge Gravity Wells

As shown in Figs. 9.1, 9.8, and 9.19, a discharge well pumping from an unconfined aquifer does not create upconing when

$$D_{xu} - s_d \geq \alpha_s H_o \tag{9.38}$$

where s_d is the drawdown at any radius r from the well. If we draw a horizontal plane BB distance $\alpha_s H_o$ above sea level, then its intersection with the water table will lie above the toe of the natural interface (assuming the validity of the Ghyben-Herzberg approach). If a well is operating landward beyond the toe, upconing does not necessarily occur, as ex-

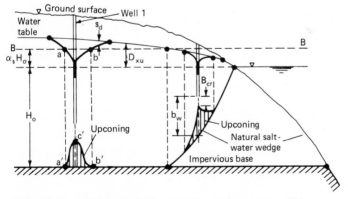

Figure 9.19 Upconing of salt water beneath gravity wells.

plained earlier. If we assume that well I (Fig. 9.19) is pumping at a high capacity in such a way that a portion of its drawdown curve is lower than line BB, then upconing will occur. The upward flow of the salt water will be very slow, and therefore, the upconing may not develop immediately and may not be encountered if the operation if intermittent. In order to err on the side of safety, in water-resources management the time lag for upconing should be neglected; otherwise efficient control should be exercised in well operation coupled with careful field observations.

If the drawdown curve intersects line BB (Fig. 9.19) at points a and b, upconing develops only below ab, as shown by the mound $a'c'b'$. If the Ghyben-Herzberg approach is applied, the upconing curve can be obtained once the drawdown curve is determined by the available methods for gravity wells whether in the steady or nonsteady states of flow (Chap. 8). Thus upconing takes place when

$$D_{xu} - s_d < \alpha_s H_o \tag{9.39}$$

The main issue is not only to find the shape of the upconing mound but also to find the critical conditions of upconing beyond which salt water enters the well. Schmorak and Mercado (1969) assumed a horizontal interface (or approximately so below the well) and determined the critical height B_{cr} of the upconing beneath a partially penetrating well.

$$B_{cr} = \frac{Q_{cr}}{2\pi K_h b_w} \frac{K_v t_{cr}}{2nb_w + \alpha_s K_v t_{cr}} \tag{9.40}$$

where Q_{cr}, t_{cr} = the critical values of Q and t
$\qquad\quad b_w$ = the height of the bottom of the well above the interface
$\qquad\quad n$ = the porosity of the aquifer
K_h, K_v = the hydraulic conductivities in the horizontal and vertical directions (anisotropic aquifer)

If the aquifer under consideration is isotropic, then

$$B_{cr} = \frac{Q_{cr}}{2\pi b_w (2nb_w/t_{cr} + \alpha_s K)} \tag{9.41}$$

At $t = \infty$ (steady-state case),

$$B_{cr} = \frac{Q_{cr}}{2\pi \alpha_s K b_w} \quad \text{(steady state)} \tag{9.42}$$

The value of B_{cr} was found by the same authors to vary between $0.4b_w$ and $0.6b_w$ before a sudden rise in the saltwater mound takes place, leading to well deterioration.

When pumping stops, the upconing mound undergoes decay as well as freshwater displacement (seaward) until it vanishes. It has been reported (Bear, 1979) that the transition zone becomes thicker when upconing takes place, and brackish water may then contaminate the well before the criitcal stage is reached. Bear (1979) also gave solutions for an array of wells pumping at the same rate and located above the interface.

He also gave a qualitative description of the formation of a groundwater divide in confined and unconfined aquifers, allowing the flow of a sheet of water to the sea between the divide and the interface. The solutions are generally complex (mostly numerical) and involve several simplified assumptions.

Strack (1976) determined the groundwater divide of a well pumping from an unconfined aquifer, but he followed a conservative approach in order to avoid the upconing zones. He indicated that a well pumping beyond the natural interface with drawdowns lower in level than plane *BB* (Fig. 9.19) produces no upconing (see Fig. 9.20). (This point has been already discussed, and upconing may occur.) The groundwater divide is shown in Fig. 9.20*b*, indicating the location of the stagnation point (SP) at which the velocity is zero. The toe has been disturbed from a straight line passing through *G* to a curved line passing through the same point to satisfy the new configuration of the flow lines. The dashed line passing through the SP (Fig. 9.20*b*) is an equipotential line. The SP coordinates

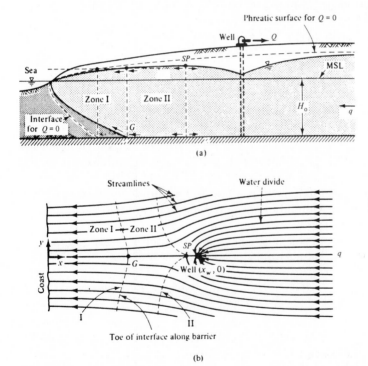

Figure 9.20 Location of the interface for a single well near the coast. (*a*) Vertical section; (*b*) plan. (*O. D. L. Strack, "A Single-Potential Solution for Regional Interface Problems in Coastal Aquifers," Water Resources Res. vol. 12, no. 6, 1976, p. 1168.*)

x_s, $y_s = 0$ are given by (Strack, 1976):

$$x_s = x_w \sqrt{1 - \frac{Q}{\pi q x_w}} \tag{9.43}$$

where q is natural steady flow per unit length (in m²/min) and x_w and $y_w = 0$ are coordinates of the well location.

The equation of the toe of the interface is given by

$$\frac{\alpha_s(1 + \alpha_s)H_o^2 K}{q} = 2x + \frac{Q}{2\pi q} \ln \frac{(x - x_w)^2 + y^2}{(x + x_w)^2 + y^2} \tag{9.44}$$

Pumping creates water-table levels lower than that above point G. Strack (1976) considered that as long as SP is landward from G, no upconing will take place "even when water levels are lower than sea level."* Strack's conservative approach considered that the disturbance of the original interface takes place when x_s decreases, until it becomes less than L. This occurs whenever x_w decreases and/or Q increases. A limiting condition is established when $x_s = L$ or when the interface passes by SP. By substituting the x value given in Eq. (9.44) with the x_s value given in Eq. (9.43) and setting $y = 0$, the limiting condition is obtained:

$$\frac{\alpha_s(1 + \alpha_s)H_o^2 K}{q x_w} = 2\frac{x_s}{x_w} + \frac{Q_{cr}}{\pi q x_w} \ln \frac{x_w - x_s}{x_w + x_s} \tag{9.45}$$

where Q_{cr} is the critical rate of pumping.

A well should then be placed at a certain distance x_w and pumped at a certain rate $Q < Q_{cr}$ in such a way that all the flowing water entering the well is beyond the natural saltwater wedge. It should be noted also that some common simplifying assumptions are introduced in this approach (that is, the flow is assumed steady and the well is assumed to completely penetrate the aquifer).

As long as the approach satisfies the condition $x_s \geq L$ and Q_{cr} takes place when $x_s = L$, then from Eq. (9.43),

$$L = x_w \sqrt{1 - \frac{Q_{cr}}{\pi q x_w}}$$

and
$$Q_{cr} = \pi q \left(x_w - \frac{L^2}{x_w} \right) \tag{9.46}$$

L is determined from Eq. (9.12b).

---

* It has been shown that this statement is not valid.

9.8 Control of Saltwater Intrusion

It has been illustrated that a natural interface develops between salt water and fresh water in confined and unconfined aquifers under the effect of natural flow within the aquifer, assuming that the two fluids are immiscible. It has also been indicated that the transitional zone, within which dispersion and diffusion take place, is relatively small, except probably when natural flow is decreasing or under exceptional circumstances such as in Hawaii and Florida (Cooper et al., 1964). Thus from the management point of view, this transitional zone may be disregarded at least until more satisfactory approaches are developed. Some cases have also been examined in which the natural interfaces are disturbed by natural and artificial causes. The study of all these factors and the understanding of the hydrologic and geometric boundaries should be the first step in planning a successful control method.

Depending on water usage, control methods may have one or more of the following purposes:

1. Preventing the complete or partial escape of the fresh water to the sea

2. Increasing the freshwater pressures in order to increase the rate of flow within the aquifer or the size of the freshwater lens, thus reducing the size of the saltwater wedge

3. Regulating the activities of water withdrawal in a certain zone in order to maintain a state of saltwater intrusion that will not lead to critical upconing beneath wells and drains

Potential Methods of Control

Barriers must be constructed in order to stop or minimize the seaward flow of fresh water. These barriers may be either *physical* or *hydraulic* (Kashef, 1971; Todd, 1974, 1980; Zielbauer, 1966; Peters and Cuming, 1967). A physical barrier should be constructed near the coast to an optimum design depth using relatively impervious material. A trench backfilled with compacted silty clay material will serve the purpose. Long impervious curtains, such as sheet piles, coffer dams, concrete panel walls constructed with the aid of slurries, or grout curtains formed by injecting the soil with clay slurries, cement grout, or chemicals can be used. The methods of constructing such physical barriers are well-known geotechnical engineering techniques (see, for example, Terzaghi and Peck, 1967). However, barriers in general have numerous construction and economic problems associated with them. When a barrier must extend through the entire depth of a deep artesian aquifer, the cost is prohibitive, although the resulting increase in the piezometric heads is the desired effect and can be achieved with such a barrier. If soil grouting is used, there should be sufficient effective weight above the upper boundary of the artesian aquifer to counterbalance the increasing upward pressures produced during injection of the grout. If the depth of the barrier covers part of the outflow surface in artesian aquifers, the freshwater flow will emerge from underneath the barrier with very high velocities that may lead to backward erosion (Sec. 5.4) and ultimately to land subsidence. However, if a physical barrier is constructed in an unconfined aquifer (the uppermost formation), it may be economically feasible but still create other problems. The water table will rise to a new surface behind the barrier similar to the backwater curve behind dams across open channels or rivers. Although water conservation is necessary to retard saltwater encroachment, it can lead to waterlogging of the coastal agricultural lowlands. The depths, materials, and layout of a solid barrier will then be subject to the normal decision-making techniques used in any major engineering work.

Recharge methods are characterized not only by recharge wells or horizontal drains but also by water spreading directly on the ground surface or through artificial or natural ponds or channels (Water Research Association, 1971; Baumann, 1965; Brown and Signor, 1972, 1974; International Association of Scientific Hydrology, 1967; Peters and Rose, 1968). The water is allowed to infiltrate through the soil beds freely by gravity. Recharge methods are relatively recent and are accompanied by several problems. The common problems in recharge wells are water turbidity, clogging of wells, accumulated air problems, effects on injection and shutdown periods, and the effectiveness of the procedure (Sternau, 1967). Gravel packs should be properly designed and constructed and water withdrawal should be tested before injection. In some cases, it was found necessary to withdraw water from injected wells during certain periods to improve recharge capacity. Recharging a well with free-falling water can produce unsatisfactory results; air being sucked into the well during the process may cause inflow stoppage (Sternau, 1967). Air and water will then be violently emitted from the well casing. Rebhun and Schwartz (1968) explained the problem of recharge-well clogging and contamination of water pumped in dual-purpose wells (wells used for both recharge and discharge purposes). They attributed these problems to small concentrations of suspended organic matter in recharge water, and they suggested as a solution the redevelopment of recharge wells by small-volume pumping. Geotechnical engineering aspects of recharge wells should also be studied. The sudden drop in pressure due to a shutdown of pumping (increasing the effective pressure) can lead to excessive land subdsidence. In addition, during the recharging operation, the magnitudes of the water pressure at the upper boundary of an artesian aquifer should be lower than the downward effective stress at the same plane; the ratio of these values should be based on an appropriate safety factor.

Hydraulic barriers can also be constructed by a battery of discharge wells parallel and close to the shoreline. The wells pump out a mixture of salt water and fresh water, thus stemming up the salt water and changing its inland flow direction. Fresh water can then be pumped out safely from the inland side of this battery by any producing wells from the trough formed between the battery and the producing wells. This control method is known as the *pumping trough* (Todd, 1980). A field experiment using this approach was conducted in Ventura County, California. Plans were also made to test the pumping trough in Los Angeles, which consists of 30.5-cm diameter (12-in) wells spaced at 198 m (650 ft), pumped at a rate of 1.89 to 3.4 m³/min (500 to 900 gal/min) per well, and drilled 76 m (250 ft) within the aquifer. It is apparent that a saltwater body would be trapped at the inland side next to the well battery. In this method, a considerable amount of fresh water is lost to the sea, degrading the existing storage capacity of the aquifer. Todd (1980) concluded that in general the method is not economically feasible and suggested its temporary use only in urgent cases until more appropriate procedures are planned.

A method has been suggested (Kashef, 1971) to overcome the disadvantages of the pumping-trough method. In this method, a battery of discharge wells is installed along the coast *to pump out only pure salt water.* The location of the screens, the pumping capacity of each well, and the spacings are designed to prevent any fresh water from entering these wells. With this approach, the depth of the fresh water should be deepened gradually inland to the bottom of the aquifer as a result of increases in water pressure and retreat of the initial saltwater wedge. The success of the method can be tested continuously by checking the water salinity at various depths in monitoring wells. Unfortunately, this method has not yet been tested despite the high costs of field tests of lesser efficiency. The method is promising and has a substantial advantage over both hydraulic barriers and pumping troughs. It does not require imported fresh water or the withdrawing of fresh water from the aquifer.

In Orange County, California, field tests have been conducted using a combination of hydraulic barriers (Kashef, 1972). It is claimed that use of the pumping-trough method in conjunction with recharge wells minimizes the effects of subsidence and waterlogging and allows more flexibility in operation (Bruington and Seares, 1965). However, it is obvious that such a combination will result in the waste of substantial amounts of fresh water. Furthermore, the costs can be expected to be prohibitive. In cases where there is real danger of land subsidence, recharge wells must be used.

Control by Recharge Wells

As previously explained, most of the analyses of control methods to alleviate the detrimental effects of saltwater intrusion are still in the research stage, although some of the methods have been used in the field. No definite practical guidelines have been established that will be of benefit to managers of groundwater resources in coastal areas. Some analytic approaches have been introduced recently for the design of recharge wells that either retard saltwater intrusion (Kashef, 1976a) or form hydraulic barriers (Kashef, 1975, 1977). Only the technical aspects of this method are explained here. The economics of the method should also be investigated in order to evaluate how this method compares to other methods of control.

If a battery of recharge wells is arranged parallel to the shoreline, under certain conditions it will produce a combined flow effect equivalent to a unidirectional uniform increase in the natural flow, allowing a uniform retreat of the original saltwater wedge.

If the concepts already discussed in connection with discharge wells are applied here and the same grid pattern (Fig. 9.14) and the same symbols are adopted, the following equations describe the effects of a single recharge well:

$$H^\circ = \sqrt{\bar{x}} + \bar{\beta}W\{\bar{\delta}[(\bar{x}_w - \bar{x})^2 + (\bar{y}_w - \bar{y})^2]\} \qquad (\bar{x} \leqslant 1.0) \qquad (9.47)$$

and $H^\circ = \frac{1}{2}(1 + \bar{x}) + \bar{\beta}W\{\bar{\delta}[(\bar{x}_w - \bar{x})^2 + (\bar{y}_w - \bar{y})^2]\} \qquad (\bar{x} \geqslant 1.0) \quad (9.48)$

If H° calculated from the preceding equations is greater than 1.0, the artesian aquifer will be entirely in a freshwater zone. However, when $0 < H^\circ < 1.0$ within a certain zone, there will be both fresh and salt waters. The line $H^\circ = 1.0$ therefore gives the boundary between these two zones. The effects based on the preceding equations for two cases of recharge wells are summarized in Figs. 9.21 and 9.22.

The method of images has to be applied whenever $L < r_{e,t}$, and the following equation should be used for a *single* recharge well:

$$H^\circ = \sqrt{\bar{x}} + \bar{\beta}W\{\bar{\delta}[(\bar{x}_w - \bar{x})^2 + (\bar{y}_w - \bar{y})^2]\} - \bar{\beta}W\{\bar{\delta}[(\bar{x} + \bar{x}_w)^2 + (\bar{y}_w - \bar{y})^2]\}$$
$$(9.49)$$

If we apply Eq. (9.49) to a point at the shoreline ($\bar{x} = 0$), then H° will be zero. Thus the residual total head at the shoreline will always remain zero, irrespective of the location of the recharge well or the pumping

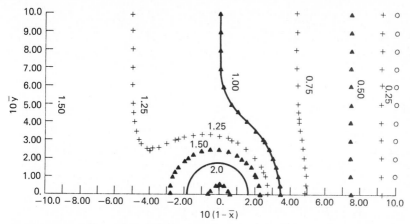

Figure 9.21 Effect of one recharge well; lines $H°$ corresponding to $\bar{x}_w = 0.0$, $\bar{\beta} = 1.0$, and $\bar{\delta} = 9$. *(A. I. Kashef, "Management of Retardation of Salt Water Intrusion in Coastal Aquifers," Office of Water Research and Technology Report, U.S. Department of Interior, Washington, 1975.)*

period. However, the intermediate head values between the well and the shoreline will be greater when $r_{e,t}$ increases, maintaining the same pumping capacity. The increase in these heads is produced through the effect of prolonged periods of pumping.

If we consider next a battery of recharge wells placed parallel to the shoreline with equal spacings and equal pumping rates and assume that

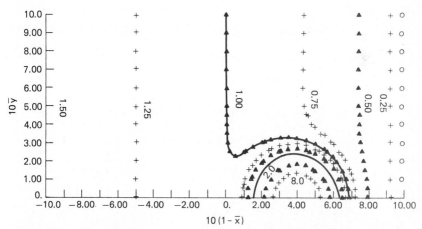

Figure 9.22 Effect of one recharge well; lines $H°$ corresponding to $\bar{x}_w = 0.6$, $\bar{\beta} = 1.0$, and $\bar{\delta} = 25$. *(A. I. Kashef, "Management of Retardation of Salt Water Intrusion in Coastal Aquifers," Office of Water Research and Technology Report, U.S. Department of Interior, Washington, 1975.)*

all wells start pumping at the same instant, then Q, t, and $\bar{x}_w$ will be the same for all wells. If n_w is the number of wells, then $\bar{y}_w$ of a certain well n will be

$$(\bar{y}_w)_n = \frac{2n}{n_w - 1} \tag{9.50}$$

where
$$n = 1, 2, 3, \ldots, \frac{n_w - 1}{2} \tag{9.51}$$

The different u values ($u = r^2 S/rTt$) for the various locations of wells and their images correspond to the following:

u_o: a well located along the $\bar{x}$ axis

u_{oi}: a well image located along the $\bar{x}$ axis

u_u: a well located above the $\bar{x}$ axis

u_{ui}: a well image located above the $\bar{x}$ axis

u_d: a well located below the $\bar{x}$ axis

u_{di}: a well image located below the $\bar{x}$ axis

Thus,

$$
\begin{aligned}
u_o &= \bar{\delta}[(\bar{x} - \bar{x}_w)^2 + \bar{y}^2] & (9.52a) \\
u_{oi} &= \bar{\delta}[(\bar{x} - \bar{x}_w)^2 + \bar{y}^2] & (9.52b) \\
u_u &= \bar{\delta}[(\bar{x} - \bar{x}_w)^2 + (\bar{y} - \bar{y}_{wn})^2] & (9.52c) \\
u_{ui} &= \bar{\delta}[(\bar{x} + \bar{x}_w)^2 + (\bar{y} - \bar{y}_{wn})^2] & (9.52d) \\
u_d &= \bar{\delta}[(\bar{x} - \bar{x}_w)^2 + (\bar{y} + \bar{y}_{wn})^2] & (9.52e) \\
u_{di} &= \bar{\delta}[(\bar{x} + \bar{x}_w)^2 + (\bar{y} + \bar{y}_{wn})^2] & (9.52f)
\end{aligned}
$$

The total dimensionless heads H° (with respect to datum AA in Fig. 9.13) at any nodal point (Fig. 9.14) is as follows:

$$
H^\circ = \sqrt{\bar{x}} + \bar{\beta}\{W(u_o) - W(u_{oi}) + \sum_{n=1}^{(n_w-1)/2} [W(u_u) \\
+ W(u_d) - W(u_{ui}) - W(u_{di})]_n\} \qquad (\bar{x} \leqslant 1.0) \tag{9.53}
$$

$$
H^\circ = \tfrac{1}{2}(1 + \bar{x}) + \bar{\beta}\{W(u_o) - W(u_{oi}) + \sum_{n=1}^{(n_w-1)/2} [W(u_u) \\
+ W(u_d) - W(u_{ui}) - W(u_{di})]_n\} \qquad (\bar{x} \geqslant 1.0) \tag{9.54}
$$

Equations (9.53) and (9.54) have been solved on a computer for 192 cases consisting of batteries of $n_w = 3, 5, 6$, and 9 wells (Kashef, 1975). In each case, various well locations $\bar{x}_w$ were considered as well as four values of $\bar{\delta}$, that is, 56.25, 25, 14,0625, and 9, and three values of $\bar{\beta}$, that is, 0.1, 1, and 10 (Table 9.4). It has been found that values of $\bar{\delta}$ greater than 56.25 and values of $\bar{\beta}$ less than 0.1 produce insignificant effects (Kashef,

TABLE 9.4 Performance of Recharge-Well Batteries and Percent of Saltwater Retardation Corresponding to Continuous Lines $H° = 1.0$

$\bar{x}_w$ (location of wells)	n_w (no. of wells)	$\bar{\beta}$ = 0.1 $\bar{\delta}$ a	b	c	d	$\bar{\beta}$ = 1.0 $\bar{\delta}$ a	b	c	d	$\bar{\beta}$ = 10 $\bar{\delta}$ a	b	c	d
1.0	3	10.00†	X	X	X	23.50†	16.00†	X	X	43.00†	29.75†	18.00†	X
	5	16.00†	12.50†	8.25†	X	35.00	28.00	20.00†	11.50†	52.00	42.00	32.00†	19.50†
	7	19.50	15.75	12.00	6.50†	37.50	30.25	23.50	16.75†	54.75	44.50	34.75	22.50
	9	21.00	17.50	13.75	8.50	39.00	31.75	25.25	18.00	56.00	46.00	35.75	25.10
0.7	3	X	X	X	X	†	X	X	X	66.50†	45.50†	X	X
	5	25.00†	X+	X	X	57.50	51.00†	42.50†	X	77.50	67.50	56.50†	X+
	7	37.50	X+	X	X	61.00	55.25	48.25	40.00†	79.50	70.00	60.00	49.25
	9	41.75	37.50	HY	X	63.25	56.75	49.50	42.75	81.25	71.00	60.05	50.25
0.4	3	X	X	X	X	X+	X	X	X	X+	X	X	X
	5	X	X	X	X	HY	HY	X	X	96.00	92.50	X+	X
	7	X	X	X	X	85.50	HY	HY	X	97.50	95.00	HY	HY
	9	X	X	X	X	87.00	HY	HY	HY	97.75	95.25	HY	HY
0.1	3	X	X	X	X	X	X	X	X	X	X	X	X
	5	X	X	X	X	X	X	X	X	HYW	HYW	X	HYW
	7	X	X	X	X	HYW	HYW	X	X	HYW	HYW	X	HYW
	9	HYW	X	X	X	HYW	HYW	HYW	X	HYW	HYW	HYW	HYW

Note: a, b, c, and d correspond, respectively, to $\bar{\delta}$ values 9, 14.0625, 25, and 56.25; X corresponds to lines $H° = 1.0$ are not continuous or form closed loops of salt water; X+ corresponds to conditions forming crooked lines $H° = 0$ in addition to closed loops; HY corresponds to conditions forming hydraulic barriers; and HYW corresponds to conditions forming hydraulic barriers with freshwater waste.

† Crookedness of line $H° = 1.0$ is greater than $0.05L$.

‡ Severe crookedness in line $H° = 1.0$

1975). In each case, the center well is placed along the x axis and the battery of wells is equally spaced parallel to the shore in such a way that it occupies a distance of $2L$. In some cases, the well batteries indicated a certain degree of saltwater retardation and in others the formation of hydraulic barriers, depending on their hydraulic and geometric conditions.

Before pumping starts, line $H° = 1.0$ should be parallel to the shoreline and at a distance $\bar{x} = 1.0$. Thus the ideal approach of saltwater retardation is to develop the means to advance line $H° = 1.0$ parallel to itself toward the shoreline. This means that the original length of intrusion L will be shortened to a new position as a result of a uniform increase in the rate of flow. Thus *the degree of saltwater retardation P,* measured as a percentage, can be defined as the ratio of the added extension in the entirely freshwater zone divided by the original length L of the natural saltwater intrusion multiplied by 100, or P is equal to $100(1 - \bar{x}_r)$, where $\bar{x}_r$ is the dimensionless average coordinate of the new position of $H° = 1.0$ resulting from the effects of a battery of recharge wells. In order to produce a successful result, the recharge wells should be capable of producing an equipotential line $H° = 1.0$ *almost* parallel to the shoreline. The resulting interface will be equivalent to a new rate of flow resulting from the natural flow plus the added uniform rate of flow produced by the combined effects of recharge wells. An almost perfect straight line $H° = 1.0$ parallel to the shoreline will be produced under certain favorable conditions of well spacing, well-pumping capacities, distance of the battery from the shoreline, and duration of pumping under the prevailing geometric and boundary conditions. The solution of the various cases (Table 9.4) revealed, as expected, that line $H° = 1.0$ cannot always be a perfect straight line parallel to the shoreline. The solutions result in one of the following (Kashef, 1975 and 1976a):

1. Line $H° = 1.0$ is parallel or almost parallel to the coastline. This condition necessitates checking the degree of crookedness of the resulting line $H° = 1.0$ by determining the maximum difference between its extreme $\bar{x}_r$ coordinates. If this difference does not exceed 0.05 (that is, 5 percent of L), the line is considered parallel or almost parallel to the shoreline (curve a in Fig. 9.23b).

2. The degree of crookedness of line $H° = 1.0$ exceeds 0.05 (curve a in Fig. 9.23a). In this case, the retardation of saltwater intrusion is considered unsatisfactory. Increasing the pumping factor $\bar{\beta}$ or decreasing well spacing or the time factor $\bar{\delta}$ or all these combined will change this condition to conditon 1.

3. Line $H° = 1.0$ forms a loop around the well (curves c and d in Fig. 9.23c) or in some cases away from the well (loop b in Fig. 9.23c), forming

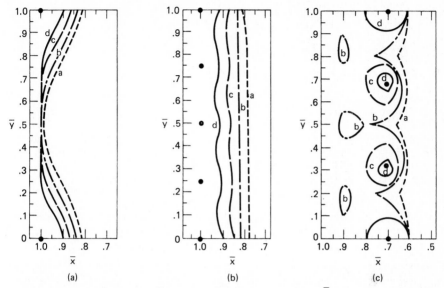

Figure 9.23   Performance of batteries of recharge wells when $\bar{\beta} = 0.1$. Curve a, $H°$ at $\bar{\delta} = 9.0$; curve b, $H°$ at $\bar{\delta} = 14.0625$; curve c, $H°$ at $\bar{\delta} = 25.0$; curve d, $H°$ at $\bar{\delta} = 56.25$. (a) $n_w = 3$; (b) $n_w = 9$; (c) $n_w = 7$. (A. I. Kashef, "Management of Retardation of Salt Water Intrusion in Coastal Aquifers," Office of Water Research and Technology Report, U.S. Department of Interior, Washington, 1975.)

trapped salt water within the freshwater zones. This condition is considered undesirable. Again, increasing $\bar{\beta}$ or the well spacings and/or decreasing $\bar{\delta}$ will change this condition gradually to condition 2 and eventually to condition 1. However, the trapped salt water must be pumped out. When the wells are at or closer to line $\bar{x} = 1.0$, curve $H° = 1.0$ may not form a complete loop and touches line $\bar{x} = 1.0$ (curves b, c, and d in Fig. 9.23a).

4. Two curves of the same intensity $H° = 1.0$ develop on both sides of the well battery, especially when $\bar{x}_w$ is very small. These curves may be almost straight lines (curves b and c in Fig. 9.24) or crooked (curve d in Fig. 9.24). These cases provide perfect conditions for forming hydraulic barriers. In practice, if the width of the barrier is relatively large, alternate shutdown of pumping and repumping should be conducted at the proper time to maintain the formation of these barriers.

5. When the well battery is very close to the shoreline, a hydraulic barrier can develop, forming only one line $H° = 1.0$ on the inland side of the recharge wells (Fig. 9.25). This means that the injected water is partially wasted seaward from the other side. This condition is consid-

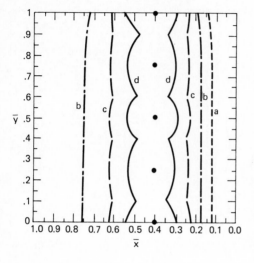

Figure 9.24 Formation of hydraulic barriers, type **HY** (Table 9.4): $\bar{x}_w = 0.4$, $n_w = 9$, $\bar{\beta} = 1.0$. Curves a, b, c, and d correspond, respectively, to curves $H° = 1.0$ at $\bar{\delta} = 9$, 14.0625, 25, and 56.25. (A. I. Kashef, "Management of Retardation of Salt Water Intrusion in Coastal Aquifers," Office of Water Research and Technology Report, U.S. Department of Interior, Washington, 1975.)

ered undesirable and similar — in the opposite sense — to the pumping trough.

A design curve for condition 1 is given in Fig. 9.26 (Kashef, 1976a), and it shows the relationship between the degree of saltwater retardation P as a percentage and $(Q/q_s)^{\sqrt{1/\delta}}$, where $q_s = qS_p$. The well spacing S_p is equal to $2L/(n_w - 1)$ in a battery of n_w wells equally spaced over a distance $2L$ parallel to the shoreline. Thus if Eq. (9.12a) is applied, q_s becomes equal to $K'H_o^2/(n_w - 1)$. The solid curves in Fig. 9.26 are plotted using the solved cases (Table 9.4) corresponding to $\bar{x}_w = 1.0$ and $\bar{x}_w = 0.7$. The two dashed curves corresponding to $\bar{x}_w = 0.8$ and 0.9 are

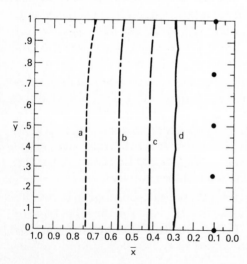

Figure 9.25 Formation of hydraulic barriers, type **HYW** (Table 9.4): $\bar{x}_w = 0.1$, $n_w = 9$, $\bar{\beta} = 10$. Curves a, b, c, and d correspond, respectively, to curves $H° = 1.0$ at $\bar{\delta} = 9$, 14.0625, 25, and 56.25. (A. I. Kashef, "Management of Retardation of Salt Water Intrusion in Coastal Aquifers," Office of Water Research and Technology Report, U.S. Department of Interior, Washington, 1975.)

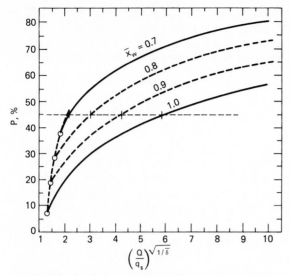

Figure 9.26 Design curves for saltwater retardation.
(*A. I. Kashef, "Control of Salt Intrusion by Recharge Wells," Proc. Am. Soc. Civ. Eng., vol. 102, no. IR4, 1976a.*)

plotted by interpolation. The steep dashed curve to the left in the same figure passes through the limiting ends of the four curves. For example, if the desired P is 30 percent, $\bar{x}_w$ has to be greater than 0.7, and to obtain $P = 25$ percent, $\bar{x}_w$ should be greater than 0.8, and so on. The use of Fig. 9.26 in designing a well battery to achieve a certain degree of retardation is illustrated in the following example (Kashef, 1976a).

EXAMPLE 9.2 In an artesian coastal aquifer, $q = 0.2789$ m² per day (3 ft² per day), $H_o = 91.44$ m (300 ft), $K = 122$ m per day, and $S = 9 \times 10^{-4}$. Determine L, S_p, and the number of equally spaced wells in a battery over 10 km corresponding to $P = 45$ percent. Assume $\alpha_s = 0.025$ and $Q = 2729$ m³ per day.

Solution

$$L = \frac{\alpha_s K H_o^2}{2q} = 45.72 \text{ km}$$

$$v = \frac{T}{S} = \frac{K H_o}{S} = 1.24 \times 10^7 \text{ m}^2 \text{ per day}$$

$$\bar{\delta} = \frac{L^2}{4vt} = \frac{42.14}{t} \qquad (t \text{ in days})$$

Therefore, $\dfrac{Q}{q_s} = \dfrac{9785}{S_p}$ (S_p is in meters)

From Fig. 9.26, $(Q/q_s)^{\sqrt{1/\delta}} = 2.15, 3.00, 4.25,$ and 5.8 when $\bar{x}_w = 0.7, 0.8,$ 0.9, and 1.0, respectively. Four alternatives are given in Table 9.5. It is recommended that the well battery be placed at $\bar{x}_w = 1.0$ in order to avoid any trapped salt water and obtain a better performance. Thus if 10 days of continuous pumping is allowed, the wells should be spaced at 265 m, whereas if 90 days is allowed, the well spacings should be increased to 2938 m (Table 9.5).

Hydraulic Barriers

Hydraulic barriers produce the same general effects as physical barriers by either vertical recharge or discharge wells arranged along a line parallel and very close to the shoreline. In several cases, the aquifer boundaries in contact with the saltwater bodies extend beyond the shoreline and the wells must be drilled offshore. Hydraulic barriers also may be formed by means of horizontal drains with vertical shafts (wells) at properly designed locations. Pumping in or out of these vertical shafts yields effects similar to those produced by vertical recharge and discharge

TABLE 9.5   Saltwater Retardation by Recharge Wells: Results of Solved Example 9.2

t, day	$\dfrac{Q}{q_s}$	S_p, m	n_w°
At $\bar{x}_w = 1.0$ and $x_w = 45.72$ km:			
5	165	59	169
10	37	265	38
90	3.33	2938	3
At $\bar{x}_w = 0.9$ and $x_w = 41.15$ km:			
5	67	146	68
10	20	501	20
90	2.69	3637	3
At $\bar{x}_w = 0.8$ and $x_w = 36.58$ km:			
5	24	402	25
10	9.55	1024	10
90	2.12	4615	2
At $\bar{x}_w = 0.7$ and $x_w = 32$ km:			
5	9.24	1059	9
10	4.82	2030	5
90	1.69	5790	2

° Number of wells per 10 km parallel to the shoreline (rounded).
SOURCE: A. I. Kashef, "Control of Salt Intrusion by Recharge Wells," *Proc. Am. Soc. Civ. Eng.*, vol. 102, no. IR4, 1976, pp. 445–457.

wells. In many cases, horizontal drains are preferred to vertical wells. However, upconing of the interface due to water withdrawal is less disastrous than in the case of direct pumping out of vertical discharge wells (Schmorak and Mercado, 1969). Placing the drains at the upper boundary of a deep artesian aquifer can be very costly. Thus the use of drains should only be considered practical in shallow unconfined aquifers. At the present time, information concerning optimum design recommendations is scarce.

Before the Los Angeles Counties project was started, it was estimated that 1200 million m³ (1 million acre-feet) of water was withdrawn from storage, lowering the water levels more than 31 m (100 ft) below sea level (Bruington and Seares, 1965). In 1965, the total annual water requirement was estimated at 719 million cubic meters. This demand was partially met by importing 376 million cubic meters for direct use and withdrawing the balance from the aquifers. The principal capital cost of the freshwater barrier was $15 million (excluding the costs of water-rights litigation), and the annual costs total $4.4 million for administration (1965), watermaster costs, the operation and maintenance of the spreading grounds and the freshwater barrier, and the purchase of supplemental water supplies. The last item alone costs $3.3 million.

Nassau County, Long Island, New York is another major area in the United States where hydraulic barriers have been constructed using injection wells. Urban development in this area greatly increased the runoff that would otherwise have been infiltrated in the upper sand and gravel beds. Recharge basins have been built to salvage storm water. The recharge water consists of reclaimed or renovated water from sewage treatment plant effluent (Baumann, 1952).

A preliminary attempt was made to establish some rules about the proper development of the hydraulic barriers (Kashef, 1975). It has been seen in the previous section that a battery of recharge wells parallel to the shoreline may under certain conditions develop hydraulic barriers when two curves of the same intensity $H^\circ = 1.0$ develop on both sides of the well battery. The straight lines given in Fig. 9.24 (curves b and c) and the crooked line d in the same figure provide perfect conditions for forming hydraulic barriers. The following conclusions were drawn from the solutions given in the previous section for various values of x_w, $S_p = 2L/(n_w - 1)$, and a factor $F = QTt/qSL^3$, which combines both the time factor $\bar{\delta}$ and the pumping factor $\bar{\beta}$:

1. Hydraulic barriers cannot be formed when $x_w > 0.4L$ and $S_p > 0.5L$.

2. When $S_p = L/2$, hydraulic barriers are formed within a range of F between 0.45 and 0.70.

3. For a range of F between 0.11 and 0.45 and a value of $S_p = L/3$, hydraulic barriers develop.

4. By increasing the pumping capacities Q and/or reducing the period of pumping t so that a high range of F between 1.05 to 2.50 is obtained, hydraulic barriers develop for a range of S_p between $L/3$ and $L/2$.

5. When $x_w = 0.1L$, $S_p \leqslant L/2$, and $F \geqslant 1.05$, hydraulic barriers with one side form, wasting most of the injected fresh water.

In all these cases, several combinations of Q, t, x_w, and S_p can be obtained. In each case when the higher limit of F is reached, pumping should be stopped until the lower limit is reached, at which time pumping should be resumed. The preceding results were obtained for specific predetermined data and should be considered approximate, but the procedure can be expanded to get more refined design procedures.

Control by Regulation of Activities

Saltwater intrusion control can be accomplished by limiting the ongoing water withdrawal and prohibiting any future water withdrawal in a certain area. Such a procedure requires extensive legal and administrative planning, including an appropriate technical monitoring system. This approach is not recommended because it is a conservative one based mainly on lack of technical skill or fear of assuming major responsibilities.

A more suitable approach includes the regulation of water withdrawal, allowing more water development without any detrimental effects on the water quality using appropriate control methods. This type of approach requires a great deal of technical, legal, and administrative expertise, well-planned monitoring systems, and sophisticated models (usually mathematical). Redistribution of producing wells and regulation of their capacities should be planned in such a way that saltwater intrusion remains in equilibrium. The method implies simply the prevention of further saltwater intrusion beyond an acceptable limit and at the same time the development of the aquifer to its maximum possible yield.

PROBLEMS AND DISCUSSION QUESTIONS

9.1 In a coastal unconfined aquifer, the lower impervious boundary is 95 m below the mean sea level. The average darcian natural velocity is 4.5 cm per year. If the unit weights of fresh and salt water are, respectively, 1.0

and 1.030 g/cm³, find the water-table elevation above sea level at the apex of the saltwater wedge using the Ghyben-Herzberg principle.

9.2 If the slope of the shore in Problem 9.1 is about $30°$ with the horizontal, plot the free surface by means of the Schaffernack and Iterson method (Chap. 5), assuming that the flow lines are horizontal along the plane of mean sea level. (*Hint:* Determine the length of the intruded salt wedge from the darcian velocity given in Problem 9.1 and an average hydraulic conductivity K of 2.5×10^{-2} cm/s.)

9.3 Plot the Ghyben-Herzberg interface on the basis of the data and results obtained from Problems 9.1 and 9.2. Write the ordinates of this curve at six equal intervals.

9.4 The natural flow in a coastal artesian aquifer has a hydraulic gradient of 0.002. If the aquifer is 55 m deep and the elevation of its upper boundary is 20 m below sea level, determine the length of the intruded wedge. The unit weight of salt water is 1.025 g/cm³, the hydraulic conductivity K is 2.5×10^{-2} cm/s, the ground surface elevation is $+5.00$ m, and the thickness of the upper impervious layer is 25.0 m (assume the outflow surface as vertical).

9.5 Plot the interface in Problem 9.4 using all possible methods given in the text. Compile the results of the various methods in tabulated form.

9.6 The Ghyben-Herzberg principle was proved for unconfined aquifers. Prove that it is also valid for artesian aquifers.

9.7 Equation (9.12a and b) indicates that the length of the intruded wedge does not depend on the height η of sea level above the upper boundary of the artesian aquifer. However, in deriving the Ghyben-Herzberg's equation, this height η was taken into consideration. Explain the effect of changes in height η, comparing the results when $\eta = 20$ m (Problem 9.4) and $\eta = 10$ m while other data remain unchanged.

9.8 Apply the *modified* Ghyben-Herzberg method and determine the saltwater-freshwater interface corresponding to the case given in Problem 9.1 assuming that the outflow face is
(a) Vertical
(b) Horizontal
(c) Sloping $45°$ with the horizontal (use interpolation similar to that used in earth dams for this case)
The free surface is considered the same as that determined from Problem 9.2.

9.9 Using the data and results or Problems 9.1 and 9.2, plot the interface according to Eq. (9.10) and Table 9.1 assuming that the outflow surface is horizontal. Also determine the length of the outflow surface in each case.

9.10 Rework Problem 9.9 assuming that the outflow surface is vertical. Also determine the length of the outflow surface in each case.

9.11 Tabulate the ordinates of the interfaces determined in Problems 9.3, 9.8, 9.9, and 9.10 in one table (at least six ordinates in each case).

9.12 Discuss the advantages and disadvantages of the viscous flow models used in connection with saltwater-intrusion problems.

9.13 Write an article about 10 pages long on hydrodynamic dispersion compiled from DeWiest (1969), Bear (1979), and the summary given in this text. Refer to any other paper given in these references that may further clarify your article.

9.14 Discuss the practicality of the methods used to find the interface in layered systems [see Eqs. (9.20) through (9.24)].

9.15 Define the critical depths of upconing beneath water wells and drains. Compile the various methods used in the text and indicate what type of data are required to determine these depths.

9.16 By means of a sketch, explain how a saltwater mound develops beyond the natural interface.

9.17 By means of a sketch, indicate the sequence of steps required to determine the upconing curve beneath an artesian well in a coastal aquifer.

9.18 Explain the main assumptions and drawbacks in the methods represented by Eqs. (9.10), (9.12), (9.14), (9.15), and (9.43) through (9.46).

9.19 What are the major methods of controlling saltwater intrusion? Explain using sketches, whenever necessary, of the various types of hydraulic barriers.

9.20 Rework Example 9.2 using P values of 20, 25, and 30 percent (P is the degree of saltwater retardation).

REFERENCES

Baumann, P.: "Technical Development in Ground Water Recharge," in V. T. Chow, (ed.), *Advances in Hydroscience*, vol. 2, Academic Press, New York, 1965, pp. 209–279.

Baumann, P.: "Groundwater Movement Controlled through Spreading," *Trans. Am. Soc. Civ. Eng.*, vol. 117, 1952, p. 1024.

Bear, J.: "Scales of Viscous Analogy Models for Ground Water Studies," *Proc. Am. Soc. Civ. Eng.*, vol. 86, no. HY2, 1960, pp. 11–23.

Bear, J.: *Hydraulics of Groundwater*, McGraw-Hill, New York, 1979.

Bear, J., and G. Dagan: "Some Exact Solutions of Interface Problems by Means of the Hodograph Method," *J. Geophys. Res.*, vol. 69, no. 8, 1964a, p. 1563.

Bear, J., and G. Dagan: "Moving Interface in Coastal Aquifers," *Proc. Am. Soc. Civ. Eng.*, vol. 90, no. HY4, 1964b, p. 193.

Braithwaite, F.: "On the Infiltration of Salt Water into the Springs of Wells under London and Liverpool," *Proc. Inst. Civ. Eng.*, vol. 14, 1855, p. 507.

Bredehoeft, J. D.: "Response of Well-Aquifer Systems to Earth Tides," *J. Geophys. Res.*, vol. 72, no. 12, 1967, p. 3075.

Brown, F. F., and D. C. Signor: "Groundwater Recharge," *Water Res. Bull.*, vol. 8, 1972, pp. 132–149.

Brown, F. F., and D. C. Signor: "Artificial Recharge: State of the Art," *Ground Water*, vol. 12, 1974, pp. 152–160.

Brown, J. S.: "Study of Coastal Ground Water with Special Reference to Connecticut," U.S. Geological Survey Water Supply Paper 537, Washington, 1925.

Bruington, A., and Seares, F.: "Operating a Sea Water Barrier Project," *Proc. Am. Soc. Civ. Eng.*, vol. 91, no. IR1, 1965, p. 17.

Charmonman, S.: "A Solution of the Pattern of Fresh-Water Flow in an Unconfined Coastal Aquifer," *J. Geophys. Res.*, vol. 70, 1965, pp. 2813–2819.

Columbus, N.: "Viscous Model Study of Sea Water Intrusion in Water Table Aquifers," *Water Resources Res.*, vol. 1, no. 2, 1965, p. 313.

Cooper, H. H., Jr., F. A. Kohout, H. R. Henry, and R. E. Glover: "Sea Water in Coastal Aquifers," U.S. Geological Survey Water Supply Paper 1613, Washington, 1964.

Cooper, H. H., Jr., et al.: "The Response of Well-Aquifer Systems to Seismic Waves," *J. Geophys. Res.*, vol. 70, 1965, pp. 3915–3926.

d'Andrimont, R.: "Notes sur l'hydrologie du littoral Belge" (Notes on the Hydrology of the Belgian Coast), *Ann. Soc. Geol. Belg.*, vol. 29, no. M129, 1902.

d'Andrimont, R.: "L'allure des nappes aquifére continues dans des terrains perméables en petit, an voisinage de la mere" (The Nature of Ground Water Contained in Freely Pervious Aquifers Adjacent to the Sea), *Ann. Soc. Geol. Belg.*, vol. 32, no. M101, 1905a.

d'Andrimont, R.: "Notes préliminaire sur une nouvelle méthode pour étudier experimentalement l'allure des nappes aquiféres dans les terrains perméables en petit" (Preliminary Notes on a New Method of Studying Experimentally the Phenomenon of Ground Water in Freely Pervious Aquifers), *Ann. Soc. Geol. Belg.*, vol. 32, no. M115, 1905b.

de Josselin de Jong, G.: "Longitudinal and Transverse Diffusion in Granular Deposits," *Trans. Am. Geophys. Union*, vol. 39, 1968, pp. 67–74.

DeWiest, R. J. M. (ed.): *Flow through Porous Media*, Academic Press, New York, 1969.

Dvoracek, M. J., and V. H. Scott: "Ground-Water Flow Characteristics Influenced by Recharge Pit Geometry," *Trans. Am. Soc. Agric. Eng.*, vol. 6, 1963, pp. 262–265, 267.

Gelhar, L. W., and M. A. Collins: "General Analysis of Longitudinal Dispersion in Nonuniform Flow," *Water Resources Res.*, vol. 7, 1971, pp. 1511–1521.

Ghyben, E. W.: "Nota in verbandmet de voorgenomen put boring nabij Amesterdam" (Notes on Probable Results on the Proposed Well Drilling near Amstedam), *Tijdschr. K. Inst. Ingen.*, vol. 21, 1888–1889.

Glover, R. E.: "The Pattern of Fresh-Water Flow in a Coastal Aquifer," in *Sea Water in Coastal Aquifers*, U.S. Geological Survey Water Supply Paper 1613-C, Washington, 1964, pp. C32–C35.

Hantush, M. S.: "Unsteady Movement of Fresh Water in Thick Unconfined Saline Aquifers," *Bull. Int. Assoc. Sci. Hydrol.*, vol. 13, no. 2, 1968, p. 40.

Hele-Shaw, H. S.: "Investigation of the Nature of the Surface Resistance of water and Stream-Line Motion under Certain Experimental Conditions," *Trans. Inst. Naval Architects*, vol. 40, 1898, p. 21.

Hele-Shaw, H. S.: "Stream-Line Motion of a Viscous Film," *Br. Assoc. Adv. Science*, vol. 68, 1899, p. 136.

Henry, H. R.: "Saltwater Intrusion into Fresh Water Aquifers," *J. Geophys. Res.*, vol. 64, no. 11, 1959, p. 1911.

Herzberg, B.: "Die Wasserversorgung einiger Nordseebäder," (The Water Supply on Parts of the North Sea Coast), *J. Gabeleucht. Wasservorsorg. (Munich)*, vol. 44, 1901, p. 815.

Hubbert, M. K.: "Theory of Ground-Water Motion," *J. Geol.*, vol. 48, no. 8, 1940, p. 785.

Hubbert, M. K.: *The Theory of Ground-Water Motion and Related Papers*, Hafner, New York, 1969.

International Association of Scientific Hydrology: "Artificial Recharge and Management of Aquifers," Publication 72, 1967.

Jacob, C. E.: "Flow of Ground Water," in H. Ronse (ed.), *Engineering Hydraulics*, Wiley, New York, 1950, chap. 5.

Kashef, A. I.: "Salt-Water Enroachment in the Eastern North Carolina Coastal Region," in *Proceedings of the Symposium on Hydrology of Coastal Regions of North Carolina*, vol. 5, Water Resources Research Institute, University of North Carolina, Raleigh, 1967, pp. 24–38.

Kashef, A. I.: "Salt-Water Intrusion in Coastal Well Fields," in *Proceedings of the Symposium on Ground-Water Hydrology*, vol. 4, American Water Resources Association, 1967, pp. 235–258.

Kasehf, A. I.: "Dispersion and Diffusion in Porous Media: Salt-Water Mounds in Coastal Aquifers," Water Resources Research Institute Report 11, University of North Carolina, Raleigh, 1968a.

Kashef, A. I.: "Fresh-Salt Water Interface in Coastal Ground-Water Basins," in Proceedings of the International Association of Hydrogeologist, Vol. 8, Turkish Committee of the International Association of Hydrogeologists, Istanbul, 1968b, pp. 369–375.

Kashef, A. I.: "Model Studies of Salt Water Intrusion," Water Res. Bull., vol. 6, no. 6, 1970, pp. 944–967.

Kashef, A. I.: "On Ground Water Management in Coastal Aquifers," Ground Water, vol. 9, no. 2, 1971, pp. 12–20.

Kashef, A. I.: "What Do We Know about Salt Water Intrusion?" Water Res. Bull., vol. 8, no. 1, 1972, pp. 282–293.

Kashef, A. I.: "Management of Retardation of Salt Water Intrusion in Coastal Aquifers," Office of Water Research and Technology Report, U.S. Department of Interior, Washington, 1975.

Kashef, A. I.: "Control of Salt Intrusion by Recharge Wells," Proc. Am. Soc. Civ. Eng., vol. 102, no. IR4, 1976a, pp. 445–457.

Kashef, A. I.: "Theoretical Developments and Practical Needs in the Field of Salt Water Intrusion," in Advances of Groundwater Hydrology, American Water Resources Association, 1976b, pp. 228–242.

Kashef, A. I.: "Management and Control of Salt-Water Intrusion in Coastal Aquifers," CRC Crit. Rev. Environ. Control, vol. 7, no. 3, 1977, pp. 217–275.

Kashef, A. I.: "Harmonizing Ghyben-Herzberg Interface with Rigorous Solutions," Ground Water, vol. 21, no. 2, 1983, pp. 153–159.

Kashef, A. I., and J. C. Smith: "Expansion of Salt-Water Zone due to Well Discharge," Water Resour. Bull., vol. 11, no. 6, 1975, pp. 1107–1120.

Kohout, F. A.: "Cyclic Flow of Salt Water in the Biscayne Aquifer of Southeastern Florida," J. Geophys. Res., vol. 65, no. 7, 1960, pp. 26–30.

Lusczynski, N. J., and W. V. Swarzenski: "Salt-Water Encroachment in Southern Nassau and Southeastern Queens Counties, Long Island, New York," U.S. Geological Survey Water Supply Paper 1613-F, Washington, 1966.

Marino, M. A.: "Hele-Shaw Model Study of the Growth and Decay of Ground-Water Ridges," J. Geophys. Res., vol. 72, 1967, pp. 1195–1205.

Muskat, M.: The Flow of Homogeneous Fluids through Porous Media, McGraw-Hill, New York, 1937; second printing, 1946, by J. W. Edwards, Ann Arbor, Mich.

Newport, B. D.: "Salt Water Intrusion in the United States," Report EPA-600/8-77-011, U.S. Environmental Protection Agency, Ada, Okla., 1977.

Oakes, D. B., and K. J. Edworthy: "Field Measurements of Dispersion Coefficients in the United Kingdom," Paper no. 12, International Conference on Groundwater Quality, Measurement, Prediction, and Protection, Water Research Center, Reading, England, 1976.

Ogata, A.: "Theory of Disperson in a Granular Medium," U.S. Geological Survey Professional Paper 411-1, Washington, 1970, pp. 11–134.

Peters, J. H., and D. Cuming: "Water Conservation by Barrier Injection," Water Sewage Works, vol. 114, 1967, p. 63.

Peters, J. H., and J. L. Rose: "Water Conservation by Reclamation and Recharge," Proc. Am. Soc. Civ. Eng., vol. 94, no. SA4, 1968, pp. 625–639.

Pinder, G. F., and H. H. Cooper: "A Numerical Technique for Calculating the Transient Position of Salt-Water Front," Water Resources Res., vol. 6, no. 3, 1970, p. 875.

Polubarinova-Kochina, P. Y.: Theory of Groundwater Movement, Princeton University Press, Princeton, N.J., 1962; translated by R. J. M. DeWiest from the Russian edition, 1952.

Rebhun, M., and Schwartz, J.: "Clogging and Contamination Processes in Recharge Wells," Water Resources Res., vol. 4, no. 6, 1968, p. 1207.

Rumer, R. R., Jr., and Harleman, D. F.: "Intruded Salt-Water Wedge in Porous Media," Proc. Am. Soc. Civ. Eng., vol. 89, no. HY6, 1963, p. 193.

Santing, G.: "A Horizontal Scale Model, Based on the Viscous Flow Analogy, for Studying

Ground-Water Flow in an Aquifer Having Storage," International Associaton of Scientific Hydrology, Toronto, 1957, p. 105.

Scheidegger, A. E.: "General Theory of Dispersion in Porous Media," *J. Geophys. Res.*, vol. 66, 1961, pp. 3273–3278.

Schmorak, S., and A. Mercado: "Upconing of Fresh Water-Sea Water Below Pumping Wells: Field Study, *Water Resources Res.*, vol. 5, no. 6, 1969, p. 1290.

Sternau, R.: "Artificial Recharge of Water Through Wells: Experience and Techniques," in *Proceedings of the Symposium on Artificial Recharge and Aquifer Management,* International Association of Scientific Hydrology, Toronto, 1967, p. 91.

Strack, O. D. L.: "Some Cases of Interface Flow Toward Drains," *J. Eng. Math. (Holland)*, vol. 6, 1972, pp. 175–191.

Strack, O. D. L.: "A Single-Potential Solution for Regional Interface Problems in Coastal Aquifers," *Water Resources Res.*, vol. 12, no. 6, 1976, pp. 1165–1174.

Terzaghi, K., and Peck, R. B.: *Soil Mechanics in Engineering Practice*, 2d ed., Wiley, New York, 1967.

Todd, D. K.: "Salt Water Intrusion and Its Control," *J. Am. Water Works Assoc.*, vol. 66, 1974, pp. 180–187.

Todd, D. K.: *Ground Water Hydrology,* 2d ed., Wiley, New York, 1980.

U.S. Geological Survey: "Salt-Water Encroachment in Southern Nassau and Southeastern Queens Counties, Long Island, New York," Water Supply Paper 1613-F, Washington, 1966.

Verruijt, A.: "A Note on the Ghyben-Herzberg Formula," *Bull. Inst. Assoc. Sci. Hydrol.*, vol. 13, no. 4, 1968, p. 43.

Verruijt, A.: "Steady Dispersion across an Interface in a Porous Medium," *J. Hydrol.*, vol. 14, 1971, p. 337.

Water Research Association: *Proceedings of the Artificial Groundwater Recharge Conference,* Medmenham, England, 1971.

Zielbauer, E. J.: "Sea Water Intrusion and the Barrier Projects," in *Engineering Geology in Southern California, Glendale, California,* special publication of the Association of Engineering Geologists, Los Angeles, Calif., 1966.

APPENDIX

A

SYMBOLS AND ABBREVIATIONS

A	Cross-sectional area
A_q	Coefficient in Eq. (7.9) and Table 7.1
A_s	Coefficient in Eq. (9.10) and Table 9.1
A_t	Amplitude of tide
A_v	Sectional area of voids
Å	Angstrom
AGU	American Geophysical Union
APHA	American Public Health Association
ASCE	American Society of Civil Engineers
ASTM	American Society for Testing and Materials
AWRA	American Water Resources Association
AWWA	American Water Works Association
a	Cross-sectional area of burette [Eq. (3.18)]
B	Base width of a solid dam or rectangular earth section; the length of the horizontal projection of free surface between its entrance and exit points in earth dams
B_1	Factor in Eqs. (5.26) and (5.27); $B_1 = f(b_1, b_2, t_s)$
B_2	Factor in Eqs. (5.26) and (5.27); $B_2 = f(b_1, b_2, t_s)$
B'	Distance between two bore holes
B_{cr}	Critical depth between bottom of a well and the highest point of an upconing curve
B_e	B of an equivalent rectangular earth section of a dam with sloping downstream

B'_e	Equal to $B_e + \Delta B_e$
B_f	Coefficient in Eqs. (8.100) and (8.101)
B_k	Leakage factor; $B_k = \sqrt{TD_a/K_a}$
B_{re}	Breadth of a recharge basin
B_s	Coefficient in Eq. (9.10) and Table 9.1
B_w	Width of the upper water surface in a stream
BE	Barometric efficiency
BOD	Biochemical oxygen demand
b_1	Portion of base length of a dam or weir from left end of base to the location of a sheet pile
b_2	Equal to $B - b_1$
b_{cr}	Critical depth between a subdrain and the highest point of an upconing curve
b_w	Height of a well bottom above an interface
C	Constant in Eq. (3.2)
C_1, C_2, C_3	Constants in Eqs. (6.53), (6.54), (8.71), and (8.72)
C'	Constant in Eq. (3.1)
C_c	Factor equal to $D_{30}^2/(D_{10} \times D_{60})$; see Table 1.9
C_{co}	Compression index of undisturbed soils
C'_{co}	Compression index of remolded soils
C_f	Coefficient in Eqs. (8.100) and (8.101)
C_h	Factor in Eq. (3.8)
C_i	Constant in Eqs. (5.63) through (5.66)
C_k, C'_k	Factors in Eq. (6.60) and Fig. 6.18
C_q	Constant in Eqs. (5.41), (5.42), and (5.43)
C_r	Coefficient in Eq. (8.111), determined from Fig. 8.51
C_s	Shape factor in Eq. (3.7a and b)
C_{st}	Relative concentration of salt; $C_{st} = c/c_o$
C_u	Uniformity coefficient; $C_u = D_{60}/D_{10}$ [Eq. (1.1)]
C_w	Constant in Eqs. (8.8) through (8.12)
C_ϕ	Constant in Eqs. (8.22), (8.23), and (8.24)
C_ψ	Constant in Eqs. (8.25) and (8.26)
CAE	Carbon-alcohol extract method
CCE	Carbon-chloroform extract method
CH	Clay of high compressibility (Table 1.9)
CL	Clay of low compressibility (Table 1.9)
c	Average concentration of salt in water
c_o	Initial concentration of salt in water
c_v	Coefficient of consolidation
D	Diameter of a circular pipe; diameter of spherical soil particles
D_{10}	Effective size; size at which 10 percent of soil aggregate is finer
D_{15}	Size at which 15 percent of soil aggregate is finer
D_{30}	Size at which 30 percent of soil aggregate is finer
D_{50}	Size at which 50 percent of soil aggregate is finer
D_{60}	Size at which 60 percent of soil aggregate is finer
D_{10}^f	D_{10} of filter
D_{15}^f	D_{15} of filter
D_{50}^f	D_{50} of filter
D_a	Depth of an aquitard or depth of a soil layer

D_d	Height of exit point of free surface above the impervious base in two-dimensional flow systems
D_{gu}	D_{xu} above point g in Fig. 9.10
D_o	Length of outflow surface of fresh water *below* sea level in unconfined aquifers
D_{ou}	Length of outflow surface of fresh water *above* sea level in unconfined aquifers
D_q	Depth of an artesian aquifer
D_r	Height of free surface of a gravity well above the impervious boundary at radius r
$D_{r,t}$	D_r at time t
D_s	Average diameter size of soil; $D_s \approx D_{10}$ or D_{15}
D_t	Equal to $\sqrt{t_d T/S_y}$
D_w	Height of exit point of free surface in a gravity well above the impervious boundary; depth below the upper impervious boundary in the case of saltwater upconing in confined aquifers
D_x	Height of free surface above the impervious boundary at x in a two-dimensional flow system; depth of freshwater lens below mean sea level in unconfined coastal aquifers
D_{xa}	Thickness of freshwater lens in coastal artesian aquifers
D_{xh}	Equivalent to D_x when the outflow surface is assumed horizontal in saltwater-intrusion analysis
D_{xo}	D_x at the origin
D_{xu}	Height of free surface above mean sea level in coastal aquifers
D_{xv}	Equivalent to D_x when the outflow surface is assumed vertical in saltwater intrusion analysis
D_x'	Depth of freshwater zone below sea level in coastal unconfined aquifers according to the Ghyben-Herzberg method
$\overline{D}_q$	Average depth of saturated zone in unconfined aquifers
$\overline{D}_r$	Equal to D_r/h_u
$\overline{D}_w$	Equal to D_w/h_u
d_c	Maximum height of water in a stream
d_p	Depth of a borehole or well below water table
d_s	Height of perforated, screened, or uncased sections in wells
d_v	Depth of vadose zone
E_s	Bulk modulus of compression
E_w	Bulk modulus of water ($\approx 20{,}000$ kg/cm^2); $E_w = 1/\beta_w$
E_{wf}	Well efficiency
$-Ei(-u)$	Exponential integral; $-Ei(-u) = W(u)$
EIS	Environmental impact statements
EPA	U.S. Environmental Protection Agency
ET	Rate of evapotranspiration in a hydrologic budget
$(ET)_g$	Rate of groundwater evapotranspiration in a groundwater budget
E_h	Redox potential of water
e	Void ratio of soil; $e = V_v/V_s = n/(1-n)$
e_o	Natural (in situ) void ratio of soil
F	Dimensionless factor; $F = QTt/qSL^3$

F_B	Hydraulic boundary force
F_s	Seepage force
$F(\alpha, \beta)$	Function used in Eq. (7.10) (infiltration of recharge basins)
FS	Factor of safety
G	Specific gravity of solid matters; $G = \gamma_s/\gamma_w = W_s/V_s\gamma_w$
GC	Gravel type (Table 1.9)
GM	Gravel type (Table 1.9)
GP	Gravel type (Table 1.9)
g	Acceleration due to gravity
H°	Equal to $(H_b)_{x,y}/\alpha_s D_q$
$(H_b)_x$	Total head along the upper boundary of an artesian aquifer (in saltwater-intrusion analysis
H_e	Equal to $[h_u - D_q - \eta(1 + \alpha_s)]/\alpha_s$
H_o	Depth of fresh water below mean sea level in unconfined coastal aquifers; $H_o = D_q$ of unconfined aquifers (in saltwater-intrusion analysis)
H_{ot}	Stage of tide
H_{ou}	D_{xu} at the apex of the saltwater wedge
H_r	Hardness of water in milligrams per liter as $CaCO_3$
H_p	Maximum flow path of drainage due to compressibility in a consolidation test or in a clay layer in the field
h	Total hydraulic head
h_{av}	Average h
h_b	Base pressure head (along the impervious boundary)
h_{bB}	h_{bx} at $x = B$
h_{br}	h_b at radius r
$h_{b(r,t)}$	h_b at radius r and time t
h_{bx}	h_b at x
h_d	Depth of downstream water (tailwater)
h_e	Excess hydraulic head; $h_e = h_u - h_d$
h_m	Distance between the mean of two values of h
h_0	h at time t_0
$h_{0,0}$	$h_{x,y}$ at origin
h_{oxt}	Amplitude due to tidal wave of water-level fluctuation in an observation well distance x from shore
h_r	h at radius r
$h_{r,t}$	h at radius r and time t
h_t	h at time t
h_u	Depth of upstream water; depth of undisturbed water table; depth of undisturbed piezometric surface
h_w	Height of water level in a well above the impervious boundary under steady-state conditions ($h_{w,t}$ at $t = \infty$)
$h_{w,t}$	Height of water level in a well above the impervious boundary at time t
$h_{x,y}$	h at point (x, y)
$\bar{h}$	Height of Dupuit's curve above the impervious boundary
I	Rate of inflow into a groundwater basin
IASH	International Association of Hydrological Sciences
IASH	International Association of Scientific Hydrology

i	Equal to $\sqrt{-1}$
i_g	Hydraulic gradient
$i_{g,av}$	Average i_g
$i_{g,c}$	Critical hydraulic gradient
$i_{g,e}$	Exit gradient
i_{gp}	Pressure gradient $= i_g \gamma_w =$ seepage force per unit volume
i_{gx}, i_{gy}	Components of i_g along x and y axes
K	Hydraulic conductivity (coefficient of permeability)
K_f	Curve e versus $\log \bar{\sigma}$ for soil in field
K_h	K_{max}, which is the hydraulic conductivity along direction of bedding (stratification)
K_0	Modified Bessel function of the second kind and zero order
K_p	Physical permeability (intrinsic)
K_q	K of soil layer in Fig. 5.20 in Eq. (5.88)
K_r	Curve e versus $\log \bar{\sigma}$ for a remolded soil
K_u	Curve e versus $\log \bar{\sigma}$ for an undistorbed soil
K_v	K_{min}, which is the hydraulic conductivity in the normal direction to bedding (stratification)
K_x	K along x axis
K_y	K along y axis
K_z	K along z axis
K_α	K along a direction making an angle α with the direction of bedding (stratification)
K'	Equal to $\alpha_s K$
L	Creep length in weirs (approximately length of the uppermost flow line); base width of an earth dam measured from the projection of the entrance point of the free surface; length of intruded saltwater wedge
L_{re}	Length of a recharge basin
L_s	Dimension shown in Fig. 5.20 [given in Eq. (5.88)]
L'	Average length of flow path [Fig. 5.19 and Eq. (5.87)]
L''	Average value of L_1 and L_2 [Fig. 9.10 and Eqs. (9.23) and (9.24)]
$\bar{L}$	Equal to L/H_o
l	Length of a soil specimen
MH	Silt type (Table 1.9)
ML	Silt type (Table 1.9)
MPC	Maximum permissible concentrations of radionuclides
MPN	Most probable number of Coliform group organisms
m	Equal to B/h_u
m_v	Vertical soil compressibility (coefficient of volume compressibility)
m_w	Equal to h_w/h_u; $m_w = h_{w,t}/h_u$
N	Coefficients in Eqs. (6.52), (6.53), and (6.55); a number in Tables 8.3 and 8.8
N_i	Number of image wells
N_s	Equal to $(S + S_y)/S$
NAE	National Academy of Engineering
NAS	National Academy of Sciences

NAVFAC	Naval Facilities Engineering Command, U.S. Department of the Navy
NEPA	National Environmental Policy Act
n	Constant in Eqs. (3.1) and (8.100); distance normal to a flow line; porosity as a fraction; a power of a parameter
$n(\%)$	Equal to $(V_v/V)100 = 100e/(1 + e)$
n_E	Number of equipotential drops in a flow net
n_F	Number of flow channels in a flow net
n_w	Number of wells
O	Rate of outflow from a groundwater basin
OH	Type of organic soil (Table 1.9)
OL	Type of organic soil (Table 1.9)
P	Degree of saltwater retardation measured as a percentage; $P = 100(1 - \bar{x}_r)$
P_a	P_x at $x = 0$ in earth dams
P_b	P_x at $x = B$ in earth dams
P_e	P_r at $r = r_e$
P_I	Plasticity index; $P_I = W_L - W_p$
P_i	Rate of precipitation (rainfall and snow) on a groundwater basin
P_o	P_r at $r = r_o$
P_r	Area of pressure-head diagram across vertical sections (in water wells)
$P_{r,t}$	P_r at time t
P_w	P_r at $r = r_w$
P_x	Area of pressure-head diagram across a vertical section in a two-dimensional flow system (at x)
P_u	Uplift force
PL	Public law
Pt	Peat soil (Table 1.9)
p	Pore-water pressure
p_b	Water pressure on a base of a solid dam (or weir)
p_s	Water pressure on a sheet pile
p_{st}	Pore saltwater pressure
pH	Hydrogen-ion concentration
p/γ_w	Pressure head
Q	Rate of flow; steady rate of well pumping
Q_{cr}	Q at $t = t_{cr}$ (in saltwater-intrusion analysis)
Q_g	Rate of groundwater runoff
Q_{rs}	Portion of well pumping Q that is withdrawn from a stream
$Q_{r,t}$	Rate of flow at a cylindrical surface of a radius r and at time t due to well pumping: $Q_{r,t} < Q$
Q_{st}	Rate of stream flow in a hydrologic budget
Q_{sub}	Rate of subsurface underflow in both hydrologic and groundwater budgets
q	Rate of flow per unit length, or per unit depth
q_a	Steady rate of flow in a confined aquifer; the portion of the total flow rate q below the plane coinciding with sea level in a coastal unconfined aquifer

q_b Rate of flow below tailwater level; $q_b = q - q_u$

q_{dr} Rate of flow in a subdrain (per unit length)

q_{re} Reduced rate of flow to accommodate the effect of upstream slopes in earth dams

q_{rs} Portion of q of a well that is withdrawn from a stream (Q_{rs} per unit depth)

q_s Equal to $q \times S_p$

q_u Rate of flow above tailwater level; $q = q_u + q_b$

q' Rate of seepage from the bottom of a stream

R_c Critical radius of upconing (artesian wells)

R_d Dial reading in a consolidation test

Re   Reynold's number

R_f Radius of a spherical grain of a filter (idealized)

R_g Rate of groundwater recharge in a groundwater basin

R_s Radius of a spherical grain of a soil (idealized)

R_{15} Equal to D_{15}^f/D_{15}

R_{50} Equal to D_{50}^f/D_{50}

r Coordinate; $r = \sqrt{x^2 + y^2}$

r_b Radius of uncased borehole

r_c Internal radius of casing

r_e Radius of influence of a well at steady state of flow

$r_{e,t}$ r_e at time t

$r_{e,t_{cr}}$ r_e at time $t = t_{cr}$

r_o Radius of the boundary between artesian and gravity flow conditions in the case of an overpumped artesian well

r_w Effective radius of a well

$\bar{r}$ Equal to r/L

$\bar{r}_e$ Equal to r_e/h_u

S Coefficient of storage of an aquifer

S_a Coefficient of storage of an aquitard

S_e Length of outflow surface in earth dams

S_g Groundwater storage in a groundwater budget

S_m Soil moisture storage in a hydrologic budget

S_p Uniform well spacing in a battery of recharge wells

S_r Degree of saturation measured as a percentage; $S_r = 100 V_w/V_v$

S_s Coefficient of specific storage

S_t Sensitivity of clays

S_{ty} Equal to $S_{ya} + S$, which is approximately equal to S_y in unconfined aquifers

S_u Ultimate settlement

S_y Specific yield

S_{ya} Intermediate specific yield

SC   Sand type (Table 1.9)

SM   Sand type (Table 1.9)

SP   Sand type (Table 1.9); stagnation point

s Distance along a flow line

s_d Drawdown at radius r; $s_d = h_u - h_r$

s_{da} Drawdown in an aquitard [Eq. (6.36)]

s_{df} Final drawdown (at equilibrium)

s_{dj}	Adjusted drawdown; $s_{dj} = s_d - (s_d^2/2h_u)$
s_{dw}	s_d at radius r_w
s_i	s_d for a well image
s_p	Correction parameter used in partially penetrating wells [Eqs. (8.47) and (8.106)]
s_R	Equal to $s_r \pm s_i$
s_r	s_d for a real well
s_{wf}	Drawdown s_{dw} for fully penetrating well
s_{wp}	Drawdown s_{dw} for partially penetrating well
s_x	Drawdown of natural flow at x
s_d'	Residual drawdown in an aquifer
T	Coefficient of transmissivity; $T = KD_q$ (or Kh_u)
T_c	Coefficient of transmissivity in unconfined aquifers; $T_c = K\overline{D}_q$
T_v	Time factor in Eq. (6.14) and Fig. 6.2
TDS	Total dissolved solids
TE	Tidal efficiency
t	Time
t_b	Depth of borehole below the ground surface
t_{cr}	Critical time to reach saltwater upconing in artesian wells
t_d	Delay index
t_{fi}	Thickness of filter material
t_i	Time corresponding to image wells
t_L	Time lag between the change of the stage of tide and the response in water level in an observation well
t_{ot}	Tidal period
t_r	Time corresponding to real well (when dealing with images)
t_s	Depth of sheet pile; depth of a soil prism adjacent to a sheet pile
t'	Time measured after pumping stops or after infiltration ceases (in a recharge basin)
U	Total buoyancy; $U = V\gamma_w$
U_c	Degree of consolidation
U_g	Buoyancy on the grains; $U_g = V_s\gamma_w$
USBPR	U.S. Bureau of Public Roads
USBR	U.S. Bureau of Reclamation
USDA	U.S. Department of Agriculture
USDC	U.S. Department of Commerce
USDI	U.S. Department of Interior
USGS	U.S. Geological Survey
u	Equal to $r^2S/4Tt$
u_i	u for image wells
u_r	u for real wells when dealing with images
u_w	Equal to $r_w^2S/4Tt$
$u_{x,w}$	Equal to $x_w^2S/4Tt$
u_y	Equal to $r^2S_y/4Tt$
V	Total volume
V_s	Volume of voids
V_w	Volume of water
v	Velocity; darcian velocity; rate of leakage

v_{am}	Arrival recharge rate (per unit area) at water table of infiltrated water from a recharge basin
v_{ap}	Velocity of approach
v_c	Lower critical velocity
v_o	Darcian velocity of natural flow
$v_{r,t}$	Radial velocity at radius r and at time t
v_s	Seepage velocity (true velocity); $v_s = v/n$
$v_{s,t}$	Velocity along direction s_t
$v_x, v_{x'}$	Velocity component along x direction
$v_{x,t}$	Component of $v_{r,t}$ along x direction
$v_y, v_{y'}$	Velocity component along y direction
$v_{y,t}$	Component of $v_{r,t}$ along y direction
W	Total weight of soil specimen; $W = W_s + W_w$
W_{di}	Designation for a discharge well
W_{fi}	Dry weight of filter material
W_r	Constant rate of rainfall, evaporation, or evapotranspiration
W_{re}	Designation for a recharge well
W_s	Weight of solids in a soil specimen
$W(u)$	Well function; $W(u) = -Ei(u)$
W_w	Weight of water in a soil specimen
WATSTORE	National Water Data Storage and Retrieval System
WPCF	Water Pollution Control Federation
WPRS	Water and Power Resources Service, U.S. Department of Interior (a title change for U.S. Bureau of Reclamation for a brief period)
w	Water content of soil (as a percentage); $w = 100W_w/W_s$
w_L	Liquid limit of soil
w_h	Natural moisture content (w in field)
w_s	Shrinkage limit of soil
$\overline{w}$	Complex plane; $\overline{w} = \phi + i\psi$
X	Shortened x axis; $X = x\sqrt{K_y/K_x}$
x_r	Coordinate of line $H° = 1.0$ (Chap. 9)
x_s	x coordinate of stagnation point
x_w	x coordinate of well location
$\overline{x}$	Equal to x/L
$\overline{x}_r$	Equal to x_r/L
$\overline{x}_w$	Equal to x_w/L
y_e	Elevation head
y_g	Maximum y value of groundwater divide (Fig. 8.6)
y_o	Height of free surface at $x = 0$ (Fig. 5.16a)
y_s	y coordinate of stagnation point
y_w	y coordinate of well location
$\overline{y}$	Equal to y/L
$\overline{y}_w$	Equal to y_w/L
$Z(x, y, t)$	Function in Eq. (7.11)
$Z(x, y, t - t')$	Function in Eq. (7.11)
z_c	Equal to $\alpha_s D_q + \eta(1 + \alpha_s)$; Fig. 9.3
$\overline{z}$	Complex plane; $\overline{z} = x + iy$

Greek Symbols

α	Angle of slope of piezometric surface, tangents, etc.
α_s	Equal to $(\gamma_{st} - \gamma_w)/\gamma_w$
α_1, α_2	Factors in Eq. (7.10)
α ray	Type of radiation due to nuclear-decay reactions
β	Downstream slope angle of an earth dam
β_w	Water compressibility ($\approx 4.69 \times 10^{-8}$ cm^2/g); $\beta_w = 1/E_w$
β_1, β_2	Factors in Eq. (7.10)
β ray	Type of radiation due to nuclear-decay reactions
$\bar{\beta}$	Pumping factor; $\bar{\beta} = Q/(4\pi\alpha_s KD_q^2)$
γ_d	Dry unit weight of soil; $\gamma_d = W_s/V$
γ_{fi}	Unit weight of filter material
γ_m	Bulk (mass) unit weight; $\gamma_m = W/V$
γ_s	Unit weight of soil particles; $\gamma_s = W_s/V$
γ_{st}	Unit weight of salt water
γ_{sat}	Saturated unit weight of soil
γ_w	Unit weight of water (fresh water) at 4°C; $\gamma_w = 1$ g/cm$^3 \approx$ 62.4 lb/ft^3
γ ray	Type of radiation due to nuclear-decay reactions
γ'	Submerged unit weight; $\gamma' = (\gamma_{\text{sat}} - \gamma_w) = (W - V\gamma_w)/V$
Δ	Differential
Δa	Vertical projection of Δs (Fig. 5.6)
ΔB_e	Additional width of the equivalent rectangular earth section to compensate for the sloping upstream surface
$(\Delta s_d)_o$	Δs_d value over one log cycle of t or over one log cycle of r^2/t or over one log cycle of r
$(\Delta s_d')_o$	$\Delta s_d'$ value over one log cycle of t'/t
$\Delta\left(\dfrac{1}{Q}\right)_o$	$\Delta\left(\dfrac{1}{Q}\right)$ over one log cycle of t [Eq. (8.68)]
δ	Slope angle
$\bar{\delta}$	Time factor; $\bar{\delta} = L^2S/4Tt = uL^2/r^2$
ϵ_c	Value of dispersion, assumed constant in all directions
ϵ_m	Molecular diffusion
ϵ_n	Lateral dispersion
ϵ_s	Longitudinal dispersion
η	Height of mean sea level above the upper boundary of the artesian aquifer
θ	$\tan^{-1}(y/x)$; also angle shown in Fig. 8.6
λ	Constant in Eq. (5.61); $\lambda = 4$ when $\cot \beta > 1$ and $\lambda = 3$ when $\cot \beta < 1$
λ_n, λ_s	Intrinsic dispersivity along the n and s directions, respectively
μ	Absolute or dynamic viscosity
ν	Soil diffusivity; $\nu = T/S = K/S_s$
ν'	Kinematic viscosity of a fluid; $\nu' = \mu/\rho$
ρ	Mass density of fluid (or water)
ρ_o	Mass density of fluid (or water) at atmospheric pressure
σ_t	Total stress

$\bar{\sigma}$	Effective stress
$\bar{\sigma}_o$	Effective stress in the field at the location from which a soil specimen is taken (related to consolidation of soils)
ϕ	Velocity potential
ϕ_o	Velocity potential, natural flow
ϕ_{wd}	Velocity potential, discharge wells
ϕ_{wr}	Velocity potential, recharge wells
ψ	Stream function
ψ_k	Coefficient in Eq. (8.81); $\psi_k = r\sqrt{K_a S_a}/4\sqrt{TSD_a}$
ψ_o	Stream function, natural flow
ψ_s	Stream function of the groundwater divide
ψ_{wd}	Stream function, discharge well
ψ_{wa}	Stream function, recharge well

CONVERSION TABLES

Length

Angstrom unit (Å)	=	1.0×10^{-7}	millimeter (mm)
	=	3.93701×10^{-9}	inch (in)
1.0 foot (ft)	=	12.0	inches (in)
	=	0.3048	meter (m)
1.0 kilometer (km)	=	1000	meters (m)
	=	3280.84	feet (ft)
	=	0.62137	mile (mi)
1.0 meter (m)	=	1000	millimeters (mm)
	=	100	centimeters (cm)
	=	39.37008	inches (in)
	=	1.09361	yards (yd)
	=	3.28083	feet (ft)
1.0 mile (mi)	=	1.60934	kilometers (km)
1.0 yard (yd)	=	3.0	feet (ft)
	=	0.9144	meter (m)

Area

1.0 acre	=	43,560	square feet (ft^2)
	=	4046.8564	square meters (m^2)

1.0 hectare	=	10,000	square meters (m²)
	=	2.471	acres
1.0 square kilometers (km²)	=	100	hectares
	=	0.3861	square mile (mi²)
1.0 square meter (m²)	=	10.76391	square feet (ft²)
1.0 square mile (mi²)	=	2.58999	square kilometers (km²)

Volume

1.0 acre-foot (acre-ft)	=	1233.482	cubic meters (m³)
	=	43,560	cubic feet (ft³)
1.0 cubic centimeter (cm³)	=	0.001	liter (L)
	=	0.033814	fluid ounce (fl. oz)
1.0 cubic meter (m³)	=	1000	liters (L)
	=	264.1721	gallons (gal)
	=	35.31467	cubic feet (ft³)
1.0 cubic foot (ft³)	=	28.31685	liters (L)
	=	7.48052	gallons (gal)
1.0 cubic kilometer (km³)	=	10^9	cubic meters (m³)
	=	0.81071×10^6	acre-feet (acre-ft)
	=	0.23991	cubic mile (mi³)
1.0 liter (L)	=	0.26417	gallon (gal)
	=	33.814	fluid ounces (fl. oz)
1.0 gallon (gal)	=	0.13368	cubic foot (ft³)
	=	128.0	fluid ounces (fl. oz)
	=	3.78541	liters (L)

Velocity (or Units of Hydraulic Conductivity)

1.0 foot per second (ft/s)	=	0.3048	meter per second (m/s)
1.0 meter per second (m/s)	=	3.60	kilometers per hour (km/h)
	=	3.28084	feet per second (ft/s)
	=	2.23694	miles per hour (mi/h)
1.0 gallon per square foot per day	=	0.0407458	meter per day
	=	0.13368	foot per day

Force

1.0 pound (lb)	=	4.4482	newtons (N)
1.0 kilogram (kgf)	=	9.80665	newtons (N)
	=	2.2046	pounds (lb)
1.0 dynes (dyn)	=	1.0×10^{-5}	newton (N)

Mass

1.0 gram (g)	=	0.001	kilogram (kg)
1.0 ounce (avdp)	=	28.34952	grams (g)
	=	0.0625	pound (avdp)
1.0 pound (avdp)	=	16.00	ounces (avdp)
	=	0.45359	kilogram (kg)
1.0 kilogram (kg)	=	2.20462	pounds (avdp)
	=	0.06852	slug
	=	0.10197	kilogram (force) – second squared per meter (kgf-s^2/m)
1.0 short ton	=	907.1847	kilograms (kg)
	=	0.90718	metric ton (t)
	=	2000	pounds (avdp)
1.0 metric ton (t)	=	1000	kilograms (kg)
(tonne or megagram)	=	2204.62	pounds (avdp)
	=	1.10231	short tons
1.0 long ton	=	1016.047	kilograms
	=	2240	pounds (avdp)
	=	1.01605	metric tons
	=	1.120	short tons

Pressure

1.0 meter-head (column of water measured at 4°C)	=	9.80636	kilopascals
	=	9806.36	newtons per square meter (N/m^2)
	=	1000	kilograms (f) per square meter (kgf/m^2)
	=	73.554	mmHg (measured at 0°C)
	=	3.28084	feet of water (measured at 4°C)
	=	204.81	pounds per square foot
	=	0.096781	standard atmosphere
1.0 foot of water (measured at 4°C)	=	2.98898	kilopascals (kPa)
	=	2988.98	newtons per square meter (N/m^2)
	=	304.79114	kilograms (f) per square meter (kgf/m^2)
	=	0.3048	meters-head (measured at 4°C)
	=	22.4193	mmHg (measured at 0°C)
	=	62.4261	pounds per square foot
	=	0.029499	standard atmosphere

1.0 kilopascal (kPa)	=	1000	newtons per square meter (N/m^2)
	=	10	millibars (mbar)
	=	0.01	bar
	=	7.50064	mmHg (measured at $0\,°C$)
	=	0.10197	meter-head (measured at $4\,°C$)
	=	20.8854	pounds per square foot
	=	9.8692×10^{-3}	standard atmosphere
1.0 bar	=	100	kilopascals (kPa)
	=	1000	millibars (mbar)
1.0 standard atmosphere	=	760	mmHg (measured at $0\,°C$)
	=	33.90	feet of water (measured at $4\,°C$)
	=	10.3327	meters-head (measured at $4\,°C$)
	=	1.03327	kilograms (f) per square centimeter (kgf/cm^2)
	=	14.70	pounds per square inch
1.0 short ton per square foot	=	95.76052	kilopascals (kPa)
	=	9765.091	kilograms (f) per square meter (kgf/m^2)
	=	0.9765091	kilogram (f) per square centimeter (kgf/cm^2)
1.0 kilogram (f) per square centimeter (kgf/cm^2)	=	1.024056	short tons per square foot
	=	0.96780	standard atmosphere
	=	2048.112	pounds per square foot
	=	98.0665	kilopascals (kPa)
	=	10.0	meters-head (measured at $4\,°C$)
	=	32.8083	feet of water (measured at $4\,°C$)
	=	735.54	mmHg (measured at $0\,°C$)

Temperature T: ($°C$, degrees Celsius; $°F$, degrees Fahrenheit; K, degrees Kelvin)

($T°C$)	=	$\frac{5}{9}[(T°F) - 32°]$
($T°F$)	=	$\frac{9}{5}(T°C) + 32°$
(T K)	=	$T°C + 273.18° = \frac{5}{9}[(T°F) + 459.67]$

Standard Gravitational Acceleration g

| g | = | 9.80665 | meters per second squared (m/s^2) |
| | = | 32.17405 | feet per second squared (ft/s^2) |

Symbols of the International System of Units (SI)*

Quantity	Name of unit	SI symbol
Length	Meter	m
Mass	Kilogram	kg
Force	Newton	N (kg $\cdot$ m/s^2)
Pressure	Pascal	Pa (N/m^2)
Time	Second	s
Temperature	Kelvin	K
Plane angle	Radians	rad

* SI = Système International d'Unités.

Prefixes for SI Units

Prefix	Symbol	Multiplication factor	Prefix	Symbol	Multiplication factor
tera	t	10^{12}	centi	c	10^{-2}
giga	G	10^{9}	milli	m	10^{-3}
mega	M	10^{6}	micro	μ	10^{-6}
kilo	k	10^{3}	nano	n	10^{-9}
hecto	h	10^{2}	pico	p	10^{-12}
deka	da	10^{1}	femto	f	10^{-15}
deci	d	10^{-1}	atto	a	10^{-18}

C

PROOF OF EQ. (5.14)

The natural sediments in a flow medium are commonly composed of stratified soil layers of various thicknesses and hydraulic conductivities. Therefore, the transmissivity T of each layer usually differs depending on the specific values of K and D.

If water flows in the direction of stratification (x axis), the average K_x is given by

$$K_x = \frac{1}{\sum\limits_{i=1}^{n} D_i} \sum_{i=1}^{n} K_i D_i \tag{C.1}$$

When the flow is normal to the direction of stratification, the average K_y is given by

$$K_y = \frac{\sum\limits_{i=1}^{n} D_i}{\sum\limits_{i=1}^{n} D_i/K_i} \tag{C.2}$$

Equations (C.1) and (C.2) are based on Darcy's law; their proof can be

found in any standard book on soil mechanics. As was discussed in Sec.
3.7 K_x is greater than K_y (except in loessial soils). For a certain number of
layers, $\sum_{i=1}^{n} D_i$ and $\sum_{i=1}^{n} K_i$ have constant values. Thus Eqs. (C.1) and
(C.2) can be written as

$$K_x = \left(\sum_{i=1}^{n} K_i \right) \sum_{i=1}^{n} \left(\frac{K_i}{\sum_{i=1}^{n} K_i} \times \frac{D_i}{\sum_{i=1}^{n} D_i} \right)$$

$$= \left(\sum_{i=1}^{n} K_i \right) \sum_{i=1}^{n} (A_i B_i) \tag{C.3}$$

and
$$K_y = \left(\sum_{i=1}^{n} K_i \right) \frac{1}{\sum_{i=1}^{n} (B_i/A_i)} \tag{C.4}$$

where
$$A_i = \frac{K_i}{\sum_{i=1}^{n} K_i} < 1.0$$

and
$$B_i = \frac{D_i}{\sum_{i=1}^{n} D_i} < 1.0$$

Since $1/A_i$ is greater than A_i, the value of K_x [Eq. (C.3)] is greater than K_y
[Eq. (C.4)].

An anisotropic soil formation can be visualized as being composed of
very thin stratifications that form a *more or less* homogeneous (yet hydraulically anisotropic) formation. Therefore in anisotropic formations,
$K_x > K_y$ and K_α along a direction making an angle α with the x axis has an
intermediate value expressed by the following equation (Casagrande,
1937):

$$K_\alpha = \frac{K_x K_y}{K_x \sin^2 \alpha + K_y \cos^2 \alpha} \tag{C.5}$$

[This equation is the same as Eq. (3.24) on page 99.] It was explained in
Sec. 5.1 that in an anisotropic medium a flow net can be drawn on a
transformed section (similar to that shown in Fig. 5.4a) that has been
shortened along the x axis to a new coordinate $X = x\sqrt{K_y/K_x}$ in accordance with Eqs. (5.9) and (5.10). The K_α value changes continuously along
a flow line in the natural-scale section (except along an impervious
boundary). It is therefore convenient to plot a flow net on a transformed
section rather than on a natural-scale section. The actual rate of flow q

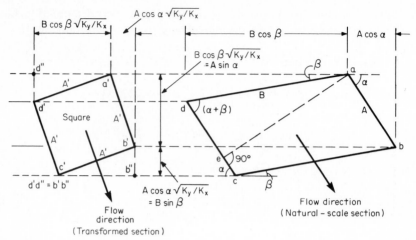

Figure C.1   An inclined square field in a flow net drawn on a transformed section and its projection on a natural-scale section.

can be determined from the flow net drawn on a transformed section using the following equation from Chap. 5.

$$q = \sqrt{K_x K_y} \, h_e \frac{n_F}{n_E} \tag{5.14}$$

In both the original and transformed sections h_e, n_F and n_E are the same. Since the transformed section represents a homogeneous and isotropic medium, its average hydraulic conductivity is the same at all points and in all directions. Its value K_{tr} is given by

$$K_{tr} = \sqrt{K_x K_y} \tag{C.6}$$

and is included in Eq. (5.14). Taylor (1948) proved this equation for flow in a horizontal direction. His procedure can be followed to prove Eq. (C.6) [and hence Eq. (5.14)] for flow in a vertical direction. This author (Kashef, 1954) extended the proof for flow in a general direction.

Figure C.1 shows an inclined square field in a flow net drawn on a transformed section. Its projection on a natural-scale section is shown in the right-hand side of the figure (parallelogram *abcd*). The values of the projections along the *x* and *y* axes are in accordance with the transformation procedure given in Sec. 5.1. From these projections in the figure, it is obvious that

$$\cos \beta = \frac{A}{B} \frac{\sin \alpha}{\sqrt{K_y/K_x}}$$

and
$$\sin \beta = \frac{A}{B} \cos \alpha \sqrt{\frac{K_y}{K_x}} \qquad (C.7)$$

If Darcy's law is applied to find the rate of flow Δq through the field, then

$$\Delta q = K_{tr} A' \frac{\Delta h}{A'} = K_{tr} \Delta h \qquad \text{(transformed section)} \qquad (C.8)$$

and
$$\Delta q = K_\alpha \Delta h \frac{\overline{ae}}{ab} \qquad \text{(natural-scale section)} \qquad (C.9)$$

In both these equations, Δq and Δh have the same magnitudes; therefore

$$K_{tr} = K_\alpha \frac{\overline{ae}}{ab} = K_\alpha \frac{\overline{ae}}{A} \qquad (C.10)$$

But
$$\overline{ae} = B \sin(\alpha + \beta)$$
$$= B(\sin \alpha \cos \beta + \cos \alpha \sin \beta)$$

If this value of $\overline{ae}$ and K_α [Eq. (C.5)] are substituted into Eq. (C.10),

$$K_{tr} = \frac{K_x K_y B(\sin \alpha \cos \beta + \cos \alpha \sin \beta)}{(K_x \sin^2 \alpha + K_y \cos^2 \alpha)A}$$

If the values for $\cos \beta$ and $\sin \beta$ from Eq. (C.7) are then substituted in this equation,

$$K_{tr} = \frac{K_x K_y(\sin^2 \alpha/\sqrt{K_y/K_x} + \cos \alpha \sqrt{K_y/K_x})}{K_x \sin^2 \alpha + K_y \cos^2 \alpha}$$
$$= \frac{K_x \sqrt{K_x K_y} \sin^2 \alpha + K_y \sqrt{K_x K_y} \cos^2 \alpha}{K_x \sin^2 \alpha + K_y \cos^2 \alpha}$$

or
$$K_{tr} = \sqrt{K_x K_y}$$

If this value is substituted in Eq. (C.8) and it is noted that $q = \Delta q \, n_F$ and $\Delta h = h_e/n_E$, Eq. (5.14) is obtained.

REFERENCES

Casagrande, A.: "Seepage through Dams," *J. N. Engl. Water Works Assoc.*, vol. 51, 1937, pp. 131–172.

Kashef, A. I.: "Seepage through Anisotropic Soils," *Civ. Eng. Mag. (Cairo)*, vol. 2, no. 2, 1954, pp. 61–66.

Taylor, D. W.: *Fundamentals of Soil Mechanics*, Wiley, New York, 1948.

AUTHOR INDEX

SUBJECT INDEX